Recommended Reference Materials
for the Realization of
Physicochemical Properties

IUPAC Commission on Physicochemical Measurement and Standards and IUPAC Subcommittee of The Physical Chemistry Division on Physicochemical Measurements and Standards

The membership of the Commission on Physicochemical Measurements and Standards and its Subcommission on Calibration and Test Materials and the subsequent Subcommittee of the Physical Chemistry Division, during the period 1971 to 1986 when the initial and revised sections were prepared, was as follows:*

Titular members

Chairmen: H. Kienitz (Germany), D. Ambrose (U.K.),
P. Cali (U.S.A.), K. N. Marsh (Australia, U.S.A.)

Vice-Chairman and Secretary: E. Brunner (Germany),
G. A. Uriano (U.S.A.)

Members: I. Brown (Australia), L. Crovini (Italy), H. Feuerberg (Germany), J. Franc (Czechoslovakia), R. P. Graham (Canada), E. Juhász (Hungary), J. E. Lane (Australia), Y. Mashiko (Japan), G. Milazzo (Italy), T. Plebanski (Poland), R. Sabbah (France), S. Saeki (Japan), J. Terrien (France)

Associate members

S. Bukówiecki (Switzerland), J. C. G. Calado (Portugal), A. J. Head (U.K., Secretary-Subcommittee), A. Ishitani (Japan), W. Künzel (G.D.R.), B. Le Neindre (France), M. Matrka (Czechoslovakia), L. Molle (Belgium), A. Newton (U.K.), W. Simon (Switzerland), W. M. Smit (Netherlands), L. A. K. Stavely (U.K.), D. R. Stull (U.S.A.), O. Suschny (Austria), W. Trabczyński, K. Wandleburg (Germany), H. F. van Wijk (Netherlands)

Members of subcommittee and/or subcommission

J. D. Cox (U.K.), G. Girard (France), D. V. S. Jain (India),
S. D. Rasberry (U.S.A.), W. Thomas (Germany), H. Ziebland (U.K.).

National representatives

A. B. Biswas (India), C. Černý (Czechoslovakia), M. Milone (Italy),
J. N. Mukerjee (India)

* Members listed in the various categories have served in other offices and categories during the period 1971 to 1986.

International Union of Pure and Applied Chemistry

Recommended Reference Materials for the Realization of Physicochemical Properties

EDITED BY
K. N. MARSH
Thermodynamics Research Center
Texas Engineering Experiment Station
The Texas A & M University System
College Station
Texas, U.S.A.

BLACKWELL SCIENTIFIC PUBLICATIONS

OXFORD LONDON EDINBURGH

BOSTON PALO ALTO MELBOURNE

© 1987 International Union
of Pure and Applied Chemistry
and published for them by
Blackwell Scientific Publications
Editorial offices:
Osney Mead, Oxford OX2 0EL
 (*Orders*: Tel. 0865 240201)
8 John Street, London WC1N 2ES
23 Ainslie Place, Edinburgh EH3 6AJ
52 Beacon Street, Boston
 Massachusetts 02108, USA
667 Lytton Avenue, Palo Alto
 California 94301, USA
107 Barry Street, Carlton
 Victoria 3053, Australia

All rights reserved. No part of this publication
may be reproduced, stored in a retrieval system,
or transmitted, in any form or by any means,
electronic, mechanical, photocopying, recording
or otherwise without the prior permission of the
copyright owner.

First published 1987

Printed in Great Britain
at the Alden Press, Oxford

DISTRIBUTORS

USA and Canada
Blackwell Scientific Publications Inc
PO Box 50009, Palo Alto
California 94303
(*Orders*: Tel. (415) 965-4081)

Australia
Blackwell Scientific Publications
(Australia) Pty Ltd
107 Barry Street,
Carlton, Victoria 3053
(*Orders*: Tel. (03) 347 0300)

British Library
Cataloguing in Publication Data
Recommended reference materials for the
 realization of physicochemical properties.
 1. Materials—Analysis—Standards
 I. Marsh, K.N.
 620.1'12'0218   QD131

ISBN 0-632-01718-X

Library of Congress
Cataloging-in-Publication Data
Recommended reference materials for the
 realization of physicochemical properties.
 At head of title: International Union of Pure
and Applied Chemistry
 1. Chemistry, Physical and theoretical—
Standards.
I. Marsh, K. N. II. International Union of Pure
and Applied Chemistry.
QD455.2.R43 1987   541.3   87-11724
ISBN 0-632-01718-X

Contents

Collators, vii

Preface, ix

1 Introduction, 1

2 Density, 5
Collator: G. GIRARD

3 Surface tension, 39
Collator: D. V. S. JAIN

4 Viscosity, 45
Collators: W. KÜNZEL, H. F. VAN WIJK AND K. N. MARSH

5 Pressure-volume-temperature, 73
Collator: D. AMBROSE

6 Distillation-testing of columns, 115
Collator: E. F. G. HERINGTON

7 Relative humidity of air, 157
Collator: W. KÜNZEL

8 Temperature, 163
Collator: D. AMBROSE

7 Enthalpy, 219
Collators: A. J. HEAD AND R. SABBAH

10 Thermal conductivity, 321
Collator: B. LE NEIDRE

11 Electrolytic conductivity, 371
Collators: E. JUHÁSZ AND K. N. MARSH

12 Permittivity, 379
Collator: K. N. MARSH

13 Potentiometric ion activities, 401
Collator: E. JUHÁSZ

14 Optical rotation, 419
Collator: B. COXEN

15   Optical refraction, 427
     Collators: A. FELDMAN AND
     I. H. MALITSON

16   REFLECTANCE, 447
     Collators: H. FEUERBURG, D. GUNDLACH
     AND H. TERSTIEGE

17   Wavelength and transmittance, 465
     Collators: R. W. BURKE AND
     R. A. VELAPOLDI

18   Relative molecular mass, 491
     Collator: K. N. MARSH

Collators

D. Ambrose
Department of Chemistry
University College London
20 Gordon Street,
London, WC1H OAJ (U.K.)

R. W. Burke
Center for Analytical Chemistry
National Bureau of Standards
Gaithersburg, MD 20899 (U.S.A.)

B. Coxon
National Bureau of Standards
U.S. Department of Commerce
Gaithersburg, MD 20899 (U.S.A.)

L. Crovini
Instituto de Metrologia del Consiglio Nazionale
delle Richerche
Strada delle Cacce 73
I-10135 Torino (Italy)

A. Feldman
National Bureau of Standards
Gaithersburg, MD 20899 (U.S.A.)

H. Feuerburg
Bundesanstalt fur Materialprüfung
Unter den Eichen 87
D-1000 Berlin 45
(Federal Republic of Germany)

G. Girard
Bureau International des Poids et Mesures
Pavillon de Breteuil
F-92310 Sèvres (France)

D. Gundlach
Bundesanstalt fur Materialprüfung
Unter den Eichen 87
D-1000 Berlin 45
(Federal Republic of Germany)

A. J. Head
Division of Quantum Metrology
National Physical Laboratory
Teddington, Middlesex TW11 OLW (U.K.)

E. F. G. Herington
29 Seymour Road
East Molesey
Surrey, KT8 OPB (U.K.)

D. V. S. Jain
Department of Chemistry
Panjab University
Chandigarh - 160014 (India)

E. Juhász
National Office of Measurements
Németölgyi út 37
H-1124 Budapest XII (Hungary)

W. Künzel
Amt für Standardisierung, Messwesen
und Warenprüfung (ASMW)
Fürstenwalder Damm 388, DDR-1162 Berlin
(German Democratic Republic)

B. Le Neindre
L.I.M.H.P.-C.N.R.S.
Université Paris-Nord
93430-Villetaneuse (France)

K. N. Marsh
Thermodynamics Research Center
Texas A&M University
College Station, TX 77843 (U.S.A.)

I. H. Malitson (retired)
National Bureau of Standards
Gaithersburg, MD 20899 (U.S.A.)

R. Sabbah
Centre de Thermodynamique et
de Microcalorimétrie
Centre National de la Recherche Scientifique
26, rue du 141è R.I.A.
13003 Marseille (France)

H. Terstiege
Bundesanstalt fur Materialprüfung
Unter den Eichen 87
D-1000 Berlin 45
(Federal Republic of Germany)

H. F. van Wijk
Hoofgroep Maatschappelijke Technologie
Organisatie voor Toegepast
Natuurwetenschappelijk Onderzoek
POB 108, Utrechtseweg 48
NL-3700 AC Zeist (Netherlands)

R. A. Velapoldi
Centers for Radiation Research and
Analytical Chemistry
National Bureau of Standards
Gaithersburg, MD 20899 (U.S.A.)

Preface

In 1953 the Commission on Physicochemical Measurements and Standards (Commission I.4) of the International Union of Pure and Applied Chemistry adopted terms of reference that included the promotion and encouragement of "the use of standard substances for calibrating and checking in physicochemical measurements." In the years immediately following, however, the Commission was more concerned with the determination of purity, a prerequisite for the establishment of satisfactory reference materials, and it was not until 1972 (Ref. 1), with a second edition in 1976 (Ref. 2), that "Physicochemical Measurements: Catalogue of Reference Materials from National Laboratories" appeared. This was exactly what its name implies, a catalogue, and it had no explanatory comments. More was needed, and in 1969 a Task Group on Standard Calibration Substances had been formed under the chairmanship of Professor Hermann F. K. Kienitz. On behalf of BASF AG he invited the Group to meet in Ludwigshafen in 1970, and at this meeting and a second one in Zeist in 1972 at the Institute of Physical Chemistry, TNO, the guiding principles were decided that were put into practice in the first Recommendations (Ref. 3).

It is difficult now to remember the uncertainties we had to begin with about the direction in which we wanted to go, and about the decisions we needed to take. A first question was nomenclature. At that time what are now known as reference materials were variously referred to in different parts of the world as standard samples, standard substances, standard reference materials, reference samples, and test samples. There was no agreed classification, nor were the functions of different types of *standards* agreed. The first step was to remove *standard* from the vocabulary of the subject on the grounds that it had too many different meanings and served no useful purpose here. *Substance* was too restrictive because that is a specific form of material, a single defined chemical entity; the properties of a purified sample of a specified substance are fixed and may be known with as high an accuracy as current techniques of measurement allow. A *material*, on the other hand, may not be, indeed usually is not, a single chemical entity and different batches may be of different compositions; the properties of each batch must therefore be determined if it is to be used for reference purposes. Use of a purified substance is to be preferred but it is necessary to use a wider range of materials because in some measurements, *e. g.* in the determination of viscosity, the range of values required cannot be provided easily by the use of single substances only. Moreover, outside the field of physical chemistry, there are many technological measurements made on materials that have no relevance to the behaviour of single substances, and it is desirable that the nomenclature used in this work should not conflict with the nomenclature needed in other fields. So we arrived at *Reference materials*, but how many classes of reference materials were there? No section caused more difficulties than the comparatively short and apparently straightforward "General Introduction" by J. P. Cali (Ref. 3) which he had to rewrite several times to bring it into accord with the changing and conflicting views expressed by different meetings, and even by the same meeting at different stages of its proceedings. The arguments are to a large extent lost because there are limits to what can be preserved on record in the reports of meetings, but, even if our initial classification is not quite identical

Preface

with what is used in this book, I think the Task Group can claim that it provided the first forum for discussion of the nomenclature now used in the reference material field, and that its ideas permeated subsequent developments both within IUPAC and without, particularly in OIML and ISO-REMCO (thanks largely to those of our members who were also members of these bodies).

There were two more lively special meetings in Warsaw in 1974 and in Rome in 1976, and then the renamed Task Group on Calibration and Test Materials, which had become a Subcommission, effectively became the Commission itself, meeting at the General Assemblies of IUPAC until 1983 when it became a Subcommittee of the Division of Physical Chemistry for the production of this volume. It is the fruit of a project created by Kienitz, and I think that, but for his belief in the needs it sets out to meet and his consequent enthusiasm that inspired, or goaded, its many participants, it never would have come to fruition. Kienitz backed his enthusiasm with considerable support from BASF AG, which included the services of Dr. E. Brunner as secretary in the crucial initial stages when doubts and uncertainties were clarified and a goal was defined. Producing these Recommendations was the most important activity of Commission I.4 while I was a member and it is both a duty and a pleasure to record here the part played by Kienitz in its initiation. It is sad that he did not live to see its completion.

A firm editorial hand is necessary in a publication such as this if uniform presentation that does not conflict with the manifold recommendations of other IUPAC bodies is to be achieved, not to speak of the difficulties of ensuring we do not contravene what we ourselves previously decided, and this responsibility was first undertaken by Dr. E. F. G. Herington. He has been ably followed by Dr. K. N. Marsh who, as Chairman of the Subcommittee and Editor, has produced this volume.

D. Ambrose
Chairman, Commission I.4, 1975-1979

REFERENCES

1. *Pure and Appl. Chem.* **1972**, *29*, 597.
2. *Pure and Appl. Chem.* **1976**, *48*, 503.
3. *Pure and Appl. Chem.* **1974**, *20*, 393.

1

SECTION: INTRODUCTION

Science and technology is maintained, in large part, through the application of numerical data obtained by measurement. The accuracy of the data depends, to a large extent, on the reliability of measurement and a common frame of reference. The common frame of reference for the base system of units recognized internationally is the *Système International d'Unités* (SI) (Ref. 1) and the means to realize this system and the definitions of units, as well as the methods used to realize them, are the responsibility of the Comité International des Poids et Mesures (CIPM) working with various international scientific unions and national standards laboratories. It is necessary to be able to transfer these units from standards laboratories to working laboratories which contain various measuring systems and instruments. Such apparatus often yield results whose uncertainty of measurement and limits of error cannot be established without the employment of materials with known properties. The materials used for this purpose are called *reference materials*.

The Reference Materials Committee of the International Organization for Standardization (ISO-REMCO) have considered the use of terms and definitions in connection with reference materials (Ref. 2) and their recommendations have been considered by the four international organizations concerned with metrology – the International Bureau of Weights and Measures (BIPM), the International Electrotechnical Commission (IEC), the International Organization for Standardization (ISO) and the International Organization for Legal Metrology (OIML). The agreed definitions of these terms are contained in the International Vocabulary of Basic and General Terms in Metrology (Ref. 3). The terms and definitions which are of concern in this publication are based on their definitions, except that the word 'standard' has been replaced by 'reference material' in accordance with the recommendations of ISO-REMCO.

In considering the position of different types of reference materials in the metrological hierarchy, we include in the classification the descriptives *primary*, *secondary*, and *working* as used for the designation of the classes of reference materials in this publication. A *primary reference material* has the highest metrological quality in a specified field. A *secondary reference* material is one whose value is fixed by comparison with a primary reference material. The terms primary and secondary denote the position in the metrological hierarchy and a secondary reference material is not necessarily of inferior quality to a primary reference material. A *working reference material* is one which, usually calibrated against a primary or secondary reference material, is used routinely to calibrate or check measuring instruments. A primary or secondary reference material may be used as a working reference material when high accuracy is required. Working reference materials are primarily used for three purposes; (1) to calibrate instruments to ensure that the numerical values obtained for a measured property are accurate within the uncertainty limits, (2) to check the performance of an instrument or its operator, and (3) provide a means of quality control.

Some materials have their properties certified by, or their certification is traceable to (as defined in reference 3), an authoritative national or international organization, agency, or laboratory and these are called *certified reference materials* for the purpose of these recommendations. When available, certified reference materials have advantages over uncertified reference materials because uncertainties in the property value are defined.

Certified reference materials also provide a convenient method for the transfer of a value of a property from a standardizing laboratory to a working laboratory and for intercomparisons between standardizing laboratories and between working laboratories. Some pure materials which are not certified are used for the realization of certain properties. We have sometimes recommended the use of such uncertified materials which may either require purification by the user or be available in adequate purity from reliable sources. In these cases the estimated uncertainty given in the recommendation is based upon an analysis of the published data, but the actual value may have a higher uncertainty depending on the specific nature of the impurities present in the particular sample. A number of recommendations have been published previously in the IUPAC journal Pure and Applied Chemistry (Refs. 4 to 19). This publication brings together most of these recommendations: some of them remain unchanged while others have been revised or expanded, and a few deleted. The recommendations are based on information from the literature, from correspondence with scientists in each discipline, and from comments solicited from metrological bodies, national standards laboratories, manufacturers and suppliers. The values given result from a careful evaluation by the collators and contributors and the recommendations have been referred to the appropriate IUPAC bodies and national standards laboratories for their comment.

The introduction to each section is not intended to be exhaustive but reflects the state-of-the-art for the measurement under consideration at the completion date. Many of the recommendations include a brief outline of the experimental methods available and the advantages and disadvantages of each method. The majority of the recommendations include the following information for each recommended reference material: physical property, units, range of variables, physical state(s) within range, the names of contributors, intended usage, sources of supply, methods of preparation, pertinent physicochemical data, and references.

In a publication of this form it is impossible for all the recommendations to be completed at the same time and to be current at the time of publication. The following is a tabulation of the completion date of each section.

1.	Introduction	1985
2.	Density	1983
3.	Surface Tension	1983
4.	Viscosity	1984
5.	Pressure-Volume-Temperature	1983
6.	Distillation – Testing of Columns	1983
7.	Relative Humidity of Air	1984
8.	Temperature	1985
9.	Enthalpy	1983
10.	Thermal Conductivity	1986
11.	Electrolytic Conductivity	1986
12.	Permittivity	1986
13.	Potentiometric Ion Activities	1985
14.	Optical Rotation	1986
15.	Optical Refraction	1984
16.	Reflectance	1984
17.	Wavelength and Transmittance	1986
18.	Relative Molecular Mass	1986

The symbols, terminology and units used are those recommended by IUPAC (Ref. 20).

The following provisos apply to the information on sources of supply: (a) the recommended materials, in most instances, have not been checked independently by the IUPAC; (b) the quality of material may change with time; (c) the quoted sources of supply may not be exclusive sources because no attempt has been made to seek out all possible alternatives; (d) the IUPAC does not guarantee any material that is recommended.

This book was produced using the computerized typesetting program TeX developed by Donald Knuth for the American Mathematical Society. The program was adapted to run on a Hewlett – Packard 9000 Series 500 computer under the HPUX operating system by Kenneth N. Marsh and Gregory Marriott. The assistance of Hermalinda Ryan, Beverly Boyd, Cathryn L. Marsh and Lui Hsaio–Ping in interpreting TeX and Barbara M. Marsh in proofreading is gratefully acknowledged.

Kenneth N. Marsh

REFERENCES

1. *Le Système International d'Unités (SI)*. 5th edition: B.I.P.M., **1985**.
2. *Terms and definitions used in connection with reference materials, Certification of reference materials–General and statistical principles*. ISO Guide 30, International Organization for Standardization: Geneva, **1981**.
3. *International Vocabulary of basic and general terms in metrology*, International Organization for Standardization: Geneva, **1984**.
4. Cali, J. P. *Pure and Appl. Chem.* **1974**, *40*, 393.
5. Cox, J. D. *Pure and Appl. Chem.* **1974**, *40*, 399.
6. Brown, I.; Lane, J. E. *Pure and Appl. Chem.* **1974**, *40*, 451.
7. Brown, I.; Lane, J. E. *Pure and Appl. Chem.* **1974**, *40*, 457.
8. Brown, I.; Lane, J. E. *Pure and Appl. Chem.* **1974**, *40*, 463.
9. Brown, I.; Lane, J. E. *Pure and Appl. Chem.* **1976**, *45*, 1.
10. Dietz, R.; Green, J. H. S. *Pure and Appl. Chem.* **1976**, *48*, 241.
11. Milazzo, G. *Pure and Appl. Chem.* **1977**, *49*, 661.
12. Ambrose, D. *Pure and Appl. Chem.* **1977**, *49*, 1437.
13. Feuerberg, H.; Terstiege, H. *Pure and Appl. Chem.* **1978**, *50*, 1477.
14. Durst, R. A.; Cali, J. P. *Pure and Appl. Chem.* **1978**, *50*, 1485.
15. Brunner, E.; Herington, E. F. G.; Newton, A. *Pure and Appl. Chem.* **1979**, *51*, 2421.
16. Plebanski, T. *Pure and Appl. Chem.* **1980**, *52*, 2393.
17. Juhasz, E.; Marsh, K. N. *Pure and Appl. Chem.* **1981**, *53*, 1841.
18. Kienitz, H.; Marsh, K. N. *Pure and Appl. Chem.*, **1981**, *53*, 1847.
19. Ziebland, H. *Pure and Appl. Chem.* **1981**, *53*, 1863.
20. *Manual of Symbols and Terminology for Physicochemical Quantities and Units*, 1979 Edition, International Union of Pure and Applied Chemistry, *Pure and Appl. Chem.* **1979**, *51*, 1.

2

SECTION: DENSITY

COLLATOR: G. GIRARD

CONTENTS:

2.1. Introduction

2.2. Methods of measurement
  2.2.1. Hydrostatic weighing
  2.2.2. Pycnometric method
  2.2.3. Vibrating tube method
  2.2.4. Other methods

2.3. Water as a reference material
  2.3.1. Standard mean ocean water (SMOW)
  2.3.2. Recent improvements in the results of isotopic analysis
  2.3.3. Variation in the density of water as a function of isotopic composition
  2.3.4. Preparation of water having a density as near as possible to that of SMOW
  2.3.5. The absolute density of SMOW as a function of temperature
  2.3.6. Relative density of water
  2.3.7. Absolute density of SMOW
  2.3.8. Recommended table of the absolute density of SMOW for the temperature range 0 to 40 °C
  2.3.9. Problems due to the effect of pressure and dissolved gases on the density of water

2.4. Mercury as a reference material
　　2.4.1. Preparation of a suitable sample of mercury
　　2.4.2. Effect of impurities and isotopic composition on the density of mercury
　　2.4.3. Absolute density of mercury

2.5. Silicon as a reference material

2.6. Other reference materials

2.7. Reference materials for density measurement
　　2.7.1. Water
　　2.7.2. Mercury
　　2.7.3. Crystal of silicon
　　2.7.4. 2,2,4-trimethylpentane
　　2.7.5. Cyclohexane
　　2.7.6. *trans*-bicyclo(4,4,0)decane
　　2.7.7. 100 Neutral solvent
　　2.7.8. Flugene 112

2.8. Contributors

2.9. List of Suppliers

2.1. Introduction

The symbol for density is ρ and the SI unit is kg m^{-3}. The symbol for relative density is d, where d is the ratio of the absolute density of a material to the absolute density of a specified reference material. This is a unitless quantity. Conditions such as pressure and temperature may be different for the two materials. In the section on water, d refers to the ratio of the density, ρ, of a given sample of water at a temperature t °C, relative to its maximum value ρ_{max} which is reached at a temperature close to 4 °C, the water being free from dissolved gases and under a pressure of 101 325 Pa. Temperatures refer to the International Practical Temperature Scale of 1968 (IPTS-68).

The U.S. National Bureau of Standards (NBS) has determined the densities of four silicon objects with a standard deviation of $\pm 1 \times 10^{-6}$. These objects are used as in-house reference standards in the determination of densities of liquid reference materials in the range of about 800 to 2000 kg m^{-3}. Silicon pieces of certified density are at present made available to the public by NBS. The NBS, co-operating with other national standards laboratories, will compare the densities of suitable solid artifacts with the single-crystal, in-house silicon density

2. Density

standards of NBS. The relative uncertainty in such a comparison is no more than $\pm 3 \times 10^{-6}$. This number represents three standard deviations plus $\pm 1.8 \times 10^{-6}$ systematic uncertainty in the silicon crystal standards. However, in the measurements of density, a liquid reference material will frequently be required. Water and mercury are highly suitable substances for the calibration of apparatus used for density measurement and the determination of the volume of apparatus. Samples of a small number of other liquids of certified density are also required for use when water or mercury are unsuitable.

The density of almost all liquids and solids is determined in relation to the density of water, hence the metrological importance of this liquid. When water is to be used as a reference material, it must be remembered that the density of water is affected by certain factors which should or should not be considered, depending on the degree of accuracy desired. These factors include the absolute value of the density at a reference temperature, the variations of this value as a function of temperature, the isotopic composition, the presence of dissolved atmospheric gases, and the compressibility of water. This discussion will refer only to the use of natural waters (that is, water not made artificially) for the temperature range 0 to 40°C and at a pressure close to the standard atmosphere.

Table 2.1.1. Maximum uncertainties of liquid water density values tabulated in table 2.7.1.1

Source of uncertainty	$\dfrac{10^3 \, \Delta\rho}{\text{kg m}^{-3}}$		
	15 °C	25 °C	40 °C
Uncertainty in ρ_{max} (V-SMOW)	3	3	3
Differences between basic density tables of Thiesen and Chappuis	1	2	5
Uncertainty in temperature scale	1	2	4
Uncertainty in isotopic composition	0.5	0.5	0.5
Uncertainty in gas content	1	1	
Uncertainty in compressibility	-	-	-
Total maximum uncertainty	6.5	8.5	12.5

Table 2.1.2. Temperature control requirements for an uncertainty of $\pm 1 \times 10^{-3}$ kg m^{-3} in the density of water

$t/^\circ$C	$\Delta t/$K
15	0.0066
25	0.0038
40	0.0026

The different factors affecting the density of water will be discussed later; it will be shown that each factor contributes a rather large maximum uncertainty. Table 2.1.1 gives these uncertainties in order of magnitude.

Thus at the present time (1983) under fixed physical conditions, the density of water is known to not better than $\pm 1 \times 10^{-2}$ kg m^{-3}. If a lesser accuracy is desired, the isotopic composition or dissolved gas factors, or the compressibility of water can be disregarded; in this case, an additional uncertainty of about $\pm 1 \times 10^{-2}$ kg m^{-3} is introduced. Since temperature is an important factor, table 2.1.2 gives the accuracy with which temperature measurements must be made in order to obtain an uncertainty in the density of water of $\pm 1 \times 10^{-3}$ kg m^{-3}.

2.2. Methods of measurement

A good general description of the apparatus and of methods for the measurement of density is given by Bauer and Lewin (Ref. 1) and by Bowman and Schoonover (Ref. 2). The most precise methods are the pycnometric (Refs. 1, 3), hydrostatic weighing (Refs. 1, 2, 4), magnetic float (Refs. 1, 5), and temperature flotation (Ref. 6) methods. At the present time, methods are available for the determination of the density of a liquid compared to that of a reference liquid with a uncertainty of $\pm 1 \times 10^{-3}$ kg m^{-3} or better. However, it is not possible to take full advantage of this precision in density determinations because of the uncertainty in the density of water, and the uncertainty in the meniscus volume of mercury resulting from variations in its surface properties.

2.2.1. Hydrostatic weighing

The density of solids can be determined by the hydrostatic method, using the density of water as the reference material. This same method can be used to determine the density of liquids provided that a solid body of known density (in relation to water or obtained by an absolute method) is available.

The measurement is performed in two stages:

2. Density

1. The tare, having a mass m_t, equilibrates a standard mass of mass m_2 placed on the pan of a balance and an assembly cradle (submerged in water) and suspension wire, of mass m_k, suspended from the same pan of the balance.

2. The mass standard of mass m_2 is removed and equilibrium with the unchanged tare is re-established by placing another mass standard, having a mass of m_1, on the pan of the balance and placing the object to be measured, of mass m (determined beforehand) and a density of ρ', in the cradle suspended from the pan and submerged in water. The densities of water, air and the objects of mass m_1 and m_2 being respectively ρ, a and λ, we have the following equations of equilibrium:

$$m_t = m_2(1 - a/\lambda) + m_k, \tag{2.2.1.1}$$

$$m_t = m - (m/\rho')\rho + m_1(1 - a/\lambda) + m_k, \tag{2.2.1.2}$$

whence

$$\rho' = \frac{m\rho}{m - (m_2 - m_1)(1 - a/\lambda)} \tag{2.2.1.3}$$

To determine the density of a liquid, the same procedure is followed, with ρ'' now being the density of the liquid and V' being the known volume (from m and ρ') of the submerged body. The following equations result:

$$m_t = m_2(1 - a/\lambda) + m_k, \tag{2.2.1.4}$$

$$m_t = m - V'\rho'' + m_1(1 - a/\lambda) + m_k, \tag{2.2.1.5}$$

$$m - V'\rho'' = (m_2 - m_1)(1 - a/\lambda), \tag{2.2.1.6}$$

whence

$$\rho'' = [m - (m_2 - m_1)(1 - a/\lambda)]/V'. \tag{2.2.1.7}$$

To obtain a minimum uncertainty, a certain number of precautions must be taken when using this method. For example, the effect of surface tension is theoretically eliminated. However, care must be taken to cover, with a layer of platinum black, the suspension wire (made of either platinum or stainless steel) that is located in the air-water interface region (Ref. 7). The suspension wire submerged in water must remain the same length to avoid any variation in the hydrostatic buoyancy. The water must be degassed beforehand, since water density is currently defined under these conditions. Moreover, this operation prevents the formation of bubbles on the surface of the submerged body. The dissolved-gas content of water must be determined at the time of measurement by means of auxiliary measurements. Furthermore, depending on the accuracy required, the deuterium and oxygen-18 isotopic composition of the water should be determined in order to apply subsequently a calculated correction factor using equation (2.3.3.1). The variations in the volume of the object due to temperature must also be known with sufficient accuracy.

These precautions apply to the measurement of density itself. However, one must also take into account uncertainties in the values of the mass standards used and of the differential correction for air buoyancy, this correction being calculated from the densities of the solid objects under consideration and of the mass reference standards. An equation for the determination of the density of moist air has been recommended by the Comité International des Poids et Mesures (CIPM) in 1981 (Ref. 8). In applying this equation, it is important to distinguish between the uncertainty due to the equation itself and the uncertainty due to the quantities measured.

(i) *Uncertainty due to the equation itself:* If the quadratic sum of the random uncertainties ($1\,\sigma$) is calculated, an average relative uncertainty of $\pm 2.5 \times 10^{-5}$ is obtained. If, in the same way, the systematic uncertainties are combined, an average relative uncertainty ($1\,\sigma$) of $\pm 6 \times 10^{-5}$ is obtained. However, if the arithmetic sum of these same uncertainties is calculated, an average relative value of $\pm 10 \times 10^{-5}$ is obtained.

(ii) *Uncertainty of the density of air due to the quantities measured after applying the given equation:* The uncertainties in ρ due to uncertainties in the measurements of pressure (p), temperature (T), relative humidity (h) or temperature of the dew point (t_r) and the mole fraction of CO_2 (x_{CO_2}) are given with good accuracy, under normal conditions, by:

$$\delta\rho/\rho(p) \simeq +1 \times 10^{-5}\ \text{Pa}^{-1}\ \delta p, \qquad (2.2.1.8)$$

$$\delta\rho/\rho(T) \simeq -4 \times 10^{-3}\text{K}^{-1}\ \delta T, \qquad (2.2.1.9)$$

$$\delta\rho/\rho(h) \simeq -9 \times 10^{-3}\ \delta h, \qquad (2.2.1.10)$$

$$\delta\rho/\rho(t_r) \simeq -3 \times 10^{-4}\text{K}^{-1}\ \delta t_r, \qquad (2.2.1.11)$$

$$\delta\rho/\rho(x_{CO_2}) \simeq +0.4\delta x_{CO_2} \qquad (2.2.1.12)$$

Using the hydrostatic weighing method and following the necessary precautions, it is possible to determine the relative density of a solid with a reproducibility of $\pm 1 \times 10^{-6}$. Be that as it may, at the present time the absolute value of this density may be determined to not better than $\pm 1 \times 10^{-2}$ kg m^{-3}, because of our insufficient knowledge of water density, as mentioned above.

Specially calibrated standard reference silicon can be used to find the density of either liquids or other solids. The necessary techniques have been well documented (Refs. 9, 10). To measure the density of liquids, one measures the buoyant loss in weight of the silicon in the liquid. Then

$$\rho = \rho_s(m_s - \sigma_s)/m_s, \qquad (2.2.1.13)$$

where ρ is the unknown liquid density, ρ_s is the silicon density, m_s is the mass of the silicon, and σ_s is the apparent mass of the silicon when submerged in the fluid. Once the fluid density is known, the same fluid can immediately be used to determine the density of an unknown solid (Refs. 9, 10, 11). Using more than one silicon object in the same series of measurements increases the statistical degrees of freedom. This can be useful in analyzing experimental results.

2. Density

2.2.2. Pycnometric method

The use of pycnometers is widespread for determining the density of liquids and solids. Different forms of pycnometers may be employed depending on the liquid used for the determination (Ref. 1). When determining the density of a liquid, the pycnometer must be calibrated with reference to water.

Let m be the mass of the pycnometer, m' the mass of the pycnometer filled with water and m_1 its mass when filled with the liquid under study. If ρ is the density of the water, under the physical conditions of the measurement, the unknown density of the liquid, ρ_1, is:

$$\rho_1 = \rho(m_1 - m)/(m' - m). \qquad (2.2.2.1)$$

If one is determining the density of a solid having a mass of m_c, its density, ρ_c, would be given by:

$$\rho_c = m_c \rho/(m' + m_c - m_2), \qquad (2.2.2.2)$$

m_2 being the mass of the pycnometer containing the solid and the water. If water is not compatible with the solid, another appropriate liquid must be used, the density of which must be determined beforehand.

When using this method, various precautions should be taken depending on the degree of accuracy required. To obtain density values with an accuracy of ± 1 kg m^{-3} the calculation of the air buoyancy correction may be disregarded. For an accuracy of $\pm 1 \times 10^{-1}$ kg m^{-3}, one can assume the air density to be 1.2 kg m^{-3}. The ambient air density must be calculated to obtain a higher accuracy.

The following important factors should be kept in mind: The mass of a glass object may vary appreciably depending on the amount of water adsorbed. Erratic behaviour of the balance occurs if the vessels become electrostatically charged by friction; this can be eliminated by placing a safe amount of radioactive material in the balance case. The pycnometer must be well cleaned both inside and out. It must be maintained at a temperature as close as possible to the balance temperature. Also after fabrication, the volume of a glass pycnometer will vary with time. When using a graduated pycnometer which has not been standardized beforehand, successive menisci must be brought back to the same graduation. This will avoid errors which may result if, for example, the internal surface of the tube is not regular or if the graduations are not exactly equidistant. In order to avoid the formation of air bubbles, particular care must be taken when filling the pycnometer with water or another liquid; any air bubbles formed must be eliminated. An error may result if the solid appears in powder form, since microbubbles may adhere to the surface of the grains. If such is the case, a degassing operation should be considered, if possible. During the manipulation of a liquid or a liquid mixture (for example, hygroscopic or volatile liquids), any variation in composition must be avoided.

2.2.3. Vibrating tube method

An elegant method to measure liquid densities is that developed by Kratky et al. (Ref. 12) which measures the natural vibration frequency of a tube containing the liquid investigated. Because the oscillating frequency is related to the mass of the tube, it also is a simple function of the density of the liquid contained in the tube. Among the advantages we note are the small volume requirements, large operating temperature range and a digital readout mode. However, it suffers from some important drawbacks. To achieve density measurement within $\pm 3 \times 10^{-3}$ kg m^{-3} care must be taken in the filling procedure, an integrating time constant of about 10 minutes is required, in addition to a long initial thermal equilibration period.

The liquid studied is circulated at a constant flow rate through the vibrating tube during the frequency (or period) measurement. Provided that this circulation is not perturbed, the natural vibration period of the tube will be related to the density of the liquid in the static case:

$$\rho = A + B\tau^2 \qquad (2.2.3.1)$$

where ρ is the liquid density, τ the oscillating period, and A and B are the constants of the apparatus (determined from frequency measurements on two liquids of accurately known densities, or more simply determined from air and water, if the required accuracy is not high).

The tube is a V-shaped stainless steel tube anchored (soldered) through a brass plate. Its volume is less than a cubic centimeter. This plate is close to a copper vessel thermally regulated by a water jacket. The resonant vibration of the V-shaped tube is activated and sustained with a magnetic pick-up, the coil of which is incorporated in an electronic circuit. The period of the natural oscillation of the tube is measured using a digital frequency meter.

An improved version of this densitometer has been realized by A. Poisson (Ref. 13). The temperature of the liquid studied is regulated by an independent circuit and measured very close to the input and output of the tube. Another circuit maintains the temperature of the cell as closly as possible to that of the liquid studied. The calibration is made using two fluids, the densities of which bracket the one to be measured. To eliminate the drift inherent to the densitometer, a measurement is made on a reference liquid between every measurement on the liquid under study. The calculation of the density is then reduced to

$$\rho = \rho_o + B(\tau^2 - \tau_o^2) \qquad (2.2.3.2)$$

where ρ_o and τ_o correspond to the reference liquid. It is possible to obtain a uncertainty of $\pm 1 \times 10^{-3}$ kg m^{-3} if the apparatus is used with care and adequate precautions are taken.

2.2.4. Other methods

Other methods may be used in certain special cases and may be adapted to specific needs. These include:

2. Density

(i) *The flotation method:* The float is completely submerged in the liquid and equilibrium is attained when its density is equal to that of the liquid. This method may be used to study small variations of the density of the liquid (for example, changes in isotopic composition), compensating the variations by changing the temperature or pressure. The uncertainty obtained for $\Delta \rho$ can be at best $\pm 1 \times 10^{-5}$ kg m^{-3}.

(ii) *The magnetic float method:* A magnetic float is submerged in a liquid and suspended by a magnetic field. The position of the float is controlled by varying the field, and the current passing through the solenoid is a function of the density. Once again, this method is used to study small variations of density (Ref. 14). An uncertainty of $\pm 1 \times 10^{-4}$ kg m^{-3} can be obtained when measuring density differences.

2.3. Water as a reference material

In spite of the problems of isotopic variation between samples of carefully purified water from various sources, water is still the most suitable material for use as a density reference material. Recent work at Bureau International des Poids et Mesures (Girard) and the Institute of Oceanography (Menaché) in France has enabled variations in the densities of samples of water to be related to the isotopic compositions.

2.3.1. Standard mean ocean water (SMOW)

To allow for small variations in the density between samples of purified water due to variations in their isotopic composition it is necessary to have a reference datum. It is noted that sea water has a very stable isotopic composition, such that density changes of samples of sea water resulting solely from differences in isotopic composition do not exceed $\pm 1 \times 10^{-3}$ kg m^{-3}. It has thus been recommended by Girard and Menaché (Ref. 15) that Standard Mean Ocean Water (SMOW) be used as a reference datum for precise density determinations. Standard Mean Ocean Water is a pure water which was prepared from ocean water by Craig (Ref. 16), who proposed it as a universal isotopic standard for reporting deuterium and ^{18}O concentrations in natural waters. SMOW is kept by the International Atomic Energy Agency which is responsible for its distribution in small quantities to laboratories who specialize in isotope ratio measurements by mass spectrometry.

The isotopic composition of SMOW has been determined by Craig (Ref. 16) (^{18}O) and Hagemann *et al.* (Ref. 17) (D); the values found are as follows:

$$r_{18}(\text{SMOW}) = [^{18}\text{O}]/[^{16}\text{O}](\text{SMOW}) = (1\ 993.4 \pm 2.5) \times 10^{-6} \qquad (2.3.1.1)$$
$$r_{\text{D}}(\text{SMOW}) = [\text{D}]/[\text{H}](\text{SMOW}) = (155.76 \pm 0.05) \times 10^{-6} \qquad (2.3.1.2)$$

where $[^{18}\text{O}]/[^{16}\text{O}]$ and $[\text{D}]/[\text{H}]$ are the ratios of the numbers of specified atoms in the sample.

The isotopic composition of any sample of water is usually determined by comparison with SMOW using mass spectrometry. The results of this comparison are expressed by the relative differences δ_{18} and δ_D, which are defined by the following relations:

$$\delta_{18} = [r_{18}(\text{sample})/r_{18}(\text{SMOW}) - 1] \times 10^3, \quad (2.3.1.3)$$

$$\delta_D = [r_D(\text{sample})/r_D(\text{SMOW}) - 1] \times 10^3. \quad (2.3.1.4)$$

2.3.2. Recent improvements in the results of isotopic analyses

The accuracy obtained in recent isotopic analyses and the consistency of results for the same sample analyzed in different laboratories have been improved appreciably following the preparation, at the request of the International Atomic Energy Agency (IAEA) in Vienna, of two new reference waters. These samples allow the definition of a scale of values for $\delta(\delta_{18}$ and $\delta_D)$ which includes virtually all values found in nature. These two reference liquids are kept at the IAEA who, on request, provides small quantities to laboratories specializing in isotopic analyses of water.

(i) *First reference water:* The first reference sample is new SMOW, the 'Vienna SMOW' (V-SMOW), prepared by Craig who obtained a liquid having practically the same isotopic composition as the former reference liquid: the same value of r_{18} and a new value of r_D, lower by 0.2 per cent. The Vienna-SMOW is the *fundamental reference liquid* for isotopic analyses of water. The absolute values of its isotopic ratios determined by Hagemann et al. (Ref. 17) for r_D and by Baertschi (Ref. 18) for r_{18} are as follows:

$$\begin{aligned} r_D(\text{V-SMOW}) &= (155.76 \pm 0.05) \times 10^{-6} \\ r_{18}(\text{V-SMOW}) &= (2\,005.20 \pm 0.45) \times 10^{-6} \end{aligned} \quad (2.3.2.1)$$

(ii) *Second reference water:* Standard Light Antarctic Precipitation (SLAP). The absolute value of r_D determined by Hagemann et al. (Ref. 17) is:

$$r_D(\text{SLAP}) = (89.02 \pm 0.05) \times 10^{-6} \quad (2.3.2.2).$$

It is the lowest value found in nature to date. SLAP is a water obtained from melted ice of the Antarctic, taken by Professor E. Picciotto at Plateau Station, at an altitude of 3700m and at 79°15′ S and 40°30′ E.

Scale of values for δ_{18} and δ_D relative to V-SMOW; consistency of new results:

(i) Point of origin V-SMOW:

$$\delta_{18} = \delta_D = 0 \text{ by definition} \quad (2.3.2.3)$$

(ii) SLAP: The limit point which corresponds to water having the lowest isotopic concentrations of ^{18}O and D. The values (Ref. 19) in comparison to V-SMOW are:

$$\begin{aligned} \delta_{18}(\text{SLAP}) &= -55.5 \\ \delta_D(\text{SLAP}) &= -428 \end{aligned} \quad (2.3.2.4).$$

2. Density

These two values have been universally adopted.

For these samples the determination entails three analyses: that of the sample and those of the two reference points: V-SMOW and SLAP. In principle, the measured value of $\delta_m(\text{SLAP})$ should coincide, on the scale of δ, with the value assigned to $\delta_o(\text{SLAP})$, that is, -55.5 for ^{18}O and -428 for deuterium. If not, the scale obtained by observation is in error and must be normalized in order to coincide with the scale corresponding to the assigned values. The normalization for the sample is obtained by multiplying the measured value δ_m (sample) by the ratio $\delta_o(\text{SLAP})/\delta_m(\text{SLAP})$, recalling that the values for δ are calculated with respect to V-SMOW.

The normalized value δ_o (sample) for a sample of water, which is the result of analysis, is equal to:

$$\delta_o(\text{sample}) = \delta_m(\text{sample}) \times \delta_o(\text{SLAP})/\delta_m(\text{SLAP}) \qquad (2.3.2.5)$$

where

$$\delta_m(\text{sample}) = \{[r(\text{sample}) - r(\text{V-SMOW})]/r(\text{V-SMOW})\} \times 10^3 \qquad (2.3.2.6)$$
$$\delta_m(\text{SLAP}) = \{[r(\text{SLAP}) - r(\text{V-SMOW})]/r(\text{V-SMOW})\} \times 10^3 \qquad (2.3.2.7)$$
$$\delta_o(\text{SLAP}) = -55.5 \text{ for } ^{18}O \qquad (2.3.2.8)$$

and

$$\delta_o(\text{SLAP}) = -428 \text{ for } D \qquad (2.3.2.9)$$

are the assigned values.

The relation $\delta_o(\text{SLAP})/\delta_m(\text{SLAP})$ is called the normalization factor. At this time, all isotopic analyses of water are done with V-SMOW as the reference material. This liquid, for simplification, will be referred to as SMOW throughout the remainder of this document.

2.3.3. Variation in density of water as a function of isotopic composition

The variation in the density of water with isotopic composition has been investigated by Girard and Menaché (Refs. 20, 22) over a density interval of 2×10^{-2} kg m^{-3}. The isotopic density correction may be determined from the following relation which was obtained from their experimental measurements (Ref. 23).

$$[(\rho(\text{sample}) - \rho(\text{SMOW})] \times 10^3/(\text{kg m}^{-3}) = 0.233\delta_{18} + 0.016\,6\delta_D \qquad (2.3.3.1).$$

This relation applies to densities at temperatures from 0 to 40 °C for samples prepared from natural waters. It is estimated that the uncertainty introduced by the use of this relation is less than $\pm 1 \times 10^{-3}$ kg m^{-3}. This error can be reduced by using samples of water having an isotopic composition as close as possible to that of SMOW.

2.3.4. Preparation of a water having a density as near as possible to that of SMOW

The stability of the isotopic composition of sea water is such that, by means of classical distillation, a pure water can be obtained having a density varying very little from that of SMOW. However, such a water cannot replace SMOW as a reference liquid to carry out accurate density measurements for other water samples. Such measurements can only be done by using, as reference materials, standard pure water whose composition is known exactly. Cox, McCartney and Culkin (Ref. 24) have proposed a triple distillation method for the purification of sea water and natural waters which, if followed rigourously, does not alter the deuterium and ^{18}O isotopic concentration. This fact has been verified by Menaché (Ref. 25) employing the mass spectrometric determinations of the D/H and $^{18}O/^{16}O$ ratios by the method described by Nief *et al.* (Refs. 26, 27). However, it has also been verified that this triple distillation method of sea water too often results in deception. Only after very extensive experience can satisfactory results be obtained. Thus, when using this method, it is strongly recommended, especially for accurate density determinations, to conduct an isotopic control analysis of the sample as soon as possible before or after its density determination.

Obtaining a pure water with a density near that of SMOW remains a useful exercise. Provided that its isotopic composition is determined by mass spectrometry then this water can serve as an auxiliary reference liquid to replace SMOW in isotopic analysis. Further, the isotopic corrections in density measurements are minimized.

For density measurements of natural pure waters with reference to SMOW, the difference in the density of a sample from that of SMOW can be calculated at any temperature, using equations (2.3.1.3), (2.3.1.4) and (2.3.3.1). The density of the sample can then be obtained at the required temperature once the density of SMOW is known at that temperature.

2.3.5. The density of SMOW as a function of temperature

Since the beginning of this century a large number of tabulations of the relative and absolute densities of water in the temperature range 0 to 40 °C have been published which give appreciably different values for any particular temperature. These tables are often given without reference to their origin, without adequate definition of the particular temperature scale to which they refer and, for absolute values, without reference to the isotopic composition of the water used. This leads to considerable confusion and doubt in the assessment of precise density data because scientists may use a particular, but unspecified tabulation in the belief that its contents are both unique and accurate. This unfortunate situation has been reviewed by Menaché and Girard (Ref. 28).

2.3.6. Relative density of water

All the published tables of relative density, $d_t = \rho_t/\rho_{max}$, are based on the direct observations of the thermal expansion of water made by Thiesen *et al.* (Ref. 29) published in 1900 and the values published by Chappuis (Ref. 30) in 1907. More recently Steckel and Szapiro (Ref. 31) have made new determinations which have yielded values closer to those reported by Thiesen *et al.* than to those recorded by Chappuis. The tables published by Mendeleev (Ref. 32), Tilton and Taylor (Ref. 33), Kell (Ref. 34), Bigg (Ref. 35), Aleksandrov and Trakhtengerts (Ref. 36), and Wagenbreth and Blanke (Ref. 37) which can be designated 'derived tables' are all based on reassessments of the basic measurements made by Thiesen *et al.* and by Chappuis.

Menaché and Girard (Ref. 28) have emphasised that the values of relative density d shown in these basic and derived tables, are independent of the isotopic composition of the water used, as both t_{max} and the thermal expansion coefficient show negligible variation within the range of isotopic concentrations found for natural waters.

At the present time, our knowledge of the values of the relative density of water as a function of temperature in the range 0 to 40 °C is, unfortunately, subject to appreciable uncertainties. The values given in the two basic tables by Thiesen *et al.* and Chappuis show differences that increase in absolute value with the temperature, Thiesen's values being generally lower than those reported by Chappuis. Above 16 °C these differences exceed 1×10^{-3} kg m^{-3} and rapidly reach an unacceptable level of 6×10^{-3} kg m^{-3} at 25 °C rising to 9×10^{-3} kg m^{-3} at 40 °C. In addition to this, there are problems due to the definitions of temperature scales. The data reported by Thiesen *et al.* and by Chappuis related to the 'Echelle Normale' of the hydrogen thermometer. There are two schools of thought (Ref. 28), one which believes this temperature scale to coincide practically (in the range 0 to 40 °C) with the International Practical Temperature Scale of 1968 (IPTS-68) and another which believes it to coincide practically with the International Practical Temperature Scale of 1948 (IPTS-48). In spite of the thorough investigations made by Hall (Ref. 38), this difference of opinion cannot be resolved until new precise density determinations relative to IPTS-68 have been made. At 40 °C the difference between IPTS-68 and IPTS-48 is 0.01 K. The measurements made by Steckel and Szapiro (Ref. 31) were related to IPTS-48.

2.3.7. Density of SMOW

To calculate the density of SMOW as a function of temperature from values of relative density we require a value of ρ_{max}(SMOW). Absolute determinations of ρ_{max} for water made at the beginning of the century (Ref. 39) led to the mean value $\rho_{max} = 999.972$ kg m^{-3}, but we do not know the isotopic composition of the water corresponding to this value. Girard and Menaché (Refs. 15, 28) have proposed to take for the maximum density of SMOW, which occurs at a temperature close to 4 °C under a pressure of one standard atmosphere (101 325 Pa) in the absence of dissolved atmospheric gases, the value:

$$\rho_{max}(\text{SMOW}) = 999.975 \text{ kg m}^{-3} \quad (2.3.7.1).$$

This value is regarded as exact, with an uncertainty determined to be $\pm 3 \times 10^{-3}$ kg m^{-3}. If, as recommended, this absolute value is regularly used in all future density determinations, the results obtained will be consistent. When a more accurate absolute value is determined in the future, these results will easily be corrected.

2.3.8. Recommended table of the density of SMOW for the temperature range 0 to 40 °C

To reduce confusion and to promote uniformity in the determination of liquid densities, it is recommended that a single table of provisional values of the density of SMOW, a water of known isotopic composition, be recommended pending new precise determinations of the density of water. The table of recommended values attached to this report uses ρ_{max}(SMOW) equal to 999.975 kg m^{-3} and is based on the table given by Bigg (Ref. 35) in which the figures are the weighted means of the values presented by Thiesen *et al.* and by Chappuis. It has been assumed that the Echelle Normale coincides with IPTS-68 and therefore that the temperatures in reference 35 refer to the currently used scale. Account has been taken of the possible effects of the uncertainties in the temperature scale of the original measurements in the estimates now given of the maximum uncertainties in the tabulated values of density.

These uncertainties can only be reduced by precise redeterminations of the density, relative to IPTS-68, of samples of water of known isotopic composition. Measurements have already commenced in Australia and Japan and it is hoped that other countries will also contribute.

2.3.9. Problems due to the effect of pressure and dissolved gases on the density of water

The effect of these two factors on the density of water has been discussed by Menaché (Ref. 40). A change in pressure of 133.3 Pa changes the density of water by approximately 0.066×10^{-3} kg m^{-3}. Values of the compressibility of water from -20 to 110 °C are also given by Kell (Ref. 34). The effect of dissolved gases is now under study. At 4 °C air-saturated water is believed to be about 4×10^{-3} kg m^{-3} less dense than air-free water and at 20 °C this difference should be about 2 to 3×10^{-3} kg m^{-3}.

2.4. Mercury as a reference material

The density of mercury at 20 °C has been measured with great accuracy by Cook (Refs. 41, 42), its thermal expansion is known over a wide range of temperature, probably more exactly than that of any other liquid (Refs. 43, 44), and a table of values of the density of mercury in the range -20 to $+300$ °C (IPTS-48) based on these measurements has been published (Ref. 45). Nevertheless, it is questionable whether mercury will often be the most suitable calibration or test material for use in pycnometric measurement of densities because it suffers from the following disadvantages.

(i) As the density of mercury is so much greater than the density of most other liquids, difficulties may arise because of the relatively large mass that must be determined when the pycnometer contains mercury.

2. Density

(ii) The interpretation of measurements may be difficult because the contact angle of mercury with the walls of the vessel (about 140° when the vessel is glass) differs greatly from that of other liquids and is affected by the cleanliness of the apparatus and the purity of the mercury.

However the density of mercury is particularly important because of the use of mercury as a manometric fluid. The following paragraph is inspired by the report on the IPTS-68 (Ref. 46).

In practice pressures are usually determined by means of a mercury column. The mean density of pure mercury at the temperature t_{68} in a barometric column supported by the pressure p being measured is given, with sufficient accuracy over the temperature range from 0 to 40 °C and for the pressures relevant to these measurements, by the relation

$$\rho(t_{68}, p/2)/(\text{kg m}^{-3}) = \rho(20\ °\text{C}, p_\circ)/\{[1 + A(x - 20) + B(x - 20)^2] \\ \times [1 - \chi(p/2 - p_\circ)/\text{Pa}] \text{kg m}^{-3}\} \qquad (2.4.1)$$

where

$$x = t/°\text{C},$$
$$A = 18\ 115 \times 10^{-8},$$
$$B = 0.8 \times 10^{-8},$$
$$\chi = 4 \times 10^{-11},$$
$$\rho(20\ °\text{C}, p_\circ) = 13\ 545.87\ \text{kg m}^{-3}.$$

This last value is the density of pure mercury at $t_{68} = 20\ °\text{C}$ under a pressure $p_\circ = 101\ 325$ Pa (1 standard atmosphere). A sufficiently accurate value of the local gravity may be obtained by using the Reséau Gravimétrique International Unifié 1971 (IGSN-71) of the International Union of Geodesy and Geophysics.

A simple relationship between density and the compressibility has been assumed in the above relation. Mercury is also used for confining the sample and transmitting the pressure in volumetric measurements on fluids at high pressures, and under these conditions a more complex equation for the compressibility may be required (Ref. 47). When mercury is used as a manometric fluid, the effect of capillary depression must also be taken into account (Refs. 48 to 51).

Mercury is also frequently used for the determination of the volumes of vessels by weighing. For this purpose the large mass involved may not always be a disadvantage, and mercury may, in many instances, be the most suitable substance because of its low volatility and non-wetting character. Care must be taken to account for the dilation of the vessel.

2.4.1. Preparation of a suitable sample of mercury

The most widely used method of purifying mercury, after it has been filtered to remove gross contamination, is treatment by dilute nitric acid (say, 1 part by volume of concentrated acid in 20 parts of the solution) to remove base metals, followed by washing with water, and by distillation to remove the higher-boiling metals (Refs. 41, 52, 53). It has been shown that the nitric acid treatment is more effective if the mercury is shaken for a few minutes with the acid rather than for the mercury to fall in drops through a column of the acid (although the latter procedure has often been used, and commercial apparatus for carrying it out conveniently is available). Distillation is most effective if carried out in the presence of clean air at a pressure of about 3 kPa.

The purified mercury is best stored in soft glass or polythene bottles (it may be more difficult to pour cleanly from the latter because of the development of electrostatic charges) (Ref. 54) and the bottles, their interior surfaces free of scratches, should be cleaned with nitric acid before use. If base metals, such as zinc or copper, are present in mercury a film quickly forms on the bright surface, but slight surface contamination of even the purest samples is often observed after prolonged storage in air, and it is therefore preferable for air to be displaced from the bottles by an inert gas or for the mercury to be stored *in vacuo*. Small amounts of impurity on the surface of mercury may be removed by filtration, or a pure sample may be drawn off through a tube dipping below the surface.

2.4.2. Effect of impurities and isotopic composition on the density of mercury

The following table (Ref. 41) gives the mass fractions of impurities which produce a change of 1×10^{-2} kg m^{-3} in the density of mercury.

Table 4.2.1. Influence of impurities on the density of mercury

Metal	$10^{-6} \times$ mass fraction	Metal	$10^{-6} \times$ mass fraction
platinum	2.7	tin	1.1
gold	3.4	iron	1.4
zinc	1.5	sodium	0.13
copper	1.9	calcium	0.13
lead	4.3	aluminium	0.25

Furthermore, mercury is composed of isotopes of mass numbers 196, 198, 199, 200, 201, 202 and 204 and variations of 0.005 per cent in the abundances of one or two isotopes would change the density of a sample of mercury by 1×10^{-2} kg m^{-3}. The sensitivities of analysis of impurities or of the determination of isotopic abundances available at the time of the investigation by Cook and Stone (Refs. 41, 42) were not adequate to ensure the absence of impurities or of isotopic constancy to a degree sufficient for the density to be specified more closely than $\pm 10 \times 10^{-2}$ kg m^{-3}. Nevertheless, four samples of mercury of different origins were purified by at least two successive shakings with nitric acid and at least two distillations in the presence of air, and the values obtained for their densities agreed within $\pm 1 \times 10^{-2}$ kg m^{-3}. It therefore appears probable that mercury can be prepared with a lower concentration of impurities than could then be assayed and that the methods of purification do not have a significant effect on the isotopic composition.

2.4.3. Density of mercury

A table of the absolute density of mercury from -20 to $+300$ °C, based on the table given by Bigg (Ref. 45) is attached. In this table due allowance has been made for the change in the temperature scale to IPTS-68. The likely uncertainties in the values are also shown.

2.5. Silicon as a reference material

The density of a perfect single crystal of silicon ρ_s is given by

$$\rho_s = nM/N_A a_\circ^3 \qquad (2.5.1)$$

where n is the number of atoms per unit cell (eight), M is the average molar mass of silicon atoms, $a_\circ$ is the spacing of a unit cell and N_A is the Avogadro constant. Such perfection is closely approximated by dislocation-free silicon grown by the float-zone technique.

Even for a perfect material ρ_s may vary from crystal to crystal because silicon has three naturally occuring isotopes. One can infer from the reported global variability in isotopic mixes (Refs. 55, 56) that ρ_s may vary by about 30×10^{-3} kg m^{-3} among crystals grown from silicon taken from different natural sources. Evidence accumulated indicates that further segregation of isotopes due to the zone refining process is much smaller than the global variability noted above and may, in fact, be negligible (Refs. 11, 56). Experiments have also shown that ρ_s/M is constant for impurity-free material (Ref. 57).

Detectable impurities which may occur in single-crystal, float-zone refined silicon are oxygen (Ref. 58) and carbon (Ref. 59), as well as a dopant (usually boron (Ref. 60) or phosphorus). For lightly doped silicon (resistivity $\geq 1\ \Omega$ cm) (Ref. 61) the effect of these impurities on ρ_s is far below the effect of isotopic variability. A certain amount of care must be taken in fabricating silicon artifacts from a larger crystal so that the finished piece is free from damage (Ref. 62).

The thermal expansion of silicon is well-characterized between 6 K and 340 K (Ref. 63). The volume coefficient of expansion at 20 °C is $(7.66 \pm 0.02) \times 10^{-6}$ K^{-1}, based on the data of reference 63, and is reasonably constant in this region. The surface of silicon oxidizes in ambient laboratory air (Refs. 64, 67) so that the surface has physical properties similar to fused silica. The silicon objects are thus easily cleaned with non-abrasive detergent or with a number of common organic solvents. Vapour degreasing in inhibited 1,1,1-trichloroethane is also a good cleaning method. Silicon is more brittle than most metals but, because of its high homogeneity, accidental chipping will not alter the density of a macroscopic piece.

A brief comment should be made about the long term stability of silicon artifacts. It is known that a thin oxide layer grows on a silicon surface in air at room temperature (Refs. 64, 67). The various published experimental results differ somewhat as to the precise nature of this film and its precise growth rate. All results, however, show a film growth of less than 6 nm after 2 years. The presence of a uniform oxide layer can easily be shown to have the following effect on ρ_s:

$$\Delta\rho/\rho = (\rho_\circ/\rho_s - 1)(A/V)L \tag{2.5.2}$$

where $\rho_\circ$ is the density of the oxide (2 200 kg m^{-3} for fused silica), A/V is the ratio of area to volume of the sample, and L is the oxide thickness. For the silicon objects described in section 7.3, A/V is less than 200 m^{-1}. Thus for an oxide which has a density within 10 per cent of fused silica, L would have to be 35 nm in order to produce a relative change in the density of 1×10^{-6}. Extrapolation from the most pessimistic data (Ref. 65) suggests that this thickness represents a growth of several hundred years. Experimentally, it has been found that a crystal with $A/V \sim 130$ m^{-1}, which was subjected to numerous boilings in distilled water and vapor degreasings in 1,1,1-trichloroethane did not change its density over a period of 30 months. The detectable limit for the relative density change was about 5×10^{-7} (Ref. 68). Thus, there is good reason to believe that silicon objects have a stable density to well within the stated accuracies.

2.6. Other reference materials

Samples of a number of hydrocarbons with certified values of density are available from National Laboratories. Samples of 2,2,4-trimethylpentane are available from suppliers (A) and (C). Samples of hexane, octane, nonane, cyclohexane, methylcyclohexane, *trans*-decahydronaphthalene and toluene are also available from supplier (A) with values of density certified to ± 0.005 kg m^{-3} at temperatures at 5 K intervals from 20 to 50 °C. Samples of methylcyclohexane and toluene are available from supplier (C) with densities accurate to ± 0.05 kg m^{-3} at 20, 25, and 30 °C and samples of hexadecane, *trans*-decahydronaphthalene and 1-methylnaphthalene with an estimated uncertainty of ± 0.08 kg m^{-3} at the same temperatures are also available from supplier (C). At present there are no suitable reference materials for use over a wide temperature and pressure range.

Three of these certified reference materials have been chosen as suitable examples. Values are given for 2,2,4-trimethylpentane, cyclohexane and *trans*-bicyclo[4,4,0]decane (*trans*-decahydronaphthalene). Two other reference materials provide lower accuracies: 100 Neutral Solvent and flugene 112 (1,1,2,2-tetrachloro 1,2-difluoroethane, CCl_2FCCl_2F), both available from supplier (D).

REFERENCES

1. Bauer, N.; Lewin, S. Z. *Techniques of Chemistry*, 4th ed. Weissberger, A., editor, Vol. 1, Part IV, Interscience: New York, **1972**, p. 57.
2. Bowman, H. A.; Schoonover, R. M. *J. Res. Nat. Bur. Stand.* **1967**, *71c*, 179.

3. Washburn, E. W.; Smith, E. R. *J. Res. Nat. Bur. Stand.* **1934**, *12*, 305.
4. Forziati, A. F.; Mair, B. J.; Rossini, F. D. *J. Res. Nat. Bur. Stand.* **1945**, *35*, 513.
5. Hales, J. L. *J. Phys. E.* **1970**, *3*, 855.
6. Kozdon, A. *Proc. First International Conference on Calorimetry and Thermodynamics*, Warsaw, August **1969**, p. 831.
7. Bonhoure, A. *Procès Verbaux CIPM* **1958**, *26A*, 122.
8. *Procès Verbaux CIPM* **1981**, *49*, C1.
9. Bowman, H. A.; Schoonover, R. M.; Carroll, C. L. *J. Res. Nat. Bur. Stand.* **1973**, *78A*, 13.
10. Bowman, H. A.; Schoonover, R. M.; Carroll, C. L. *Metrologia* **1974**, *10*, 117.
11. Davis, R. S. *Metrologia* **1982**, *18*, 193.
12. Picker, P; Tremblay, E.; Jolicoeur, C. *J. Sol. Chem.* **1974**, *3*, 377.
13. Poisson, A. *Thesis Dr. Sciences*, Université P. et M. Curie Paris, **1978**.
14. Bignell, N. *J. Phys. E* **1982**, *15*, 378.
15. Girard, G.; Menaché, M. *C.R. Acad. Sci. Paris* **1972**, *274B*, 377.
16. Craig, H. *Science* **1961**, *133*, 1833.
17. Hagemann, R.; Nief, G.; Roth, E. *Tellus* **1970**, *22*(6), 712.
18. Baertschi, P. *Earth Planet. Sci. Lett.* **1976**, *31*, 341.
19. Gonfiantini, R. *Nature* **1978**, *271*(5645), 534.
20. Girard, G.; Menaché, M. *C.R. Acad. Sci. Paris* **1967**, *265B*, 709.
21. Menaché, M.; Girard, G. *C.R. Acad. Sci. Paris* **1970**, *270B*, 1513.
22. Girard, G.; Menaché, M. *Metrologia* **1971**, *7*(3), 83.
23. Menaché, M.; Beauverger, C.; Girard, G. *Annales Hydrographiques* **1978**, 5th series, No. 6, 750, 37.
24. Cox, R. A.; McCartney, M. J.; Culkin, F. *Deep-Sea Research* **1968**, *15*, 319.
25. Menaché, M. *Deep-Sea Research* **1971**, *18*, 449.
26. Nief, G.; Botter, R. *Advances in Mass Spectrometry, Joint Conference on Mass Spectrometry*, Waldron, J. D. editor, Pergamon Press: Oxford, **1958**, p. 515.
27. Majoube, M.; Nief, G. *Advances in Mass Spectrometry* **1968**, *4*, 511.
28. Menaché, M.; Girard, G. *Metrologia* **1973**, *9*, 62.
29. Thiesen, M.; Scheel, K.; Diesselhorst, H. *Wiss. Abhandl. Physik. Tech. Reichs.* **1900**, *3*(1), 1.
30. Chappuis, P. *Trav. et Mém. du BIPM* **1907**, *13*, D1.
31. Steckel, F.; Szapiro, S. *Trans. Faraday Soc.* **1963**, *59*(2), 331.
32. Mendeleev, D. I. *Vremenik Glavnoj Palati Mer i Vesov* **1897**, part 3, 133.
33. Tilton, L. W., Taylor, J. K. *J. Res. Nat. Bur. Stand.* **1937**, *18*, 205.
34. Kell, G. S. *J. Chem. Eng. Data* **1967**, *12*, 66; **1975**, *20*, 97.
35. Bigg, P. H. *Brit. J. Appl. Phys.* **1967**, *18*, 521.
36. Aleksandrov, A. A.; Trakhtengerts, M. S. *Teploenergetika* **1970**, *17*, 86; English translation in *Thermal Engineering* **1970**, *17*, 122.
37. Wagenbreth, H.; Blanke, W. *PTB-Mitteilungen* **1971**, *6*, 412.
38. Hall, J. A. *Phil. Trans.* **1929**, *229A*, 1.
39. Guillaume, Ch-Ed. *La Création du Bureau International des Poids et Mesures et son Oeuvre*, 240. Gauthier-Villars: Paris **1927**.
40. Menaché, M. *Metrologia* **1967**, *3*(3), 58.
41. Cook, A. H.; Stone, N. W. B. *Phil. Trans. Roy. Soc. London* **1957**, *250A*, 279.

42. Cook, A. H. *Phil Trans. Roy. Soc. London* **1961**, *254A*, 125.
43. Beattie, J. A.; Blaisdell, B. E.; Kaye, J.; Gerry, H. T.; Johnson, C. A. *Proc. Am. Acad. Arts Sci.* **1941**, *71*, 371.
44. Cook, A. H. *Brit. J. Appl. Phys.* **1956**, *7*, 285.
45. Bigg, P. H. *Brit. J. Appl. Phys.* **1964**, *15*, 1111.
46. *Comptes rendus des séances de la 13ᵉ Conférence Générale des Poids et Measures* **1967**, annex 2. See also *The International Practical Temperature Scale of 1968*, HMSO. London (1969); *Metrologia* **1969**, *5*, 35.
47. Bett, K. E.; Weale, K. E.; Newitt, D. M. *Brit. J. Appl. Phys.* **1954**, *5*, 243.
48. Gould, F. A.; Vickers, T. *J. Sci. Inst.* **1952**, *29*, 86.
49. *Tables of Physical and Chemical Constants*. Originally compiled by Kaye. C. W. C.; Laby, T. H., 14th ed., Longman: London, **1973**, p. 22.
50. Gould, F. A. *Dictionary of Applied Physics* Glazebrook, R., editor, Vol. 3, Macmillan: London, **1923** p. 140.
51. Brombacker, W. G.; Johnson, D. P.; Cross, J. L. *Mercury Barometers and Manometers*, NBS Monograph 8, U.S. Department of Commerce, Washington D.C., **1964**.
52. Gordon, C. L.; Wichers, E. *Ann. New York Acad. Sci.* **1957**, *65*, 369.
53. Roberts, F. G. *CI Technical Papers* No. 156, 175 and 176 MQAD. Ministry of Defence, Royal Arsenal, East Woolwich, U.K., **1958**.
54. Biram, J. G. S. *Vacuum* **1957**, *5*, 77.
55. Tilles, D. *J. Geophys. Res.* **1961**, *66*, 3003.
56. Barns, I. L.; Moore, L. J.; Machlan, L. A.; Murphy, T. J.; Shields, W. R; *J. Res. Nat. Bur. Stand.* **1975**, *79A*, 727.
57. Deslattes, R. D. *Ann. Phys. Rev. Chem.* **1980**, *31*, 456.
58. Bond, W. L.; Kaiser, W. *Phys. Chem. Solids* **1960**, *16*, 44.
59. Baker, J. A.; Tucker, T. N.; Moyer, N. E.; Buschert, R. C.; *J. Appl. Phys.* **1968**, *39*, 4365.
60. Horn, F. H. *Phys. Rev.* **1955**, *97*, 1521.
61. Thurber, W. R.; Mattis, R. L.; Liu, Y. M.; Filliben, J. J.; *Semiconductor Measurement Technology*, NBS Special Publication 400-64. **1981**.
62. Schoonover, R. M. *Notes on the Preparation of Silicon Density Artifacts*, National Bureau of Standards Internal Report 76-1019, 1976.
63. Lyon, K. G.; Salinger, G. L.; Swenson, C. A.; White, G. K. *J. Appl. Phys.* **1977**, *48*, 865.
64. Archer, R. J. *J. Electrochem. Soc.* **1957**, *104*, 619.
65. Lukes, F. *Surface Sci.* **1972**, *30*, 91.
66. Roikh, I. L.; Belitskaya, S. G.; Ordynskaya, V. V. *Izv. Akad. Nauk SSSR* (Neorg. Mat.) **1972**, *8*, 1525.
67. Raider, S. I.; Flitsch, R.; Palmer, M. J. *J. Electrochem. Soc.: Solid-State Sci. and Tech.* **1975**, *122*, 413.
68. Deslattes, R. D.; Reference wavelengths–Infrared to gamma-rays. In: Proc. Course 68, *Metrology and Fundamental Constants*, p.101, Summer School of Physics, Enrico Fermi, Varenna, Italy (1976). North Holland: Amsterdam 1980.

2.7. Reference materials for density measurement

2.7.1. Water

Physical property: Density, ρ
Unit: kg m^{-3}
Recommended reference material: Water (H$_2$O)
Range of variables: 0 to 40 °C
Physical state within the range: liquid
Contributors to the first version: D. Ambrose, G.T. Armstrong, H.A. Bowman, I. Brown, J. P. Cali, G. Girard, E.F.G. Herington, J.E. Lane, M. Menaché, T. Plebanski, J. Terrien.
Contributors to the revised version: G. Girard, M. Menaché

Intended usage: Water is a primary reference material which can be used for the determination, by weighing, of the volume of apparatus such as that used for the measurement of density. A good general description of methods is given by Bauer and Lewin (Ref. 1) and by Bowman and Schoonover (Ref. 2).

Sources of supply and/or methods of preparation: V-SMOW is the reference liquid for density determinations of water, but actually it is only used for isotopic analysis; the density can be obtained from the isotopic composition and the isotopic correction according to equation 2.3.3.1, and reference to table 2.7.1.1 which gives the density of SMOW as a function of temperature.

V-SMOW is maintained at the International Atomic Energy Agency (IAEA), Section of Isotope Hydrology, P.O. Box 100, A-1400 Vienna, Austria, who provide, on request, samples of about 100 cm^3. At the time of delivery, the isotopic composition of the sample (eqns. 2.3.1.3 and 2.3.1.4) is known with accuracy. All auxiliary reference waters could be obtained in the laboratory by distillation and degassing of sea water in order to obtain an absolute density and an isotopic composition as close as possible to those of SMOW and to minimize the risks of error when applying equation 2.3.3.1. Purification and degassing of the water samples slightly modifies their isotopic composition and, consequently, their absolute density. This also applies to evaporation. An isotopic analysis should be conducted after purification and degassing of the water sample used in the density measurements. All measurements with a water sample must then be accompanied by measurements of the auxiliary properties such as temperature, pressure, dissolved atmospheric gases, humidity and isotopic composition.

Pertinent physicochemical data: The following table of recommended values for the density of Standard Mean Ocean Water was calculated from the table presented by Bigg (Ref. 4); see also Wagenbreth and Blanke (Ref. 5). Values calculated by the relation

$$\rho(\text{SMOW})/(\text{kg m}^{-3}) = [\rho(\text{Bigg})/(\text{kg m}^{-3})] \times (999.975/999.972) \qquad (2.7.1.1)$$

were fitted within $\pm 10^{-4}$ kg m^{-3} by the following equation from which the table was generated:

$$\rho(\text{SMOW})/(\text{kg m}^{-3}) = a_0 + a_1 x + a_2 x^2 + a_3 x^3 + a_4 x^4 + a_5 x^5 \qquad (2.7.1.2)$$

where $x = t/^\circ\text{C}$ and

$$a_0 = 999.842\,594; \qquad a_1 = 6.793\,952 \times 10^{-2};$$
$$a_2 = -9.095\,290 \times 10^{-3}; \qquad a_3 = 1.001\,685 \times 10^{-4};$$
$$a_4 = -1.120\,083 \times 10^{-6}; \qquad a_5 = 6.536\,332 \times 10^{-9}.$$

If the density, $\rho(\text{sample})$, of a given specimen of purified and degassed water is required at a temperature t, it is first necessary to determine the isotopic composition of the sample. The quantities δ_{18} and δ_D are then calculated by means of equations 2.3.2.5 to 2.3.2.9 and the resulting quantities are substituted into equation 2.3.3.1 to obtain the isotopic density correction to the appropriate value of $\rho(\text{SMOW})$ taken from the table of recommended values. The table preserves the format of that from which it was derived but if the estimates of maximum uncertainties (given in table 2.1.1) are correct, the last digit of each values is not significant. Samples of purified natural waters, or preferably sea water, of undetermined isotopic concentration can be used with table 2.7.1.1 for density determinations if an accuracy of only $\pm 20 \times 10^{-3}$ kg m^{-3} is required.

Any future revision of table 2.7.1.1 will be done in co-operation with the International Association for the Properties of Steam which is currently engaged in a comprehensive review of the properties of water.

REFERENCES

1. Bauer, N.; Lewin, S. Z. *Techniques of Chemistry*, 4th ed. Weissberger, A., editor, Vol. 1, Part IV, Interscience: New York, **1972**, p. 57.
2. Bowman, H. A.; Schoonover, R. M. *J. Res. Nat. Bur. Stand.* **1967**, *71c*, 179.
3. Cox, R. A.; McCartney, M. J.; Culkin, F. *Deep-Sea Research* **1968**, *15*, 319.
4. Bigg, P. H. *Brit. J. Appl. Phys.* **1967**, *18*, 521.
5. Wagenbreth, H.; Blanke, W. *PTB-Mitteilungen* **1971**, *6*, 412.

2. Density

Table 2.7.1.1. Density of Standard Mean Ocean Water (SMOW) free from dissolved atmospheric gases, at a pressure of 101 325 Pa.

$t_{68}/°C$	.0	.1	.2	.3	.4	.5	.6	.7	.8	.9
				ρ/kg m^{-3}						
0	999.8426	8493	8558	8622	8683	8743	8801	8857	8912	8964
1	999.9015	9065	9112	9158	9202	9244	9284	9323	9360	9395
2	999.9429	9461	9491	9519	9546	9571	9595	9616	9636	9655
3	999.9672	9687	9700	9712	9722	9731	9738	9743	9747	9749
4	999.9750	9748	9746	9742	9736	9728	9719	9709	9696	9683
5	999.9668	9651	9632	9612	9591	9568	9544	9518	9490	9461
6	999.9430	9398	9365	9330	9293	9255	9216	9175	9132	9088
7	999.9043	8996	8948	8898	8847	8794	8740	8684	8627	8569
8	999.8509	8448	8385	8321	8256	8189	8121	8051	7980	7908
9	999.7834	7759	7682	7604	7525	7444	7362	7279	7194	7108
10	999.7021	6932	6842	6751	6658	6564	6468	6372	6274	6174
11	999.6074	5972	5869	5764	5658	5551	5443	5333	5222	5110
12	999.4996	4882	4766	4648	4530	4410	4289	4167	4043	3918
13	999.3792	3665	3536	3407	3276	3143	3010	2875	2740	2602
14	999.2464	2325	2184	2042	1899	1755	1609	1463	1315	1166
15	999.1016	0864	0712	0558	0403	0247	0090	9932*	9772*	9612*
16	998.9450	9287	9123	8957	8791	8623	8455	8285	8114	7942
17	998.7769	7595	7419	7243	7065	6886	6706	6525	6343	6160
18	998.5976	5790	5604	5416	5228	5038	4847	4655	4462	4268
19	998.4073	3877	3680	3481	3282	3081	2880	2677	2474	2269
20	998.2063	1856	1649	1440	1230	1019	0807	0594	0380	0164
21	997.9948	9731	9513	9294	9073	8852	8630	8406	8182	7957
22	997.7730	7503	7275	7045	6815	6584	6351	6118	5883	5648
23	997.5412	5174	4936	4697	4456	4215	3973	3730	3485	3240
24	997.2994	2747	2499	2250	2000	1749	1497	1244	0990	0735
25	997.0480	0223	9965*	9707*	9447*	9186*	8925*	8663*	8399*	8135*
26	996.7870	7604	7337	7069	6800	6530	6259	5987	5714	5441
27	996.5166	4891	4615	4337	4059	3780	3500	3219	2938	2655
28	996.2371	2087	1801	1515	1228	0940	0651	0361	0070	9778*
29	995.9486	9192	8898	8603	8306	8009	7712	7413	7113	6813
30	995.6511	6209	5906	5602	5297	4991	4685	4377	4069	3760
31	995.3450	3139	2827	2514	2201	1887	1572	1255	0939	0621
32	995.0302	9983*	9663*	9342*	9020*	8697*	8373*	8049*	7724*	7397*
33	994.7071	6743	6414	6085	5755	5423	5092	4759	4425	4091
34	994.3756	3420	3083	2745	2407	2068	1728	1387	1045	0703
35	994.0359	0015	9671*	9325*	8978*	8631*	8283*	7934*	7585*	7234*
36	993.6883	6531	6178	5825	5470	5115	4759	4403	4045	3687
37	993.3328	2968	2607	2246	1884	1521	1157	0793	0428	0062
38	992.9695	9328	8960	8591	8221	7850	7479	7107	6735	6361
39	992.5987	5612	5236	4860	4483	4105	3726	3347	2966	2586
40	992.2204									

* The leading figure decreases by 1.0.

2.7.2. Mercury

Physical property: Density ρ
Unit: kg m^{-3}
Recommended reference material: Mercury (Hg)
Range of variables: −20 to 300 °C
Contributors to the first version: D. Ambrose, I. Brown, E.F.G. Herington
No revision made

Intended usage: Mercury is a primary reference material which can be used for the determination, by weighing, of the volume of apparatus such as pycnometers and other equipment used for density determinations. A good general description of methods is given by Bauer and Lewin (Ref. 1).

Source of supply and/or methods of preparation: Samples of mercury can be purified by treatment with dilute nitric acid (1:20) followed by washing with water and by distillation under reduced pressure (about 3 kPa) in the presence of clean air. Further details on purification are given in Section 4.1.

Pertinent physicochemical data: Values for the density of mercury are given in table 2.7.2.1 which is based on data presented by Bigg (Ref. 2) corrected to bring the values to the International Practical Temperature Scale of 1968.

The table was generated from the value $\rho_0 = 13\,595.08$ kg m^{-3} by means of the following equations:

$$\rho = \rho_0/(1 + \alpha x) \tag{2.7.2.1}$$

and

$$10^8 \alpha = a_0 + a_1 x + a_2 x^2 + a_3 x^3 \tag{2.7.2.2}$$

where $x = t/°$ C and

$$a_0 = 18\,152.53$$
$$a_1 = 0.589\,8$$
$$a_2 = 31.507 \times 10^{-4}$$
$$a_3 = 2.405 \times 10^{-6}$$

REFERENCES

1. Bauer, N.; Lewin, S. Z. *Techniques of Chemistry*, 4th ed. Weissberger, A., editor, Vol. 1, Part IV, Interscience: New York, **1972**, p. 57.
2. Bigg, P. H. *Brit. J. Appl. Phys.* **1964**, *15*, 1111.

2. Density

Table 2.7.2.1. Density of Mercury $\rho/\text{kg m}^{-3}$ at a pressure of 101 325 Pa for temperatures $t/°C$ on the International Practical Temperature Scale of 1968

$t/°C$					Density/kg m^{-3}						$\Delta\rho/\Delta t$ kg m^{-3} K^{-1}	Uncertainty of ρ $\pm\delta\rho$ / kg m^{-3}
$t/°C$	−20	−19	−18	−17	−16	−15	−14	−13	−12	−11		
13	644.59*	642.11	639.62	637.14	634.66	632.18	629.70	627.22	624.75	622.27	2.48	0.3
$t/°C$	−10	−9	−8	−7	−6	−5	−4	−3	−2	−1		
13	619.79	617.32	614.85	612.37	609.90	607.43	604.96	602.49	600.02	597.55	2.47	0.04
$t/°C$	0	1	2	3	4	5	6	7	8	9		
13	595.08	592.62	590.15	587.68	585.22	582.76	580.29	577.83	575.36	572.91	2.46	0.02
13	570.44	567.98	565.52	563.07	560.60	558.15	555.69	553.24	550.78	548.33	2.46	0.01
13	545.87	543.42	540.97	538.52	536.06	533.62	531.17	528.71	526.26	523.82	2.45	0.01
13	521.37	518.92	516.47	514.03	511.58	509.14	506.70	504.26	501.82	499.37	2.44	0.02
13	496.93	494.49	492.05	489.62	487.18	484.73	482.29	479.86	477.42	474.99	2.44	
13	472.55	470.11	467.68	465.25	462.82	460.38	457.96	455.52	453.09	450.67	2.43	0.04
13	448.23	445.81	443.37	440.95	438.52	436.10	433.67	431.24	428.82	426.39	2.43	
13	423.97	421.55	419.13	416.70	414.28	411.86	409.43	407.01	404.60	402.18	2.42	0.06
13	399.77	397.34	394.92	392.50	390.09	387.67	385.26	382.85	380.42	378.01	2.42	
13	375.60	373.18	370.77	368.36	365.94	363.54	361.12	358.72	356.30	353.89	2.41	0.08
$t/°C$	100	110	120	130	140	150	160	170	180	190		
13	351.5	327.4	303.4	279.4	255.4	231.5	207.6	183.7	159.8	136.0	2.40	0.1
$t/°C$	200	210	220	230	240	250	260	270	280	290		
13	112.1	088.3	064.5	040.7	016.9	993.0#	969.2#	945.4#	921.5#	897.6#	2.38	0.2
$t/°C$	300											
12	873.7											

* Leading figures are 13 except where otherwise indicated.
\# Leading figures are 12 in each instance.

2.7.3. Crystal of Silicon

Physical property: Density ρ
Unit: kg m^{-3}
Recommended reference material: Crystal of silicon
Range of variables: 20 to 40 °C
Physical state within the range: Solid
Contributor: R. S. Davis

Intended usage: A crystal of silicon is a primary reference material which can be used for the determination of the density of liquids or solids by means of hydrostatic weighing. A good general description of methods is given by Bauer and Lewin (Ref. 1) and by Bowman and Schoonover (Ref. 2)

Sources of supply and/or methods of preparation: Samples of 100 g (SRM 1840) or 200 g (SRM 1841) with certified values for density are available from supplier (B).

Pertinent physicochemical data: The following certified values for density at $t_{68} = 20$ °C are directly traceable to the units of mass and length as defined in the International System of Units (SI).

SRM 1840 (100 g) $\rho = (2\,329.074 \pm 0.019)$ kg m^{-3}
SRM 1841 (200 g) $\rho = (2\,329.075 \pm 0.017)$ kg m^{-3}

The uncertainties are at the 99 per cent confidence level and include traceability to the SI definitions of mass and length. The coefficient of volume expansion between -3 °C and 37°C is given by the equation:

$$10^8 \alpha/\text{K} = 3(236.28 + 1.030\,6x - 0.003\,35x^2) \qquad (2.7.3.1)$$

where $x = t/$°C (Ref. 3). The compressibility of silicon is approximately 1.0×10^{-11} Pa^{-1} (Ref. 4).

REFERENCES

1. Bauer, N.; Lewin, S. Z. *Techniques of Chemistry*, 4th ed. Weissberger, A., editor, Vol. 1, Part IV, Interscience: New York, **1972**, p. 57.
2. Bowman, H. A.; Schoonover, R. M. *J. Res. Nat. Bur. Stand.* **1967**, *71c*, 179.
3. Lyon, K. G.; Salinger, G. L.; Swenson, C. A.; White, G. K. *J. Appl. Phys.* **1977**, *48*, 865.
4. Deslattes, R. D.; Reference wavelengths-Infrared to gamma-rays. In: Proc. Course 68, *Metrology and Fundamental Constants*, Summer School of Physics, Enrico Fermi, Varenna, Italy, 1976. North Holland-Elsevier: Amsterdam, 1980, p. 101.

2.7.4. 2,2,4-Trimethylpentane

Physical property: Density ρ
Unit: kg m^{-3}
Recommended reference material: 2,2,4-trimethylpentane (C$_8$H$_{18}$)
Range of variables: 20 to 50 °C
Physical state within the range: Liquid
Contributors to the first version: I. Brown, J.P. Cali, E.F.G. Herington, T. Plebanski.
Contributor to the revised version: G. Girard

Intended usage: Samples of 2,2,4-trimethylpentane can be used for the determination, by weighing, of the volume of apparatus such as is used for density determinations. A good general description of methods is given by Bauer and Lewin (Ref. 1) and by Bowman and Schoonover (Ref. 2).

Sources of supply and/or methods of preparation: Samples with certified values of density are available from suppliers (A) and (C).

Pertinent physicochemical data: The following values of density, which apply to air-saturated material available from supplier (A), were determined using precise pycnometric and flotation methods (Ref. 3), the pycnometers were calibrated with water (Refs. 4 to 6) and mercury (Ref. 7).

Table 2.7.4.1. Density of 2,2,4-trimethylpentane from supplier (A), mole fraction purity of 0.9975, air saturated

$\dfrac{t_{68}}{°C}$	$\dfrac{\rho}{\text{kg m}^{-3}}$	$\dfrac{t_{68}}{°C}$	$\dfrac{\rho}{\text{kg m}^{-3}}$
20	691.959	40	675.348
25	687.849	45	671.124
30	683.711	50	666.871
35	679.543		

Uncertainty limit at 99 per cent confidence level $\pm 5 \times 10^{-3}$ kg m^{-3}. The density values were calculated from the equation

$$\rho/(\text{kg m}^{-3}) = \sum_{0}^{3} c_i (t/°C)^i \qquad (2.7.4.1)$$

where $c_0 = 708.113$, $c_1 = -7.962 \times 10^{-1}$, $c_2 = -5.77 \times 10^{-4}$, $c_3 = 6.7 \times 10^{-8}$.

REFERENCES

1. Bauer, N.; Lewin, S. Z. *Techniques of Chemistry*, 4th ed. Weissberger, A., editor, Vol. 1, Part IV, Interscience: New York, **1972**, p. 57.
2. Bowman, H. A.; Schoonover, R. M. *J. Res. Nat. Bur. Stand.* **1967**, *71c*, 179.
3. Kozdon, A. *Proc. 1st Intern. Conf. Calorimetry and Thermodynamics*, Vol. 4, Warsaw, **1969**, p. 831.
4. Kell, G. S. *J. Chem. Eng. Data* **1967**, *12*, 66.
5. Franks, F., *Water, a Comprehensive Treatise*, Plenum Press: New York, **1972**, p. 373.
6. Girard, G.; Menaché, M. *Metrologia* **1971**, *7*, 83.
7. Bigg, P. H. *Brit. J. Appl. Phys.* **1964**, *15*, 1111.

2.7.5. Cyclohexane

Physical property: Density ρ
Unit: kg m^{-3}
Recommended reference material: Cyclohexane (C_6H_{12})
Range of variables: 20 to 50 °C
Physical state within the range: Liquid
Contributor to the first version: T. Plebanski
No revision made

Intended usage: Samples of cyclohexane can be used for the determination, by weighing, of the volume of apparatus such as that used for density measurement. A good general description of methods is given by Bauer and Lewin (Ref. 1) and by Bowman and Schoonover (Ref. 2).

Sources of supply and/or method of preparation: Samples with certified values of density are available from supplier (A).

Pertinent physicochemical data: The following values of density, which apply to air-saturated material available from supplier (A) were determined using precise pycnometric and flotation methods (Ref. 3), the pycnometers were calibrated with water (Refs. 4 to 6) and mercury (Ref. 7).

2. Density

Table 2.7.5.1. Density of cyclohexane, supplier (A) mole fraction purity of 0.9998, air-saturated.

$\dfrac{t_{68}}{°C}$	$\dfrac{\rho}{\text{kg m}^{-3}}$	$\dfrac{t_{68}}{°C}$	$\dfrac{\rho}{\text{kg m}^{-3}}$
20	778.583	40	759.624
25	773.896	45	754.805
30	769.172	50	749.960
35	764.414		

Uncertainty limits at 99 per cent confidence level are $\pm 5 \times 10^{-3}$ kg m^{-3}. These density values were calculated from the equation

$$\rho/(\text{kg m}^{-3}) = \sum_{0}^{3} c_i (t/°C)^i \qquad (2.7.5.1)$$

where $c_0 = 7\,796.922$, $c_1 = -8.989 \times 10^{-1}$, $c_2 = -9.67 \times 10^{-4}$, $c_3 = 3.19 \times 10^{-6}$.

REFERENCES

1. Bauer, N.; Lewin, S. Z. *Techniques of Chemistry*, 4th ed. Weissberger, A., editor, Vol. 1, Part IV, Interscience: New York, **1972**, p. 57.
2. Bowman, H. A., Schoonover, R. M. *J. Res. Nat. Bur. Stand.* **1967**, *71c*, 179.
3. Kozdon, A. *Proc. 1st Intern. Conf. Calorimetry and Thermodynamics*, Vol. 4, Warsaw, **1969**, p. 831.
4. Kell, G. S. *J. Chem. Eng. Data* **1967**, *12*, 66.
5. Franks, F., *Water, a Comprehensive Treatise*, Plenum Press: New York, **1972**, p. 373.
6. Girard, G., Menaché, M. *Metrologia* **1971**, *7*, 83.
7. Bigg, P. H. *Brit. J. Appl. Phys.* **1964**, *15*, 1111.

2.7.6. *trans*-Bicyclo [4,4,0]decane

Physical property: Density ρ
Unit: kg m^{-3}
Recommended reference material: *trans*-bicyclo [4,4,0]decane ($C_{10}H_{18}$)
(*trans*-decahydronaphthalene)
Range of variables: 20 to 50 °C
Physical state within the range: Liquid
Contributors to the first version: I. Brown, J.P. Cali, E.F.G. Herington, T. Plebanski.
No revision made

Intended usage: Samples of *trans*-bicyclo [4,4,0]decane can be used for the determination, by weighing, of the volume of apparatus such as that used for density measurement. A good general description of methods is given by Bauer and Lewin (Ref. 1) and by Bowman and Schoonover (Ref. 2).

Sources of supply and/or method of preparation: Samples with certified values of density are available from suppliers (A) and (C).

Pertinent physicochemical data: The following values of density, which apply to air-saturated material available from supplier (A) were determined using precise pycnometric and flotation methods (Ref. 3), the pycnometers were calibrated with water (Refs. 4 to 6) and mercury (Ref. 7).

Table 2.7.6.1. Density of *trans*-bicyclo[4,4,0]decane from supplier (A) with an estimated mole fraction purity of 0.970, air-saturated.

$\dfrac{t_{68}}{°C}$	$\dfrac{\rho}{\text{kg m}^{-3}}$	$\dfrac{t_{68}}{°C}$	$\dfrac{\rho}{\text{kg m}^{-3}}$
20	869.623	40	854.693
25	865.895	45	850.945
30	862.165	50	847.185
35	858.432		

Uncertainty limit at 99 per cent confidence level $\pm 5 \times 10^{-3}$ kg m^{-3}. These density values were calculated from the equation

$$\rho/(\text{kg m}^{-3}) = \sum_{0}^{3} c_i (t/°C)^i \qquad (2.7.6.1)$$

where $c_0 = 884.579$, $c_1 = -7.513 \times 10^{-1}$, $c_2 = 2.440 \times 10^{-4}$, $c_3 = -3.519 \times 10^{-6}$

The following values of density apply to air-saturated material available from supplier (C). The density values were determined by the American Petroleum Institute Research Project 6 at the Carnegie-Mellon University.

Table 2.7.6.2. Density of *trans*-bicyclo[4,4,0]decane, sample 561-x, mole fraction purity of $(0.999\ 8 \pm 0.000\ 2)$

$\dfrac{t_{68}}{°\text{C}}$	$\dfrac{\rho}{\text{kg m}^{-3}}$
20	869.71 ± 0.08
25	865.92 ± 0.08
30	862.22 ± 0.08

REFERENCES

1. Bauer, N.; Lewin, S. Z. *Techniques of Chemistry*, 4th ed. Weissberger, A., editor, Vol. 1, Part IV, Interscience: New York, **1972**, p. 57.
2. Bowman, H. A., Schoonover, R. M. *J. Res. Nat. Bur. Stand.* **1967**, *71c*, 179.
3. Kozdon, A. *Proc. 1st Intern. Conf. Calorimetry and Thermodynamics*, Vol. 4, Warsaw, **1969**, p. 831.
4. Kell, G. S. *J. Chem. Eng. Data* **1967**, *12*, 66.
5. Franks, F., *Water, a Comprehensive Treatise*, Plenum Press: New York, **1972**, p. 373.
6. Girard, G., Menaché, M. *Metrologia* **1971**, *7*, 83.
7. Bigg, P. H. *Brit. J. Appl. Phys.* **1964**, *15*, 1111.

2.7.7. 100 Neutral Solvent

Physical property: Density ρ
Unit: kg m^{-3}
Recommended reference material: 100 Neutral solvent
Range of variables: 10 to 50 °C
Physical state within the range: Liquid
Contributor: G. Girard

Intended usage: Samples of this liquid can be used for the determination, by weighing, of the volume of apparatus such as that used for the measurement of density. A good general description of methods is given by Bauer and Lewin (Ref. 1) and by Bowman and Schoonover (Ref. 2).

Sources of supply and/or method of preparation: Samples with certified values of density are available from supplier (D).

Pertinent physicochemical data: The following value for the density was determined using hydrostatic or pycnometric methods.

  The density of water was taken from the National Bureau of Standards tables.
  The temperature is relative to IPTS-68.
  Density at 20 °C is 861.80 kg m^{-3}.
  Cubic thermal expansion at 20 °C is 7.4×10^{-4} K^{-1}
  Relative uncertainty of the determination is $\pm 5 \times 10^{-5}$.
  Variation of the density after keeping at room temperature for six months is less than 3×10^{-2} kg m^{-3}.

REFERENCES

1. Bauer, N.; Lewin, S. Z. *Techniques of Chemistry*, 4th ed. Weissberger, A., editor, Vol. 1, Part IV, Interscience: New York, **1972**, p. 57.
2. Bowman, H. A., Schoonover, R. M. *J. Res. Nat. Bur. Stand.* **1967**, *71c*, 179.

2.7.8. Flugene 112

Physical property: Density ρ
Unit: kg m^{-3}
Recommended reference material: Flugene 112 (1,1,2,2-tetrachloro 1,2-difluoroethane) (CCl$_2$FCCl$_2$F)
Range of variables: 40 to 60 °C
Physical state within the range: Liquid
Contributor: G. Girard

Intended usage: Samples of flugene 112 can be used for the determination, by weighing, of the volume of apparatus such as that used for the measurement of density. A good general description of methods is given by Bauer and Lewin (Ref. 1) and by Bowman and Schoonover (Ref. 2).

Sources of supply and/or method of preparation: Samples with certified values of density are available from supplier (D).

Pertinent physicochemical data: The following value for the density was determined using hydrostatic or pycnometric methods.

    The density of water was taken from the National Bureau of Standards tables.
    Samples are solid below 30 °C.
    The temperature is IPTS-68.
    Density at 50°C is 1595.10 kg m^{-3}.
    Cubic thermal expansion at 50 °C is 8×10^{-4} K^{-1}.
    Relative uncertainty of the determination is $\pm 5 \times 10^{-5}$.
    Variation of the density after keeping at room temperature for six months is less than 3×10^{-2} kg m^{-3}.

REFERENCES

1. Bauer, N.; Lewin, S. Z. *Techniques of Chemistry*, 4th ed. Weissberger, A., editor, Vol. 1, Part IV, Interscience: New York, **1972**, p. 57.
2. Bowman, H. A., Schoonover, R. M. *J. Res. Nat. Bur. Stand.* **1967**, *71c*, 179.

2.8. Contributors to the revised versions

R. S. Davis,
National Bureau of Standards,
U.S. Department of Commerce,
Gaithersburg, MD 20899 (U.S.A)

G. Girard,
Bureau International des Poids et Mesures,
Pavillon de Breteuil,
F-92310 Sèvres (France)

M. Menaché
7 rue de Reims
F-75013 Paris (France)

2.9. List of Suppliers

(A) Polish Committee for Standardization and Measures
Division of Physicochemical Metrology,
Ul. Elektoralna 2, PL 00-139 Warsaw (Poland).

(B) Office of Standard Reference Materials,
U.S. Department of Commerce,
National Bureau of Standards,
Gaithersburg, MD 20899 (U.S.A.)

(C) API Standard Reference Materials,
Carnegie-Mellon University,
Schenley Park, Pittsburgh,
PA 15213 (U.S.A.)

(D) Service de Matériaux de Référence,
Laboratoire National d'Essais,
1 rue Gaston Boissier,
F-75015 Paris (France)

3

SECTION: SURFACE TENSION

COLLATOR: D. V. S. JAIN

CONTENTS:

3.1. Introduction

3.2. Methods of measurement

3.3. Reference materials for surface tension
    3.3.1. Water
    3.3.2. Benzene
    3.3.3. Heptane

3.4. Contributors

3.1. Introduction

Ever since Young (Ref. 1) (1805) and Laplace (Ref. 2) (1806) demonstrated that the pressure difference across a point on a surface is determined by the surface tension and the curvature of the surface at that point, it has been possible to measure surface tension precisely, and without the need for any form of calibration. However, for at least a century after Young's and Laplace's work, very few reliable measurements of surface tension were made. The major sources of error were (and these sources are often still present in modern measurements) the use of imprecise measuring techniques, the use of inappropriate mathematical expressions to calculate values of surface tension from the experimental observations and failure to exclude surface active impurities from the apparatus. Experimental work since 1910 by Richards and Carver (Ref. 3) and by Harkins *et al.* (Refs. 4, 5) has shown how to avoid the errors arising from the first and last of these sources while correct mathematical expressions, and convenient methods for using them, have been developed by Laplace (Ref. 2), Rayleigh (Ref. 6), Bashforth and Adams (Ref. 7), and Sugden (Refs. 8, 9). Notwithstanding the earlier comment that the calibration of apparatus is unnecessary, nevertheless for many of the measuring methods calibration is convenient and often desirable because of difficulties in obtaining accurate values for some dimensions of the apparatus. Water, benzene, and heptane are recommended as suitable calibration materials. It is recommended that the symbol γ be used to denote surface tension, and that the unit be mN m^{-1} (millinewton per metre). This unit gives the same number as the older c.g.s. unit of dyne cm^{-1}.

3.2. Methods of measurement

Excellent reviews of available methods have been given by Harkins and Alexander (Ref. 5), Adam (Ref. 10) and Adamson (Ref. 11).

The capillary rise technique is the favoured absolute method and experimental details have been given by Richards and Carver (Ref. 3). Sugden (Ref. 8) has given an accurate and convenient method for the calculation of values of the surface tension from capillary rise, and this method has recently been improved by Lane (Ref. 12). Absolute values of surface tension of common liquids can be measured to ± 0.02 mN m^{-1} by the capillary rise method but this requires very accurate measurement of the capillary radius which is difficult to achieve. The problem can be avoided by the determination of the capillary radius by calibration with a liquid of known surface tension.

The maximum bubble pressure method is described by Sugden (Ref. 9), and for properly formed tips has the advantage over the capillary rise method of being independent of contact angle. If the contact angle is $< 90°$, the bubble becomes detached from the outside of the tip; otherwise it becomes detached from the inside of the tip. The appropriate radius must be used for the calculation of the surface tension. Correction factors for the non-sphericity of the bubble have been given by Sugden (Ref. 9), and in a more convenient and accurate form by Johnson and Lane (Ref. 13). The tip radii can also be determined by calibration. In this manner it is possible to obtain values of the surface tension reliable to ± 0.1 mN m^{-1}.

A third reliable method employs the Wilhemly vertical plate balance with a smooth plate (Refs. 14 to 16). Problems associated with rough plates have been discussed by Lane and Jordan (Ref. 17) and also by Princen (Ref. 18). With smooth plates no special corrections are required for the calculation of surface tension. With care, absolute measurements of surface tension to ± 0.05 mN m^{-1} can be achieved. No advantage is obtained by calibration with a liquid of known surface tension.

Many laboratories have used the du Nouy balance and a horizontal platinum ring but this method is not recommended, as uncertain correction factors have to be introduced, and the maximum force necessary to raise the ring is difficult to measure. These problems are discussed by Adam (Ref. 10). The drop-weight (or drop-volume) method is based on a dynamic measurement and thus lacks the precision of the other methods given above, all of which are static. However, it is particularly convenient for measuring the tension at a liquid/liquid interface. Suitable experimental arrangements are described by Harkins and Alexander (Ref. 5). Harkins and Brown (Ref. 4) have determined empirically the values of the correction factors needed to calculate surface tension from the experimental data.

REFERENCES

1. Young, T. *Phil. Trans.* **1805**, *95*, 65.
2. Laplace, P. S. *Mecanique celesté*, suppl. 10th vol., **1806**.
3. Richards T. W.; Carver, E. K. *J. Amer. Chem. Soc.* **1921**, *43*, 827.
4. Harkins W. D.; Brown, F. E. *J. Amer. Chem. Soc.* **1919**, *41*, 499.
5. Harkins W. D.; Alexander, A. E. *Techniques of Organic Chemistry*, 3rd ed. Weissberger, A., editor, Vol. I, Part 1, Interscience: New York, **1963**, p. 757.
6. Lord Rayleigh, *Proc. Roy. Soc.* **1915-16**, *A92*, 184.
7. Bashforth F.; Adams, J. C. *An attempt to test the theories of capillary action.* Cambridge University Press: Cambridge, **1883**.
8. Sugden, S. *J. Chem. Soc.* **1921**, *119*, 1483.
9. Sugden, S. *J. Chem. Soc.* **1922**, *121*, 858; **1924**, *125*, 27.
10. Adam, N. K. *The Physical Chemistry of Surfaces*, 2nd ed., Oxford University Press: Oxford, **1941**, p. 363.
11. Adamson, A. W. *Physical Chemistry of Surfaces*, 2nd ed., Interscience: New York, **1967**, p. 9.
12. Lane, J. E. *J. Coll. Interface Sci.* **1973**, *42*, 145.
13. Johnson C. H. J.; Lane, J. E. *J. Coll. Interface Sci.*, in press.
14. Dervichian, D. G. *J. Phys. Radium*, **1935**, *6*, 221, 429.
15. Harkins W. D.; Anderson, T. F. *J. Amer. Chem. Soc.* **1937**, *59*, 2189.
16. Jordan, D. O.; Lane, J. E. *Austral. J. Chem.* **1964**, *17*, 7.
17. Lane, J. E.; Jordan, D. O. *Austral. J. Chem.* **1970**, *23*, 2153; **1971**, *24*, 1297.
18. Princen, H. M. *Austral. J. Chem.* **1970**, *23*, 1789.

3.3. Reference materials for surface tension measurements

3.3.1. Water

Physical property: Surface tension γ
Unit: $mN\ m^{-1}$
Recommended reference material: Water (H_2O)
Range of variables: 0 to 60 °C
Physical state within the range: liquid
Contributors to the first version: I. Brown, J. Franc, J. E. Lane
No revision made

Intended usage: Water can be used for the calibration and for the testing of apparatus used to measure surface tension.

Sources of supply and/or methods of preparation: Purification by distillation, redistillation from alkaline permanganate and a third distillation from a trace of sulphuric acid is recommended by Harkins and Brown (Ref. 1). Because water is easily contaminated by surface active materials that are often present in air, the apparatus and the water should be prepared immediately before measurements are made.

Pertinent physicochemical data: Harkins and Alexander (Ref. 2) have examined the surface tension data for the water/air interface as a function of temperature and find that values of surface tension can be calculated to within experimental error (± 0.02 mN m^{-1}) by means of the following equation

$$\gamma_{\text{H}_2\text{O}/\text{air}}/(\text{mN m}^{-1}) = 75.680 - 0.138x - 3.56 \times 10^{-4}x^2 + 4.7 \times 10^{-7}x^3 \qquad (3.3.1.1)$$

where $x = (t/^\circ\text{C})$ and the temperature is restricted to the range 0 to 60 °C.

This equation which gives the values 72.78 mN m^{-1} for 20 °C and 72.01 mN m^{-1} for 25 °C for the surface tension of the water/air interface can be used to calculate values for water prepared by the recommended method (Ref. 1).

REFERENCES

1. Harkins, W. D.; Brown, F. E. *J. Amer. Chem. Soc.* **1919**, *41*, 499.
2. Harkins W. D.; Alexander, A. E. *Techniques of Organic Chemistry*, 3rd ed. Weissberger, A., editor, Vol. I, Part 1, Interscience: New York, **1963**, p. 757.

3.3.2. Benzene

Physical property: Surface tension γ
Units: mN m^{-1}
Recommended reference material: Benzene (C$_6$H$_6$)
Range of variables: 10 to 60 °C
Physical state within the range: liquid
Contributors to the first version: I. Brown, J. Franc, J. E. Lane
No revision made

Intended usage: Benzene can be used for the calibration and for the testing of apparatus used to measure surface tension.

Sources of supply and/or method of preparation: Samples can be purified by shaking them with mercury and then with sulphuric acid. The samples should next be distilled and then should be fractionally frozen five times (Refs. 1 to 3).

Pertinent physicochemical data: The value of the surface tension of the benzene/air interface at 20 °C is well established (Refs. 1 to 4) as (28.88 ± 0.01) mN m^{-1}. A considerable amount of data for temperatures from 10 to 60 °C has been collated by Young and Harkins (Ref. 5), who produced the equation

$$\gamma_{\text{benzene}/\text{air}}/(\text{mN m}^{-1}) = 31.58 - 0.137x + 0.0001x^2 \qquad (3.3.2.1)$$

where $x = (t/°C)$. This equation gives values 28.88 mN m^{-1} for 20 °C and 28.22 mN m^{-1} for 25 °C. The values for 30 °C given by Koefoed and Villadsen (Ref. 6) and for 60 °C given by Marechal (Ref. 7) agree with values calculated from the equation to within ±0.03 mN m^{-1}. The data recorded by Donaldson and Quale (Ref. 8) appear to be too low for temperatures above 60 °C. The equation given above should produce values with uncertainty better than ±0.05 mN m^{-1} and can be used to calculate values for samples of benzene purified as described (Refs. 1 to 3).

REFERENCES

1. Harkins, W. D.; Brown, F. E. *J. Amer. Chem. Soc.* **1919**, *41*, 499.
2. Richards, T. W.; Shipley, J. W. *J. Amer. Chem. Soc.* **1914**, *36*, 1825.
3. Richards, T. W.; Carver, E. K. *J. Amer. Chem. Soc.* **1921**, *43*, 827.
4. Richards, T. W.; Coombs, L. B. *J. Amer. Chem. Soc.* **1915**, *37*, 1656.
5. Young T. F.; Harkins, W. D. *International Critical Tables*, Washburn, E. W., editor in chief, Vol. IV, McGraw-Hill: New York, **1928**, p. 454.
6. Koefoed, J.; Villadsen, J. V. *Acta. Chem. Scand.* **1958**, *12*, 1124.
7. Marechal, J. *Bull. Soc. Chem. Belges* **1952**, *61*, 149.
8. Donaldson, R. E.; Quale, O. R. *J. Amer. Chem. Soc.* **1950**, *72*, 35.

3.3.3. Heptane

Physical property: Surface tension γ
Unit: mN m^{-1}
Recommended reference material: Heptane (C$_7$H$_{16}$)
Range of variables: 20 to 50°C
Physical state within the range: liquid
Contributor: D. V. S. Jain

Intended usage: Heptane can be used for the calibration and for testing of apparatus used to measure surface tension.

Sources of supply and/or methods of preparation: Commercially available heptane (having mole fraction purity > 0.99) distilled three times from sodium using a fractionating column is suitable. The fraction distilling between 97.5 to 98.5 °C at atmospheric pressure should be stored over molecular sieves prior to use.

Pertinent physicochemical data: The value of the surface tension of the heptane/air interface has been well established (Refs. 1 to 3) as (20.30 ± 0.02) mN m^{-1} at 20 °C. Values over the range of temperature from 20 to 50 °C have been reported from several laboratories (Refs. 1 to 4). The results can be fitted to the equation

$$\gamma_{\text{heptane/air}}/(\text{mN m}^{-1}) = 23.298 - 0.2045 \times 10^{-4}x + 0.340 \times 10^{-10}x^2$$
$$- 0.353 \times 10^{-16}x^3 \tag{3.3.3.1}$$

where $x = (t/°C)$. The equation gives values of surface tension in agreement with the results from three different laboratories within ± 0.02 mN m^{-1} in the temperature range 20 to 50 °C.

REFERENCES

1. Quayle, O. R.; Day, R. A.; Bown, G. M. *J. Amer. Chem. Soc.* **1944**, *66*, 935.
2. Koefoed, J.; Villadsen, J. V. *Acta Chem. Scand.* **1958**, *12*, 1124.
3. Aveyard, R. *Trans. Faraday Soc.* **1967**, *63*, 2778.

3.4. Contributor to the revised versions

D. V. S. Jain,
Department of Chemistry,
Panjab University,
Chandigarh - 160014 (India)

4

SECTION: VISCOSITY

COLLATORS: W. KÜNZEL, H. F. VAN WIJK, K. N. MARSH

CONTENTS:

4.1. Introduction

4.2. Methods of measurement
- 4.2.1. Capillary viscometers
- 4.2.2. Falling body viscometers
- 4.2.3. Rotational viscometers
- 4.2.4. Other viscometers
- 4.2.5. Calibration using reference materials

4.3. Reference materials for viscometers
- 4.3.1. Water
- 4.3.2. Certified reference materials

4.4. Contributors

4.5. List of Suppliers

4.1. Introduction

The recommended (ISO, IUPAC) symbol for viscosity (still sometimes referred to as dynamic viscosity, though this term is now commonly reserved for a frequency dependent property) is η (Ref. 1). The SI unit for viscosity is the pascal second (Pa s). Alternatives are the newton

second per square metre (N s m^{-2}) and the kilogram per metre per second (kg m^{-1} s^{-1}).*
The viscosity is 1 pascal second for a fluid for which there is a tangential force of 1 newton on 1 square metre of either of two infinite parallel planes 1 metre apart when (a) the space between those planes is filled with the fluid, (b) one of the planes moves with a velocity of 1 metre per second in its own plane relative to the other, and (c) the flow of the fluid is laminar.

The recommended (ISO, IUPAC) symbol of kinematic viscosity is ν (Ref. 1). The SI unit for kinematic viscosity is the square metre per second (m^2 s^{-1}) where 1 square metre per second is the kinematic viscosity of a fluid having a viscosity of 1 pascal second and a density of 1 kilogram per cubic metre.**

Additional ways of representing the viscosity of dilute solutions have been defined (Ref. 2). The relative viscosity η_r is defined as the ratio between the viscosity of the solution and the viscosity of the pure solvent at the same temperature and pressure, while the relative viscosity increment η_i, formerly called the specific viscosity is defined by $\eta_i = \eta_r - 1$. Both these quantities are dimensionless.

This recommendation concerns only reference liquids which are Newtonian; that is their viscosity is independent of the rate of shear. A detailed discussion of Newtonian and non-Newtonian behaviour of fluids is given in reference 2.

The viscosity η is given by
$$\tau_{xz} = F_x/A = \eta(\mathrm{d}v_x/\mathrm{d}z). \tag{4.1.1}$$

The kinematic viscosity is given by
$$\nu = \eta/\rho, \tag{4.1.2}$$

where ρ is the density, Equation (4.1.1) can be used to determine the viscosity of a fluid directly by measuring the shear stress τ_{xz} required to give laminar flow with a velocity gradient $\mathrm{d}v_x/\mathrm{d}z$. In practice, such a measurement is difficult to make. There are a variety of other methods which can be used to make measurements of the viscosity (Refs. 3 to 5, 7). One such method is to measure the time taken for a known volume of liquid to flow through a capillary of known dimensions under a known pressure difference (Ref. 3). This method is not routinely used because it is difficult to measure the area of the capillary with sufficient

---

* A commonly used unit for the viscosity the poise (P) has been defined (Ref. 1) in terms of SI units as P = 10^{-1} Pa s. A commonly used subunit is the centipoise (cP) where cP = 10^{-3} Pa s. This unit does not belong to the International System of Units and its use should be progressively discouraged.

** A commonly used unit for kinematic viscosity the stokes (St) has been defined (Ref. 1) in terms of SI units as St = 10^{-4} m^2 s^{-1}. A commonly used subunit is the centistokes (cSt) where cSt = 10^{-6} m^2 s^{-1}. This unit does not belong to the International System of Units and its use should be progressively discouraged. An author who uses either the poise or the stokes must define it in terms of the SI unit in each publication in which it is used.

precision and it is also necessary to apply corrections for various kinetic effects which occur at the ends of the capillary. Thus a reference material is convenient for the calibration of viscometers. The internationally recognized reference material is water in equilibrium with air at 293.15 K and atmospheric pressure (Ref. 6).

The viscosity of fluids can range from less than 10^{-6} Pa s to more than 10^6 Pa s. For organic liquids the viscosity ranges generally from 10^{-3} Pa s to 10^3 Pa s. Hence viscometers which are designed to give a suitable measured quantity when calibrated with water will have unsuitable values of the measured quantities when used to determine the viscosities of highly viscous liquids. For this reason reference materials with certified values are used to calibrate viscometers. The certified values of these reference materials are always determined by comparison with the viscosity of water, either directly or indirectly, through a chain of intermediate reference liquids and master class instruments. In each step of the comparison a part of the accuracy is lost, so that with increasing viscosity the uncertainty increases. The viscosities of a number of reference materials relative to that of water are often given with an uncertainty considerably less than the known uncertainty of the absolute value for water (Refs. 6, 7). The consistency of the various national viscosity scales is checked by a continuing programme of direct international comparisons made by the exchange of both master viscometers and reference materials between the various suppliers. This comparison suggests that the viscosity of reference materials can, at best, be certified to ± 0.2 per cent with reference to water, even though in many cases agreement between laboratories is better than ± 0.1 per cent (Ref. 8).

Generally, selected pure materials are not used as reference materials because the viscosity usually depends significantly and indeterminately on the purity of the material. Suppliers use a wide range of reference materials which are usually of indeterminate composition, the materials being petroleum oils, polyisobutenes, silicone oils, undefined polymers and asphalts, and molten glasses. It is necessary that the materials show Newtonian behaviour, that they are non-corrosive, and, except for the molten glasses, have a high solubility in at least two readily available organic solvents so as to enable appropriate cleaning. In the case of viscosities greater than 10^2 Pa s, special care must be taken to ensure that the liquids are Newtonian in behaviour under the conditions of calibration.

The viscosity range covered by the reference materials available from various laboratories is given in section 3.2. There are a considerable number of reference materials for use within the viscosity range 10^{-3} to 30 Pa s, while in the range 30 to 10^4 Pa s there is an inadequate number of reference materials and there are no reference materials for viscosity greater than 10^4 Pa s at 298 K. The coverage of the temperature scale is not adequate in that only a few reference materials are certified below 273 K and there are no materials for use within the temperature range 410 to 850 K. A variety of glasses are certified for use between 737 and 1818 K. One supplier provides reference materials with certified numerical values for an equation that enables viscometers to be calibrated at temperatures other than those at which the viscosity values are certified. In the viscosity range 10^{-3} to 10^{-1} Pa s it is generally necessary for the temperature to be controlled and known to within ± 0.01 K of the temperature of certification in order to use the reference material within its specified uncertainty, which is usually ±0.2 per cent with reference to water. In the higher viscosity range, temperature control is still important. In the viscosity ranges 10^{-1} to 10 Pa s and greater than 10 Pa s the uncertainty of reference materials are of the order of ±0.3 per cent and ±1 per cent respectively. However, the temperature dependence of the viscosity of high viscosity fluids is much greater than that for low viscosity fluids, so that temperature control and measurement must still be of the order of ±0.02 K.

Table 4.1.1. Temperature dependence of viscosity of selected certified reference materials

Polish CRM		Cannon CRM S8000		Cannon CRM S3		NBS SRM-710	
T/K	η/Pa s	T/K	η/Pa s	T/K	$10^3 \eta$/Pa s	T/K	η/Pa s
293.15	589	293.15	33	219.26	260	850.1	10^{11}
323.15	59	298.15	20	233.15	70	930.6	10^9
353.15	10	313.15	5.9	293.15	3.9	979.3	10^7
373.15	4.05			313.15	2.4	1094.7	10^5
				373.15	0.9	1292.2	10^3
						1707.5	10

Most reference materials must be used within a specified time period otherwise certification is not valid. In general, materials with viscosities in the range of 10^{-3} to 10^3 Pa s must be used within times varying from 2 to 12 months. This is because the viscosity of reference materials in the medium to high viscosity range made from petroleum oils increases at a rate of from 0.01 to 0.03 per cent per month. For petroleum oils, the viscosity changes with pressure seldom exceed 0.003 per cent per kilopascal so that changes in atmospheric pressure can be neglected in most cases. The use of pure fluids with $\nu < 10^{-6}$ m^2 s^{-1} as reference materials for viscosity has been discussed in detail by Bauer and Meerlender (Ref. 9).

No reference materials are available for calibrating and checking viscometers to be used for measuring the viscosity of gases. Hanley (Refs. 10 to 12) has published several reviews on the viscosity of a variety of gases over a temperature and pressure range.

REFERENCES

1. *Manual of Symbols and Terminology for Physicochemical Quantities and Units*, **1979** Edition, International Union of Pure and Applied Chemistry, *Pure and Appl. Chem.* **1979**, *51*, 1.
2. *Manual of Symbols and Terminology for Physicochemical Quantities and Units. Appendix II: Definitions, Terminology, and Symbols in Colloid and Surface Chemistry. Part I.13. Selected Definitions, Terminology and Symbols for Rheological Properties.* International Union of Pure and Applied Chemistry, *Pure Appl. Chem.* **1979**, *51*, 1213.
3. Swindells, J. R.; Coe, J. R.; Godfrey, T. B. *J. Res. Nat. Bur. Std.* **1952**, *48*, 1.
4. White, N. S.; Kearsley, E. A. *J. Res. Nat. Bur. Std.* **1971**, *75A*, 541.
5. Roscoe, R.; Bainbridge, W. *Proc. Phys. Soc. (Lond.)* **1958**, *72*, 585.
6. *International Organization for Standardization, Technical Report 3666* **1977**.
7. Marvin, R. S. *J. Res. Nat. Bur. Std.* **1971**, *75A*, 535.
8. Daborn, J. E.; Kuhlborsch, G. *Report on the Second BCR Viscosity Audit* Commission of the European Communities, BCR information, Report EUR 8359 EN, **1983**.
9. Bauer, H.; Meerlender, G. *Rheologica Acta* **1984**, *23*, 514.
10. Hanley, H. J. M. *J. Phys. Chem. Ref. Data* **1973**, *2*, 619.
11. Hanley, H. J. M.; Ely, J. R. *J. Phys. Chem. Ref. Data* **1973**, *2*, 735.
12. Hanley, H. J. M.; McCarty, R. D.; Haynes, W. H. *J. Phys. Chem. Ref. Data* **1974**, *3*, 979.

4.2. Methods of measurement

The most important methods used for measuring the viscosity of liquids include the capillary tube method, the falling body method, the rotating cylinder method and the use of a slit between cone and plate. A variety of other methods have been used and some are outlined at the end of the section.

The majority of viscometers have to be calibrated with reference materials.

4.2.1. Capillary viscometers

Table 4.2.1 lists the most important types of viscometers based on laminar flow through a capillary. The equation of Hagen and Poiseuille for Newtonian liquids

$$\eta = \frac{\pi r^4 t}{8Vl}\Delta p, \qquad (4.2.1.1)$$

can be modified for gravity flow type capillary viscometers

$$\nu = \frac{\pi g r^4 h t}{8Vl}, \qquad (4.2.1.2)$$

where r is the radius of the capillary tube, Δp is the average pressure difference between the inlet and outlet of the capillary, t is the efflux time for the volume V, l is the length of the capillary and h is the average vertical distance between the upper and lower menisci (Ref. 1).

Table 4.2.1. Types of capillary viscometers

capillary viscometers	gravity flow type	Modified Ostwald viscometers
		suspended-level viscometers
		reverse-flow viscometers
	pressure flow type	high pressure viscometers
		process viscometers

Gravity flow type capillary viscometers

These viscometers are based on the principle of a liquid flowing through a capillary under gravity and the efflux time t for a constant volume to flow between two marks is a function of kinematic viscosity ν. The various terms in equation 4.2.1.2 which are constant for a particular viscometer are grouped into the viscometer constant C:

$$\nu = Ct. \tag{4.2.1.3}$$

Equation 4.2.1.3 is not correct since it neglects end effects and the kinetic energy correction term. A better relation between the kinematic viscosity and the flow time is

$$\nu = Ct - E/t^2, \tag{4.2.1.4}$$

where E is a second constant for a particular viscometer, and this second term becomes insignificant when the flow time is greater than a minimum value.

The International Standard ISO 3105-1976 (Ref. 2) describes these corrections and the conditions under which the correction terms become insignificant. The corrections are also discussed by Hardy (Ref. 3). In practice, it is necessary to have a range of capillary viscometers, which differ from each other by the diameter of the capillary, in order to cover a wide range of fluid viscosities.

4. Viscosity

Table 4.2.2 lists the properties of the three main types of capillary viscometers: the modified Ostwald viscometer, the suspended-level viscometer and the reverse-flow viscometer.

Except for the suspended-level viscometer, it is necessary to ensure that the viscometer contains a fixed and constant volume of liquid. The main advantage of the suspended-level viscometer is that the suspension guarantees a uniform middle driving head of liquid, independent of the volume of the sample charged into the viscometer. The reverse-flow viscometer has the advantage that it can be used both for transparent and opaque liquids but at a lower precision than the direct flow pattern.

Table 4.2.2. Types and range of capillary viscometers

Type	Range $\dfrac{10^6 \nu}{m^2\ s^{-1}}$	Number of viscometers to cover range	Ratio of constant between two neighbouring sizes	length of capillary $\dfrac{l}{mm}$
Modified Ostwald viscometers	0.5 to 2×10^4	5 to 12	2 to 5	72 to 130
Suspended-level viscometers	0.3 to 1×10^6	7 to 16	2 to 3	80 to 210
Reverse-flow viscometers	0.4 to 3×10^6	6 to 12	2 to 4	150 to 210

Master viscometers

Master viscometers are used to establish the viscosity scale by the so-called "step-up" procedure to ensure the smallest possible uncertainty (Ref. 2) in calibration. They are mostly the suspended-level type but sometimes the modified Ostwald type has been used. They differ from the usual capillary viscometers with regard to the length of the capillary. The capillaries measure at least 400 mm in order to reduce the correction terms due to kinetic energy and surface tension effects.

"Conventional" viscometers

A "conventional" viscometer is one where the capillary has been "degenerated" to an orifice or to a very short tube. They are also known as "viscosity cups" or "efflux type viscometers" (Ref. 9). Examples are the Ford viscosity cup, Zahn cup, Redwood cup, ISO cup, and Engler viscometer. The results from these viscometers are only meaningful if the convention for them is maintained. A comparison of results obtained with these "conventional" viscometers and with the usual capillary viscometers is not always possible.

Automatic Capillary Viscometers

Commonly used capillary viscometers and master viscometers have been developed so that the flow-time between the two marks can be measured automatically. In some cases the liquid is displaced automatically, after a measurement, back to the starting position by means of a peristaltic pump and electrical valves (Ref. 5, 6). Automatic time measurement is usually based on the following: light from a source, guided by an image fibre, passes through the glass wall of the viscometer at the position of the upper and lower reference mark and illuminates two photo-transistors. At the moment when the liquid is replaced by air, the light is scattered at the internal glass-air interface and the intensity of the light is reduced. This effect is used to start and to stop a counter. Automatic capillary viscometers allow routine repeat measurements which usually result in an increase in timing precision. They are not suitable for liquids which foam.

Pressure flow capillary viscometers

In these viscometers, the liquid is forced through a capillary at a known constant flow rate and the difference in pressure at the two ends of the capillary is measured (Ref. 7).

High pressure viscometers

These viscometers are often used to determine the behaviour of non-Newtonian liquids at different shear rates in capillaries (Ref. 8). However, the shear rate may not be constant over the length of the capillary. The change of the shear relations is determined by altering the pressure at the top of the capillary.

Process viscometers

When the viscosity of a fluid in a process stream has to be monitored continuously, a capillary viscometer (Ref. 9) can be mounted in a by-pass. One method is to use a gear pump which draws up a fluid sample from the process pipe line and ensures a constant volume flow rate. The differential pressure developed across the capillary is measured by a differential pressure transducer and the temperature of the liquid in the capillary is measured by a temperature transmitter.

4. Viscosity

4.2.2. Falling body viscometers

Table 4.2.2.1 lists the most important designs of falling body viscometers.

Table 4.2.2.1. Types of falling body viscometers

Falling body viscometers	Viscometers with vertical falling bodies, normally a falling ball
	Viscometers with a rolling ball in an inclined tube
	Forced ball viscometers

The underlying principle of falling ball viscometers is based on the law of laminar flow around a ball, known as Stokes law,

$$W = 6\pi\eta rv, \qquad (4.2.2.1)$$

which relates the viscous drag W on the ball to the radius r and the constant velocity v attained when falling in an infinite and homogeneous liquid.

At a constant velocity, the viscous drag W is equal to the apparent weight of the sphere

$$W = \frac{4}{3}\pi r^3 (\rho_s - \rho_l)g, \qquad (4.2.2.2)$$

where ρ_s and ρ_l are the densities of the ball and the liquid respectively. From these two equations

$$\eta = 2(\rho_s - \rho_l)\frac{gr^2}{9v}. \qquad (4.2.2.3)$$

Other falling bodies such as cylinders have been used in the design of special viscometers. For such bodies other equations can be derived.

For all falling ball viscometers the condition of an infinite amount of liquid around the falling ball is not realized. Corrections for the finite size of the tube have been summarized by Peter (Ref. 10) and Flude and Daborn (Ref. 11). For a ball rolling in an inclined tube (Höppler viscometers), an empirical relation has been given by Weber (Ref. 12).

Normal falling ball viscometers

Viscometers with either falling or rolling balls in either vertical or inclined calibrated glass or metal tubes are most common. The following equation is valid for both falling or rolling ball viscometers;

$$\eta = k(\rho_s - \rho_l)t, \qquad (4.2.2.4)$$

where k is the constant for the body (sphere or other shaped body) and t is the time for which the sphere falls or rolls a fixed distance.

In the design of forced ball viscometers, the ball does not fall freely, but its motion (either up or down) is realized by attaching it to a special balance. The velocity of the ball can be varied by changing the effective weight. If the difference in the buoyancy of the ball is small, the equation for the viscosity with this viscometer is

$$\eta = k' f t, \qquad (4.2.2.5)$$

where f is the effective weight and k' is the constant for a particular ball. It is possible to characterize the non-Newtonian behaviour of liquids with these viscometers but it is not possible to express the exact relation between the shear stress and the velocity gradient. Table 4.2.2.2 lists the various types of falling ball viscometers.

Bubble viscometers

Bubble viscometers are based on Stokes' law and they are used for routine measurements. A series of identical tubes are filled with reference fluids, varying stepwise in viscosity and one tube is filled with the liquid of unknown viscosity. Bubbles of the same size are introduced into the various tubes and the bubble speeds are compared by inverting the tubes and measuring the time for the various bubbles to pass between two fixed marks.

Falling ball viscometers for special tasks

Falling ball viscometers are frequently used to measure the viscosity at high static pressure (up to 500 MPa) since it is easy to measure the falling time in these special viscometers by electronic devices. Many papers have described special high static pressure viscometers of the falling ball type (Refs. 13, 14). The falling bodies are either balls or cylinders. The constant k is determined by calibration with reference materials at atmospheric pressure and the variation of the constant k with pressure is calculated. Falling ball viscometers have been used up to 1600 °C.

4.2.3. Rotational viscometers

Table 4.2.3.1 summarizes the most important designs of rotational viscometers.

The fluid under investigation is placed between two rotationally-symmetric and coaxially arranged surfaces and one of the surfaces rotates with a constant angular velocity. The angular

4. Viscosity

velocity ω of the surface can be controlled, in many cases by means of a microcomputer and the velocity gradient and the shear stress τ can be determined. The relationship $\tau = f(D)$ is calculated and if the relation is linear, the liquid is Newtonian.

Table 4.2.2.2. Types of falling ball viscometers

Type	Range Pa s	Number of balls with different diameters	Special names
Viscometers with vertical falling ball	0.01 to 40	3	Portable viscometers (without thermostatic bath)
Viscometers with rolling ball in an inclined tube	5×10^{-4} to 1×10^2	6 to 7 (Material: both glass and corrosion-resistant steel alloy)	Höppler viscometer
Forced ball viscometers	1×10^{-5} to 10^5	2 to 5	Viscobalance Rheoviscometer, Consistometer

Table 4.2.3.1. Types of rotational viscometers

Rotational viscometers	Coaxial-cylinder viscometers	Couette
		Searle
	Cone-plate viscometers	
	Plate viscometers	

Rotational viscometers have advantages over other types of viscometers for the investigation of non-Newtonian behaviour since the relation between the velocity gradient D and the shear stress τ can be readily determined.

A detailed description of the measurements of viscosity of Newtonian liquids by means of rotational viscometers is contained in DIN 53018, part 1 (Ref. 15). The second part of that document describes possible sources of systematic errors in coaxial-cylinder rotational viscometers.

Coaxial-cylinder viscometers

The flow of Newtonian liquids in a coaxial-cylinder viscometer can be described with the modified Margules equation

$$T = \frac{4\pi r_a^2 r_i^2}{r_a^2 - r_i^2} \omega \eta (l + \Delta l). \tag{4.2.3.1}$$

In this equation r_a and r_i are the radii of the outer and inner cylinders respectively. T is the torque, ω is the angular velocity and l is the length of the cylinder on which the torque is measured. The influence of the end surfaces can be corrected with the additional length Δl. The constant k of a particular coaxial cylinder of a viscometer is

$$k = \left(\frac{1}{4\pi(l + \Delta l)}\right)\left(\frac{r_a^2 - r_i^2}{r_a^2 r_i^2}\right), \tag{4.2.3.2}$$

and it is usually determined by means of reference materials with Newtonian behaviour.

Coaxial-cylinder viscometers with rotating outer cylinders are called Couette viscometers, those with rotating inner cylinders are called Searle viscometers. In a Searle viscometer, inertia forces disturb the stability of the flow at relative high angular velocity and relatively small slits.

There are many modifications of coaxial-cylinder viscometers which enable them to be used under the following conditions:

a) for normal use ($\sim 10^{-3}$ to 10^4 mPa s, 0 to 100 °C);

b) for use at high temperatures (up to 1700 °C) for the determination of the viscosity of molten glass and molten slags;

c) for measurement of viscosity under pressure (to 100 MPa);

d) as continuously measuring instrument in a by-pass of a pipe line;

e) for the measurement of very small quantities (~ 0.4 cm^3) and at very small angular velocity gradients (7×10^{-2} to 4.5 s^{-1});

f) for measurement of very high ($> 10^5$ s^{-1}) angular velocity gradients.

Cone-plate viscometers

In many cases coaxial-cylinder viscometers can be modified to cone-plate viscometers. To a first approximation the following equation is valid for Newtonian flow;

$$\eta = Tk/\omega, \qquad (4.2.3.3)$$
$$k = \frac{3\alpha}{2\pi r_o^3}, \qquad (4.2.3.4)$$

where α is the angle between the cone and the plate and r_o is the radius of the cone. The coefficient k of a cone-plate viscometer can be determined by means of viscosity reference materials. This viscometer has the advantage that for small values of α, the rate of shear is essentially constant for constant ω.

Plate viscometers

Plate viscometers are used primarily for the investigation of substances having very high viscosities, i.e., $\eta > 10^{10}$ Pa s. The flow equation is:

$$\eta = Tk/\omega, \qquad (4.2.3.5)$$

where

$$k = \frac{2h}{\pi r_o^4}, \qquad (4.2.3.6)$$

and r_o is the radius of the plates and h is the distance between the rotating and fixed plate. A calibration with viscosity reference materials is usually not possible because of the lack of suitable reference fluids having a high viscosity.

4.2.4. Other viscometers

Specialized viscometers have been developed for specific requirements. These include the oscillating-disk viscometers and quartz oscillating viscometers. A survey of these instruments is given in reference 16. Viscometers based on these principles are described in reference 17.

4.2.5 Calibration using reference materials

The calibration of viscometers consists of, in general, the determination of the constant(s) of the viscometers. It is usual to calibrate viscometers with at least two reference fluids having different values of viscosity or by direct calibration against a previously calibrated viscometer using a transfer fluid.

Non-Newtonian behaviour is observed when the relation $\tau = f(D)$ is not linear. Hence it is necessary to determine τ and D seperately. It is also necessary to take into account correction terms for the particular viscometer. For the estimation of these correction terms, existing reference materials can be useful. Since rotational viscometers are often used for the investigation of non-Newtonian liquids these viscometers are frequently calibrated with reference materials or the dimensions of the viscometer are determined. Most other viscometers are calibrated with reference fluids.

Factors influencing the accuracy of the measured values of viscosity

(i) The viscosity scale is based on the value of the viscosity of water.
(ii) The uncertainty of the value of viscosity increases as the difference from the value of the viscosity of water increases. If all conditions of viscosity measurement represent an optimum, the additional uncertainty of viscosity values can be estimated as follows (Ref. 4):

low viscosity	±0.1%
viscosity up to 1 Pa s	±0.2%
viscosity up to 100 Pa s	±0.6%

(iii) Control and measurement of temperature

Factors influencing the precision of the measured values of viscosity

To obtain a precise viscosity measurement requires the precise control of temperature. In addition, time measurement influences the precision of viscosity measurement by the capillary and falling ball methods while the measurement of the torque is important for rotational viscometers. The highest precision is reached with automatic capillary viscometers where the repeatability can reach one part per million.

The results of interlaboratory tests indicate that, on identical test materials, over a long period of time, the repeatability of a measured value was ±0.35 per cent (95 per cent confidence level) and the reproducibility ±0.7 per cent (95 per cent confidence level).

REFERENCES

1. ISO, International Standard 3104-1976, Petroleum products–Transparent and opaque liquids–Determination of kinematic viscosity and calculation of dynamic viscosity.
2. ISO, International Standard 3105-1976, Glass capillary kinematic viscometers-specification and operating instructions.
3. Hardy, R. C. *NBS Viscometer Calibrating Liquids and Capillary Tube Viscometers*, National Bureau of Standards Monograph 55, U.S. Dept. of Commerce: Washington, **1962**.
4. Daborn, J. E. *Metrologia* **1975**, *11*, 79.
5. Smith, J. S.; Irving, H. M. N. H.; Simpson, R. B. *Analyst* **1970**, *95*, 743.
6. Priel, Z. *J. Phys. E: Sci. Instr.* **1980**, *13*, 814.
7. Müller, H. G. *Instrum. Rev.* **1967**, *14*, 102.
8. Klein, J.; Müller, H. G.; Leidigkeit, G. *Rheol. Acta* **1982**,
9. Karam, H. J.; *Ind. and Eng. Chem.* **1963**, *55*(2), 38.
10. Peter, S. *Chem. Ing. Tech.* **1960**, *32*, 437.
11. Flude, M. J. C.; Daborn, J. E. *J. Phys. E. Sci. Instrum.* **1982**, *15*, 1313.
12. Weber, von W. Kolloid-Z **1956** *147*, 14.
13. Irving, J. B.; Barlow, A. J. *J. Phys. E: Sci. Instru.* **1971**, *4*, 232.

4. Viscosity

14. Künzel, W. *Exp. Tech. Phys.* **1965**, *13*, 468.
15. DIN 53018, Teil 1 und 2, 1976: Messung der dynamischen Viskosität newtonischer Flüssigkeiten mit Rotationasviskosimetern
16. Touloukian, Y. S. *Viscosity*, Vol. 11 of the TPRC Data Series, IFI, Plenum: New York, **1974**.
17. Van Wazer, J. R. et al. *Viscosity and Flow Measurement*, New York: Interscience Publishers, **1963**.

4.3. Reference materials for viscosity

4.3.1. Water

Physical property: Viscosity η and kinematic viscosity ν
Units: Pa s, $m^2 \, s^{-1}$
Recommended reference material: Water (H_2O)
Range of variables: 293.15 K, atmospheric pressure
Physical state within the range: liquid
Contributors: T. Plebanski, K. N. Marsh

Intended usage: Water is used for calibration of viscometers intended for the measurement of the viscosity of liquids.

Sources of supply and/or methods of purification: Purification by double distillation and passing through a fine sintered glass frit (to saturate the sample with air) just prior to use is recommended (Ref. 1).

Pertinent physicochemical data: In 1958 the International Organization for Standardization, Technical Commission 66, circulated a draft proposal to accept water at 293.15 K as a reference material for the calibration of viscometers (Ref. 9). A value of 0.001002 Pa s was recommended, based on a number of determinations (Refs. 1 to 3) but primarily based on the value of 0.0010019 Pa s reported by Swindells, Coe, and Godfrey (Ref. 1). The values in references 1 and 3 were obtained using capillary viscometers, with possible systematic errors which could not be estimated.

In many cases the true value of the viscosity is not as important as consistency with other measurements throughout the world. Since the value of 0.001002 Pa s was already used rather widely by 1958, a change could only have been justified by convincing evidence that this value differed from the true value by more than its uncertainty. An estimate of this uncertainty was published in 1971 (Refs. 4 to 7). The viscosity of a liquid was measured by two independent absolute methods involving different types of flow. The results differed by more than 0.5 per cent, five times the uncertainty (including both the variability and estimate of systematic errors) assigned to either measurement. Based on all the above results it is recommended (Refs. 8, 9) that viscometer calibrations be based on the following value for the viscosity of water

$$\eta(H_2O, 293.15 \text{ K}, 101.325 \text{ kPa}) = 0.001002 \text{ Pa s}$$

with the proviso that when only agreement between measurements made in different laboratories is important, the uncertainty associated with this value be ignored but that when a true value of the viscosity is required, an uncertainty of ±0.25 per cent be assigned to this value. This value for the viscosity of water is recommended by ISO, OIML, and IUPAC.

The kinematic viscosity of water is derived from the recommended value for the dynamic viscosity and the recommended value (Ref. 10) for the density of air-free water at 293.15 K and atmospheric pressure. At 293.15 K air-saturated water is about 1×10^{-3} kg m^{-3} less dense than air-free water (Ref. 10). This difference is insignificant.

The recommended value is

$$\nu(H_2O, 293.15 \text{ K}, 101.325 \text{ kPa}) = 1.0038 \times 10^{-6} \text{ m}^2 \text{ s}^{-1}$$

with the proviso that when only agreement between measurements made in different laboratories is important, the uncertainty associated with this value be ignored but that when a true value of the kinematic viscosity is required, an uncertainty of ±0.25 per cent be assigned to this value. This value for the kinematic viscosity is recommended by OIML, ISO, and IUPAC.

The viscosity of water at other temperatures relative to the value at 293.15 K given in the table below have essentially been derived from the measurements and the estimation of Collings and Bajenov (Ref. 11), Kestin *et al.* (Ref. 12), and James *et al.* (Ref. 13).

The International Association of Steam (IAPS) adopted in 1982 a new formulation for the properties of water (Ref. 14). Their recommended values for the dynamic and kinematic viscosity are not consistent with the recommendations of ISO, OIML, and IUPAC.

Table 4.3.1.1. Viscosity of water at various temperatures

T/K	$10^3 \eta/\text{Pa s}$	T/K	$10^3 \eta/\text{Pa s}$	T/K	$10^3 \eta/\text{Pa s}$
273.15	1.791	303.15	0.7975	343.15	0.404$_2$
283.15	1.307	313.15	0.6530	353.15	0.355$_1$
293.15	1.002	323.15	0.5469	363.15	0.315$_0$
298.15	0.8902	333.15	0.4665	373.15	0.282$_1$

REFERENCES

1. Swindells, J. F.; Coe, J. R.; Godfrey, T. B. *J. Res. Nat. Bur. Std.* **1952** *48*, 1.
2. Roscoe, R.; Bainbridge, W. Proc Phys. Soc. (Lond.) **1958**, *72*, 585.
3. Maliarov, G. A. *Trudy Nauchno Issled. Inst. Metrologi* **1959**, *37*(97), 125.
4. Kawata, H.; Kurase, K.; Yoshida, K. *Proc. Fifth Inst. Congress on Rheology 1*, 453 ed. S. Onogi, University of Tokyo Press: Tokyo, **1969**.
5. White, H. S.; Kearsley, E. A. *J. Res. Nat. Bur. Std.* **1971**, *75A*, 541.
6. Penn, R. W.; Kearsley, E. A. *J. Res. Nat. Bur. Std.* **1971**, *75A*, 553.
7. Marvin, R. S. *J. Res. Nat. Bur. Std.* **1971**, *75A*, 535.
8. Marvin, R. S. *The Calibration of Viscometers*, National Bureau of Standards Special Publication 300, *Precision Measurement and Calibration*, Vol. 8. Washington, DC, **1972**.
9. *International Organization for Standardization, Technical Report 3666*, **1977**.
10. Brown, I.; Lane, J. E. *Pure Appl. Chem.* **1976**, *45*, 1.
11. Collings, A. F,; Bajenov, N. *Metrologia* **1983**, *19*, 61.
12. Kestin, J.; Sololov, M.; Wakeham, W. A. *J. Phys. Chem. Ref. Data* **1978**, *7*, 947.
13. James, C. J.; Mulcahy, D. E.; Steel, B. J. *J. Phys. D: Appl. Phys.* **1984**, *17*, 225.
14. Sengers, J. V.; Kamgar-Parsi, B. *J. Phys. Chem. Ref. Data* **1984**, *13*, 185.

4.3.2. Certified reference materials

Physical property: Viscosity η and kinematic viscosity ν.
Units: Pa s, $m^2\ s^{-1}$
Recommended reference materials: Variety of oils, polyisobutenes, and glasses.
Range of variables: 220 to 1696 K, 101.325 kPa
Physical state within the range: liquid
Contributors: T. Plebanski, K. N. Marsh

Intended usage: These certified reference materials are used to calibrate and to check viscometers intended for measurements on liquids. Methods for measuring the viscosity of liquids have been summarized by Johnson, Martin, and Porter (Ref. 1). For silicone oils with viscosity greater than 10^3 Pa s care should be taken to ensure that the liquids are Newtonian under the conditions of calibration. Silicone oils can effect glass surfaces.

Sources of supply and/or methods of preparation: Samples are available from supplies A, B, C, D, E, F, G, H, I, J, K with certified values of (dynamic or kinematic) viscosity, with or without density. In general the uncertainties are at an 95 per cent confidence level or better.

Pertinent physicochemical data: Supplier A (CSIRO National Measurement Laboratory, Australia) produces six reference materials listed in table 4.3.2.1.

The uncertainty values given assume no uncertainty in the value of the viscosity of water. The value given in table 4.3.2.1 are nominal values and should not be taken to be actual values of the reference materials supplied. Sample size 275 cm^3.

Supplier B (Laboratoire National d'Essais, France) produces sixteen reference materials listed in table 4.3.2.2.

The uncertainty for the viscosity is ±0.2 per cent or higher. No uncertainty for the viscosity of water is assumed. The sample size 125 cm^3. The values given in table 4.3.2.2 are nominal values and should not be taken to be actual values of the reference materials supplied.

Supplier C (Physikalisch-Technische Bundesanstalt, Federal Republic of Germany) produces 25 reference materials listed in table 4.3.2.3.

The uncertainty is ±0.2 per cent for viscosities up to 2 Pa s and ±0.3 per cent or higher for higher viscosities. No uncertainty in the value for the viscosity of water is assumed. On demand the materials can be certified at additional temperatures.

Sample sizes are 100 cm^3, 250 cm^3, and 500 cm^3. Samples marked by * indicate that amounts of 1000 cm^3 are also available. The viscosity of the samples increases with time but remains within 0.2 per cent over three months. The values given in table 4.3.2.3 are nominal values only.

Supplier D (National Office of Measures, Hungary) produces a variety of reference materials with viscosities ranging from 2×10^{-3} to 1000 Pa s. The materials are certified at 293.15 K for viscosity, kinematic viscosity and density. Their materials are divided into two groups. Group 1 are mineral oils with viscosities from 2×10^{-3} to 5 Pa s having an uncertainty of ±0.1 per cent and a certification period of six months. Group 2 are polymers with viscosities in the range of 100 to 1000 Pa s having an uncertainty of ±1 per cent and a certification period of one year. The materials are said to be proved as Newtonion liquids.

The uncertainty values for the viscosity assume no uncertainty in the viscosity of water.

Supplier E (National Research Laboratory of Metrology, Japan) produces 13 reference materials as listed in table 4.3.2.4.

The highest viscosity material is also certified at 298.15 K (not included in table 4.3.2.4). The sample size is 500 cm^3. The values given in table 4.3.2.4 are nominal values only. Uncertainties are not given.

Supplier F (Institute of Applied Chemistry TNO, The Netherlands) produces a variety of reference materials with viscosities ranging from 2.5×10^{-3} to 1000 Pa s. The materials are certified at 293.15 K for viscosity and listed in five groups.

Group 1 are mineral oils with viscosities in the range 2.5×10^{-3} to 1 Pa s having an uncertainty of ±0.2 per cent. Group 2 are polymers with viscosities in the range 1 to 20 Pa s having an uncertainty of ±0.3 per cent.

Group 3 are polymers with viscosities in the range 20 to 30 Pa s having an uncertainty of ±0.5 per cent.

4. Viscosity

Group 4 are polymers with viscosities in the range 50 to 500 Pa s having an uncertainty of ±1 per cent. Group 5 are polymers with viscosities in the range 500 to 1000 Pa s having an uncertainty of ±2 per cent.

The period of validity of certification for samples with viscosity less than 1 Pa s is 1 year while for the high viscosity materials it is six months. Samples certified at higher and lower temperatures can be supplied on request. The sample size is 250 cm^3. The materials are stated to be Newtonian.

Supplier G (Research and Development Centre for Standard Reference Materials, Poland) produces 18 reference materials with viscosities ranging from 1.5×10^{-3} to 600 Pa s. The 11 materials (mineral oils) in the viscosity range 1.5×10^{-3} to 4 Pa s are certified at 293.15, 323.15 and 353.15 K for viscosity, kinematic viscosity and density. The 7 materials (polyisobutenes) in the viscosity range 4.1 to 589 Pa s are certified at 293.15, 323.15 and 313.15 K for viscosity only. The uncertainty limits increase from ±0.1 per cent in the low viscosity range to ±2 per cent in the high viscosity range. An equation is supplied to allow the material to be used at any desired temperature within the range.

Supplier H (National Physical Laboratory, U.K.) produces eleven reference materials, with viscosities ranging from 1.5×10^{-3} to 70 Pa s. The seven materials in the range 1.5×10^{-3} to 2 Pa s are mineral oils; the four materials in the range 5 to 70 Pa s are polybutenes. The materials are certified at 293.15, 298.15, and 313.15 K. In addition, two oils with approximate viscosities of 3.2×10^{-3} and 10.1×10^{-3} Pa s at 310.15 K for haematology are available. Sample size is 100 cm^3. See table 4.3.2.5.

Supplier I (Cannon Instrument Company, U.S.A.) produces 19 reference materials with viscosities ranging from 0.3×10^{-3} to 900 Pa s. The majority of these materials are certified at 293.15 K for viscosity, kinematic viscosity, and density. Some of the materials are certified at a variety of temperatures ranging between 219.2 K and 403 K. Five of the materials specifically reproduce asphalt viscosities up to 900 Pa s between 293.15 K and 408 K and two reproduce viscosities of standard engine lubricants at 219.25 and 233.15 K. These reference materials are sponsored by ASTM D-2, RDDVII-A. See table 4.3.2.6.

Certified samples can be supplied with certification at specified temperatures between 273 and 373 K on request. Custom viscosity samples are also supplied. The values given in this table are nominal values only. Sample size is 470 to 500 cm^3.

Supplier J (National Bureau of Standards, U.S.A.) produces 3 certified reference materials (glasses) with viscosities ranging from 10 to 10^{11} Pa s. The materials are certified for viscosity in the temperature range 737 K to 1818 K. See table 4.3.2.7. These materials are designed for testing and calibrating rotating cylinder instruments and fibre elongation equipment.

Supplier K (All Union Scientific Research Institute of Metrology, U.S.S.R) produces 18 references materials with kinematic viscosities ranging from 2.5×10^{-6} m^2 s^{-1} to 10^2 m^2 s^{-1}.

Table 4.3.2.1. National Measurement Laboratory, Australia. Certified reference viscosity materials

designation	$\dfrac{10^3 \eta}{\text{Pa s}}$		$\dfrac{10^6 \nu}{\text{m}^2\text{ s}^{-1}}$		$\dfrac{\rho}{\text{kg m}^{-3}}$		uncertainty $\pm\%$	
	298.15 K	313.15 K	298.15 K	313.15 K	298.15 K	313.15 K	viscosity*	density*
AS 2.5	2.8	2.0	3.4	2.5	820	810	0.10	0.02
AS 7.5	11	6.5	12	7.6	860	850	0.15	0.02
AS 25	43	22	50	26	860	850	0.3	0.02
AS 75	160	66	180	76	880	870	0.3	0.02
AS 200	480	180	540	200	900	890	1.0	0.02
AS 600	1800	550	1900	600	920	910	1.0	0.02

* The uncertainties are at the 99 per cent confidence level. The certification periods for the four lower viscosity samples and the two higher viscosity samples are 12 months and 9 months respectively.

4. Viscosity

Table 4.3.2.2. Laboratoire National d'Essais, France. Certified reference viscosity materials

designation	$\dfrac{10^6 \nu}{\mathrm{m^2\,s^{-1}}}$				$\dfrac{\rho}{\mathrm{kg\,m^{-3}}}$			
	293.15 K	313.15 K	323.15 K	373.15 K	293.15 K	313.15 K	323.15 K	373.15 K
s-2	3.3	2.2	1.8	0.97	816	806	795	761
s-5	7.4	4.3	3.4	1.5	834	820	814	781
s-10	17.7	8.9	6.7	2.8	836	823	816	785
A	51	21	15	4.2	862	849	843	811
B	76	30	20	5.4	854	842	836	805
C	106	38	25	6.1	874	861	855	833
D	215	70	44	8.9	881	868	863	832
E	273	85	53	10.2	885	872	867	836
F	377	111	68	12.6	888	876	867	838
200	639	174	102	16.3	892	879	872	844
350	1220	292	166	30	898	885	880	849
G	2100	475	260	31	903	891	885	856
H	4300	790	399	37	932	920	913	884
	253.15 K							
BT 200	217				—			
BT 200A	245				—			
BT 300A	350				—			

Table 4.3.2.3. PTB, Federal Republic of Germany. Certified reference viscosity materials

designation	$\dfrac{10^3 \eta(293.15\ \text{K})}{\text{Pa s}}$	$\dfrac{10^6 \nu(293.15\ \text{K})}{\text{m}^2\ \text{s}^{-1}}$
1B	0.98	1.26
2A	2.1	2.6
5B	4.6	5.7
10A	8.5	10.2
10B	12.9	15.2
10D*	14.7	16.7
20C	21.2	24.9
20E*	45.7	49.8
50C	88	100
100D	131	149
100C	157	177
200A	223	254
200G*	339	382
500B	430	485
500F	650	725
500E*	875	970
2000C	1790	1990
2000E	3040	3500
2000F	1940	2260
5000A	4200	4800
10000A	8900	10200
10000D	13600	15400
10000B	17500	20000
20000D	31400	35600

Table 4.3.2.4. National Research Laboratory of Metrology, Japan. Certified reference viscosity materials

designation	$\dfrac{10^3 \eta}{\text{Pa s}}$			$\dfrac{10^6 \nu}{\text{m}^2\text{ s}^{-1}}$		
	293.15 K	303.15 K	313.15 K	293.15 K	303.15 K	313.15 K
JS 2.5	2	1.6	1.3	2.5	2.1	1.7
JS 5	4	3	2.5	5	4	3
JS 10	8	6	5	10	7	6
JS 20	17	11	8	20	14	10
JS 50	42	25	18	50	30	21
JS 100	85	50	30	100	60	35
JS 200	170	90	52	200	110	60
JS 500	450	220	120	500	250	140
JS 1000	900	440	220	1000	500	250
JS 2000	1800	800	390	2000	900	440
JS 20H	16000	6000	3000	18000	7000	3000
JS 60H	50000	19000	8000	56000	21000	9000
JS 200H	150000	–	–	170000	–	–

Table 4.3.2.5. NPL, United Kingdom. Certified reference viscosity materials

designation	$\dfrac{10^3 \eta}{\text{Pa s}}$			
	293.15 K	298.15 K	310.15 K	313.15 K
V1	1.4	1.3	—	1.0
V2	5.0	4.3	3.1	3.0
V3	20	16	10	9
V4	57	43	—	22
V5	220	160	—	69
V6	780	530	—	200
V7	2900	1900	—	600
V8	8000	5100	—	1500
V9	15000	9500	—	2700
V10	36000	23000	—	6600
V11	110000	70000	—	20000

designation	$\dfrac{10^6 \nu}{\text{m}^2 \text{ s}^{-1}}$			
	293.15 K	298.15 K	313.15 K	373.15 K
V1	1.9	1.7	—	1.3
V2	6.0	5.3	3.9	3.7
V3	23	19	12	11
V4	66	50	—	25
V5	250	180	—	79
V6	880	600	—	230
V7	3200	2000	—	640
V8	9000	5800	—	1700
V9	17000	11000	—	3100
V10	41000	26000	—	7300
V11	120000	78000	—	22000

Table 4.3.2.6. Cannon (U.S.A.). Certified reference viscosity materials

designation	$\dfrac{10^3 \eta}{\text{Pa s}}$			
	293.15 K	298.15 K	313.15 K	373.15 K
N.4	0.41	0.29	0.26	
N.8	0.73	0.68	0.56	
N1.0	0.92	0.85	0.69	
S3	3.9	3.3	2.4	0.9
S6	9.4	7.6	4.8	1.4
S20	38	29	15	3.1
S60	150	110	46	5.9
S200	560	390	150	14
S600	2100	1400	460	28
S2000	7600	4900	1500	62
S8000	33000	20000	5900	
S30000		72000	20000	

designation	$\dfrac{10^6 \nu}{\text{m}^2\,\text{s}^{-1}}$			
	293.15 K	298.15 K	313.15 K	373.15 K
N.4	0.47	0.45	0.40	
N.8	0.95	0.89	0.75	
N1.0	1.3	1.2	0.97	
S3	4.6	4.0	2.9	1.2
S6	11	8.9	5.7	1.8
S20	44	34	18	3.9
S60	170	120	54	7.2
S200	640	450	180	17
S600	2400	1600	520	32
S2000	8700	5600	1700	75
S8000	37000	23000	6700	
S30000		81000	23000	

Table 4.3.2.6 (cont.)

designation	Viscosity
N600	160×10^{-6} m^2 s^{-1} (333.15 K); 12×10^{-6} m^2 s^{-1} (408.15 K)
N2000	470×10^{-6} m^2 s^{-1} (333.15 K); 26×10^{-6} m^2 s^{-1} (408.15 K)
N8000	1700×10^{-6} m^2 s^{-1} (333.15 K);
N30000	120 Pa s (293.15 K); 24 Pa s (310.93 K)
N190000	900 Pa s (293.15 K); 190 Pa s (310.93 K)
N27B	low temperature viscosity reference material
N115B	low temperature viscosity reference material

4. Viscosity

Table 4.3.2.7. National Bureau of Standards certified reference viscosity materials

η/Pa s	10	10^2	10^3	10^4	10^5	10^6
designation	\multicolumn{6}{c}{T/K}					
SRN 710	1707.5	1459.9	1292.2	1178.5	1094.9	1030.3
SRN 711	1600.3	1346.0	1182.2	1067.9	983.6	918.8
SRN 717	1818.3	1522.0	1332.6	1201.1	1104.4	1030.3

η/Pa s	10^7	10^8	10^9	10^{10}	10^{11}
designation	\multicolumn{5}{c}{T/K}				
SRN 710	979.3	937.9	903.6	874.7	850.1
SRN 711	867.5	825.9	791.4	762.4	737.7
SRN 717	971.8	924.3	885.1	852.2	824.1

REFERENCE

1. Johnson, J. F.; Martin, J. R.; Porter, R. S.; *Physical Methods of Chemistry*, Pt VI, Weissberger, A.; Rossiter, B. W., editors, Interscience: New York, **1977**, p. 63..

4.4. Contributors

T. Plebanski
Research and Development Centre
for Standard Reference Materials,
UL. Elecktoralna 2,
PL 00-139, Warsawa (POLAND)

K. N. Marsh
Thermodynamics Research Center
The Texas A&M University System,
College Station, TX   77843
(U.S.A.)

H. F. van Wijk
Hoofdgroep Maatschappelijke Technologie,
Organisatie voor Toegepast
POB 108, Utrechtseweg 48
NL-3700 AC Zeist (NETHERLANDS)

W. Künzel
Amt für Standardisierung, Messwesen
und Warenprüfung (ASMW)
Fürstenwalder Damm 388, DDR-1162 Berlin
(GERM. DEM. REP.)

4.5. List of Suppliers

A. Commonwealth Scientific and Industrial
   Research Organization,
   National Measurement Laboratory,
   P.O. Box 218,
   Lindfield, N.S.W. 2070 (AUSTRALIA)

B. Laboratoire National d'Essais,
   1, Rue Gaston Boissier,
   75015 Paris (FRANCE)

C. Physikalisch-Technische Bundesanstalt,
   Braunschweig, 3300,
   Bundesalle, 100
   (FEDERAL REPUBLIC OF GERMANY)

D. Hungarian National Office of Measures,
   Orszagos Meresugyi Hivatal,
   1224 Budapest XII,
   Nemetvolgyi ut 37-39 (HUNGARY)

E. National Research Laboratory of Metrology,
   Agency of Industrial Science and Technology,
   Ministry of International Trade and Industry,
   1-4, 1-Chome, Umezono, Sakura-Mura,
   Niihari-Gun, Ibaraki 305 (JAPAN)

F. Institute of Applied Chemistry
   Utrechtseweg 48,
   P.O. Box 108,
   Zeist (NETHERLANDS)

G. Research and Development Centre
   for Standard Reference Materials,
   UL. Elecktoralna 2,
   PL 00-139, Warsawa (POLAND)

H. National Physical Laboratory,
   Teddington, Middlesex,
   TW11 0LW (U.K.)

I. Cannon Instrument Company,
   P.O. Box 16,
   State College, PA   16801 (U.S.A.)

J. Office of Standard Reference Materials,
   U.S. Department of Commerce
   National Bureau of Standards,
   Gaithersburg, MD   20899 (U.S.A)

K. All Union Scientific Research
   Institute of Metrology, Sverdlousk Branch
   Centre of the State Service for
   Standard Samples (U.S.S.R.)

5

SECTION: PRESSURE–VOLUME–TEMPERATURE

COLLATOR: D. AMBROSE

CONTENTS:

5.1. Introduction

5.2. Reference materials for vapour pressures at temperatures up to 770 K, critical temperatures and critical pressures.

    5.2.1. Carbon dioxide

    5.2.2. Pentane

    5.2.3. Benzene

    5.2.4. Hexafluorobenzene

    5.2.5. Water

    5.2.6. Mercury

    5.2.7. Ice

    5.2.8. Naphthalene

5.3. Reference materials for vapour pressures at temperatures above 600 K.

    5.3.1. Cadmium

    5.3.2. Silver

    5.3.3. Gold

5.4. Reference materials for orthobaric volumes (or densities) and critical volume (or density).

    5.4.1. Nitrogen

    5.4.2. Methane

5.4.3. Carbon dioxide

5.4.4. Benzene

5.4.5. Water

5.5. Reference materials for Pressure–Volume–Temperature behaviour (gases).

5.5.1. Helium (second virial coefficients)

5.5.2. Benzene (second virial coefficients)

5.5.3. Air

5.5.4. Nitrogen

5.5.5. Oxygen

5.5.6. Methane

5.5.7. Ethylene

5.6. Reference materials for Pressure–Volume–Temperature behaviour (liquids)

5.6.1. Water

5.7. Contributors

5.8. List of Suppliers

5.1. Introduction

The properties displayed on a Pressure–Volume–Temperature (PVT) diagram may be considered as including those of the co-existent condensed and vapour phases (properties (i) to (iii)), and those of the individual phases (properties (iv) and (v)) listed below:

(i) vapour pressure,

(ii) liquid - vapour critical temperature and critical pressure,

(iii) orthobaric volumes of liquid and vapour, including the critical volume,

(iv) PVT properties of the unsaturated vapour or gas, and

(v) PVT properties of the compressed liquid.

(The part of the diagram relating to a solid condensed phase is not considered here except with reference to vapour pressure.)

Recommendations are made in this section on reference materials for use in measurements involving all these properties.

5. Pressure-Volume-Temperature

Vapour pressures at temperatures up to 770 K, critical temperatures and critical pressures.

Measurements of vapour pressure have been reported ranging from 10^{-10} Pa up to 151 MPa (the critical pressure of mercury). The accuracy attainable varies greatly, and the present recommendations are restricted to the range of pressure from just below 0.1 Pa to 7.38 MPa (the critical pressure of carbon dioxide) and of temperature from 90 to 770 K, in which reliable and accurate methods of measurements are available, *viz.* static measurement of temperature and pressure, or ebulliometry.

Whereas ebulliometric measurements of vapour pressures near atmospheric may, if the sample is of high enough purity, be accurate within a part in 7×10^4, corresponding to ± 0.001 K, both above and below this range the attainable accuracy is much lower. In the range 500 kPa to the critical pressure the vapour pressures of some liquefied gases such as oxygen, methane and carbon dioxide have been measured very accurately, but there are few sets of values for compounds normally liquid, with the exception of water, that may be relied on to better than 1 part in 10^3 of the pressure; the vapour pressures of benzene and hexafluorobenzene are probably the best known for this class of compound because precise and consistent measurements have been made on them in more than one laboratory. Water falls into a category of its own; there have been many investigations of its thermophysical properties, including vapour pressure, and extensive assessment of their reliability.

Below 1 kPa the attainable accuracy falls off sharply and errors of $\pm 10\%$ in such measurements are to be expected. In general, measurements of the vapour pressure of liquids can be expected to be of higher accuracy than those of solids, but ice and naphthalene are recommended as reference materials because it is doubtful whether the vapour pressure of any liquid having such a low vapour pressure as these two solids is known as accurately.

A triple point, at which solid, liquid and vapour phases are in equilibrium, being invariant, is in principle particularly suitable for reference purposes, and the temperatures of several triple points are defining or secondary fixed points of the International Practical Temperature Scale of 1968. The triple-point pressure of water has been suggested as a reference pressure for low-pressure manometers, which are usually dependent upon elastic deformation of a diaphragm and require calibration (Ref. 1). The exact value of this triple-point pressure (611.2 Pa) has been in doubt (Ref. 2) but an authoritative investigation has recently been published (Ref. 3). Of other triple-point pressures only that of carbon dioxide (517.95 kPa) has been measured sufficiently accurately for the value to be used as a reference pressure.

The temperature range in which a vapour pressure is exerted is experimentally important, and materials with a wide range of vapour pressures and a wide range of volatilities have been included. Most of the vapour-pressure tables in these recommendations have been generated from Chebyshev polynomials fitted to the data specified. The equation used (Ref. 4) is

$$(T/K)\log(p/p^\circ) = \frac{a_o}{2} + \sum_{s=1}^{n} a_s E_s(x) \qquad (5.1.1)$$

where T is the thermodynamic temperature (in practice, T_{68}, the International Practical Temperature Scale 1968), p is the pressure, p° is a reference pressure, $E_s(x)$ is the Chebyshev polynomial in x of degree s, and $x = [2T - (T_{max} + T_{min})]/(T_{max} - T_{min})$, T_{min} and T_{max} being two temperatures respectively just below and just above the extreme temperatures of the measured values. The series of Chebyshev polynomials $E_o(x) = 1$, $E_1(x) = x$, $E_2(x) = 2x^2 - 1$, $E_3(x) = 4x^3 - 3x$, $E_{s+1}(x) = 2xE_s(x) - E_{s-1}(x)$ may be summed by forming the coefficients $b_n, b_{n-1}, \ldots, b_s, \ldots, b_o$ from $b_s = 2xb_{s+1} - b_{s+2} + a_s$ with $b_{n+1} = b_{n+2} = 0$. Then

$$(T/\text{K})\log(p/p^\circ) = \frac{(b_o - b_2)}{2} \qquad (5.1.2).$$

An approximate solution may be obtained as

$$(T/\text{K})\log(p/p^\circ) \approx \frac{(a_o + 2xa_1)}{2} \qquad (5.1.3)$$

and if the value of T for a given value of p is required, the approximate values of T and dT/dp obtainable from equation 5.1.3 allow rapid convergence of the iterative solution of equation 5.1.1 necessary for calculation of the exact value of T. Coefficients of the equations are given in the table below; by their use values of the pressure in kilopascals are obtained as logarithms to the base 10.

For values of vapour pressure other than those tabulated, simple linear interpolation will often be adequate; more accurate values will be obtained by linear interpolation of $\log(p/p^\circ)$ against $1/T$. More exact interpolation still will be obtained by fitting an Antoine equation (Ref. 5) to three of the tabulated points in the range of interest if this form is preferred to that of equation 5.1.1.

The vapour-pressure curve terminates at the critical point, which is frequently determined by observation of the temperature of disappearance of the meniscus between liquid and vapour in a sample contained in a thick-walled sealed glass tube. Carbon dioxide, pentane, benzene and hexafluorobenzene, for which vapour pressures up to the critical temperature and pressure are given here, are suitable as reference materials in measurements of critical temperature and pressure. Naphthalene, for which low vapour pressures only are given here but which is stable at its relatively high critical temperature, may also be used for this purpose. Water, on the other hand, otherwise attractive because of its easy availability in a pure state, is inconvenient because of its high critical pressure. Reproducibility in measurements of critical temperature T_c of ± 0.05 K or better and of critical pressure p_c of $\pm 10^{-3} p_c$ may be looked for if the samples are sufficiently pure (purity is particularly important here as the position of the critical point is often very sensitive to amounts of impurities which would cause only a small change in the vapour pressure measured just below the critical temperature).

5. Pressure-Volume-Temperature

Table 5.1.1.1. Coefficients of equation 5.1.1 for $p_o = 1$ kPa

	$\dfrac{T_{\min}}{\text{K}}$	$\dfrac{T_{\max}}{\text{K}}$	a_0	a_1	a_2	a_3	a_4	a_5	a_6
Carbon dioxide (solid)	93	217	60.124	564.125	−3.929	0.744	0.692	−	−
Carbon dioxide (liquid)	218	305	1779.489	291.773	0.034	0.508	0.067	0.033	−
Pentane	147	470	1171.153	1093.595	−30.010	9.705	−0.092	−	−
Benzene	285	563	2332.342	917.602	−9.000	4.577	−0.106	0.173	0.061
Hexafluoro-benzene	278	517	2001.998	823.086	−10.576	4.406	−0.010	0.168	0.051
Water	273	648	2794.027	1430.604	−18.234	7.674	−0.022	0.263	0.146
	a_7 0.055	a_8 0.033	a_9 0.015	a_{10} 0.013					
Mercury	400	1765	8745.7706	4708.7980	13.2829	29.7025	11.7077	4.5180	−
Ice	180	275	−989.461	453.879	0.240	−0.137	0.009	−	−
Naphthalene	230	344	301.62	791.49	−8.25	0.40	−	−	−

Note: for ice $a_0 = 375.539$, $a_1 = 596.379$ and for naphthalene $a_0 = -1420.38$, $a_1 = 620.49$ with other coefficients unchanged gives $p/$Pa.

Vapour pressures at temperatures above 600 K.

A special group of vapour-pressure reference materials comprises the metals cadmium, silver, gold, platinum and tungsten, which will cover the temperature range 350 to 3000 K and pressure range 10^{-7} to 200 Pa. The cadmium, silver and gold are supplied by the National Bureau of Standards with a certificate of purity of each sample and its vapour pressure over a range of temperature; certification of the platinum and tungsten is not yet completed. International co-operation in this project is being organized by Commission II.3 (High Temperatures and Refractories) of IUPAC. At these temperatures, errors in vapour-pressure measurement may be large (up to 100% of the pressure) and use of certified reference materials will allow investigators to check whether the methods they use in this range, which include the Knudsen, torque-Knudsen, Langmuir, and mass-spectrometric methods, are subject to systematic errors.

Orthobaric volumes (or densities) and critical volume (or density).

The volumes of the saturated liquid and vapour, the orthobaric volumes, are usually related through their reciprocals, the orthobaric densities ρ_l, ρ_g, by the well known "law of rectilinear diameters" due to Cailletet and Mathias according to which the sum of the orthobaric densities varies linearly with temperature. This may be expressed as

$$\rho_l + \rho_g = 2\rho_c + a(1 - T_r) \tag{5.1.4}$$

where ρ_l, ρ_g are the densities of liquid and vapour, ρ_c is the critical density, T_r is the reduced temperature T/T_c, and a is a constant for each substance. In addition, the difference between the two densities may be expressed as

$$\rho_l - \rho_g = b(1 - T_r)^n \tag{5.1.5}$$

where b is a constant for each substance and $n \approx 1/3$. An equation similar to equation (5.1.5) was proposed by Verschaffelt (Ref. 6) and was applied by Goldhammer (Ref. 7) and by Jüptner (Ref. 8); it has since been rediscovered more than once and has been extensively discussed (Refs. 9, 10). Equation 5.1.4 allows calculation of ρ_c, and addition and subtraction of equations (5.1.4) and (5.1.5) give ρ_l and ρ_g. However, these equations are not exact; they are empirical and extra terms must often be added for the representation of exact measurements, which are difficult in the region close to T_c. As a result there must always be some uncertainty in the extrapolation to $T_r = 1$ from the region where observations are made, and there is always an uncertainty of about $\pm 3\%$ in ρ_c because of the physical indefiniteness of this property. Except when $T_r \to 1$, $\rho_g \ll \rho_l$, and the orthobaric density of the vapour cannot be determined with the same accuracy as that of the liquid. For T_r less than about 0.8, ρ_g is very small and equation (5.1.4) and (5.1.5) cease to be useful for the representation of the variation of ρ_g with temperature although they may still be satisfactory for ρ_l; in these conditions ρ_g is more usually obtained by extrapolation to the vapour pressure of an equation of state representing the behaviour of the unsaturated vapour.

Pressure–Volume–Temperature behaviour of unsaturated vapours and gases.

Different reference materials and ways of expressing their properties are needed for different methods of study of the PVT properties of gases and vapours. These methods may be broadly divided into three groups:

(i) gases and vapours at pressures up to twice atmospheric,

(ii) gases at higher pressures over wide ranges of temperature, and

(iii) vapours of substances liquid at room temperature, at pressures up to, usually, not more than 10 MPa.

Experiments in group (i) are usually carried out in glass apparatus, often with the prime aim of determining coefficients of the virial equation of state; whereas experiments in groups (ii) and (iii), which involve higher pressures, are carried out in metal apparatus (possibly with a glass liner). Few have been carried out at temperatures greater than 800 K. The results of these measurements are frequently presented in the first instance as isotherms either in terms of p and V or of the compression factor $Z = pV/RT$, and are then frequently represented by

5. Pressure-Volume-Temperature

an equation of state. Equations of state and the experimental procedures have been surveyed in several reviews and books (Refs. 11 to 15).

A need additional to the standardisation of measurements arises in this section of the recommendations, *viz.* that there are occasions when, in testing large plant, large quantities of a fluid of known density must be used. The question of importance is therefore not what fluids have the best known properties, but what are the properties of the fluids that must be used because of practical considerations. Accordingly, with these needs in mind, the recommendations include the following:

(i) Air is included in addition to nitrogen, for which more accurate data are available, because it is unlikely that nitrogen will be used on a large scale in preference to air except in the air-separation industry where nitrogen is freely available.

(ii) Oxygen is included because operators of plant containing this gas or, more particularly, liquid, who need to conduct tests on the plant or calibrate instruments *in situ* may be reluctant to admit other fluids because of the necessity for thorough purging.

(iii) Ethylene is included because this is an industrially important gas that is often transported by pipeline in which it is metered by volume but paid for by mass. The particular value adopted for the density of ethylene in these transactions is therefore directly convertible into money terms. Metering of ethylene with flowmeters calibrated by use of other fluids has been shown not to be sufficiently reliable for this conversion, and exact and agreed values of the density of ethylene itself are needed.

In this field, the greatest effort has been put into the study of the properties of water substance, from which have arisen the International Skeleton Steam Tables and various more detailed tables based upon them. These are referred to in recommendation 5.6.1 relating to water as a reference material for the Pressure-Volume-Temperature behaviour of liquids.

A special aspect of PVT properties is that they frequently cannot be reduced to compact sets of numbers conforming to the pattern set elsewhere in these recommendations and in some instances, therefore, the recommendations here are of the best sets of tables available.

Pressure-Volume-Temperature behaviour of liquids.

Many of the comments made concerning PVT behaviour of unsaturated vapours and gases also apply here. Whenever a liquid is required in large quantities, water is the obvious choice.

Units: As far as possible SI units have been used in these recommendations, but in some instances, where reference is made to other publications, other units are specified. The SI unit of pressure is the newton per square metre N m^{-2} which has been given the name pascal, symbol Pa. The following conversion factors apply to the units likely to be used in this field:

$1 \text{ Pa} = 10^{-5}$ bar exactly $= 9.869\ 23 \times 10^{-6}$ atm $= 7.500\ 62 \times 10^{-3}$ mmHg
$= 1.019\ 716 \times 10^{-5}$ kgf cm^{-2} $= 1.450\ 38 \times 10^{-4}$ lbf in^{-2}

1 atm (standard atmosphere) $= 101\ 325$ Pa exactly
$= 1.013\ 25$ bar exactly $= 760.000$ mmHg
$= 1.033\ 227$ kgf cm^{-2} $= 14.695\ 95$ lbf in^{-2}

1 mmHg $= 133.3224$ Pa $= 1.315\ 79 \times 10^{-3}$ atm
$= 1.359\ 51 \times 10^{-3}$ kgf cm^{-2} $= 1.933\ 68 \times 10^{-2}$ lbf in^{-2}

1 kgf cm^{-2} $= 98\ 066.5$ Pa exactly $= 735.559$ mmHg
$= 0.967\ 841$ atm $= 14.223\ 34$ lbf in^{-2}

1 lbf in^{-2} $= 6894.76$ Pa $= 6.804\ 60 \times 10^{-2}$ atm
$= 51.7149$ mmHg $= 7.030\ 70 \times 10^{-2}$ kgf cm^{-2}

1 kg $= 2.204\ 62$ lb; 1 lb $= 0.453\ 592\ 37$ kg exactly

1 in $= 2.54$ cm exactly; 1 cm $= 0.393\ 701$ in

1 ft^3 $= 2.831\ 68 \times 10^{-2}$ m^3; 1 m^3 $= 35.3147$ ft^3

1 g cm^{-3} $= 62.4280$ lb ft^{-3}; 1 lb ft^{-3} $= 0.016\ 0185$ g cm^{-3}.

REFERENCES

1. van Hook, W. A. *Metrologia* **1971**, *7*, 30.
2. Ambrose, D.; Lawrenson, I. J. *J. Chem. Thermodynamics* **1972**, *4*, 755.
3. Guildner L.; Johnson, D. P.; Jones, F. E. *J. Res. Nat. Bur. Stand. A* **1976**, *80*, 505.
4. Ambrose, D.; Counsell, J. F.; Davenport, A. J. *J. Chem. Thermodynamics* **1970**, *2*, 283.
5. Thomson, G. W. *Chem. Rev.* **1946**, *38*, 1.
6. Vershaffelt, J. E. *Comm. Leiden* **1896**, *28*, 12.
7. Goldhammer, D. A. *Z. physik. Chem.* **1910**, *71*, 555.
8. Jüptner, H. *Z. physik. Chem.* **1913**, *85*, 1.
9. Riedel, L. *Chemie-Ing.-Tech.* **1954**, *26*, 259.
10. Bowden S. T.; Costello, J. M. *Rec. Trav. Chim.* **1958**, *77*, 28, 32, 36, 803.
11. Ellington, R. T.; Eakin, B. E. *Chem. Eng. Prog.* **1963**, *59*, (11), 80.
12. Cox, J. D.; Lawrenson, I. J. *Specalist Periodical Report: Chemical Thermodynamics*, McGlashan, M. L. senior reporter, Vol. 1, The Chemical Society: London, **1973**, 162.
13. Martin, J. J. *Ind. Eng. Chem.* **1967**, *59*, (12), 34.

14. Mason, E. A.; Spurling, T. H. *The Virial Equation of State*, Pergamon: London, **1969**.
15. Le Neindre, B.; Vodar, B. (Eds.) *Experimental Thermodynamics Vol. II, Experimental Thermodynamics of Non-reacting Fluids*, Butterworths: London, **1975**.

Note added in press

In the period since these Recommendations were completed in 1983 it has become clear that a polynomial equation, as used here, is not the best form for representation of vapour pressures up to the critical point. The contributor of the Recommendations for vapour-pressure reference materials now favours equations of the type proposed by Wagner (Refs. 1 to 5), but has not found it practical to revise these recommendations again. The equations given fit the experimental results well, and it is unlikely that new equations of the preferred form would give rise to changes in the tabulated values of vapour pressure greater than those arising from the present revision of the original Recommendations (Ref. 6), which are in general less than the probable experimental error. However, it should be noted that the properties of water substance are the concern of the International Association on the Properties of Steam, and those seeking the most up-to-date conclusions of IAPS about the vapour pressure of water should consult the NBS/NRC Steam Tables (Ref. 7), where the vapour pressure is tabulated and a Fortran program for its calculation is given. Another equation for the vapor pressure of water has been proposed by Wagner and Saul (Ref. 8). A comparison of different equations for the vapour pressure of water has been made by Somayajulu (Ref. 9).

REFERENCES

1. Wagner, W *Cryogenics* **1973**, *13*, 470.
2. Wagner, W. *Bull. Inst. Int. Froid, Annexe* **1973**, *4*, 65.
3. Wagner, W. Habilitationschrift, TU Braunschweig 1973, Forschr.–Ber VDI–Z. Reihe 3, Nr. 39, **1974**.
4. Wagner, W. *A new correlation method for thermodynamic data applied to the vapour-pressure curve for argon, nitrogen, and water*, IUPAC Thermodynamic Tables Project Center, PC/T 15: London, **1977**.
5. Ambrose, D. *J. Chem. Thermodynamics* **1986**, *18*, 45.
6. Ambrose, D. (collator) *Pure and Appl. Chem.* **1977**,*49*, 1437.
7. Haar, L.; Gallagher, J. S.; Kell, G. *NBS/NRC Steam Tables*, Hemisphere Pub. Inc.: New York, **1984**.
8. Wagner, W.; Saul, A. *Proc. Xth Int. Conf. Properties of Steam*, **1984**.
9. Somayajulu, G. R. *J. Chem. Eng. Data* accepted.

5.2. Reference materials for vapour pressures at temperatures up to 770 K, critical temperatures and critical pressures

5.2.1. Carbon dioxide

Physical property: Vapour pressure
Units: kPa, K
Recommended reference material: Carbon dioxide
Range of variables: 1 to 7380 kPa, 120 to 304 K
Physical states within the range: solid and vapour, liquid and vapour
Contributor: D. Ambrose

Intended usage: Carbon dioxide is recommended as a test substance for vapour-pressure measurements on relatively high-boiling gases because its vapour pressure is well established and it may easily be obtained in a state of high purity. Vapour-pressure measurements in the temperature range from 120 K up to room temperature are normally made by the static method, frequently in the course of calorimetric studies, or in apparatus the design of which is based upon calorimetric practice (*i.e.* the thermal isolation of the equilibrium vessel by means of vacuum jackets and the elimination by careful design of unwanted flows of heat through connecting tubes) (Refs. 1 to 3). The vapour pressure of carbon dioxide at 273.15 K or 273.16 K, which is accurately known, has often been used as a standard pressure for the calibration of pressure balances and, although calibration by a standards laboratory (against a standard instrument, itself calibrated by exact measurements of the diameters of the piston and cylinder) is to be preferred, measurements of the vapour pressure at 273.15 K and 273.16 K by means of a pressure balance provides a useful check upon the experimental technique. An apparatus for making this measurement by a dynamic, *i.e.* ebulliometric, method has been described which could be adapted for use with other gases (Ref. 4).

Sources of supply and/or method of preparation: At one time pure samples of carbon dioxide were prepared chemically (Ref. 5) but this is probably not now necessary. Carbon dioxide of very high purity may be bought in cylinders and the probable residual impurity is air. This may be removed by pumping in the apparatus (Ref. 6). If air is present to an appreciable extent in the cylinder, which may be checked by passing the gas into an aqueous solution of potassium hydroxide and determination of the fraction not absorbed, slow venting of a large proportion of the contents of the cylinder, *i.e.* a Rayleigh distillation, will reduce the residual air in the cylinder to a very low level. A more effective and less wasteful method of purification is to distill the material under reflux at 273 K (Refs. 7, 8), as is done in the apparatus mentioned above (Ref. 4).

5. Pressure-Volume-Temperature

Pertinent physicochemical data:

Table 5.2.1.1. Vapour pressure p of carbon dioxide

Solid				Liquid			
T_{68}/K	p/kPa	T_{68}/K	p/kPa	T_{68}/K	p/kPa	T_{68}/K	p/kPa
120	0.004	175	16.86	216.58*	517.95	270	3203
125	0.012	180	27.62	220	599.6	273.15	3485
130	0.032	185	44.02	225	735.7	275	3698
135	0.080	190	68.44	230	893.5	280	4160
140	0.187	194.674	101.325	235	1075.2	285	4710
145	0.409	195	104.07	240	1283.0	290	5315
150	0.848	200	155.11	245	1519.0	295	5979
155	1.674	205	227.07	250	1785.6	300	6710
160	3.158	210	327.17	255	2084.9	304.21**	7382
165	5.721	215	464.78	260	2419.4		
170	9.987	216.58*	517.95	265	2791		

*Triple-point temperature.
**Critical temperature.

The tables have been computed from the coefficients given in table 5.1.1.1. For the solid these coefficients were obtained from the values tabulated by Meyers and van Dusen (Ref. 5), by a fit that was constrained to pass through the normal sublimation temperature 194.674 K, which is a secondary reference point on the IPTS-68 (Refs. 9, 10).

Angus, Armstrong and de Reuck have critically evaluated the data for the thermodynamic properties of liquid carbon dioxide (Ref. 11), and the coefficients in table 5.1.1.1 were obtained by a fit to the values tabulated by these authors; They differ slightly from those given originally in these recommendations (Ref. 12) but the consequent changes in vapour pressure tabulated here are nowhere greater than a relative change in pressure of 3×10^{-4}. Sengers and Chen concluded from a survey of their own and other accurate measurements that the best values of p at 273.15 K lie in the range 3485 ± 0.3 kPa (Ref. 6). Values below 1 kPa are included above because they were included in the work on which the table is based (Ref. 5), but they are probably not of high accuracy; at higher pressures the measurements are among the most accurate that have been made and reasonable estimates of their uncertainty are $\pm 5 \times 10^{-4} p$ for 10 kPa $< p <$ 518 kPa and $\pm 2 \times 10^{-4} p$ for $p >$ 518 kPa.

REFERENCES

1. Goodwin, R. D. *J. Res. Nat. Bur. Std.* **1961**, *65C*, 231.
2. Hoge, H. J. *J. Res. Nat. Bur. Std.* **1950**, *44*, 321.
3. Straty, G. C.; Prydz, R. *Rev. Sci. Instrum.* **1970**, *41*, 1223.
4. Edwards, J. L.; Johnson, D. P. *J. Res. Nat. Bur. Std.* **1968**, *72C*, 27.
5. Meyers, C. H.; van Dusen, M. S. *J. Res. Nat. Bur. Std.* **1933**, *10*, 381.
6. Levelt Sengers, J. M. H.; Chen, W. T. *J. Chem. Phys.* **1972**, *56*, 595.
7. Keulemans, A. I. M. *Gas Chromatography*, 2nd ed., Reinhold: New York, **1959**, p. 227.
8. Wilkins, M.; Wilson, AERE Report 3244. Atomic Energy Research Establishment, Harwell, **1960**.
9. *The International Practical Temperature Scale of 1968.* HMSO, London **1976**; *Metrologia* **1969**, *5*, 35; **1976**, *12*, 7.
10. Barber, C. R. *Brit. J. Appl. Phys.* **1966**, *17*, 391.
11. Angus, S.; Armstrong, B.; de Reuck, K. M. *Carbon Dioxide, International Thermodynamic Tables of the Fluid State*, Pergamon Press: Oxford, **1977**.
12. Ambrose, D. *Pure Appl. Chem.* **1977**, *49*, 1437.

5.2.2. Pentane

Physical property: vapour pressure
Units: kPa, K
Recommended reference material: Pentane
Range of variables: 7.5 to 3370 kPa, 250 to 469.7 K
Physical states within the range: liquid and vapour
Contributor: D. Ambrose

Intended usage: Pentane is recommended as a test substance for use in any method of vapour-pressure measurements when liquids of comparable volatility are under investigation.

Sources of supply and/or methods of preparation: Pentane is normally purified by fractional distillation of a suitable starting material and the process of purification may be monitored by gas chromatography. The acceptability of any sample for the present purpose may be confirmed by comparative ebulliometric measurements against water. Samples of pure pentane are available from suppliers (A), (C), (D) and (E).

Pertinent physicochemical data: Table 5.2.2.1 has been computed from the coefficients given in table 5.1.1.1 which were determined by a fit to the measurements by Willingham, Taylor, Pignocco and Rossini (Ref. 1) and by Beattie, Levine and Douslin (Ref. 2); the critical temperature is that selected for the TRC Thermodynamic Tables – Hydrocarbons (Ref. 3). It is desirable that values recommended for reference purposes should have been confirmed by more than one investigation; in this instance, confirmation at pressures above atmospheric is not as satisfactory as is desirable since the only other investigation [by Young (Ref. 4)]

5. Pressure-Volume-Temperature

of this substance in the high-pressure range, while of outstanding merit for its time, did not approach the accuracy now attainable. Pentane is recommended however because there appears to be no other liquid of comparable volatility that is more suitable; it has been preferred over diethyl ether (Ref. 5), which would otherwise have been chosen, because of its much lower solvent capacity for water and its consequent relative ease of handling without contamination by atmospheric moisture.

Table 5.2.2.1. Vapour pressure p of pentane

T_{68}/K	p/kPa	T_{68}/K	p/kPa	T_{68}/K	p/kPa
250	7.60	320	144.3	400	1038
260	12.98	330	145.7	410	1256
270	21.15	340	260.1	420	1507
280	33.11	350	339.4	430	1793
290	50.01	360	435.9	440	2120
300	73.170	370	551.5	450	2490
309.21	101.325	380	688.8	460	2910
310	104.07	390	850.2	469.7*	3368

*Critical temperature.

REFERENCES

1. Willingham, C. B.; Taylor, W. J.; Pignocco, J. M.; Rossini, F. D. *J. Res. Nat. Bur. Std.* **1945**, *35*, 219.
2. Beattie, J. A.; Levine, S. W.; Douslin, D. R. *J. Am. Chem. Soc.* **1951**, *73*, 4431.
3. TRC Thermodynamic Tables – Hydrocarbons, Thermodynamics Research Center, Texas A&M University, College Station, Texas. Loose-leaf data sheets extant **1985**.
4. Young, S. *J. Chem. Soc.* **1897**, *71*, 446; *Sci. Proc. Roy. Dublin Soc.* **1910**, *12*, 374.
5. Ambrose, D.; Sprake, C. H. S.; Townsend, R. *J. Chem. Thermodynamics* **1972**, *4*, 247.

5.2.3. Benzene

Physical property: vapour pressure
Units: kPa, K
Recommended reference materials: Benzene
Range of variables: 5 to 4898 kPa, 280 to 562.1 K
Physical state within the range: liquid and vapour
Contributor: D. Ambrose

Intended usage: Benzene is recommended as a test substance for any method of vapour-pressure measurement in the range 5 kPa to the critical pressure 4898 kPa. It has been used as the reference substance in the comparative ebulliometric method (Ref. 1) (see recommendation 5.2.5 relating to the use of water as a reference substance).

Sources of supply and/or method of preparation: Benzene may be purified by fractional freezing of material uncontaminated by sulphur-containing compounds. This substance has been extensively used as a test material in the development of the method for testing purity by freezing point measurements; however, confirmation of the vapour pressure of a sample by comparative ebulliometric measurements against water provides an alternative way of confirming its acceptability for the present purpose. Samples of pure benzene are available from suppliers (A), (C), (D), and (E).

Pertinent physicochemical data:

Table 5.2.3.1. Vapour pressure p of benzene

T_{68}/K	p/kPa	T_{68}/K	p/kPa	T_{68}/K	p/kPa
280	5.148	370	165.2	470	1366
290	8.606	380	215.9	480	1602
300	13.816	390	277.7	490	1868
310	21.389	400	353.2	500	2164
320	32.054	410	441.0	510	2494
330	46.656	420	545.5	520	2861
340	66.152	430	667.6	530	3267
350	91.609	440	808.8	540	3717
353.24	101.325	450	971.1	550	4216
360	124.192	460	1156	560	4770
				562.16*	4898

*Critical temperature.

5. Pressure-Volume-Temperature

The table has been computed from the coefficients given in table 5.1.1.1 which are based upon surveys (Refs. 2 to 4) of the available data. There is good agreement between values published by different authors; the normal boiling temperature should be reproducible within ±0.01 K, and pressures above atmospheric up to the critical may be expected to be within 0.1% of the values given.

REFERENCES

1. Osborn, A. G.; Douslin, D. R. *J. Chem. Eng. Data* **1969**, *11*, 501.
2. Ambrose, D. *J. Phys. E* **1968**, *1*, 41.
3. Ambrose, D.; Broderick, B. E.; Townsend, R. *J. Chem. Soc.* **1967**, *A*, 633.
4. Ambrose, D. *J. Chem. Thermodynamics* **1981**, *13*, 1161.

5.2.4. Hexafluorobenzene

Physical property: Vapour pressure
Units: kPa, K
Recommended reference material: Hexafluorobenzene
Range of variables: 4 to 3273 kPa, 280 to 516.7 K
Physical states within the range: liquid and vapour
Contributor: D. Ambrose

Intended usage: Hexafluorobenzene may be used as a test substance for any method of vapour-pressure measurement in the range 4 kPa to the critical pressure 3273 kPa. This range is substantially the same as is covered by benzene but, if it is available, hexafluorobenzene is suggested as an alternative to the latter because of the very close agreement existing between very precise investigations carried out at the Bartlesville Energy Research Center, USA (Refs. 1, 2) and at the National Physical Laboratory, UK (Refs. 3, 4). Hexafluorobenzene has the advantage in comparison with benzene of being non-toxic.

Sources of supply and/or methods of preparation: Anaesthetic grade hexafluorobenzene from supplier (B) has proved to be of suitably high purity. The purity may be monitored by gas chromatography and confirmed by freezing point measurements; however, confirmation of the vapour pressure of a sample by comparative ebulliometric measurements against water provides an alternative way of confirming its acceptability for the present purpose.

Pertinent physicochemical data: Table 5.2.4.1 has been computed from the coefficients given in table 5.1.1.1, which are based on all the sources quoted. The coefficients in table 5.1.1.1 differ slightly from those given originally in these recommendations (Ref. 5) but the consequent changes in vapour pressure tabulated here are nowhere greater than $3 \times 10^{-4} p$; reasons for the change are discussed in reference 4. Normal boiling temperatures should be reproducible within ±0.01 K, and pressures above atmospheric up to the critical may be expected to be within 0.1% of the values given.

Table 5.2.4.1. Vapour pressure p of hexafluorobenzene

T_{68}/K	p/kPa	T_{68}/K	p/kPa	T_{68}/K	p/kPa
280	4.322	360	124.816	450	1080
290	7.463	370	168.4	460	1297
300	12.328	380	223.0	470	1547
310	19.576	390	290.4	480	1833
320	30.009	400	372.6	490	2159
330	44.578	410	471.5	500	2530
340	64.380	420	589.3	510	2954
350	90.664	430	728.3	516.73*	3273
353.402	101.325	440	890.9		

*Critical temperature.

REFERENCES

1. Douslin, D. R.; Osborn, A. *J. Sci. Instrum.* **1956**, *42*, 369.
2. Douslin, D. R.; Harrison, R. H.; Moore, R. T. *J. Chem. Thermodynamics* **1969**, *1*, 305.
3. Counsell, J. F.; Green, J. H. S.; Hales, J. L.; Martin, J. F. *Trans. Faraday Soc.* **1964**, *60*, 700.
4. Ambrose, D. *J. Chem. Thermodynamics* **1981**, *13*, 1161.
5. Ambrose, D. *Pure Appl. Chem.* **1977**, *49*, 1437.

5.2.5. Water

Physical property: Vapour pressure
Units: kPa, K
Recommended reference material: Water
Range of variables: 3.66 to 22106 kPa, 273.16 to 647.31 K
Physical states within the range: liquid and vapour
Contributors: D. Ambrose

Intended usage: If the sample of the substance of which the vapour pressure is being measured is sufficiently pure, and temperature measurements are good to ±0.005 K, comparative ebulliometry (*i.e.* the comparison of the boiling or condensation temperature of the substance with that of a reference substance, preferably water, boiling at the same pressure applied by means of a buffer gas) yields values of normal boiling points that are reproducible within ±0.01 K, and variations in the vapour pressure of the substance at a given temperature ranging from about ±0.010 kPa at 20 kPa to ±0.020 kPa at 200 kPa (Refs. 1 to 4). The pressure for each observation is calculated from the known relationship between the temperature and pressure of the reference substance. For determinations of the highest accuracy corrections must be applied for the effect of the hydrostatic heads of vapour in the apparatus (Refs. 4, 5), but for more moderate accuracy (if, for example, temperatures are measured by means of mercury-in-glass thermometers) these may be ignored. Smooth boiling is essential, and special steps to ensure this are necessary at pressures below 15 kPa: water, especially, tends to bump in these conditions. Many determinations have been made in ebulliometers of the type designed by Swietoslawski (Refs. 3, 6) but an improved apparatus for this application is that designed by Ambrose (Ref. 1), and with this smooth boiling is obtained at pressures down to 3 kPa, the lowest at which it is practicable to operate an ebulliometer containing water. There is less difficulty in obtaining smooth boiling at pressures above atmospheric, and other apparatus has been described for use up to 800 kPa (Refs. 2, 7), and up to the critical pressure of organic compounds (Ref. 8). The ebulliometric method is well established for use at pressures up to 200 kPa, and is applicable for measurements on materials boiling, at the prevailing pressure, in the range 250 to 750 K. The ebulliometric method appears to be satisfactory at pressures above 200 kPa but it has not been widely used under such conditions, and for the most accurate measurements there may be objections in principle because of the presence of the buffer gas (Ref. 4).

Since the vapour pressure of liquid water is known more exactly than that of any other substance and adequate purification of a sample is straightforward, water is also recommended as a test material for other methods of vapour–pressure measurement provided there is no likelihood of reaction with materials of construction of the apparatus.

Sources of supply and/or method of purification: Purification of water has been extensively studied (Refs. 9 to 11). For the present purpose either distilled or de-ionised water may be used; excessive fractionation should be avoided as this would lead to a change in the isotopic ratio, and a subsequent measureable change in the vapour pressure.

Pertinent physicochemical data: Table 5.2.5.1 has been computed from the coefficients given in table 5.1.1.1. They differ from those given originally in these recommendations (Ref. 12), which were based on a critical survey of all the reliable measurements available (Ref. 13), because it now appears from the measurements made at the National Bureau of Standards (Ref. 14) that the triple-point pressure is higher than had previously been adopted for the internationally agreed Steam Tables. The previous equation and table gave 0.6111 kPa as the triple-point pressure; for the present revision the same data as were selected in reference 13 were refitted with substitution of the value 0.611657 kPa as the triple-point pressure. The change affects the value in the range 273.15 to 290 K; elsewhere differences between the values given previously and the present ones do not exceed $10^{-4}p$. In the range 293.15 to 373.15 K values calculated from the coefficients given in table 5.1.1.1 are practically identical with those given in another survey restricted to the range 273.15 to 373.15 K

(Ref. 15); at 283.15 K the difference between the two formulations is about 0.2 Pa $(2 \times 10^{-4} p)$. Further amendment may eventually be required because of changes in the temperature scale. Definition of thermodynamic temperature by allocation of the value of 273.16 K to the triple-point temperature of water has made the 'steam point' a measured instead of a defined point and investigations have indicated that the thermodynamic normal boiling point of water may be as much as 30 millikelvins below the previously defined value 373.15 K (Ref. 16). This result does not affect the expression of temperatures on IPTS-68 which will remain unchanged until amendment is agreed by the Conférence Général des Poids et Mesures.

Table 5.2.5.1. Vapour pressure p/kPa of water at International Practical Kelvin Temperatures T_{68}, 273.15 K, 273.16 K, at 2 K intervals from 270 K to 646 K, and 647.31 K.

T_{68}/K		p/kPa			
273.15	0.6112				
273.16	0.61166				
	0	2	4	6	8
270	0.4848	0.5620	0.6500	0.7499	0.8632
280	0.9915	1.1363	1.2994	1.4828	1.6886
290	1.9192	2.1768	2.4643	2.7844	3.1403
300	3.5353	3.9728	4.4567	4.9910	5.5800
310	6.2282	6.9407	7.7224	8.5790	9.5162
320	10.540	11.658	12.875	14.200	15.639
330	17.202	18.896	20.730	22.713	24.856
340	27.167	29.659	32.340	35.224	38.323
350	41.647	45.211	49.028	53.112	57.477
360	62.139	67.112	72.414	78.059	84.068
370	90.453	97.238	104.44	112.08	120.17
380	128.74	137.80	147.39	157.51	168.20
390	179.48	191.37	203.90	217.08	230.96
400	245.54	260.87	276.97	293.86	311.58
410	330.15	349.60	369.98	391.29	413.59
420	436.89	461.24	486.66	513.19	540.87
430	569.73	599.80	631.13	663.75	697.69
440	732.99	769.70	807.86	847.49	888.64
450	931.36	975.69	1021.7	1069.3	1118.7
460	1169.9	1222.8	1277.7	1334.4	1393.2
470	1453.9	1516.6	1581.5	1648.5	1717.6
480	1789.0	1862.8	1938.8	2017.2	2098.1
490	2181.4	2267.3	2355.8	2446.9	2540.7
500	2637.3	2736.7	2838.9	2944.1	3052.2
510	3163.3	3277.6	3394.9	3515.5	3639.3
520	3766.4	3896.9	4030.9	4168.3	4309.3

Table 5.2.5.1. continued

T_{68}/K	p/kPa				
	0	2	4	6	8
530	4453.9	4602.2	4754.3	4910.1	5069.9
540	5233.5	5401.2	5573.0	5748.9	5929.0
550	6113.4	6302.1	6495.3	6692.9	6895.1
560	7102.0	7313.6	7530.0	7751.2	7977.4
570	8208.6	8445.0	8686.5	8933.3	9185.4
580	9443.0	9706.2	9974.9	10249	10530
590	10816	11108	11406	11711	12022
600	12339	12662	12993	13330	13674
610	14025	14384	14749	15122	15503
620	15892	16288	16693	17106	17528
630	17959	18399	18848	19307	19777
640	20256	20747	21249	21763	
647.31	22106				

REFERENCES

1. Ambrose, D. *J. Phys. E* **1968**, *1*, 41.
2. Ambrose, D.; Sprake, C. H. S.; Townsend, R. *J. Chem. Thermodynamics* **1968**, *1*, 499.
3. Osborn, A. G.; Douslin, D. R. *J. Chem. Eng. Data* **1969**, *11*, 502.
4. Ambrose, D. *Specialist Periodical Report: Chemical Thermodynamics*, McGlashan, M. L. senior reporter *Vol. 1*, The Chemical Society: London, **1972**, p. 218.
5. Beattie, J. A.; et al. *Proc. Amer. Acad. Arts Sci.* **1936**, *71*, 327, 361, 395.
6. Swietoslawski, W. *Ebulliometric Measurements*, Reinhold: New York, **1945**.
7. Ambrose, D.; Sprake, C. H. S. *J. Chem. Thermodynamics* **1972**, *4*, 603.
8. Oliver, G. D.; Grisard, J. W. *Rev. Sci. Instrum.* **1956**, *24*, 204; *J. Am. Chem. Soc.* **1956**, *78*, 561.
9. Barber, C. R.; Handley, R.; Herington, E. F. G. *Brit. J. Appl. Phys.* **1954**, *5*, 41.
10. Menaché, M. *Deep Sea Res.* **1971**, *18*, 449.
11. Hughes, R. C.; Mürau, P. C.; Gundersen, G. *Anal. Chem.* **1971**, *43*, 619.
12. Ambrose, D. *Pure Appl. Chem.* **1977**, *49*, 1437.
13. Ambrose, D.; Lawrenson, I. J. *J. Chem. Thermodynamics* **1972**, *4*, 755.
14. Johnson, D. P.; Guildner, L. A.; Jones, F. E. *J. Res. Nat. Bur. Stand. A* **1976**, *80*, 505.
15. Wexler, A. *J. Res. Nat. Bur. Stand. A* **1976**, *80*, 775.
16. Quinn, T. J.; Compton, J. P. *Rept. Progr. Phys.* **1975**, *38*, 151.

5.2.6. Mercury

Physical property: Vapour pressure
Units: kPa, K
Recommended reference material: Mercury
Range of variables: 0.1 to 800 kPa, 400 to 770K
Physical states within the range: liquid and vapour
Contributor: D. Ambrose

Intended usage: The range of applicability of the comparative ebulliometric method, discussed in detail in connection with the use of water as a reference substance (5.2.5), may be extended downwards to 0.1 kPa if mercury is used as the reference substance (Ref. 1). For accurate measurements the effect of radiation in the determination of the temperature becomes important at 500 K and upwards, and for the study of high-boiling substances (up to 750 K) there may be advantage in using mercury as the reference substance, or as a test substance for check on the apparatus. For this purpose mercury may be used at pressures up to 800 kPa.

Sources of supply and/or method of preparation: The mercury should be purified by established methods, which normally include treatment with nitric acid and distillation (Refs. 2, 3). It is unlikely that variation in the isotopic ratio will cause significant change in the vapour pressure, but the possibility should be considered for the most accurate measurements. Purification of mercury is discussed in greater detail in the item dealing with its use as a reference material for density measurement.

Pertinent physicochemical data:

Table 5.2.6.1. Vapour pressure p of mercury

T_{68}/K	p/kPa	T_{68}/K	p/kPa	T_{68}/K	p/kPa	T_{68}/K	p/kPa
400	0.138	500	5.239	600	57.64	700	314.9
410	0.215	510	6.955	610	70.09	710	363.3
420	0.329	520	9.131	620	84.67	720	417.5
430	0.493	530	11.861	630	101.66	730	477.8
440	0.724	540	15.256	640	121.35	740	544.8
450	1.045	550	19.438	650	144.05	750	619.0
460	1.485	560	25.547	660	170.08	760	700.9
470	2.078	570	30.74	670	199.80	770	791.0
480	2.866	580	38.19	680	233.59		
490	3.899	590	47.09	690	271.83		

The normal boiling temperature of mercury 629.81 K is a secondary fixed point on the International Practical Temperature Scale of 1968 (Ref. 4). The table has been computed from coefficients calculated by Ambrose and Sprake (Ref. 1) and given in table 5.1.1.1.

REFERENCES

1. Ambrose, D.; Sprake, C. H. S. *J. Chem. Thermodynamics* **1972**, *4*, 603.
2. Cook, A. H.; Stone, N. W. B. *Phil. Trans. Roy. Soc. Lond.* **1957**, *250A*, 279.
3. Gordon, C. L.; Wichers, E. *Ann. New York Acad. Sci.* **1957**, *65*, 369.
4. *The International Practical Temperature Scale of 1968*, Amended edition of 1975. HMSO, London **1976**; *Metrologia* **1969**, *5*, 35; **1976**, *12*, 7.

5.2.7. Ice

Physical property: Vapour pressure
Units: Pa, K
Recommended reference materials: Ice
Range of variables: 0.1 to 611 Pa, 190 to 273.16 K
Physical states within the range: solid and vapour
Contributor: D. Ambrose

Intended usage: Vapour pressures in this range of pressure are normally determined by the static method. Accurate measurement of pressure below 1000 Pa is difficult; elastic gauges are most frequently used and the validity of the results obtained by their means depends upon assumptions that are difficult to confirm. Although the vapour pressure of ice is not known with the accuracy that is desirable (Ref. 1), it is probably better known than is that of any other substance in the same range of pressure, and ice is recommended as a reference substance either for test of a method or for calibration of a pressure gauge.

Sources of supply and/or method of preparation: See recommendation 5.2.5 relating to use of water as a vapour-pressure reference material.

Pertinent physicochemical data: Table 5.2.7.1 has been computed from the coefficients given in table 5.1.1.1, which were obtained by fitting to values selected from the table extending over the range -100 to 0 °C at intervals of 0.1 K calculated by Wexler from an equation reconciling vapour-pressure measurements with the related thermodynamic properties (Ref. 1). Wexler took as a reference point the new value by Guildner, Johnson and Jones (Ref. 2) for the triple point pressure (611.657 Pa) in place of the one previously accepted (611.2 Pa) on which the table in the previous issue of these recommendations were based (Ref. 3), and the consequent effect is that the recommended values are increased by about $10^{-3}p$.

Table 5.2.7.1. Pressure p of water vapour over ice

T_{68}/K	p/Pa	T_{68}/K	p/Pa	T_{68}/K	p/Pa
190	0.032	220	2.656	250	76.04
195	0.074	225	4.942	255	123.2
200	0.163	230	8.953	260	195.8
205	0.344	235	15.81	265	306.0
210	0.702	240	27.28	270	470.1
215	1.387	245	46.03	273.15	611.15
				273.16*	611.66

*Triple-point temperature.

REFERENCES

1. Wexler, A. *J. Res. Nat. Bur. Stand. A* **1977**, *81*, 5.
2. Guildner, L.; Johnson, D. P.; Jones, F. E. *J. Res. Nat. Bur. Stand. A* **1976**, *80*, 505.
3. Ambrose, D. *Pure Appl. Chem.* **1977**, *49*, 1437.

5.2.8. Naphthalene

Physical property: Vapour pressure
Units: Pa, K
Recommended reference materials: Naphthalene
Range of variables: 0.04 to 1000 Pa, 250 to 353 K
Physical states within the range: solid and vapour
Contributor: D. Ambrose

Intended usage: Naphthalene is recommended as an alternative to ice as a reference material for measurements below 1000 Pa when its more convenient temperature range is an advantage. Naphthalene may also be used as a reference material for the determination of critical temperature and pressure.

Sources of supply and/or method of purification: In the past naphthalene was liable to contain thionaphthen (benzo[b]thiophen) which was difficult to remove (Ref. 1), but it is now available commercially with a low sulphur content as a chemical reagent (Ref. 2). Samples of certified purity are obtainable from suppliers (A) and (C); samples of certified vapour

5. Pressure-Volume-Temperature

pressure, purified by zone-refining of micro-analytical reagent naphthalene, are available from supplier (D).

Pertinent physicochemical data:

Table 5.2.8.1. Vapour pressure p of naphthalene

T_{68}/K	p/Pa	T_{68}/K	p/Pa	T_{68}/K	p/Pa
250	0.036	290	4.918	330	182.9
260	0.144	300	13.43	340	389.9
270	0.514	310	34.15	350	792.0
280	1.662	320	81.39	353.43*	999.6

*Triple-point temperature.

The table has been computed from the coefficients given in table 5.1.1.1 calculated by Ambrose, Lawrenson and Sprake (Ref. 3); it is based on measurements made by these authors in the range 0.25 to 490 Pa (dependent on the assumption of linearity of response of an elastic diaphragm gauge) and on other values in the literature. De Kruif *et al.* have measured the vapour pressure of naphthalene (Ref. 6) since these recommendations were first published (Ref. 7); the agreement between the two sets of results (in reference 3 and 6) is so good that no change has been required in the coefficients of the equation or the table above. The enthalpy of sublimation at 298 K obtained by differentiating the equation is identical with the value given in the section on enthalpy in these Recommendations (Ref. 4), and the difference between the enthalpy of sublimation at the triple point and the enthalpy of vaporization at the same temperature obtained from an equation representing the vapour pressure of liquid naphthalene is in good agreement with the result of a calorimetric measurement of the enthalpy of fusion (Ref. 4). The accuracy of the values of p in the table is assessed as $\pm 5\%$ at $T < 280$ K and $\pm 2\%$ at $T > 280$ K.

The critical temperature is 748.4 K and the critical pressure is 4.051 MPa (Ref. 7).

REFERENCES

1. Herington, E. F. G.; Densham, A. B.; Malden, P. J. *J. Chem. Soc.* **1954**, 2643.
2. *Reference substances and reagents for use in organic micro-analysis*, Analyst **1972**, *97*, 740.
3. Ambrose, D.; Lawrenson, I. J.; Sprake, C. H. S. *J. Chem. Thermodynamics* **1975**, *7*, 1173; The vapour pressure of naphthalene. NPL Report DCS 1/75, National Physical Laboratory, Teddington, **1975**.

4. Cox, J. D. *Pure Appl. Chem.* **1974**, *40*, 399.
5. McCullough, J. P.; Finke, H. L.; Messerly, J. F.; Todd, S. S.; Kincheloe, T. C.; Waddington, G. *J. Phys. Chem.* **1957**, *61*, 1105.
6. de Kruif, C. G.; Kuipers, T.; van Miltenburg, J. C.; Schaake, R. C. F.; Stevens, G. *J. Chem. Thermodynamics* **1981**, *13*, 1081.
7. Ambrose, D.; Broderick, B. E.; Townsend, R. *J. Chem. Soc.* **1967**, *A*, 633.

5.3. Reference materials for vapour pressures at temperatures above 600 K

5.3.1. Cadmium

Physical property: Vapour pressure
Units: Pa, K
Recommended reference material: Cadmium
Range of variables: 2.5×10^{-6} to 15 Pa, 350 to 594 K
Physical states within the range: solid and vapour
Contributor: R. C. Paule

Intended usage: The vapour pressure of cadmium (CRM 746) has been determined from a statistical analysis of measurements carried out in six laboratories where several different techniques were used (Knudsen (weight loss), torque-Knudsen and calibrated mass spectrometric methods) and the values are recommended for checking other measurements of this type (Ref. 1). The certification was carried out at the request of Commission II.3.

Source of supply: Certified Reference Material 746 is available from supplier (C) in the form of rods 6.35 mm diameter and 63.5 mm long; it is homogeneous and has a purity of 99.999%.

Pertinent physicochemical data:

Table 5.3.1.1. Vapour pressure p of cadmium

T_{68}/K	p/Pa	T_{68}/K	p/Pa
350	2.52×10^{-6}	500	2.34×10^{-1}
400	3.01×10^{-4}	550	2.59
450	1.22×10^{-2}	594*	15.3

*Melting point.

REFERENCE

1. Paule, R. C.; Mandel, J. *National Bureau of Standards Special Publication*, 260-21. U.S. Government Printing Office: Washington, D.C. **1971**.

5.3.2. Silver

Physical property: Vapour pressure
Units: Pa, K
Recommended reference material: Silver
Range of variables: 1.4×10^{-7} to 130 Pa, 800 to 1600 K
Physical state within the range: solid and vapour, liquid and vapour
Contributor: R. C. Paule

Intended usage: The vapour pressure of silver (CRM 748) has been determined from a statistical analysis of measurements carried out in nine laboratories where several different techniques were used (Knudsen, both weight loss and condensation methods, torque-Knudsen and calibrated mass spectrometric methods) and the values are recommended for checking other measurements of this type (Ref. 1). The certification was carried out at the request of Commission II.3.

Source of supply: Certified Reference Material 748 is available from supplier (C) in the form of rods 6.35 mm diameter and 63.5 mm long; it is homogeneous and has a purity of 99.999%.

Pertinent physicochemical data:

Table 5.3.2.1. Vapour pressure p of silver

T_{68}/K	p/Pa	T_{68}/K	p/Pa
800	1.37×10^{-7}	1235*	3.76×10^{-1}
900	1.50×10^{-5}	1300	1.37
1000	6.36×10^{-4}	1400	7.90
1100	1.35×10^{-2}	1500	35.8
1200	1.71×10^{-1}	1600	133

*Melting point.

REFERENCE

1. Paule, R. C.; Mandel, J. *National Bureau of Standards Special Publication*, 260-21. U.S. Government Printing Office: Washington, D.C. **1971**.

5.3.3. Gold

Physical property: Vapour pressure
Units: Pa, K
Recommended reference material: Gold
Range of variables: 1.0×10^{-3} to 190 Pa, 1300 to 2100 K
Physical state within the range: solid and vapour, liquid and vapour
Contributor: R. C. Paule

Intended usage: The vapour pressure of gold (CRM 745) has been determined from a statistical analysis of measurements carried out in eleven laboratories where several different techniques were used (Knudsen, both weight loss and condensation methods, torque-Knudsen and calibrated mass spectrometric methods) and the values are recommended for checking other measurements of this type (Ref. 1). The certification was carried out at the request of Commission II.3.

Sources of supply: Certified Reference Material 745 is available from supplier (C) in the form of wire 1.4 mm diameter and 152 mm long; it is homogeneous and has a purity of 99.999%.

Pertinent physicochemical data:

Table 5.3.3.1. Vapour pressure p of gold

T_{68}/K	p/Pa	T_{68}/K	p/Pa
1300	1.01×10^{-3}	1700	1.93
1338*	2.59×10^{-3}	1800	7.35
1400	1.02×10^{-2}	1900	24.5
1500	7.46×10^{-2}	2000	71.6
1600	4.19×10^{-1}	2100	189

*Melting point.

REFERENCE

1. Paule, R. C.; Mandel, J. *National Bureau of Standards Special Publication*, 260-19. U.S. Government Printing Office: Washington, D.C. **1970**.

5.4. Reference materials for orthobaric volumes (or densities) and critical volume (or density)

5.4.1. Nitrogen

Physical property: Orthobaric volume and density
Units: $m^3\ mol^{-1}$, $kg\ m^{-3}$, K
Recommended reference material: Nitrogen (relative molecular mass 28.0134)
Range of variables: 63.148 to 126.2 K
Physical states within the range: saturated liquid and vapour
Contributor: S. Angus

Sources of supply and/or method of preparation: Gas of high purity is available commercially but it is probably always advisable for it to be dried before use, for example, by passage through a bed of molecular sieve.

Pertinent physicochemical data:

Table 5.4.1.1. Molar volume V_m and density ρ of nitrogen

$\dfrac{T_{68}}{K}$	Saturated liquid		Saturated vapour	
	$\dfrac{10^6\ V_m}{m^3\ mol^{-1}}$	$\dfrac{\rho}{kg\ m^{-3}}$	$\dfrac{10^6\ V_m}{m^3\ mol^{-1}}$	$\dfrac{\rho}{kg\ m^{-3}}$
63.148*	32.28	867.77	41 500	0.6751
70	33.32	840.77	14 760	1.898
77.348**	34.64	808.61	6 071	4.614
80	35.18	796.25	4 596	6.096
90	37.55	746.00	1 857	15.085
100	40.68	688.65	877.2	31.94
110	45.18	620.02	448.8	62.41
120	53.35	525.1	225.0	124.5
124	61.02	459.1	156.4	179.1
126.20+	89.2	314	89.2	314

*Triple-point temperature.
**Boiling temperature at 101.325 kPa.
+Critical temperature.

This table is extracted from a more detailed table (Ref. 1) which is based on a critical survey and correlation of the available experimental results; it is consistent with the pressure–volume–temperature properties recommended for nitrogen (5.4). Values obtained by interpolation between 124 K and the critical temperature are subject to large uncertainties.

REFERENCE

1. Angus, S.; de Reuck, K. M.; Armstrong, B.; Jacobsen, R. T.; Stewart, R. B. *Nitrogen, International Thermodynamic Tables of the Fluid State*, Pergamon: Oxford, **1979**.

5.4.2. Methane

Physical property: Orthobaric volume and density
Units: $m^3\,mol^{-1}$, $kg\,m^{-3}$, K
Recommended reference material: Methane (relative molecular mass 16.0426)
Range of variables: 90.68 to 190.555 K
Physical states within the range: saturated liquid and vapour
Contributor: S. Angus

Sources of supply and/or methods of preparation: Methane may be purified by chromatographic procedures (*e.g.* by passage through a column containing activated charcoal (Ref. 1) or by distillation (Ref. 2). Supplies of the gas of purity 99.995% or better are available commercially.

Pertinent physicochemical data: Table 5.4.2.1 is extracted from a more detailed table (Ref. 3); it is consistent with the pressure–volume–temperature properties recommended for methane (5.6). Values obtained by interpolation between 190 K and the critical temperature are subject to large uncertainties.

REFERENCES

1. Douslin, D. R.; Harrison, R. H.; Moore, R. T.; McCullough, J. P. *J. Chem. Eng. Data* **1964**, *9*, 358.
2. Hummel, R. W.; Hearne, J. A. *Chem. Ind. Lond.* **1961** 1827.
3. Angus, S.; Armstrong, B.; de Reuck, K. M. *Methane, International Thermodynamic Tables of the Fluid State*, Pergamon: Oxford, **1977**.

Table 5.4.2.1. Molar volume V_m and density ρ of methane

	Saturated liquid		Saturated vapour	
$\dfrac{T_{68}}{\text{K}}$	$\dfrac{10^6 \, V_m}{\text{m}^3 \, \text{mol}^{-1}}$	$\dfrac{\rho}{\text{kg m}^{-3}}$	$\dfrac{10^6 \, V_m}{\text{m}^3 \, \text{mol}^{-1}}$	$\dfrac{\rho}{\text{kg m}^{-3}}$
90.68*	35.55	451.3	63 820	0.2514
100	36.55	438.9	23 690	0.6772
110	37.76	424.9	10 000	1.604
120	39.12	410.1	4 903	3.272
130	40.70	394.2	2 677	5.993
140	42.54	377.1	1 579	10.160
150	44.78	358.3	982.5	16.33
160	47.66	336.6	631.9	25.39
170	51.67	310.5	410.9	39.04
180	58.13	276.0	261.0	61.47
190	78.61	201.5	128.9	124.5
190.55+	98.9	162.2	98.9	162.2

*Triple-point temperature.
+Critical temperature.

5.4.3. Carbon dioxide

Physical property: Orthobaric volume and density
Units: m^3 mol^{-1}, kg m^{-3}, K
Recommended reference material: Carbon dioxide (relative molecular mass 44.0098)
Range of variables: 216.58 to 304.21 K
Physical states within the range: saturated liquid and vapour
Contributor: S. Angus

Sources of supply and/or methods of preparation: See recommendation 5.2.1 relating to use of carbon dioxide as a vapour-pressure reference material.

Pertinent physicochemical data:
Table 5.4.3.1. Molar volume V_m and density ρ of carbon dioxide

	Saturated liquid		Saturated vapour	
$\dfrac{T_{68}}{K}$	$\dfrac{10^6 \, V_m}{m^3 \, mol^{-1}}$	$\dfrac{\rho}{kg \, m^{-3}}$	$\dfrac{10^6 \, V_m}{m^3 \, mol^{-1}}$	$\dfrac{\rho}{kg \, m^{-3}}$
216.58*	37.35	1178.4	3 134	14.04
220	37.74	1166.2	2 745	16.03
230	38.98	1129.1	1 883	23.37
240	40.40	1089.4	1 318	33.37
250	42.05	1046.6	942.0	46.72
260	44.03	999.4	682.6	64.48
270	46.50	946.4	498.0	88.37
280	49.77	884.2	361.8	121.6
290	54.64	805.4	256.2	171.7
300	64.69	680.3	164.0	268.3
304.21+	94.44	466	94.44	466

*Triple-point temperature.
+Critical temperature.

This table is extracted from a more detailed table based on a critical survey and correlation of the available experimental results where equations suitable for use in a computer, which will reproduce these values, are also given (Ref. 1). Values obtained by interpolation between 300 K and the critical temperature are subject to large uncertainties.

REFERENCE

1. Angus, S.; Armstrong, B.; de Reuck, K. M. *Carbon Dioxide, International Thermodynamic Tables of the Fluid State*, Pergamon: Oxford, **1976**.

5.4.4. Benzene

Physical property: Orthobaric volume and density
Units: $m^3 \, mol^{-1}$, $kg \, m^{-3}$, K

Recommended reference material: Benzene (relative molecular mass 78.113)
Range of variables: 290 to 562.16 K.
Physical states within the range: saturated liquid and vapour
Contributor: D. Ambrose

Sources of supply and/or methods of purification: See recommendation 5.2.3 relating to the use of benzene as a vapour-pressure reference material.

Pertinent physicochemical data:

Table 5.4.4.1. Molar volume V_m and density ρ of benzene

$\dfrac{T_{68}}{\text{K}}$	Saturated liquid		Saturated vapour	
	$\dfrac{10^6\, V_m}{\text{m}^3\, \text{mol}^{-1}}$	$\dfrac{\rho}{\text{kg m}^{-3}}$	$\dfrac{10^6\, V_m}{\text{m}^3\, \text{mol}^{-1}}$	$\dfrac{\rho}{\text{kg m}^{-3}}$
290	88.54	882.3	278 710	0.2803
300	89.62	871.6	179 040	0.4363
325	92.48	844.7	68 410	1.142
350	95.60	817.1	30 760	2.539
375	99.05	788.6	15 617	5.002
400	102.94	758.8	8 686	8.993
425	107.43	727.1	5 168	15.11
450	112.8	692.7	3 368	23.19
475	119.4	654.3	2 078	37.58
500	128.1	609.7	1 356	57.60
525	141.0	554.1	874.9	89.29
550	166.1	470.4	516.7	151.2
562.16*	257	304	257	304

*Critical temperature.

The table is extracted from a more detailed table based on a critical survey and correlation of the available experimental results (Ref. 1). Values obtained by interpolation between 555 K and the critical temperature are subject to large uncertainties.

REFERENCE

1. *Thermodynamic Properties of Benzene.* Report No. 73009, Engineering Sciences Data Unit: London, **1973**.

5.4.5. Water

Physical property: Orthobaric volume and density
Units: $m^3\ mol^{-1}$, $kg\ m^{-3}$, K
Recommended reference material: Water (relative molecular mass 18.0152)
Range of variables: 273.16 to 647.126 K
Physical state within the range: saturated liquid and vapour
Contributor: S. Angus

Sources of supply and/or methods of preparation: See recommendation 5.2.5 relating to use of water as a vapour-pressure test material.

Pertinent physicochemical data:

Table 5.4.5.1. Molar volume V_m and density ρ

	Saturated liquid		Saturated vapour	
$\dfrac{T_{68}}{K}$	$\dfrac{10^6\ V_m}{m^3\ mol^{-1}}$	$\dfrac{\rho}{kg\ m^{-3}}$	$\dfrac{10^6\ V_m}{m^3\ mol^{-1}}$	$\dfrac{\rho}{kg\ m^{-3}}$
273.16*	18.019	999.8	371 130	0.004 854
323.15	18.234	988.0	216 840	0.083 08
373.15**	18.797	958.4	30 150	0.597 5
423.15	19.645	917.1	7 077	2.546
473.15	20.833	864.7	2 294	7.854
523.15	22.54	799.1	902.8	19.955
573.15	25.29	712.4	390.4	46.15
623.15	31.35	574.7	158.8	113.5
633.15	34.11	528.1	125.4	143.7
643.15	39.74	453.3	89.8	200.6
647.126+	56.9	322	56.9	322

*Triple-point temperature.
**Boiling temperature at 101.325 kPa.
+Critical temperature.

The table is extracted from a more detailed table based on a critical survey and correlation of the available experimental results (Ref. 1); it is consistent with the pressure–volume–temperature properties recommended for water (5.6.1). Values obtained by interpolation between 643.15 K and the critical temperature are subject to large uncertainties.

REFERENCE

1. Harr, L.; Gallagher, J. S.; Kell, G. S. *NBS/NRC Steam Tables*, Hemisphere Pub. Inc.: New York, **1983**.

5.5. Reference materials for Pressure–Volume–Temperature behaviour (gases)

5.5.1. Helium

Physical property: Second virial coefficient
Units: $cm^3\ mol^{-1}$, K
Recommended reference material: Helium (relative atomic mass 4.002 60)
Range of variables: 250 to 1500 K
Physical states within the range: gas
Contributors: D. Ambrose, S. Angus

Intended usage: Measurements of pVT properties of gases or unsaturated vapours are frequently made in a Burnett apparatus, in which the ratio of two volumes must be determined either from the measurements made on the substance under study (Ref. 1) or from measurements made in a separate experiment on a gas chosen because it is nearly ideal, *viz.* hydrogen (Ref. 2), or helium (Refs. 3 to 5). The values of the second virial coefficient given here will allow calculation of the volumetric properties needed for evaluation of the volume ratio of the apparatus if helium is used for the calibration.

Source of supply and/or method of preparation: Helium of adequately high purity is available commercially. Purification to a higher standard for gas chromatographic use has been described (Refs. 6,7).

Pertinent physicochemical data:

The values in table 5.5.1.1, calculated by McCarty (Ref. 8), are those appropriate to the equation,

$$pV_m = RT(1 + B_v/V_m), \quad (5.5.1.1)$$

i.e. the truncated form of the volume-explicit virial equation, which accurately represents the behaviour of helium in the temperature range given at pressures up to 10 MPa (above this pressure, a third virial coefficient C_v should be included in the equation, but values for C_v are of much lower accuracy than those for B_v). Values of B_v are available for lower temperatures, but their accuracy is less certain, and they have been omitted because they are unlikely to be required for the purpose of this recommendation. There are other assessments of the best values for B_v (Refs. 9 to 11); those given here are the ones used in the IUPAC tables of thermodynamic properties (Ref. 9). The 95% confidence levels based on a statistical analysis of the data are given in the table but the more cautious ± 0.5 cm^3 mol^{-1} for the uncertainty of the values is recommended.

Table 5.5.1.1. Second virial coefficient B_v of helium

$\dfrac{T_{68}}{\text{K}}$	$\dfrac{B_v}{\text{cm}^3\text{ mol}^{-1}}$	T_{68}/K	$\dfrac{B_v}{\text{cm}^3\text{ mol}^{-1}}$	T_{68}/K	$\dfrac{B_v}{\text{cm}^3\text{ mol}^{-1}}$
250	12.16±0.14	550	10.90±0.3	900	9.71±0.4
300	11.95±0.2	600	10.71±0.3	1000	9.41±0.4
350	11.73±0.2	650	10.53±0.3	1200	8.86±0.4
400	11.51±0.2	700	10.35±0.3	1300	8.60±0.4
450	11.30±0.3	750	10.18±0.3	1400	8.36±0.4
500	11.10±0.3	800	10.02±0.3	1500	8.12±0.4

REFERENCES

1. Butcher, E. G.; Dadson, R. S. *Proc. Roy. Soc. Lond. A* **1964**, *277*, 448.
2. Mueller, W. H.; Leland, T. W.; Kobayashi, R. *Amer. Inst. Chem. Eng. J.* **1961**, *7*, 267.
3. Anderson, L. N.; Kudchadker, A. P.; Eubank, P. T. *J. Chem. Eng. Data* **1968**, *13*, 321.
4. Kudchadker, A. P.; Eubank, P. T. *J. Chem. Eng. Data* **1970**, *15*, 7.
5. Waxman, M.; Hastings, J. R. *J. Res. Nat. Bur. Std. C* **1971**, *75*, 165.
6. Bourke, P. J.; Gray, M. D.; Denton, W. H. *J. Chromatog.* **1965**, *19*, 189.
7. Seitz, C. A.; Bodine, W. M.; Klingman, C. L. *J. Chromatog. Sci.* **1971**, *9*, 29.
8. McCarty, R. D. *Thermophysical Properties of Helium-4 from 2 to 1500 K with Pressures to 1000 Atmospheres*, National Bureau of Standards, Washington, D.C. In press.
9. Angus, S.; de Reuck, K. M.; McCarty, R. D. *Helium, International Thermodynamic Tables of the Fluid State*, Pergamon: Oxford, **1977**.
10. Levelt Sengers, J. M. H.; Klein, M.; Gallagher, J. S. *PVT Relations of Gases. Second Virial Coefficients*, in: *Handbook of Physics*, 3rd ed., McGraw-Hill: New York, **1972**, p. 4.
11. Dymond, J. H.; Smith, E. B. *The Virial Coefficients of Gases, A Critical Compilation*, 2nd ed., Clarendon Press: Oxford, **1980**.

5.5.2. Benzene

Physical property: Second virial coefficient
Units: cm^3 mol^{-1}, K
Recommended reference material: Benzene (relative molecular mass 78.1134)
Range of variables: 300 to 600 K
Physical states within the range: vapour
Contributor: M. B. Ewing

Intended usage: Measurements in different laboratories of the second virial coefficients of substances liquid at room temperature frequently differ by more than their estimated experimental errors. Large and indeterminate errors arise, for example, when the liquid is polar and tends to be adsorbed on the walls of the apparatus. The problem of adsorption is not as severe for benzene as it is for many other liquids, and benzene is one of the substances that have been studied by many investigators (Ref. 1). The following values of the second virial coefficient of benzene are recommended for use in checking apparatus for measurement of non-ideality vapours.

Sources of supply and/or methods of preparation: See recommendation 5.2.3 relating to use of benzene as a vapour–pressure reference material.

Pertinent physicochemical data:

Table 5.5.2.1. Second virial coefficient B_p of benzene

$\dfrac{T_{68}}{K}$	$\dfrac{B_p}{\text{cm}^3 \text{ mol}^{-1}}$	$\dfrac{T_{68}}{K}$	$\dfrac{B_p}{\text{cm}^3 \text{ mol}^{-1}}$	$\dfrac{T_{68}}{K}$	$\dfrac{B_p}{\text{cm}^3 \text{ mol}^{-1}}$
290	−1584	360	−901	480	−467
300	−1444	380	−790	520	−396
310	−1322	400	−700	560	−341
320	−1215	440	−564	600	−298
340	−1039				

These values have been calculated from the equation,

$$\frac{B_p}{\text{cm}^3 \text{ mol}^{-1}} = 41.5 - 1.118 \times 10^5 x^{-1} - 3.850 \times 10^7 x^{-2} - 1.289 \times 10^9 x^{-3} - 5.158 \times 10^{12} x^{-4} \tag{5.5.2.1}$$

where x is equal to (T/K). This equation was determined by Al–Bizreh and Wormald (Ref. 1): these authors estimate that the standard deviation in B_p is 2.6%.

The second virial coefficient given is that appropriate to the equation,

$$V_\text{m} = \frac{RT}{p} + B_p, \tag{5.5.2.2}$$

i.e. the truncated form of the pressure–explicit virial equation (Ref. 3).

REFERENCES

1. Dymond, J. H.; Smith, E. B. *The Virial Coefficients of Gases. A Critical Compilation*, 2nd ed., Clarendon Press: Oxford, **1980**.
2. Al–Bizreh, N. A.; Wormald, C. J. *J. Chem. Thermodynamics* **1977**, *9*, 749.
3. Cox, J. D.; Lawrenson, I. J. *Specialist Periodical Report: Chemical Thermodynamics*, McGlashan, M. L. senior reporter, Vol. 1, The Chemical Society: London, **1972**, p. 162.

5.5.3. Air

Physical property: Pressure–volume–temperature relationships
Units: bar, cm^3 g^{-1}, K
Recommended reference material: Air (effective relative molar mass 28.966)
Range of variables: 150 to 1000 K, 1 to 1000 bar
Physical states within the range: gas
Contributor: S. Angus

Intended usage: For tests requiring large quantities of gas, air is the obvious choice except at cryogenic temperatures.

Sources of supply and/or methods of preparation: For accurate measurements the air should be freed form carbon dioxide and water by, for example, passage through beds of suitable adsorbant. Removal of trace contaminants from laboratory air is discussed by Williams and Eaton (Ref. 1), but such ultra–purification should not be necessary for the present purpose.

Pertinent physicochemical data: Use of the tables by Sytchev *et al.* (Ref. 2) is recommended. These tables give values of the compression factor, density, and its derivatives, enthalpy, entropy, heat capacities, Joule–Thomson coefficient, fugacity, and speed of sound in pressure–temperature co–ordinates. Equations are given and are based on a critical survey and correlation of the available experimental data. Reference 2, which is in Russian, is not easily accessible and until an English translation is available, which it is hoped will have a wider circulation, the tables by Baehr and Schwier (Ref. 3) may be used. These give values of specific volume, enthalpy, entropy and exergy from 63 to 1077 K (in °C) up to 4500 bar.

(Exergy is the excess Gibbs energy over that under ambient conditions defined as 1 bar, 15 °C). The tables were prepared before the high-temperature high-pressure results were available, and consequently they have a larger region depending upon extrapolation than those in Ref. 2. Below 473 K the two sets of tables do not differ significantly.

REFERENCES

1. Williams, F. W.; Eaton, H. G. *Analyt. Chem.* **1974**, *46*, 179.
2. Sytchev, V. V.; Vasserman, A. A.; Kozlov, A. D.; Spiridonov, G. A.; Tsimarnii, V. A. *Thermodynamic Properties of Air*, GS SSD Monograph: Moscow, **1978**.
3. Baehr, H. D.; Schwier, K. *Die thermodynamischen Eigenschaften der Luft*, Springer-Verlag OHG: Berlin, **1961**.

5.5.4. Nitrogen

Physical property: Pressure–volume–temperature relationships
Units: bar, mol^{-1}, K
Recommended reference material: Nitrogen (relative molecular mass 28.0134)
Range of variables: Melting line (ca. 64 to 66 K) to 2000 K, 0.1 to 100 bar; Melting line (66 to 195 K) to 1200 K, 120 to 10 000 bar
Physical states within the range: liquid and gas
Contributor: S. Angus

Intended usage: For tests requiring large quantities of inert gas or liquid, nitrogen is the most readily available substance.

Sources of supply and/or methods of preparation: Gas of high purity is available commercially, but it is probably always advisable for it to be dried before use, for example by passage through a bed of molecular sieve.

Pertinent physicochemical data: Use of the tables given in reference 1 is recommended. These tables give values of the density and its derivatives and of the internal energy, enthalpy, entropy, heat capacities and speed of sound in pressure–temperature coordinates. The equation of state is given, together with its subsidiary forms necessary to calculate the above properties. The tables are based on a critical survey and correlation of the available data.

REFERENCE

1. Angus, S.; de Reuck, K. M.; Armstrong, B.; Jacobsen, R.T.; Stewart, R. B. *Nitrogen, International Thermodynamic Tables of the Fluid State*, Pergamon: Oxford, **1979**.

5.5.5. Oxygen

Physical property: Pressure–volume–temperature relationships
Units: atm, cm^3 mol^{-1}, K
Recommended reference material: Oxygen (relative molecular mass 31.998)
Range of variables: 54.351 K, 0.001 atm (triple point) to 340 K, 340 atm
Physical states within the range: liquid and gas
Contributor: S. Angus

Intended usage: When equipment is used with liquid oxygen, or when equipment is used in which a phase change may occur as a result of which oxygen liquefies, it may be necessary on grounds of safety for in situ calibrations and tests to be carried out only with oxygen.

Sources of supply and/or methods of preparation: Oxygen of high enough purity for most purposes may be bought. A more detailed consideration of the purity of the gas will be found in work relating to the use of its normal boiling point as a defining temperature of the International Practical Temperature Scale (Refs. 1 to 3).

Pertinent physicochemical data: The tables by Weber (Ref. 4) are recommended. Work is currently in progress which will result in equations and tables which are expected to supersede those in reference 4. Reference 5 may be regarded as an interim report on this work. It is intended that a final publication will form a book in the IUPAC International Thermodynamic Tables of the Fluid State.

REFERENCES

1. Ancsin, J. *Metrologia* **1970**, *6*, 53.
2. Ancsin, J. *Temperature*, Plumb, H. H. editor–in–chief, Vol. 4, Part 1, Instrument Society of America: Pittsburgh, **1972**, p. 211.
3. Tiggelman, J. L.; Durieux, M. *Temperature*, Plumb, H. H. editor–in–chief, Vol. 4, Part 1, Instrument Society of America: Pittsburgh, **1972**, p. 149.
4. Weber, L. A. *Thermodynamic and Related Properties of Oxygen from the triple point to 300 K at pressures to 1000 bar*, NASA Reference Publication 1011, NBS/R 77/865, **1977**.
5. Ewers, J.; Wagner, W. *A method for optimizing the structure of equations of state and its application to an equation of state for oxygen*, Proceeding of the 8th Symposium of Thermophysical Properties, Sengers, J. V. editor, Vol. 1, *Thermophysical Properties of Fluids*, American Society of Mechanical Engineers: New York, **1982**, p. 78.

5.5.6. Methane

Physical property: Pressure–volume–temperature relationships
Units: bar, m^3 mol^{-1}, K
Recommended reference material: Methane (relative molecular mass 16.0426)
Range of variables: 90.68 K, 0.117 bar (triple point) to 500 K, 700 bar
Physical states within the range: liquid and gas
Contributor: S. Angus

Intended usage: Testing of gas compressibility apparatus of various types – constant volume, variable volume, and expansion (Refs. 1 to 3).

Source of supply and/or methods of preparation: See recommendation 5.4.2 relating to use of methane as an orthobaric–critical volume reference material.

Pertinent physicochemical data: The tables prepared by the IUPAC Thermodynamic Tables Project are recommended (Ref. 4). They are based on a critical review of the available experimental results and extend from the triple point to 10 000 bar, up to 470 K and to 400 bar from 470 K to 620 K. The tables give values of volume, entropy, enthalpy, isobaric heat capacity, compression factor, fugacity/pressure ratio, Joule–Thomson coefficient, the ratio of the heat capacities and the speed of sound as functions of pressure and temperature; and the pressure, entropy, internal energy and isochoric heat capacity as functions of density and temperature. Subsidiary tables give the zero–pressure properties, saturation–curve properties, and melting–curve properties. Equations to reproduce the tables are also given.

REFERENCES

1. Ellington, R. T.; Eakin, B. F. *Chem. Eng. Prog.* **1963**, *59*(11), 80.
2. Mason, E. A.; Spurling, T. H. *The Virial Equation of State*, Pergamon: Oxford, **1969**.
3. Cox, J. D.; Lawrenson, I. J. *Specialist Periodical Report: Chemical Thermodynamics*, McGlashan, M. L. senior reporter, Vol. 1, The Chemical Society: London, **1973**, p. 162.
4. Angus, S.; Armstrong, B.; de Reuck, K. M. *Methane, International Thermodynamic Tables of the Fluid State*, Pergamon: Oxford, **1977**.

5.5.7. Ethylene

Physical property: Pressure–volume–temperature relationships
Units: bar, cm^3 mol^{-1}, °C
Recommended reference material: Ethylene (relative molar mass 28.1536)
Range of variables: 223.15 to 423.15 K up to 3000 bar and 223.15 to 423.15 K up to 0.0225 mol cm^{-3}
Physical states within the range: liquid and gas

Contributor: S. Angus

Intended usage: Flowmeters in use in ethylene pipelines, the standards against which they are calibrated, and proposed mass-flowmeters for ethylene, require such accuracy that calibration with other fluids is not satisfactory, and the properties of ethylene itself are required.

Sources of supply and/or methods of preparation: Ethylene of high purity may be bought; commercial grades may contain other hydrocarbons that will change its properties, especially near the critical point. Calculations based on analysis should be made to check the possible effects before the values in the tables are applied.

Pertinent physicochemical data: The tables prepared by the NBS (Ref. 1) are recommended. They are based on a critical survey of the available data, which includes the recent extensive research program designed to remedy the deficiencies noted in the previous recommended tables (Ref. 2) and which are thereby obsolete.

The tables and equations of reference 1 are inaccurate in the near-critical region. Reference 3 provides equations suitable only in this region, but in the region of overlap between the equations of reference 1 and 3 there are significant differences. A simple equation for use in the low density region is given in reference 4.

REFERENCES

1. McCarty R. D.; Jacobsen, R. T. *Thermophysical properties of ethylene*, NBS TN 960, Washington, **1981**.
2. Angus, S.; Armstrong, B.; de Reuck, K. M.; Featherstone, W.; Gibson, M. R. *Ethylene, 1972*, Butterworths: London, **1973**.
3. Levelt Sengers, J. M. H.; Gallagher, J. S.; Balfour, F. W.; Sengers, J. V. *A Thermodynamic Surface for the Critical Region of Ethylene*, Proceedings of 8th Symposium on Thermophysical Properties, Sengers, J. V. editor, Vol. 1, Amer. Soc. Mechanical Engineers: New York, **1982**, p. 368.
4. Levelt Sengers, J. M. H.; Hastings, J. R. *Int. J. Thermophysics*, **1981**, *2*, 269.

5.6. Reference materials for Pressure–Volume–Temperature behaviour (liquids)

5.6.1. Water

Physical property: Pressure–volume–temperature relationships
Units: bar, cm^3 g^{-1}, °C
Recommended reference material: Water (relative molecular mass 18.0152)
Range of variables: 273.16 K, 0.001 bar (triple point) to 653 K, 1400 bar
Physical states within the range: liquid
Contributor: S. Angus

Intended usage: For tests and calibrations requiring large quantities of liquid with well defined physical properties, water is the obvious choice for all pressures at ambient or near-ambient temperatures. The volumetric properties of water are better known than are those of any other liquid, but it is corrosive and attacks both glass and silica (although the latter more slowly) at elevated temperatures; this is of importance in relation to its use as a reference material for density.

Sources of supply and/or methods of preparation: See item 5.2.5 relating to use of water as a vapour-pressure material.

Pertinent physicochemical data: At present there exist six tables (Refs. 2 to 7) based on the 1963 International Skeleton Tables (IST 1963), (Ref. 1), which itself is based on the experimental data available up to 1960, and six more recent tables or equations (Refs. 8 to 13) based on the analysis of all the experimental data available at the time of production. In addition, there are several specialised tables available, such as the properties of steam at atmospheric pressure (Ref. 14) or at very high temperatures, when dissociation must be considered (Ref. 15). It is recommended that the tables based on the equation published by Harr, Gallagher, and Kell (Ref. 12) be used, as the most accurate representation of the most reliable experimental data, except in the near critical region. If this region is to be included in calculations, and a computer is available, reference 13 may be used. This includes the H, G, and K equation, plus an additional function which extends its range almost to the critical point and also includes equations for the transport properties.

For very high pressures, the equation of Juza, Hoffer, and Sifner (Ref. 10) which extends to 10 GPa, should be consulted.

In the references recommended, the International Practical Temperature Scale of 1968 is used by all except reference 8, which used IPTS 1948.

REFERENCES

1. *International Skeleton Tables for Steam, 1963.*
2. Bain, R. W. *Steam Tables, 1964*, HMSO, Edinburgh, **1964**.
3. Grigull, U. editor, *Properties of Water and Steam in SI units*, 3rd revision and up-dated printing, Springer: Berlin, **1982**.
4. Meyer, C. A.; McClintock, R. B.; Silvestri, G. J.; Spencer, R. C. *ASME Steam Tables, 1979*, 4th ed., American Society of Mechanical Engineers: New York, **1980**.
5. *UK Steam Tables in SI Units, 1970*, Arnold: London, **1970**.
6. Vukalovich, M. P.; Rivkin, S. L.; Alexandrov, A. A. *Tables of the Thermophysical Properties of Water and Steam*, GS SSD: Moscow, **1969**.
7. *1968 JSME Tables*, Japan Society of Mechanical Engineers: Tokyo, **1968**.
8. Keenan, J. H.; Keyes, F. G.; Hill, P. G.; Moore, J. G. *Steam Tables*, International edition, Wiley: New York, **1969**.
9. Rivkin, S. L.; Alexandrov, A. A. *Thermophysical properties of water and steam* (in Russian), Energija: Moscow, **1980**.

10. Juza, J.; Hoffer, V.; Sifner, O. *Equation of state for ordinary water between 350 and 1000 °C from 0 to 10 GPa and from 100 to 350 °C between 200 MPa and melting curve*, Acta Technica CSAV **1979**, *24*, 251.
11a. Pollak, R. *The Thermodynamic Properties of Water*, Ha. Schrift, Ruhr University, **1973**.
11b. Cullen, M. G. *CEGB Steam Tables*, revised, CEGB Report CC/N865, London, **1981**.
11c. Cullen, M. G.; Parfitt, R. J. *CEGB Steam Tables*, extended, CEGB Report CC/N883, London, **1981**.
12. Harr, L; Gallagher, J. S.; Kell, G. S. *NBS/NRC Steam Tables*, Hemisphere Pub. Inc.: New York, **1983**.
13. Kestin, J.; Sengers, J. V.; Kamgar-Parsi, B.; Levelt Sengers, J. H. M. *J. Phys. Ref. Data*, **1984**, *13*, 175. See also Kestin, J.; Sengers, J. V. *J. Phys. Chem. Ref. Data* **1986**, *15*, 305.
14. Alexandrov, A.A.; Trachtenberg, M. S. *Thermophysical propeties of water at atmospheric pressure* (in Russian), GS SSD, Izd. Standartov: Moscow, **1977**.
15. Kmonicek, V. *The Thermodynamic Functions of Dissociating Steam in the Range 1000 to 5000 K, 0.01 to 1006*, Academia: Prague, **1967**.

5.7. Contributors

D. Ambrose,
Department of Chemistry,
University College London,
20 Gordon Street,
London WC1H OAJ (UK).

M. B. Ewing,
Department of Chemistry,
University College London,
20 Gordon Street
London WC1H OAJ (UK).

S. Angus,
Department of Chemical Engineering
and Chemical Technology,
Imperial College of Science and Technology,
London, SW7 2BY (UK).

R. C. Paule,
National Bureau of Standards,
U.S. Department of Commerce,
Office of Standard Reference Materials,
Gaithersburg, MD 20899 (USA).

5.8. List of Suppliers

A. A.P.I. Samples,
Attention of A. J. Streiff,
Carnegie–Mellon University,
Schenley Park,
Pittsburgh, PA 15213 (USA).

B. Aldrich Chemical Co. Ltd,
The Old Brickyard New Road,
Gillingham, Dorset SP8 4BR,
(UK).

C. Office of Standard Reference Materials,
National Bureau of Standards,
Gaithersburg, MD 20899 (USA).

D. Office of Reference Materials,
National Physical Laboratory,
Teddington, Middlesex, TW11 OLW (UK).

E. Phillips Petroleum Company,
Bartlesville, OK 74004 (USA).

6

SECTION: DISTILLATION–TESTING OF COLUMNS

COLLATOR: E. F. G. HERINGTON

CONTENTS:

- 6.1. Introduction
- 6.2. Abbreviations and Symbols
- 6.3. General Theory
- 6.4. Design of Tests
- 6.5. Test mixtures for columns at atmospheric pressure and below atmospheric pressure
    - 6.5.1. Benzene + Toluene
    - 6.5.2. Benzene + Heptane
    - 6.5.3. *trans*-Decalin + *cis*-Decalin
    - 6.5.4. Methylcyclohexane + Toluene
    - 6.5.5. Chlorobenzene + Ethylbenzene
    - 6.5.6. Benzene + 1,2-Dichloroethane
    - 6.5.7. 2-Methylnaphthalene + 1-Methylnaphthalene
    - 6.5.8. Heptane + Methylcyclohexane
    - 6.5.9. *para*-Xylene + *meta*-Xylene
- 6.6. Test mixtures for columns at atmospheric pressure and above atmospheric pressure
    - 6.6.1. Methanol + Ethanol

6.6.2. Propene + Propane

6.7. Contributors

6.8. Supplier

6.9. Appendix
- 6.9.1. Effect of errors in the value of the relative volatility ratio.
- 6.9.2. Choice of the initial composition of the test mixture.
- 6.9.3. Effect of experimental errors in the mole fractions.
- 6.9.4. Dependence of the maximum, minimum and optimum number of theoretical plates that can be measured on the volatility ratio.
- 6.9.5. Vapour-liquid equilibrium data for test mixtures.

6.1. Introduction

Distillation is the most widely used separation process for the isolation of the components of mixtures. Very large scale distillation columns are employed in industry and small apparatus is often used in the laboratory. Separation in fractional distillation is achieved by the interchange of matter between rising vapour and falling liquid in a countercurrent process. Measurements of the theoretical plate equivalence of distillation equipment are used for the following purposes:

(i) to compare the performance of different column designs and packings,

(ii) to compare the separations achieved when distillation equipment is used under various conditions; for example to study the effect of different boil-up rates and the distributions of reflux and temperature control facilities on the separation efficiency,

(iii) to provide data by the aid of which large equipment can be built based on information obtained with small scale apparatus,

(iv) to examine the effect of the properties of the mixture on the performance of the equipment.

To facilitate intercomparisons of the performance of distillation equipment the European Working Party on Distillation, Absorption and Extraction published a report in 1969 entitled 'Recommended Test Mixtures for Distillation Columns' edited by F. J. Zuiderweg (Ref. 1). The present recommendations owe much to this earlier work. The aim of this recommendation is to update the earlier information, to provide necessary details on the

6. Distillation

purity of the components of the test mixtures and to give pertinent data in a form easy to use. To that end the requisite data on vapour-liquid equilibrium are arranged in this report so that two independent workers using the results from the same observations should have no difficulty in arriving at the same calculated number of theoretical plates. The effects of experimental errors in the observations on the derived theoretical plate values have also been examined in some detail.

The performance of a fractionating column is commonly expressed in terms of the number of theoretical plates to which the column is equivalent. The vapour arising from a theoretical plate is in equilibrium with the liquid leaving the plate so that for the computation of plate equivalence it is necessary to have vapour-liquid equilibrium data referring to a series of compositions of mixtures of the components. Experimentally measured vapour-liquid equilibrium data of this type are subject to error and it is therefore imperative that the goodness of the data be established by thermodynamic consistency tests before the data are used. All the vapour-liquid equilibrium data recommended in this Section satisfy thermodynamic consistency tests; details of the results of the application of consistency tests to the data are available in reference 2.

Although the performance of distillation columns is measured by the use of equilibrium data, the separation in fractional distillation takes place by kinetic processes so that the theoretical plate equivalence of a column depends not only on quantities such as the boil-up rate but also on the properties of the mixture used for the test. The properties of the mixture which have an influence on the separations achieved include the average relative volatility, solubilities of the liquid components, densities of the phases, surface tension, viscosity, enthalpy of the components, enthalpy of vaporization, and wetting properties of the components towards the column walls and towards the column packing. For this reason column efficiencies should be determined by the choice of mixtures similar in properties to those for which the column will be used in production. To assist in the choice of a suitable mixture, values of liquid density, viscosity, surface tension, enthalpy of vaporization and heat capacity of the liquid at the boiling temperatures of the components at the pressure listed are given in table 6.1.1.

The information in these recommendations is arranged in the following manner. The next section lists the abbreviations and symbols used. The third section sets out the general theory and this is followed by a section on general considerations governing the design of tests. The individual recommendations give information on each test mixture and include methods for the analysis of the mixtures and pertinent vapour-liquid equilibrium data. The Appendix supplies information on the effects of errors in the values of the volatility ratios and of the mole fractions, the choice of composition for the test mixtures, and the minimum, maximum and optimum theoretical plate number for mixtures.

Table 6.1.1. Properties of the components of the recommended test mixtures

Recommendation		6.5.1		6.5.2	
Quantity	Unit	Benzene	Toluene	Benzene	Heptane
Pressure	kPa	101.325	101.325	101.325	101.325
Boiling Temp.	°C	80.10	110.63	80.10	98.43
Density	kg m^{-3}	813.4	783.2	813.4	613.5
Viscosity	mPa s	0.317	0.247	0.317	0.2105
Surface Tension	mN m^{-1}	21.8	17.7	21.08	12.89
Specific Enthalpy of Vaporization	kJ kg^{-1}	393.8	361.4	393.8	316.3
Specific Heat Capacity of Liquid	kJ kg^{-1} K^{-1}	1.88	2.01	1.88	2.56

Recommendation		6.5.3		6.5.3	
Quantity	Unit	trans-Decalin	cis-Decalin	trans-Decalin	cis-Decalin
Pressure	kPa	13.33	13.33	6.666	6.666
Boiling Temp.	°C	115.90	123.65	97.10	104.64
Density	kg m^{-3}	798.2	819.1	812.5	833.3
Viscosity	mPa s	0.593	0.723	0.714	0.874
Surface Tension	mN m^{-1}	20.52	21.94	22.32	23.61
Specific Enthalpy of Vaporization	kJ kg^{-1}	313.1	322.2	321.2	330.6
Specific Heat Capacity of Liquid	kJ kg^{-1} K^{-1}	2.07	2.11	1.98	2.03

Recommendation		6.5.3		6.5.3	
Quantity	Unit	trans-Decalin	cis-Decalin	trans-Decalin	cis-Decalin
Pressure	kPa	1.333	1.333	0.666	0.666
Boiling Temp.	°C	60.89	67.99	47.82	54.75
Density	kg m^{-3}	839.9	860.8	849.6	870.6
Viscosity	mPa s	1.101	1.407	1.323	1.724
Surface Tension	mN m^{-1}	25.72	26.98	26.88	28.15
Specific Enthalpy of Vaporization	kJ kg^{-1}	338.5	348.7	345.8	356.3
Specific Heat Capacity of Liquid	kJ kg^{-1} K^{-1}	1.81	1.86	1.75	1.81

6. Distillation

Table 6.1.1 continued

Recommendation		6.5.4		6.5.4	
Quantity	Unit	Methyl-cyclohexane	Toluene	Methyl-cyclohexane	Heptane
Pressure	kPa	101.325	101.325	53.33	53.33
Boiling Temp.	°C	100.93	110.63	79.65	89.49
Density	kg m^{-3}	698.4	783.2	717.2	802.8
Viscosity	mPa s	0.30	0.247	0.36	0.292
Surface Tension	mN m^{-1}	19.37	17.7	17.1	20.3
Specific Enthalpy of Vaporization	kJ kg^{-1}	317.7	361.4	330.2	374.8
Specific Heat Capacity of Liquid	kJ kg^{-1} K^{-1}	2.223	2.01	2.121	1.93

Recommendation		6.5.4		6.5.5	
Quantity	Unit	Methyl-cyclohexane	Toluene	Chloro-benzene	Ethyl-benzene
Pressure	kPa	26.66	26.66	101.325	101.325
Boiling Temp.	°C	59.60	69.49	131.73	136.19
Density	kg m^{-3}	734.8	821.2	983	760.4
Viscosity	mPa s	0.447	0.347	0.292	0.236
Surface Tension	mN m^{-1}	14.9	22.7	20.5	17.3
Specific Enthalpy of Vaporization	kJ kg^{-1}	341.5	386.9	307.6	337.8
Specific Heat Capacity of Liquid	kJ kg^{-1} K^{-1}	2.029	1.86	1.52	0.468

Recommendation		6.5.5		6.5.5	
Quantity	Unit	Chloro-benzene	Ethyl-benzene	Chloro-benzene	Ethyl-benzene
Pressure	kPa	40.0	40.0	6.66	6.66
Boiling Temp.	°C	100.40	104.71	53.49	57.64
Density	kg m^{-3}	1019	790.4	1071	833.6
Viscosity	mPa s	0.370	0.296	0.553	0.445
Surface Tension	mN m^{-1}	24.0	20.1	29.5	24.87
Specific Enthalpy of Vaporization	kJ kg^{-1}	331.4	355.4	349.8	381.4
Specific Heat Capacity of Liquid	kJ kg^{-1} K^{-1}	1.46	0.447	1.39	0.423

Table 6.1.1 continued

Recommendation		6.5.6		6.5.7	
Quantity	Unit	Benzene	1,2-Dichloro-ethane	2-Methyl-naphthalene	1-Methyl-naphthalene
Pressure	kPa	101.325	101.325	13.33	13.33
Boiling Temp.	°C	80.10	83.51	164.68	167.78
Density	kg m^{-3}	813.4	1158	900.6	907.4
Viscosity	mPa s	0.317	0.425	0.419	0.457
Surface Tension	mN m^{-1}	21.08	24.4	23.6	24.3
Specific Enthalpy of Vaporization	kJ kg^{-1}	393.8	324.7	357.6	360.7
Specific Heat Capacity of Liquid	kJ kg^{-1} K^{-1}	1.88	1.38	1.98	2.02

Recommendation		6.5.7		6.5.7	
Quantity	Unit	2-Methyl-naphthalene	1-Methyl-naphthalene	2-Methyl-naphthalene	1-Methyl-naphthalene
Pressure	kPa	6.666	6.666	1.333	1.333
Boiling Temp.	°C	144.33	147.32	104.85	107.67
Density	kg m^{-3}	914.6	925.2	946.7	956.1
Viscosity	mPa s	0.487	0.541	0.701	0.795
Surface Tension	mN m^{-1}	25.0	26.6	29.3	31.1
Specific Enthalpy of Vaporization	kJ kg^{-1}	366.2	369.6	384.8	389.2
Specific Heat Capacity of Liquid	kJ kg^{-1} K^{-1}	1.92	1.95	1.80	1.83

Recommendation		6.5.7		6.5.8	
Quantity	Unit	2-Methyl-naphthalene	1-Methyl-naphthalene	Heptane	Methyl-cyclohexane
Pressure	kPa	0.667	0.667	101.325	101.325
Boiling Temp.	°C	90.53	93.31	98.43	100.93
Density	kg m^{-3}	959.4	967.0	613.5	698.4
Viscosity	mPa s	0.818	0.933	0.2105	0.30
Surface Tension	mN m^{-1}	30.9	32.7	12.89	14.9
Specific Enthalpy of Vaporization	kJ kg^{-1}	392.6	397.4	316.3	317.7
Specific Heat Capacity of Liquid	kJ kg^{-1} K^{-1}	1.76	1.79	2.56	2.223

6. Distillation

Table 6.1.1 continued

Recommendation		6.5.9			
Quantity	Unit	p-Xylene	m-Xylene		
Pressure	kPa	101.325	101.325		
Boiling Temp.	°C	138.35	139.10		
Density	kg m^{-3}	754.7	759.4		
Viscosity	mPa s	0.219	0.216		
Surface Tension	mN m^{-1}	16.2	16.6		
Specific Enthalpy of Vaporization	kJ kg^{-1}	340.1	345.0		
Specific Heat Capacity of Liquid	kJ kg^{-1} K^{-1}	2.11	1.99		
Recommendation		6.6.1		6.6.1	
Quantity	Unit	Methanol	Ethanol	Methanol	Ethanol
Pressure	bar	1.013	1.013	3.040	3.040
Boiling Temp.	°C	64.53	78.23	95.31	109.05
Density	kg m^{-3}	750.2	737.9	718.4	704.8
Viscosity	mPa s	0.328	0.444	0.226	0.285
Surface Tension	mN m^{-1}	18.9	17.4	16.1	14.6
Specific Enthalpy of Vaporization	kJ kg^{-1}	1121.4	842	1043.8	777
Specific Heat Capacity of Liquid	kJ kg^{-1} K^{-1}	3.00	3.00	3.44	3.44
Recommendation		6.6.1		6.6.1	
Quantity	Unit	Methanol	Ethanol	Methanol	Ethanol
Pressure	bar	5.066	5.066	10.13	10.13
Boiling Temp.	°C	111.85	125.72	137.13	151.34
Density	kg m^{-3}	699.5	683.9	666.3	647.0
Viscosity	mPa s	0.187	0.228	0.140	0.165
Surface Tension	mN m^{-1}	14.5	12.8	11.8	10.0
Specific Enthalpy of Vaporization	kJ kg^{-1}	996.0	734	914.5	662
Specific Heat Capacity of Liquid	kJ kg^{-1} K^{-1}	3.67	3.71	4.11	4.16

Table 6.1.1 continued

Recommendation		6.6.2		6.6.2	
Quantity	Unit	Propene	Propane	Propene	Propane
Pressure	bar	5.066	5.066	10.130	10.130
Boiling Temp.	°C	-4.64	2.31	19.82	27.48
Density	kg m^{-3}	552.1	525.8	514.1	488.8
Viscosity	mPa s	0.123	0.1243	0.0965	0.0952
Surface Tension	mN m^{-1}	10.8	9.57	7.70	6.60
Specific Enthalpy of Vaporization	kJ kg^{-1}	385.0	409.4	345.2	334.5
Specific Heat Capacity of Liquid	kJ kg^{-1} K^{-1}	2.33	2.01	2.60	2.71

Recommendation		6.6.2		6.6.2	
Quantity	Unit	Propene	Propane	Propene	Propane
Pressure	bar	15.20	15.20	22.09	22.09
Boiling Temp.	°C	36.36	44.47	53.35	61.91
Density	kg m^{-3}	485.5	460.1	453.4	423.3
Viscosity	mPa s	0.0810	0.0790	0.0667	0.0637
Surface Tension	mN m^{-1}	5.70	4.68	3.78	2.85
Specific Enthalpy of Vaporization	kJ kg^{-1}	311.9	301.4	267.3	260.9
Specific Heat Capacity of Liquid	kJ kg^{-1} K^{-1}	2.85	3.03	3.30	3.52

6.2. Abbreviations and Symbols

A	$A = -100\{2x(1-x)\ln[(1-x)/x]\}^{-1}$
a_0, a_1	factors in the equation $\ln \alpha = a_0 + a_1(1 - 2x)$
B_0	volume of charge put into boiler of a batch column
c_i	initial concentration of involatile component in boiler
c_h	concentration of involatile component in boiler when column is under total reflux
$HETP$	height equivalent to a theoretical plate
H_0	volume of hold-up of column
ln	logarithm to base e
$\log_{10}$	logarithm to base 10
N	number of theoretical plates
$(N+1)_{max}$	the maximum value of $(N+1)$ that should be

	evaluated with given test mixture
$(N+1)_{\min}$	the minimum value of $(N+1)$ that should be evaluated with given test mixture
$(N+1)_{\mathrm{opt}}$	the optimum value of $(N+1)$ that can be evaluated with a given test mixture
$n_\mathrm{D}(20\ °\mathrm{C})$	refractive index at 20 °C for the D sodium line
p_1	vapour pressure of pure component 1
p_2	vapour pressure of pure component 2
x	mole fraction of the more volatile component in the liquid mixture
x_c	mole fraction of the more volatile component of mixture placed in boiler to achieve a mole fraction x in the boiler when the column is under total reflux
y	mole fraction of the more volatile component in the vapour. This vapour composition may be that of the vapour in equilibrium with liquid of composition x on a plate or it may be the composition of the vapour in the still head. The meaning should be clear from the context.
α	relative volatility ratio sometimes called volatility ratio equal to $[y/(1-y)]/[x/(1-x)]$
$\rho(20\ °\mathrm{C})$	density of liquid at 20 °C

6.3. General Theory

Batch fractional distillation equipment is normally tested by boiling a binary mixture under total reflux until a steady state has been achieved. Samples are then withdrawn from the still head and boiler for analysis. Compositions are expressed as mole fractions, the mole fraction of the more volatile component in the vapour being y and in the liquid x. Throughout this report the more volatile component is listed first when the names of the two components are given. If the vapour and liquid refer to one theoretical plate the relative volatility ratio (sometimes referred to as the volatility ratio), α, is defined as follows:

$$\alpha = [y/(1-y)]/[x/(1-x)], \tag{6.3.1}$$

and hence it follows that

$$y = \alpha x/[(\alpha - 1)x + 1]. \tag{6.3.2}$$

Fractional distillation is usually carried out so that there is a temperature gradient within the distillation column but the pressure within the column remains constant to a first approximation. Thus interest centers on the values of the volatility ratio, α, for liquid mixtures containing different mole fractions of components under the same total pressure. It therefore follows that for many mixtures α is a function of x.

Two methods for recording vapour-liquid equilibria data have been adopted in this report. For a mixture of components with boiling properties that differ only slightly and which has been treated as ideal, a single value of the relative volatility ratio, α, for a given total pressure has been given. This value has been calculated from the ratio of the vapour pressures of the pure components. For such a mixture the number of theoretical plate equivalence of the column, N, can be calculated by the use of the Fenske equation,

$$N + 1 = (\ln \alpha)^{-1} \ln \left\{ [y/(1-y)]/[x/(1-x)] \right\}, \qquad (6.3.3)$$

where y is the mole fraction of the volatile component in the still head and x the mole fraction of the same component in the boiler. In continuous distillation and occasionally in batch distillation equipment the liquid leaving the bottom tray rather than the liquid in the boiler can be sampled thus avoiding the assumption that the boiler is precisely one equilibrium stage. If the mole fraction x in the liquid leaving the bottom is inserted in the right hand side of equation (6.3.3) the value of N and not the value of $N+1$ is obtained. Natural logarithms (ln) have been widely used throughout this report but logarithms to the base 10 ($\log_{10}$) have also been employed.

A method of tabulation has been used for other mixtures which records the equilibrium x–y data and enables the plate equivalence of a column to be found by a simple interpolation followed by the subtraction of two quantities. These tables have been prepared from plate to plate calculations starting with $x = 0.05$ and ending where y first exceeds 0.95. The method of tabulation and the technique for using tables of this kind will be illustrated using the table given in 5.2 (benzene + heptane). To condense the tables, values of x and y for $N+1$ equal to 1, 3, 5, 7 etc. have been tabulated although in some tables final values have been inserted for even values of $N+1$. For any value of $N+1$ the value of y represents the mole fraction of the more volatile component in the vapour in equilibrium with the liquid containing x mole fraction of the same component; hence the y, x equilibrium diagram can be constructed by plotting values of y against those of x from the table; additional values of y and x for $N+1$ of even number can also be plotted (see below).

The values of x and y for even values of $N+1$ can be written down by inspection; for example the value of x for $N+1 = 4$ from the table for benzene + heptane is 0.3175 and the value of y is 0.4741; the appropriate value of α, if required, can then be calculated from these values of x and y by the use of Equation (6.3.1). Table 6.5.2.1 can also be used to derive the plate equivalence of a column by simple calculation. For example, suppose in a test using mixture 5.2, x, the pot composition, was found to be 0.1868 and y, the still head composition was found to be 0.9238, then from the table $N+1$ for the boiler would equal 3 and $N+1$ for the still head would equal 11. Therefore $(11 - 3) = 8$ would equal the plate equivalence of the column. In a real situation the boiler and still head compositions would not coincide exactly with values in the table and therefore it will be necessary to interpolate. The following example shows that procedure using data for mixture 5.2.

6. Distillation

Example:

Test mixture 5.2 (Benzene + Heptane.)

From experiment $x = 0.452$ (boiler mole fraction),
$y = 0.931$ (still head mole fraction).

According to table 6.5.2.1 the experimental value of x lies between $N+1 = 3$ and $N+1 = 5$, and by inspection the values of $N+1$ are written down,

$N+1$	x
3	0.1868
4	0.3175
5	0.4741

Clearly the value of x lies between $N + 1 = 4$ and $N + 1 = 5$, and by proportional parts it corresponds to

$$N + 1 = 4 + (0.452 - 0.3175)/(0.4741 - 0.3175) = 4.86. \qquad (6.3.4)$$

Similarly, it may be seen that the experimental value of y lies between $N + 1 = 11$ and $N + 1 = 13$, and by inspection the values for $N + 1$ are written down.

$N+1$	y
11	0.9238
12	0.9367
13	0.9467

The value of y lies between $N + 1 = 11$ and $N + 1 = 12$, and by proportional parts it corresponds to

$$N + 1 = 11 + (0.931 - 0.9238)/(0.9367 - 0.9238) = 11.56. \qquad (6.3.5)$$

Therefore the plate equivalence of the column is $11.56 - 4.86 = 6.7$ theoretical plates.

The tables of $N + 1$, x and y for ideal mixtures where α varies with x have been constructed directly from the Antoine vapour-pressure equations (see as an example mixture 5.1). The corresponding tables for non-ideal mixtures have been calculated from values of $\ln \alpha$ derived from experimental data expressed as a power series in x where the requisite power series has been obtained by fitting experimental vapour-liquid equilibrium data by the method of least squares.

The boiling points of the pure components are listed on the data sheets for selected pressures, but the equilibrium temperatures of the mixtures have not been recorded because published values are often less accurate than the concentration data and in most instances little use is made of temperatures in testing columns.

In bubble-tray fractionating-columns the plate efficiency is defined as the number of theoretical plates determined in a test distillation divided by the number of actual trays. In packed columns the height equivalent to a theoretical plate (HETP) is obtained by dividing the packed length by the number of theoretical plates.

6.4. Design of Tests

The test mixture should be chosen so as not to be too easy nor too difficult to separate. Each data sheet shows the maxima and minimum number of plates that can be determined by the use of the mixture.

A mixture must be chosen such that the expected plate equivalence falls in the range suitable for the mixture. A mixture suitable for testing the column at the pressure required must also, of course, be chosen.

Although in the testing of the plate equivalence of fractionating columns it is not essential to determine the dynamic hold-up of the column yet a knowledge of this quantity assists in the design of the optimum conditions for the evaluation of distillation column performance. The simplest experimental method for the determination of the dynamic (operating) hold-up is to place in the boiler a known volume of a solution containing a small amount of relatively involatile material dissolved in a volatile solvent. The column is then run under total reflux at the desired boil-up rate and samples are taken from the boiler when steady state conditions have been reached. The hold-up of the column, H_0, is calculated by equation (6.4.1) from the volume, B_0, of liquid originally placed in the boiler; c_i, the initial concentration of the involatile component in the boiler and c_h, the concentration of involatile component in the boiler when the column is running under total reflux,

$$H_0 = B_0(1 - c_i c_h^{-1}). \tag{6.4.1}$$

Stearic acid determined by titration or α-bromonaphthalene determined by refractometry with heptane as the volatile component have been suggested as suitable mixtures for the measurement of hold-up. Other mixtures may be used provided the additive is of very low volatility and provided it is soluble in the mixture; for example a 10% by weight solution of stearic acid in benzene has been employed. To obtain an accurate measurement of the hold-up it is desirable to use a small volume of mixture taking care however that the volume is not so small that a phase change takes place in the boiler e.g. stearic aid must not crystallize out when the column is under total reflux. For an example and for the necessary precautions to obtain accurate results see reference 3.

In the determination of the plate equivalence of a fractionating column past experience or prior published work will usually provide a guide to the magnitude of the plate equivalence to be expected. Such knowledge will enable a suitable initial composition of the feed charge to be selected for an experiment to determine the theoretical plate equivalence. In the absence of such prior knowledge preliminary experiments may be necessary to choose conditions which will give the best accuracy in results. To obtain the best accuracy when an ideal or near ideal test mixture is employed, conditions should be chosen so that the sum of x (boiler composition) and y (still head composition) is unity *i.e.* x and y are equally disposed about the mole fraction 0.5. To satisfy this condition it is necessary that the charge put into the boiler should have a higher concentration of the more volatile component than the desired value of x under total reflux because some of the volatile component will be in the hold-up (see Appendix for discussion). The appropriate equation to calculate the desired initial mole fraction, x_c, is

$$x_c = x + H_0(0.5 - x)/B_0, \tag{6.4.2}$$

where x is the required mole fraction in the boiler when a stationary condition has been achieved under total reflux, and B_o and H_o are the volumes of the test mixture used and of the hold-up respectively.

To ensure that the components of the mixture are pure enough for use in a test mixture it is recommended that the separate components be distilled through the column before the mixture is made. The distillation should be carried out with a high reflux ratio with rejection of head and tail fractions. Only the middle fraction boiling at a steady temperature should be used. Care must be taken to eliminate volatile impurities because if they are present they will accumulate in the still head during the test and lead to erroneous values for the plate equivalence. The purity of the starting materials and the possible accumulation of impurities during distillation can best be studied by gas-liquid chromatography. As the nature of the impurities is often not known the purity of the starting materials and of the liquid and vapour samples during distillation should be investigated by the use of two gas-chromatographic columns with stationary phases of different polarities to ensure that the presence of any impurity should not pass unnoticed. Details for the purification of components are given on the data sheets, where methods recommended for analysis to obtain values of x and y are also presented.

In the past, density or refractive index measurements have been used for the analysis of liquid and vapour samples but equations connecting these properties with mole fractions are not given here because it is recommended that the property-composition curve should be determined by the analyst using values obtained from measurements made on his own instruments on synthetic mixtures prepared from samples of the components used in making the test mixture.

Gas chromatography (Refs. 4, 5) is being employed to an increasing extent for the analysis of liquid and vapour samples in the testing of columns. Numerous types of equipment are offered by a large number of manufacturers, ranging from simple routine apparatus to complex research equipment connected to a computer. For the quantitative analysis necessary here an apparatus in the average price range with a thermal conductivity detector or a combination of thermal conductivity detector and flame ionization detector is adequate. The peaks can be evaluated advantageously with a simple electronic integrator. With the simple mixtures involved in theoretical plate calculations and employing routine gas-chromatographic analysis with an internal standard and integration evaluation, a standard deviation of at least $\pm 1\%$ of the mole fraction is attainable in the range 0.05 to 0.20 mole fraction. The standard deviation may be smaller in a laboratory where comprehensive experience in gas chromatography is available. In well equipped laboratories other analytical methods such as i.r. or u.v. spectroscopy can be used.

In each investigation the effect of an analytical errors in the values of x and y on the calculated value of the plate efficiency should be examined by the use of equation (6.9.3.1) in the Appendix or by the study of the $N + 1$, x, y table appropriate to the mixture used.

For many packings the performance of the column expressed as the number of theoretical plates is closely dependent on the boil-up rate so that it is necessary to control and measure the boil-up rate whenever the number of theoretical plates is measured. Every effort must be made to ensure complete condensation at the still head and prevent leakage because even a small loss will reduce the apparent number of plates found (Ref. 6). The prevention of even very small leakages is particularly important if highly efficient columns are being tested.

The column under test must be run under total reflux for sufficient time for equilibrium to be established before samples are taken. The equilibration time will be increased if the number of theoretical plates is increased, if the hold-up is increased, if the boil-up is decreased, if the concentration of more volatile component is decreased and if the boiling temperature difference between the components is decreased (Ref. 6). When a test of plate efficiency is being made the experimenter must confirm that a stationary condition has been achieved by taking a succession of samples at increasing time intervals until no change in composition occurs. However, to avoid taking samples much too early, an equilibrium time amounting to between 1 and 1.5 h for every 10 theoretical plates may be allowed in the first instance (Ref. 3).

The data sheets giving details of test mixtures are listed in two sections (6.5 and 6.6) in accordance with the pressures at which their use is suitable and within each section mixtures for testing columns with a small number of plates appear before those mixtures for testing columns with a large number of plates. The substances discussed in these recommendations may present safety hazards because many of the liquids and vapours are flamable and toxic. When these substances are used, published information on safety must be consulted and due notice must be taken of regulations in each country concerning health and safety in laboratory and in industrial premises. Note that safety information and safety regulations are frequently updated.

REFERENCES

1. Zuiderweg, F. J., editor, *Recommended Test Mixtures for Distillation Columns*, European Federation of Chemical Engineering, Working Party on Distillation, Absorption and Extraction, Institution of Chemical Engineers: London, **1969**.
2. Herington, E. F. G. *Reference Materials for Testing Distillation Columns*, NPL Report Chem 55, National Physical Laboratory: Teddington, March, **1977**.
3. Coulson, E. A.; Herington, E. F. G. *Laboratory Distillation Practice*, Newnes: London, **1958**.
4. Ettre, L. S.; Zlatkis, A., editors, *The Practice of Gas Chromatography*, Interscience: New York, **1967**.
5. Leibnitz, E.; Struppe, H. G. *Handbuch der Gas-Chromatographie*, Verlag Chemie: Weinheim, **1967**.
6. Coulson, E. A. *J. Soc. Chem. Ind.* **1945**, *64*, 101.
7. Riddick, J. A.; Bunger, W. B. *Techniques of Chemistry*, Volume II, *Organic Solvents*, 3rd ed., Wiley-Interscience: New York, **1970**.

6.5. Test mixtures for columns at atmospheric pressure and below atmospheric pressure

6.5.1. Benzene + Toluene Test Mixture

Intended usage: This test mixture may be used to measure the plate efficiencies of columns equivalent to 1 to 6 theoretical plates at 101.325 kPa pressure.

Sources of supply and/or methods of preparation: Samples of these compounds may be purchased from many firms, from some of which high purity grades are available. However, as this mixture can only be used to test columns with a small number of plates, components of the highest purity are usually not essential. Nevertheless the components should undergo a preliminary distillation before use.

Analysis: Gas chromatography is a convenient method of analysis: the procedure should be checked by the quantitative analysis of synthetic mixtures before the specimens from the distillation experiments are analysed. Examples of columns that have been used are (i) Sil-o-cel 30-50 mesh impregnated with a mass fraction of 0.3 Apiezon N (2 m column at 130 °C with hydrogen as carrier gas); (ii) a mass fraction of 0.1 carbowax 1540 on Chromosorb W (3 m column at 80 °C). Infrared spectroscopy may also be used.

Pertinent physicochemical data: The vapour pressures of the pure compounds at $t/°C$ are given by the following equations, which yield boiling temperatures of 80.100 °C (benzene) and 110.626 °C (toluene) for a pressure of 101.325 kPa (Ref. 1):

Benzene: $\quad \log_{10}(p_1/\text{kPa}) = 6.030\,55 - 1211.033/[(t/°C) + 220.790]$, $\qquad$ (6.5.1.1)

Toluene: $\quad \log_{10}(p_2/\text{kPa}) = 6.079\,54 - 1344.800/[(t/°C) + 219.482]$. $\qquad$ (6.5.1.2)

Vapour-liquid equilibrium at 101.325 kPa: the system is treated as ideal so that

$$\ln \alpha = \ln(p_1/p_2). \qquad (6.5.1.3)$$

Table 6.5.1.1. Benzene + toluene

$N+1$	x	y	α	$N+1$	x	y	α
1	0.0500	0.1107	2.3646	5	0.6395	0.8175	2.5252
3	0.2288	0.4179	2.4196	7	0.9199	0.9674	2.5851

The methods used for the calculation of the plate equivalence are given in the general theory.

REFERENCE

1. *TRC Thermodynamic Tables–Hydrocarbons*, Thermodynamic Research Center, Texas A&M University: College Station, Texas.

6.5.2. Benzene + Heptane Test Mixture

Intended usage: This test mixture may be used to measure the plate efficiencies of columns equivalent to 1 to 13 theoretical plates at 101.325 kPa pressure.

Sources of supply and/or methods of preparation: These compounds are available from a number of chemical firms but it is essential to purchase high grade materials.

A good commercial grade of benzene can be purified further by shaking portions successively with sulphuric acid until the sample is free from thiophen, then with water, dilute sodium hydroxide and with further portions of water. The benzene sample can then be dried with anhydrous calcium chloride. Further purification can be achieved by fractional freezing but whatever pre-treatments are applied the sample should be distilled before use at a high reflux ratio with rejection of head and tail fractions.

The heptane employed must consist solely of the normal isomer and care should be taken to distil the sample before use and to employ only specimens with the correct density and/or refractive index values. A suitable grade for use in test-mixtures is heptane prepared for engine knock-rating measurements available from supplier (A).

Analysis: Determinations of refractive index (benzene, $n_D(20\ °C) = 1.501\ 12$; heptane, $n_D(20\ °C) = 1.387\ 64$) or of density (benzene, $\rho(20\ °C) = 879.01$ kg m^{-3}; heptane, $\rho(20\ °C) = 683.76$ kg m^{-3}) give accurate results. Gas chromatography can be used provided the technique is checked by the quantitative analysis of synthetic mixtures of known composition. For example a 2 m column packed with a mass fraction of 0.1 Carbowax 1540 on Chromosorb W at 80 °C has been employed. Ultraviolet spectroscopy can be used for analysis.

Pertinent physicochemical data: The vapour pressures of the pure compounds at $t/°C$ are given by the following equations (Ref. 1):

Benzene : $\qquad \log_{10}(p_1/\text{kPa}) = 6.030\,55 - 1211.033/[(t/°C) + 220.790]$, $\qquad$ (6.5.2.1)

Heptane : $\qquad \log_{10}(p_2/\text{kPa}) = 6.021\,67 - 1264.900/[(t/°C) + 216.544]$. $\qquad$ (6.5.2.2)

These equations yield boiling temperatures of 80.100 °C (benzene) and 98.425 °C (heptane) for a pressure of 101.325 kPa.

Vapour-liquid equilibrium at 101.325 kPa: The system is not ideal and the following equation based on experimental data (Ref. 2) was used to represent $\ln \alpha$ as a power series in x,

$$\ln \alpha = 0.7555 - 0.2463x - 0.0361x^2 - 0.3780x^3. \qquad (6.5.2.3)$$

Table 6.5.2.1. Benzene + heptane

$N+1$	x	y	α	$N+1$	x	y	α
1	0.0500	0.0996	2.1023	9	0.8503	0.8834	1.3332
3	0.1868	0.3175	2.0254	11	0.9067	0.9238	1.2470
5	0.4741	0.6193	1.8046	13	0.9367	0.9467	1.2002
7	0.7283	0.8017	1.5083	14	0.9467	0.9546	1.1845

The methods used for the calculation of the plate equivalence are given in the general theory. Note that the value of α falls rapidly as x increases and that when $x = 0.9467$ one plate increases the concentration of the volatile component by only 0.0079 in mole fraction. Accurate analyses are therefore required if exact plate values are to be found from still-head compositions in this concentration range.

REFERENCES

1. *TRC Thermodynamic Tables–Hydrocarbons*, Thermodynamic Research Center, Texas A&M University: College Station, Texas.
2. Zuiderweg, F. J., editor, *Recommended Test Mixtures for Distillation Columns*, European Federation of Chemical Engineering, Working Party on Distillation, Absorption and Extraction, Institution of Chemical Engineers: London, **1969**.

6.5.3. *trans*-Decalin + *cis*-Decalin Text Mixture

Intended usage: This mixtures may be used to measure the theoretical plate efficiencies of columns equivalent to 2 to 22 theoretical plates (pressure 13.33 kPa), 2 to 20 theoretical plates (pressure 6.666 kPa), 1 to 16 theoretical plates (pressure 1.333 kPa) and 1 to 15 theoretical plates (pressure 0.666 kPa).

Sources of supply and/or methods of preparation: Decalin is available commercially as a mixture of the two isomers; materials from different sources show considerable variations in the isomer ratio. Purification by repeated shaking of samples with concentrated sulphuric acid until the acid layer remains only slightly coloured is recommended. The hydrocarbon should then be washed in water, then with dilute sodium hydroxide and then with three portions of water. The product should be dried over calcium chloride or molecular sieve. The material should be fractionally distilled under reduced pressure before use as a test mixture. Decomposition has been reported when the hydrocarbon is distilled at atmospheric pressure.

Analysis: Mixtures can be analysed by gas chromatography; a 3 m column packed with silicone oil DC 200 at 150 °C or a 2 to 4 m column containing a mass fraction of 0.1 Apiezon L and a mass fraction of 0.01 phosphoric acid on Chromosorb P60-80 mesh with hydrogen as carrier gas at 130 °C and a thermal conductivity detector have been recommended.

The composition of mixtures can be determined very accurately from measurements of density (*trans*-decalin, $\rho(20\ °C) = 869.7$ kg m^{-3}; *cis*-decalin, $\rho(20\ °C) = 896.7$ kg m^{-3}) or of refractive index (*trans*-decalin, $n_D(20\ °C) = 1.469\ 32$; *cis*-decalin, $n_D(20\ °C) = 1.480\ 98$). Analysis by i.r. spectroscopy is also possible (850 to 950 cm^{-1}). Samples of the pure isomers should be used to calibrate these methods.

Pertinent physicochemical data: The vapour pressures of the pure compounds at $t/°C$ are given by the following equations (Ref. 1):

trans-Decalin : $\quad \log_{10}(p_1/\text{kPa}) = 5.981\ 71 - 1564.683/[(t/°C) + 206.259]$, (6.5.3.1)

cis-Decalin : $\quad \log_{10}(p_2/\text{kPa}) = 6.000\ 19 - 1594.460/[(t/°C) + 203.392]$. (6.5.3.2)

These equations yield boiling temperatures of 115.899, 97.101, 60.894, and 47.821 °C for *trans*-decalin and 123.653, 104.637, 67.989, and 54.748 °C for *cis*-decalin at pressures of 13.33, 6.666, 1.333 and 0.666 kPa respectively.

Vapour-liquid equilibrium: The system is treated as ideal so that

$$\ln \alpha = \ln(p_1/p_2). \qquad (6.5.3.3)$$

The methods used for the calculation of the plate equivalent are given in the general theory.

REFERENCE

1. *TRC Thermodynamic Tables–Hydrocarbons*, Thermodynamic Research Center, Texas A&M University: College Station, Texas.

6. Distillation

Table 6.5.3.1. *trans*-Decalin + *cis*-decalin

Pressure 13.33 kPa							
$N+1$	x	y	α	$N+1$	x	y	α
1	0.0500	0.0641	1.3013	13	0.5590	0.6237	1.3080
3	0.0818	0.1040	1.3017	15	0.6845	0.7397	1.3096
5	0.1313	0.1644	1.3024	17	0.7883	0.8299	1.3109
7	0.2041	0.2505	1.3034	19	0.8649	0.8936	1.3118
9	0.3035	0.3625	1.3047	21	0.9168	0.9353	1.3124
11	0.4260	0.4923	1.3063	23	0.9499	0.9614	1.3128

Pressure 6.666 kPa							
$N+1$	x	y	α	$N+1$	x	y	α
1	0.0500	0.0656	1.3344	13	0.6331	0.6986	1.3433
3	0.0857	0.1112	1.3350	15	0.7570	0.8073	1.3451
5	0.1432	0.1825	1.3359	17	0.8494	0.8836	1.3464
7	0.2298	0.2852	1.3373	19	0.9109	0.9323	1.3472
9	0.3481	0.4169	1.3391	21	0.9489	0.9616	1.3477
11	0.4894	0.5624	1.3412				

Pressure 1.333 kPa							
$N+1$	x	y	α	$N+1$	x	y	α
1	0.0500	0.0695	1.4186	11	0.6417	0.7193	1.4312
3	0.0958	0.1308	1.4197	13	0.7859	0.8404	1.4339
5	0.1761	0.2330	1.4215	15	0.8831	0.9156	1.4358
7	0.3017	0.3810	1.4242	17	0.9397	0.9572	1.4368
9	0.4674	0.5561	1.4276				

Table 6.5.3.1. continued

Pressure 0.666 kPa							
$N+1$	x	y	α	$N+1$	x	y	α
1	0.0500	0.0713	1.4583	11	0.7039	0.7780	1.4740
3	0.1007	0.1405	1.4596	13	0.8380	0.8842	1.4769
5	0.1927	0.2587	1.4620	15	0.9186	0.9435	1.4786
7	0.3381	0.4281	1.4656	16	0.9435	0.9611	1.4791
9	0.5235	0.6176	1.4699				

6.5.4. Methylcyclohexane + Toluene Test Mixture

Intended usage: This mixture may be used to measure the plate efficiencies of columns equivalent to 3 to 31 theoretical plates (pressure 101.325 kPa), 3 to 23 theoretical plates (pressure 53.33 kPa), and 3 to 24 theoretical plates (pressure 26.66 kPa).

Sources of supply and/or methods of preparation: Methylcyclohexane is available from many firms. Samples should be purified by washing with concentrated sulphuric acid, sodium carbonate solution and water. The specimen should be dried with anhydrous calcium chloride or with molecular sieve and distilled until the properties of the fraction chosen closely approach those of highly purified samples (see Analysis).

Good grades of toluene are available and specimens should be purified by distillation through an efficient column at a high reflux ratio. Thus, for example, the use of nitration grade toluene purified by distillation through a 75-plate column with a high reflux ratio has been recommended. The physical properties of the specimen used to make the test mixture should be similar to those of the pure material (see Analysis).

Analysis: The determination of the refractive index of a mixture (methylcyclohexane, $n_D(20\ ^\circ C) = 1.423\ 12$; toluene, $n_D(20\ ^\circ C) = 1.496\ 93$) is very convenient; density measurements can also be used (methylcyclohexane, $\rho(20\ ^\circ C) = 769.39$ kg m^{-3}; toluene, $\rho(20\ ^\circ C) = 866.96$ kg m^{-3}). Gas chromatography (2 m column packed with a mass fraction of 0.1 Carbowax 1540 on Chromosorb W at 90 $^\circ$C) and u.v. spectroscopy can also be employed.

6. Distillation

Pertinent physicochemical data: The vapour pressures of the pure compounds at $t/°C$ are given by the following equations (Ref. 1).

Methylcyclohexane: $\log_{10}(p_1/\mathrm{kPa}) = 5.947\,90 - 1270.763/[(t/°C) + 221.416]$. (6.5.4.1)

Toluene: $\log_{10}(p_2/\mathrm{kPa}) = 6.079\,54 - 1344.800/[(t/°C) + 219.482]$. (6.5.4.2)

These equations yield boiling temperature of 100.934, 79.646, and 59.599 °C for methylcyclohexane and 110.626, 89.485, and 69.494 °C for toluene at pressures of 101.325, 53.33 and 26.66 kPa respectively.

Vapour-liquid equilibrium: The system is non-ideal; the data used (Ref. 2) are similar to those previously provided (Ref. 3).

The following equation was used to represent $\ln \alpha$ as a power series in x for a pressure of 101.325 kPa:

$$\ln \alpha = 0.4932 - 0.4803x + 0.0332x^2. \qquad (6.5.4.3)$$

Table 6.5.4.1. Methylcyclohexane + toluene at 101.325 kPa

$N+1$	x	y	α	$N+1$	x	y	α
1	0.0500	0.0776	1.5988	19	0.8566	0.8691	1.1120
3	0.1172	0.1706	1.5486	21	0.8802	0.8899	1.1010
5	0.2370	0.3126	1.4641	23	0.8986	0.9064	1.0924
7	0.3913	0.4672	1.3639	25	0.9134	0.9197	1.0857
9	0.5361	0.5962	1.2780	27	0.9254	0.9305	1.0802
11	0.6476	0.6909	1.2166	29	0.9352	0.9395	1.0758
13	0.7275	0.7583	1.1751	31	0.9434	0.9470	1.0721
15	0.7844	0.8066	1.1467	32	0.9470	0.9504	1.0705
17	0.8257	0.8422	1.1267				

Note that the value of α decreases rapidly as the value of x increases so that an error of 0.0034 in x will lead to an error of one plate when $N+1 = 32$.

The following equation was used to represent $\ln \alpha$ as a power series in x for a pressure of 53.33 kPa:

$$\ln \alpha = 0.5558 - 0.4865x + 0.0176x^2. \qquad (6.5.4.4)$$

Table 6.5.4.2. Methylcyclohexane + toluene at 53.33 kPa

$N+1$	x	y	α	$N+1$	x	y	α
1	0.0500	0.0822	1.7015	15	0.8491	0.8679	1.1682
3	0.1305	0.1971	1.6366	17	0.8839	0.8974	1.1498
5	0.2801	0.3722	1.5233	19	0.9091	0.9191	1.1366
7	0.4636	0.5469	1.3966	21	0.9278	0.9355	1.1270
9	0.6185	0.6781	1.2990	23	0.9421	0.9480	1.1197
11	0.7269	0.7668	1.2355	24	0.9480	0.9532	1.1168
13	0.7996	0.8266	1.1949				

Again the value of α decreases rapidly as the value of x increases so that an error of 0.0052 in x will produce an error of one plate when $N+1 = 24$.

The following equation was used to represent $\ln \alpha$ as a power series in x for a pressure of 26.66 kPa:

$$\ln \alpha = 0.6782 - 0.6649x + 0.0397x^2. \qquad (6.5.4.5)$$

Table 6.5.4.3. Methylcyclohexane + toluene at 26.66 kPa

$N+1$	x	y	α	$N+1$	x	y	α
1	0.0500	0.0912	1.9061	15	0.8706	0.8845	1.1382
3	0.1569	0.2485	1.7769	17	0.8963	0.9064	1.1209
5	0.3564	0.4638	1.5625	19	0.9152	0.9229	1.1084
7	0.5581	0.6348	1.3764	21	0.9296	0.9355	1.0990
9	0.6953	0.7427	1.2650	23	0.9408	0.9455	1.0918
11	0.7801	0.8100	1.2016	25	0.9497	0.9535	1.0860
13	0.8342	0.8541	1.1632				

Again the value of α decreases rapidly as the value of x increases so that an error of 0.0038 in x will produce an error of one plate when $N+1 = 25$. The methods used for the calculation of plate equivalence are given in the general theory.

REFERENCES

1. *TRC Thermodynamic Tables–Hydrocarbons*, Thermodynamic Research Center, Texas A&M University: College Station, Texas.
2. Brunner, E. private communication.
3. Zuiderweg, F. J., editor, *Recommended Test Mixtures for Distillation Columns*, European Federation of Chemical Engineering, Working Party on Distillation, Absorption and Extraction, Institution of Chemical Engineers: London, **1969**.

6.5.5. Chlorobenzene + Ethylbenzene Text Mixture

Intended usage: This mixture may be used to measure the plate efficiency of columns equivalent to 7 to 48 theoretical plates (pressure 101.325 kPa), 5 to 42 theoretical plates (pressure 40 kPa), and 3 to 31 theoretical plates (pressure 6.66 kPa).

Sources of supply and/or methods of preparation: Both components can be purchased and the specimens purified by distillation through a column of 50 theoretical plates using a high reflux ratio. The physical properties of the samples used to make the test mixture should be similar to values listed under Analysis.

Analysis: Refractive index (chlorobenzene, $n_D(20\ °C) = 1.5248$; ethylbenzene, $n_D(20\ °C) = 1.4959$) or density (chlorobenzene, $\rho(20\ °C) = 1106.3$ kg m^{-3}; ethylbenzene, $\rho(20\ °C) = 867.02$ kg m^{-3}) measurements can be used to analyse mixtures. Alternatively gas chromatography (*e.g.* a 2 to 3 m column packed with a mass fraction of 0.05 Carbowax 20 M on Chromosorb G at 85 °C) or i.r. spectroscopy can be used.

Pertinent physicochemical data: The vapour pressures of the pure components are given by the equations below. The equation for chlorobenzene is based on the data in reference 1, and it gives values in good accord with those in reference 2; the equation for ethylbenzene was taken from reference 3.

Chlorobenzene : $\quad \log_{10}(p_1/\text{kPa}) = 6.110\,31 - 1436.135/[(t/°C) + 218.158]$, $\qquad$ (6.5.5.1)

Ethylbenzene : $\quad \log_{10}(p_2/\text{kPa}) = 6.082\,09 - 1424.255/[(t/°C) + 213.206]$. $\qquad$ (6.5.5.2)

These equations yield boiling temperatures of 131.727, 100.399, and 53.486 °C for chlorobenzene and 136.187, 104.706, and 57.636 °C for ethylbenzene at pressures of 101.325, 40.0, and 6.66 kPa respectively.

Vapour-liquid equilibrium: The system is assumed to be ideal so that

$$\ln \alpha = \ln(p_1/p_2). \qquad (6.5.5.3)$$

Table 6.5.5.1. Chlorobenzene + ethylbenzene

Pressure 101.325 kPa

$N+1$	x	y	α	$N+1$	x	y	α
1	0.0500	0.0560	1.1265	27	0.5406	0.5702	1.1278
3	0.0626	0.0700	1.1265	29	0.5994	0.6280	1.1280
5	0.0781	0.0872	1.1265	31	0.6557	0.6823	1.1281
7	0.0971	0.1081	1.1266	33	0.7079	0.7322	1.1283
9	0.1201	0.1333	1.1267	35	0.7552	0.7768	1.1284
11	0.1477	0.1634	1.1267	37	0.7971	0.8159	1.1285
13	0.1803	0.1987	1.1268	39	0.8334	0.8495	1.1286
15	0.2184	0.2394	1.1269	41	0.8644	0.8779	1.1287
17	0.2619	0.2856	1.1270	43	0.8903	0.9016	1.1287
19	0.3107	0.3369	1.1272	45	0.9118	0.9211	1.1288
21	0.3641	0.3923	1.1273	47	0.9295	0.9370	1.1288
23	0.4212	0.4507	1.1275	49	0.9438	0.9499	1.1289
25	0.4806	0.5106	1.1276	50	0.9499	0.9554	1.1289

Pressure 40.0 kPa

$N+1$	x	y	α	$N+1$	x	y	α
1	0.0500	0.0570	1.1487	23	0.5294	0.5641	1.1505
3	0.0649	0.0739	1.1488	25	0.5982	0.6314	1.1508
5	0.0840	0.0953	1.1488	27	0.6635	0.6941	1.1510
7	0.1079	0.1220	1.1489	29	0.7232	0.7505	1.1512
9	0.1377	0.1550	1.1490	31	0.7759	0.7995	1.1514
11	0.1741	0.1950	1.1492	33	0.8211	0.8409	1.1516
13	0.2178	0.2425	1.1494	35	0.8589	0.8752	1.1517
15	0.2689	0.2972	1.1495	37	0.8898	0.9029	1.1518
17	0.3271	0.3585	1.1498	39	0.9146	0.9250	1.1519
19	0.3913	0.4250	1.1500	41	0.9343	0.9424	1.1520
21	0.4595	0.4944	1.1502	43	0.9497	0.9560	1.1520

6. Distillation

Table 6.5.5.1. continued

Pressure 6.666 kPa							
$N+1$	x	y	α	$N+1$	x	y	α
1	0.0500	0.0595	1.2014	19	0.5929	0.6370	1.2049
3	0.0706	0.0836	1.2015	21	0.6790	0.7183	1.2054
5	0.0988	0.1164	1.2017	23	0.7545	0.7875	1.2059
7	0.1367	0.1599	1.2019	25	0.8172	0.8436	1.2062
9	0.1862	0.2158	1.2023	27	0.8667	0.8870	1.2065
11	0.2486	0.2846	1.2027	29	0.9045	0.9195	1.2068
13	0.3237	0.3654	1.2032	31	0.9324	0.9433	1.2069
15	0.4093	0.4548	1.2037	32	0.9433	0.9526	1.2070
17	0.5010	0.5474	1.2043				

The method used for the calculation of plate equivalence are given in the general theory.

REFERENCES

1. Brown, I. *Australian J. Sci. Research* **1952**, *5A*, 530.
2. Letcher, J. M.; Bayles, J. W. *J. Chem. Eng. Data* **1971**, *16*, 266.
3. *TRC Thermodyanmic Tables–Hydrocarbons*, Thermodynamic Research Center, Texas A&M University: College Station, Texas.

6.5.6. Benzene + 1,2-Dichloroethane Test Mixture

Intended usage: This mixture may be used to measure the plate efficiencies of columns equivalent to 7 to 56 theoretical plates at a pressure of 101.325 kPa.

Sources of supply and/or methods of preparation: The compounds can be obtained from many firms. A thiophen-free grade of benzene should be used. If a commercial grade of benzene which contains thiophen is purchased, this impurity can be removed by repeated shaking of the sample with concentrated sulphuric acid until the acid layer stays colourless. If the material still contains paraffinic hydrocarbons, as revealed by the refractive index value or by gas chromatography, they can be removed by fractional freezing of the sample. The benzene so purified should be distilled through a column before use and the physical properties of the selected fractions should be similar to those recorded under Analysis.

Commercial samples of 1,2-dichloroethane may contain other chlorinated hydrocarbons and if the sample is acidic it should be washed with water and then dried over anhydrous calcium chloride. The sample should be fractionally distilled before use and only distillate fractions with the correct physical properties (see Analysis) should be used to make the test mixture.

Analysis: Analysis by density measurement is much more accurate than analysis by refractive index although either may be used (benzene, $\rho(20\ °C) = 879.0$ kg m^{-3}; $n_D(20\ °C) = 1.5011$; 1,2-dichloroethane, $\rho(20\ °C) = 1253.1$ kg m^{-3}; $n_D(20\ °C) = 1.4448$). Ultraviolet spectroscopy has also been used for analysis.

Pertinent physicochemical data: The vapour pressures of the pure components at $t/°C$ are given by the following equations (Ref. 1):

Benzene: $\qquad \log_{10}(p_1/\text{kPa}) = 6.030\ 55 - 1211.033/[(t/°C) + 220.790],\qquad$ (6.5.6.1)

1,2-Dichloroethane: $\quad \log_{10}(p_2/\text{kPa}) = 6.283\ 56 - 1341.37/[(t/°C) + 230.05].\qquad$ (6.5.6.2)

These equations yield boiling temperatures of 80.100 °C for benzene and 83.512 °C for 1,2-dichloroethane at a pressure of 101.325 kPa.

Vapour-liquid equilibrium: The system is non-ideal, indeed the $(\ln \alpha)$ versus x curve exhibits a minimum and the following equation was used to represent $\ln \alpha$ as a power series of x. This equation was obtained by fitting an equation by least squares to the data in reference 2 combined with the numerical data which was used to construct figure 2 of reference 3.

$$\ln \alpha = 0.151\ 39 - 0.191\ 57x + 0.147\ 18x^2. \qquad (6.5.6.3)$$

Table 6.5.6.1. Benzene + 1,2-dichloroethane at 101.325 kPa

$N+1$	x	y	α	$N+1$	x	y	α
1	0.0500	0.0572	1.1528	31	0.6130	0.6340	1.0934
3	0.0653	0.0743	1.1497	33	0.6544	0.6743	1.0931
5	0.0844	0.0956	1.1460	35	0.6935	0.7122	1.0934
7	0.1078	0.1212	1.1416	37	0.7302	0.7475	1.0942
9	0.1358	0.1516	1.1367	39	0.7642	0.7802	1.0952
11	0.1685	0.1864	1.1312	41	0.7955	0.8101	1.0965
13	0.2055	0.2254	1.1255	43	0.8240	0.8371	1.0980
15	0.2463	0.2679	1.1198	45	0.8495	0.8613	1.0995
17	0.2901	0.3129	1.1143	47	0.8723	0.8827	1.1011
19	0.3361	0.3596	1.1092	49	0.8923	0.9014	1.1026
21	0.3833	0.4071	1.1047	51	0.9098	0.9176	1.1040
23	0.4309	0.4546	1.1010	53	0.9248	0.9315	1.1053
25	0.4782	0.5015	1.0979	55	0.9376	0.9433	1.1065
27	0.5246	0.5473	1.0957	57	0.9485	0.9532	1.1075
29	0.5697	0.5916	1.0942				

6. Distillation

The methods used for the calculation of the plate equivalence are given in the general theory.

REFERENCES

1. *TRC Thermodynamic Tables-Hydrocarbons*, Thermodynamic Research Center, Texas A&M University: College Station, Texas.
2. Zuiderweg, F. J., editor, *Recommended Test Mixtures for Distillation Columns*, European Federation of Chemical Engineering, Working Party on Distillation, Absorption and Extraction, Institution of Chemical Engineers: London, **1969**.
3. Coulson, E. A.; Hales J. L.; Herington, E. F. G. *Trans. Faraday Soc.* 1948, *44*, 636.

6.5.7. 2-Methylnaphthalene + 1-Methylnaphthalene Test Mixture

Intended usage: This mixture is only suitable for testing columns under reduced pressure because above 13.33 kPa thermal decomposition occurs. The mixture may be used to measure the plate efficiencies of columns equivalent to 7 to 59 theoretical plates (pressure 13.33 kPa), 7 to 54 theoretical plates (pressure 6.666 kPa), 6 to 47 theoretical plates (pressure 1.333 kPa) and 5 to 41 theoretical plates (pressure 0.667 kPa).

Sources of supply and/or methods of preparation: These compounds can be purchased from various commercial sources. The compounds can be purified by distillation under reduced pressure in a column equivalent to at least 50 theoretical plates at a high reflux ratio. 2-Methylnaphthalene which has a melting temperature of about 34.5_5 °C, can also be purified by recrystallization from a 1:1 methanol and water mixture. As some photodegradation has been reported, the compounds should be stored in the dark.

Analysis: As these compounds are liable to some thermal degradation under the test conditions, the only reliable analytical method is gas chromatography. For example a 4 m column packed with an 0.1 mass fraction Apiezon L or Silicon oil DC 550 on Chromosorb P 80-100 mesh, with hydrogen and helium as carrier gas at 180 °C with detection by thermal conductivity has been recommended. Alternatively, a 3 or 4 m column at 140 °C packed with Loc-2-R446 has been used.

Pertinent physicochemical data:: The vapour pressures of the pure liquid compounds at $t/$°C are given by the following equations (Ref. 1):

2-Methylnaphthalene: $\quad \log_{10}(p_1/\text{kPa}) = 6.193\,40 - 1840.268/[(t/°C) + 198.395]$, $\quad$ (6.5.7.1)

1-Methylnaphthalene: $\quad \log_{10}(p_2/\text{kPa}) = 6.160\,82 - 1826.948/[(t/°C) + 195.002]$. $\quad$ (6.5.7.2)

These equations yield boiling temperatures of 164.679, 144.329, 104.851, and 90.534 °C for 2-methylnaphthalene and 167.776, 147.318, 107.674, and 93.310 °C for 1-methylnaphtha-lene at pressures of 13.33, 6.666, 1.333 and 0.667 kPa.

Vapour-liquid equilibrium: The mixture is treated as ideal so that

$$\ln \alpha = \ln(p_1/p_2). \tag{6.5.7.3}$$

For each pressure the variation of $\ln \alpha$ with composition is small and symmetrical about $x = 0.5$. It is recommended that the following values of α be used in the Fenske equation, for the pressure ranges indicated, in experiments designed so that $x + y = 1$, and in which x lies between 0.05 and 0.40 and y between 0.60 and 0.95.

Table 6.5.7.1. 2-Methylnaphthalene + 1-methylnaphthalene

Pressure/kPa	Recommended value of α	α at $x = 0.05$	α at $x = 0.95$
13.33	1.104 37	1.103 81	1.104 93
6.666	1.113 59	1.112 92	1.114 26
1.333	1.138 68	1.137 68	1.139 69
0.667	1.151 12	1.149 94	1.152 30

REFERENCE

1. *TRC Thermodynamic Tables-Hydrocarbons*, Thermodynamic Research Center, Texas A&M University: College Station, Texas.

6.5.8. Heptane + Methylcyclohexane Test Mixture

Intended usage: This mixture may be used to measure the plate efficiencies of columns equivalent to 11 to 82 theoretical plates at a pressure of 101.325 kPa.

Sources of supply and/or methods of preparation: These compounds are available from a number of firms but it is essential to purchase high grade material. The heptane employed must consist solely of the normal heptane isomer and care should be taken that the sample used is suitable by establishing that it contains only one component by gas chromatographic analysis and by making measurements of density and refractive index. The results obtained should be compared with those recorded under Analysis. A suitable grade is heptane for the knock-rating of engines available from Supplier (A). A sample used to make test mixtures should be fractionally distilled before use. Methylcyclohexane samples should be purified by washing with concentrated sulphuric acid, sodium carbonate solution and water. The specimen should then be dried over anhydrous calcium chloride or molecular sieve and should

6. Distillation

be distilled until the fractions have properties approaching those of the pure compound (see Analysis).

Analysis: Mixtures can be analysed by refractive index (heptane, $n_D(20\ °C) = 1.387\ 64$; methylcyclohexane $n_D(20\ °C) = 1.423\ 12$) or density (heptane, $\rho(20\ °C) = 683.76$ kg m^{-3}; methylcyclohexane, $\rho(20\ °C) = 769.39$ kg m^{-3}). Griswold (Ref. 2) recommended the use of refractive index as leading to smaller errors. Gas chromatography has also been used (column 4 m long, a mass fraction of 0.05 by weight of Carbowax 1550 on Chromosorb G at 60 °C or a mass fraction of 0.25 squalane on Chromosorb P 60-80 mesh, at 125 °C with hydrogen as carrier gas).

Whatever method of analysis is used, synthetic mixtures of known composition should be analysed so that the error in the value of N resulting from analytical errors can be assessed (see Appendix for effect of errors in x and y on the value of N obtained). The importance of paying attention to the effect of errors will be realized when it is appreciated that if $y = 0.95$, one plate produces a change of only 0.00329 in the mole fraction.

Pertinent physicochemical data: The vapour pressures of the pure compounds at $t/°C$ are given by the following equations (Ref. 1):

Heptane : $\quad\log_{10}(p_1/\text{kPa}) = 6.021\ 67 - 1264.900/[(t/°C) + 216.544],\quad$ (6.5.8.1)

Methylcyclohexane : $\quad\log_{10}(p_2/\text{kPa}) = 5.947\ 90 - 1270.763/[(t/°C) + 221.416].\quad$ (6.5.8.2)

These equations yield boiling temperatures of 98.425 °C for heptane and 100.934 °C for methylcyclohexane at a pressure of 101.325 kPa.

Vapour-liquid equilibrium: The mixture is assumed to be ideal so that

$$\ln \alpha = \ln(p_1/p_2). \qquad (6.5.8.3)$$

The calculated value of α at 98.425 °C is 1.07381 and at 100.934 °C is 1.07582. Therefore the recommended value of α to be used in the Fenske equation is 1.07482. This value is the antilogarithm of the mean of the logarithms of the two values of α (*i.e.* the geometric mean).

REFERENCES

1. *TRC Thermodynamic Tables-Hydrocarbons*, Thermodynamic Research Center, Texas A&M University College Station, Texas.
2. Griswold, J. *Ind. Eng. Chem.* **1943**, *35*, 247.

6.5.9. *para*-Xylene + *meta*-Xylene Test Mixture

Intended usage: This mixture may be used to measure the plate efficiencies of columns equivalent to 40 to 290 theoretical plates at a pressure of 101.325 kPa.

Sources of supply and/or methods of preparation: Samples can be purchased from several sources. The most likely contaminants of *para-* and *meta-*xylenes are *ortho-*xylene and ethylbenzene but fortunately the presence of small amounts of these compounds will not interfere with the testing process because the methods of analysis recommended are specific for the *para-* and *meta-*isomers. Nevertheless the test mixture should be pre-distilled in the column using a high reflux ratio and only the middle fraction should be used for the tests.

Analysis: Infrared spectroscopy or gas chromatography can be used for analysis. For example, gas chromatography using a column packed with dinonylphthalate plus a mass fraction of 0.05 bentone 38 on Chromosorb W can be used for analysis. A high accuracy in the analysis must be achieved if a high precision in the value for the calculated number of plates is to be obtained.

Pertinent physicochemical data: The vapour pressures of the pure compounds at $t/°C$ are given by the following equations (Ref. 1):

*para-*Xylene: $\quad \log_{10}(p_1/\text{kPa}) = 6.115\,42 - 1453.430/[(t/°C) + 215.307]$, $\quad$ (6.5.9.1)

*meta-*Xylene: $\quad \log_{10}(p_2/\text{kPa}) = 6.133\,98 - 1462.266/[(t/°C) + 215.105]$. $\quad$ (6.5.9.2)

These equations yield boiling temperatures of 138.351 °C for *para-*xylene and 139.104 °C for *meta-*xylene for a pressure of 101.325 kPa.

Vapour-liquid equilibrium: The mixture is assumed to be ideal so that

$$\ln \alpha = \ln(p_1/p_2), \quad (6.5.9.3)$$

but as the following table shows the value of $\ln \alpha$ varies with pressure. However, as $\ln \alpha$ is, to a good approximation, a linear function of pressure it is recommended that a value of $\ln \alpha$ appropriate to the average pressure in the distillation column be used to calculate the number of theoretical plates by means of the Fenske equation.

Table 6.5.9.1. *para-*Xylene + *meta-*xylene

p/kPa	100	120	140	160	180	200
$\ln \alpha$	0.020 256	0.018 945	0.017 839	0.016 884	0.016 043	0.015 292

As the value of α is near unity the importance of avoiding leakage from the still head, the importance of allowing long equilibration times and the necessity for high precision in the analysis must be particularly stressed.

REFERENCE

1. *TRC Thermodynamic Tables–Hydrocarbons*, Thermodynamic Research Center, Texas A&M University: College Station, Texas.

6.6. Test mixtures for columns at atmospheric pressure and above atmospheric pressure

6.6.1. Methanol + Ethanol Test Mixture

Intended usage: This mixture can be used to measure the plate efficiencies of columns of 1 to 11 theoretical plates (1.013 bar), 1 to 14 theoretical plates (3.040 bar), 1 to 16 theoretical plates (5.066 bar), and 1 to 20 theoretical plates (10.13 bar).

Sources of supply and/or methods of preparation: The compounds are available as samples of high purity from many industrial sources. The most likely impurity is water, and every time these alcohols are transferred from one container to another the water content is likely to increase unless precautions are taken. Efficient distillation of methanol can reduce the water volume fraction to 1×10^{-4}. Absolute alcohol (ethanol) contains not more than 7×10^{-4} by volume fraction of water. The samples of both alcohols should be distilled under a high reflux ratio before use.

Analysis: Gas chromatography methods are recommended for the analysis of these mixtures because the pure compounds are hygroscopic and are easily contaminated by water. Columns that have been recommended include 1 m lengths packed with Porapak at 100 °C and 2 m lengths packed with Chromosorb 101 at 90 °C.

Pertinent physicochemical data: The vapour pressures of the pure compounds at $t/°C$ are given by the following equations (Ref. 1), which yield values over the pressure range 1 to 10 bar in good agreement with the values presented in reference 2.

Methanol: $\quad \log_{10}(p_1/\text{bar}) = 5.162\,33 - 1556.1/[(t/°C) + 237.23]$, $\qquad (6.6.1.1)$

Ethanol: $\quad \log_{10}(p_2/\text{bar}) = 4.959\,59 - 1432.0/[(t/°C) + 210.83]$. $\qquad (6.6.1.2)$

These equations yield boiling temperatures of 64.53, 95.31, 111.85 and 137.13 °C for methanol and 78.23, 109.05, 125.72 and 151.34 °C for ethanol at pressures of 1.013, 3.040, 5.066 and 10.13 bar respectively.

Vapour-liquid equilibrium: The system is not ideal; data provided by Brunner (Ref. 1), which are similar to that in reference 3 were used.

The following equation was used to represent $\ln \alpha$ as a power series in x for a pressure of 1.013 bar:
$$\ln \alpha = 0.484\,78 - 0.043\,43x + 0.210\,79x^2. \qquad (6.6.1.3)$$

Table 6.6.1.1. Methanol + ethanol at 1.013 bar

$N+1$	x	y	α	$N+1$	x	y	α
1	0.0500	0.0786	1.6212	9	0.7364	0.8312	1.7632
3	0.1215	0.1830	1.6203	11	0.8992	0.9429	1.8518
5	0.2666	0.3719	1.6294	12	0.9429	0.9688	1.8800
7	0.4935	0.6198	1.6731				

The following equation was used to represent $\ln \alpha$ as a power series in x for a pressure of 3.040 bar:
$$\ln \alpha = 0.385\,34 + 0.030\,07x + 0.047\,06x^2. \qquad (6.6.1.4)$$

Table 6.6.1.2. Methanol + ethanol at 3.040 bar

$N+1$	x	y	α	$N+1$	x	y	α
1	0.0500	0.0719	1.4725	9	0.5524	0.6517	1.5164
3	0.1025	0.1442	1.4754	11	0.7411	0.8153	1.5426
5	0.1994	0.2695	1.4817	13	0.8728	0.9148	1.5643
7	0.3543	0.4506	1.4947	15	0.9441	0.9638	1.5772

The following equation was used to represent $\ln \alpha$ as a power series in x for a pressure of 5.066 bar:
$$\ln \alpha = 0.331\,61 + 0.094\,03x - 0.043\,01x^2. \qquad (6.6.1.5)$$

Table 6.6.1.3. Methanol + ethanol at 5.066 bar

$N+1$	x	y	α	$N+1$	x	y	α
1	0.0500	0.0686	1.3996	11	0.6367	0.7181	1.4540
3	0.0936	0.1267	1.4050	13	0.7878	0.8443	1.4608
5	0.1697	0.2242	1.4139	15	0.8881	0.9207	1.4640
7	0.2910	0.3693	1.4267	17	0.9445	0.9614	1.4653
9	0.4564	0.5475	1.4413				

6. Distillation

The following equation was used to represent $\ln \alpha$ as a power series in x for a pressure of 10.13 bar:

$$\ln \alpha = 0.253\ 73 + 0.169\ 61x - 0.138\ 04x^2. \tag{6.6.1.6}$$

Table 6.6.1.4. Methanol + ethanol at 10.13 bar

$N+1$	x	y	α	$N+1$	x	y	α
1	0.0500	0.0640	1.2994	13	0.6086	0.6786	1.3578
3	0.0818	0.1042	1.3056	15	0.7413	0.7951	1.3547
5	0.1322	0.1668	1.3149	17	0.8399	0.8761	1.3483
7	0.2092	0.2598	1.3273	19	0.9049	0.9274	1.3420
9	0.3190	0.3859	1.3415	21	0.9448	0.9581	1.3375
11	0.4586	0.5341	1.3532				

REFERENCES

1. Brunner, E. private communication.
2. *Vapour pressures and critical points of pure substances, IV: C_1 to C_{20} Alcohols*, Item 74023, Engineering Sciences Data Unit: London, **1974**.
3. Zuiderweg, F. J., editor, *Recommended Test Mixtures for Distillation Columns*, European Federation of Chemical Engineers, Working Party on Distillation, Absorption and Extraction, Institution of Chemical Engineers: London, **1969**.

6.6.2. Propene + Propane Test Mixture

Intended usage: This mixture is suitable for testing columns at pressures above atmospheric and at temperatures in the approximate range −4 to 62 °C. The mixture may be used to measure the plate efficiencies of columns equivalent to 3 to 30 theoretical plates (5.066 bar), to 4 to 37 theoretical plates (10.13 bar), to 6 to 44 theoretical plates (15.20 bar) and to 6 to 56 theoretical plates (22.09 bar).

Sources of supply and/or methods of purification: Cylinders containing these compounds are available from a number of manufacturers. The individual components of the test mixtures should be purified by distillation before use and the purities of specimens employed to make test mixtures should be established by gas chromatography (see Analysis).

Analysis: Mixtures should be analysed by gas chromatography; for example, a 1 m column packed with Duropack phenyl isocyanate/Porosil C at 25 °C has been used.

Pertinent physico-chemical data: The vapour pressures of the pure compounds at $t/°C$ for the temperature ranges indicated are given by the following equations (Refs. 1 to 3):

Propene: $$\log_{10}(p_1/\text{bar}) = 4.269\,75 - 945.752/[(t/°C) + 269.92], \quad (6.6.2.1)$$

for temperatures $-20 < t/°C < 65$,

Propane: $$\log_{10}(p_2/\text{bar}) = 4.312\,93 - 998.263/[(t/°C) + 274.35], \quad (6.6.2.2)$$

for temperatures $-10 < t/°C < 70$.

These equations yield boiling temperatures of -4.64, 19.82, 36.36, and 53.35 °C for propene and 2.31, 27.48, 44.47, and 61.91 °C for propane for pressures of 5.066, 10.13, 15.20 and 22.09 bar respectively.

Vapour-liquid equilibrium: The system is not ideal, and data provided by Brunner (Ref. 1) were used.

The following equation was used to represent $\ln \alpha$ as a power series in x for a pressure of 5.066 bar:

$$\ln \alpha = 0.2539 - 0.1128x. \quad (6.6.2.3)$$

Table 6.6.2.1. Propene + propane at 5.066 bar

$N+1$	x	y	α	$N+1$	x	y	α
1	0.0500	0.0632	1.2818	17	0.6492	0.6892	1.1980
3	0.0795	0.0994	1.2775	19	0.7256	0.7585	1.1877
5	0.1233	0.1516	1.2712	21	0.7880	0.8143	1.1794
7	0.1847	0.2224	1.2625	23	0.8375	0.8581	1.1728
9	0.2644	0.3102	1.2512	25	0.8762	0.8920	1.1677
11	0.3589	0.4093	1.2379	27	0.9059	0.9181	1.1638
13	0.4603	0.5108	1.2238	29	0.9287	0.9380	1.1608
15	0.5595	0.6059	1.2102	31	0.9460	0.9531	1.1586

The following equation was used to represent $\ln \alpha$ as a power series in x for a pressure of 10.13 bar:

$$\ln \alpha = 0.2036 - 0.0834x. \quad (6.6.2.4)$$

6. Distillation

Table 6.6.2.2. Propene + propane at 10.13 bar

$N+1$	x	y	α	$N+1$	x	y	α
1	0.0500	0.0604	1.2207	21	0.6576	0.6903	1.1604
3	0.0727	0.0872	1.2184	23	0.7206	0.7486	1.1543
5	0.1041	0.1237	1.2152	25	0.7742	0.7976	1.1492
7	0.1462	0.1718	1.2110	27	0.8188	0.8380	1.1449
9	0.2004	0.2320	1.2055	29	0.8554	0.8710	1.1414
11	0.2665	0.3034	1.1989	31	0.8850	0.8975	1.1386
13	0.3423	0.3828	1.1913	33	0.9088	0.9189	1.1363
15	0.4241	0.4656	1.1832	35	0.9278	0.9358	1.1345
17	0.5067	0.5469	1.1751	37	0.9430	0.9493	1.1331
19	0.5857	0.6227	1.1674	38	0.9493	0.9550	1.1325

The following equation was used to represent $\ln \alpha$ as a power series in x for a pressure of 15.20 bar:

$$\ln \alpha = 0.1760 - 0.0793 x. \tag{6.6.2.5}$$

Table 6.6.2.3. Propene + propane at 15.20 bar

$N+1$	x	y	α	$N+1$	x	y	α
1	0.0500	0.0588	1.1877	25	0.6715	0.6980	1.1306
3	0.0691	0.0809	1.1859	27	0.7228	0.7460	1.1260
5	0.0944	0.1098	1.1835	29	0.7675	0.7874	1.1220
7	0.1273	0.1469	1.1805	31	0.8058	0.8227	1.1186
9	0.1687	0.1927	1.1766	33	0.8383	0.8526	1.1158
11	0.2190	0.2473	1.1719	35	0.8657	0.8777	1.1133
13	0.2776	0.3095	1.1665	37	0.8887	0.8987	1.1113
15	0.3427	0.3770	1.1605	39	0.9078	0.9162	1.1096
17	0.4119	0.4470	1.1541	41	0.9238	0.9307	1.1082
19	0.4819	0.5164	1.1477	43	0.9370	0.9428	1.1071
21	0.5500	0.5825	1.1416	45	0.9480	0.9527	1.1061
23	0.6137	0.6434	1.1358				

The following equation was used to represent $\ln \alpha$ as a power series in x for a pressure of 22.09 bar:

$$\ln \alpha = 0.1435 - 0.0694 x. \tag{6.6.2.6}$$

Table 6.6.2.4. Propene + propane at 22.09 bar

$N+1$	x	y	α	$N+1$	x	y	α
1	0.0500	0.0571	1.1503	31	0.6748	0.6957	1.1015
3	0.0651	0.0741	1.1491	33	0.7154	0.7342	1.0984
5	0.0841	0.0954	1.1476	35	0.7518	0.7685	1.0956
7	0.1078	0.1216	1.1457	37	0.7841	0.7988	1.0932
9	0.1368	0.1534	1.1434	39	0.8126	0.8255	1.0910
11	0.1715	0.1910	1.1407	41	0.8376	0.8489	1.0891
13	0.2120	0.2343	1.1375	43	0.8594	0.8692	1.0875
15	0.2579	0.2826	1.1338	45	0.8784	0.8869	1.0860
17	0.3084	0.3350	1.1299	47	0.8949	0.9023	1.0848
19	0.3623	0.3901	1.1256	49	0.9092	0.9156	1.0837
21	0.4181	0.4462	1.1213	51	0.9216	0.9272	1.0828
23	0.4741	0.5018	1.1170	53	0.9323	0.9371	1.0820
25	0.5289	0.5554	1.1127	55	0.9416	0.9458	1.0813
27	0.5812	0.6060	1.1087	57	0.9496	0.9532	1.0807
29	0.6300	0.6529	1.1049				

REFERENCES

1. Brunner, E. private communication.
2. *Vapour pressures of pure substances up to their critical points. I: C_1 to C_8 Alkanes*, item 72028, Engineering Sciences Data Unit: London, **1972**.
3. *Vapour pressures of pure substances up to their critical points. II: C_2 to C_6 Alkenes*, item 73008, Engineering Sciences Data Unit: London, **1973**.

6. Distillation

6.7. Contributors to the Revised Version

D. Ambrose,
Department of Chemistry,
University College London,
20 Gordon Street,
London, WC1H OAJ (UK)

E. Brunner,
BASF Aktiengesellschaft,
Ammoniak laboratorium WAA/M325,
D-6700 Ludwigshafen/Rhein,
(Federal Republic of Germany)

J. F. Counsell,
BP Research Centre,
Chertsey Road,
Sunbury-on-Thames,
Middlesex, TW16 7LN (UK)

M. W. Gilzow,
Union Carbide Corporation,
P.O. Box 836,
South Charleston, WV 25303 (USA)

J. L. Hales,
National Physical Laboratory,
Teddington, Middlesex, TW11 OLW (UK)

E. F. G. Herington,
29 Seymour Road,
East Molesey,
Surrey, KT8 OPB (UK)

J. D. Olson,
Union Carbide Coorporation,
P.O. Box 8361,
South Charleston, WV 25302 (USA)

A. Newton, formerly at
Petrochemical Division,
Imperial Chemical Industries Ltd.,
POB 90, Wilton,
Middlesbrough,
Cleveland TS6 8JE (UK)

6.8. Supplier

(A) Esso Chemicals Ltd.,
Portland House,
Stag Place,
London SW1 (UK)

6.9. Appendix

This appendix discusses aspects of the behaviour of fractional distillation columns and of test mixtures germane to the measurement of the theoretical plate equivalences of columns.

6.9.1. Effect of Errors in the Value of the Relative Volatility Ratio

This discussion will be confined to a study of mixtures for which the relative volatility ratios are constant but the conclusions reached can be usefully extended to other test materials.

The following expression is obtained by straightforward differentiation of the Fenske equation (6.3.3) by setting $(N+1) \sim N$:

$$100 dN/N \sim -100(d\alpha)/\alpha \ln \alpha. \qquad (6.9.1.1)$$

Clearly the percentage error in N (viz. $100 \, dN/N$) depends not only on the uncertainty in α (viz. $d\alpha$) but also on the actual value of α so that the percentage error in N will be greater the nearer the value of α is to unity. For example, suppose it were decided that the percentage error in N must be less than 5% (i.e. the error on a 100-theoretical-plate column must be less than 5 plates and that on a 20-plate column must be less than 1 plate) then the value of $100(d\alpha)/\alpha \ln \alpha$ must be less than 5. A mixture with $\alpha = 1.05$ might be used to test a 100-plate column, so to obtain 5% accuracy in N the value of α must be known to better than 0.0026. A mixture with $\alpha = 1.25$ might be used to test a 20-plate column, so to obtain 5% accuracy in N, α must be known to better than 0.014.

Such accuracies in the values of α are not easy to achieve, especially for mixtures used for testing columns equivalent to more than 50 theoretical plates, but every attempt has been made in these recommendations to derive accurate data. It is suggested that two investigators wishing to compare the performance of fractionating equipment would be well advised to use the same data for a mixture, and it is hoped that the information in the data sheets of this report will be used in that way.

6.9.2. Choice of the Initial Composition of the Test Mixture

If it is assumed that equal observational errors in x and y are likely to occur and that the Fenske equation (6.3.3) applies to the test mixture with α constant then it can be shown that the error in the measured plate equivalence will be least when $x + y = 1$ (i.e. when $0.5 - x = y - 0.5$ so that the liquid and vapour compositions are symmetrically arranged about a mole faction of 0.5). There is an additional advantage in choosing this condition for making measurements because when $x + y = 1$ the effect of small deviations of the mixture from ideality on the calculated plate equivalence will often be minimized. The relation between $\ln \alpha$ and x for many slightly non-ideal mixtures can often be represented by a relation of the form of equation (6.9.2.1):

$$\ln \alpha = a_0 + a_1(1 - 2x), \qquad (6.9.2.1)$$

6. Distillation

so that the value of $\ln \alpha$ at $x = 0.5$ is a_0.

As a numerical example, consider a mixture where $a_0 = 0.1823$ and $a_1 = 0$ so that the Fenske equation yields a value of $N+1$ equal to 33 plates when $x = 0.05$ and $y = 0.95$. Now suppose the mixture had in fact been such that $a_0 = 0.1823$ and $a_1 = 0.0870$; then plate-to-plate calculations yield a value of $N + 1$ equal to 36 plates when $x = 0.05$ and $y = 0.95$. Thus the value of $N + 1$ differs by only 3 plates when the value of $\ln \alpha$ is constant at 0.1823 and when $\ln \alpha$ varies from 0.2693 to 0.0953 (over the range $x = 0$ to $x = 1$), a considerable change in the value of $\ln \alpha$. If conditions had been chosen so that $x + y$ did not equal unity the effect on the value of $N + 1$ of this change in the value of $\ln \alpha$ would have been much greater.

To choose the best initial composition for a test mixture which is far from ideal the tables of $N+1$, x and y provided should be examined to find the composition range where the expected value of $N + 1$ will produce the smallest error in the calculated number of theoretical plates.

The steps in choosing the initial composition for the boiler charge for a mixture where α is constant are as follows. Guess the plate equivalence of the column N, add 1, halve the value so found and calculate by the Fenske equation the initial composition x required to produce $y = 0.5$. This is the initial composition, x, that would be required if there were no dynamic hold-up in the apparatus. The initial composition, x_c, actually to be used can be obtained by consideration of a material balance, which leads to equation (6.9.2.2),

$$x_c = x + [H_0(0.5 - x)/B_0], \tag{6.9.2.2}$$

where H_0 is the volume of the hold-up and B_0 is the volume of the test mixture used. For example, if $x = 0.258$ and $(H_0/B_0) = 0.1$, then $x_c = 0.282$.

6.9.3. Effect of Experimental Errors in the Mole Fractions

If the magnitude of likely errors in x and y are assumed to be the same, and if the Fenske equation, equation (6.3.3), is applicable with α constant, and if conditions are chosen to minimize the effect of errors in x and y (i.e. $x + y = 1$), then to find the effect of an error in the composition of the boiler contents when there is no error in the vapour composition differentiation of equation (6.3.3) with y constant followed by division by equation (6.3.3) and putting $y = (1 - x)$ and $(N + 1) \sim N$ yields:

$$100 \, dN/N \sim -100 dx \{2x(1 - x) \, \ln[(1 - x)/x]\}^{-1}. \tag{6.9.3.1}$$

By considerations of symmetry the effect of an error dy in y will be given by the similar expression obtained by writing y for x throughout in equation (6.9.3.1). The following table gives values of $-A$ where

$$A = -100\{2x(1 - x) \ln[(1 - x)/x]\}^{-1}. \tag{6.9.3.2}$$

Table 6.9.3.1. Values of the function $-A$ at various mole fractions.

x	$-A$	x	$-A$
0.01	1099	0.25	243
0.05	357	0.30	281
0.10	253	0.35	355
0.15	226	0.40	514
0.1760	223	0.45	1006
0.20	225		

A minimum in the value of this function occurs at $x = 0.1760$. This behaviour has not been remarked on previously by other workers, and it is here proposed that this optimum value of N should be termed N_{opt}. The value of the tabulated function does not change very rapidly in the range $x = 0.05$ to 0.40 but it approaches infinity as x tends to zero or to 0.50. To ensure that the errors in the values of N are not too large the range $x = 0.05$ ($y = 0.95$) provides the upper limit for the number of plates N_{max} that can be determined with a given test mixture, and the values $x = 0.40$ ($y = 0.60$) have been chosen for the lower limit for the number of plates N_{min} that can be determined with a given test mixture.

In order that the values of the function tabulated expressing $100 \, dN/N$ in terms of dx be meaningful, it is essential that the value of x should be at least $10 \, dx$. Provided this condition is satisfied the table can be used to compute the percentage error in N which will result from an estimated value of dx. For example, suppose $x = 0.05$ and $dx = 0.003$, then $100 \, dN/N$ will equal -357×0.003, $i.e.$ -1.1%. If the standard deviation of y equals that of x and both equal 0.003 then the standard deviation in N expressed as a percentage will be 1.5% ($1.1 \times \sqrt{2}$). Thus, the standard deviation on a 100-plate column would be 1.5 plates and on a 20-plate column 0.3 plates. If $x = 0.40$ ($y = 0.60$) then a standard deviation of 0.003 in x and y produce a standard deviation of 2.2% in N, ($i.e.$ a percentage standard deviation of $514 \times 0.003 \times \sqrt{2}$). However, as less than 10 plates may be under determination in these circumstances, the error will only be of the order of 0.2 plate.

Inspection of the $N + 1$, x and y values tabulated for a mixture for which α is not constant will reveal readily the error in N which will result from anticipated errors in the values of x and y.

6.9.4. Dependence of the Maximum, Minimum and Optimum Number of Theoretical Plates that can be Measured on the Relative Volatility Ratio

The values of $(N + 1)_{max}$, $(N + 1)_{min}$ and $(N + 1)_{opt}$ have been calculated for $x = 0.05$,

$y = 0.95$; $x = 0.4$, $y = 0.6$; and $x = 0.176$, $y = 0.824$ respectively, from the Fenske equation, equation (6.3.3), treating α as a constant for a given mixture.

Table 6.9.4.1. Relation between maximum, minimum and optimum number of theoretical plates and the relative volatility ratio.

α	$(N+1)_{\max}$	$(N+1)_{\min}$	$(N+1)_{\mathrm{opt}}$
1.005	1181	163	619
1.01	592	82	310
1.02	297	41	156
1.05	121	17	63
1.10	62	8.5	32
1.15	42	5.8	22
1.20	32	4.4	17
1.30	22	3.1	12
1.50	14	2.0	7.6
2.0	8.5	1.1	4.5
2.5	6.4	0.9	3.4

For test mixtures that show relatively large changes in the value of α in the range $x = 0.05$ to $y = 0.95$, rough values of $(N+1)_{\max}$, $(N+1)_{\min}$ and $(N+1)_{\mathrm{opt}}$ may be found from table 6.9.4.1 by taking an average value of α.

6.9.5. Vapour-Liquid Equilibrium Data for Test Mixtures

The relative volatility ratio α for a binary mixture is defined by equation (6.3.1), and this quantity is useful for plate-to-plate calculations because x and y are related by equation (6.3.2). Fortunately, even for non-ideal mixtures $\ln \alpha$ can usually be expressed as a polynomial in x with only a few terms. Such a power series for $\ln \alpha$ enables $(N+1)$, x, y tables to be produced readily by a computer. Since it is just as easy to use natural logarithms for the calculations as to use logarithms to the base 10 and as natural logarithms are more convenient for thermodynamic studies they have been used widely in this report.

7

SECTION: RELATIVE HUMIDITY OF AIR

COLLATOR: W. KÜNZEL

CONTENTS:

7.1. Introduction

7.2. The role of reference materials: primary and secondary standards

7.3. Recommended reference materials for the measurement of relative humidity

7.4. Reference conditions and apparatus for reproducing stable vapour-liquid equilibrium

7.5. Guarantee of the saturation of the salt solution and demands for the purity requirements

7.6. Contributors

7.7. Suppliers

7.1. Introduction

The relative humidity of air U is defined by the ratio of the mole fraction of water vapour in moist air x_v at a given temperature T and at total wet gas pressure P, to the mole fraction of saturated water vapour in the air x_{vw}, under identical conditions:

$$U = \left(\frac{x_v}{x_{vw}}\right)_{P,T} \cdot 100\% \qquad (7.1.1)$$

Relative humidity is usually expressed in per cent. Under the ideal gas assumption, the relative humidity U is equal to the ratio of the partial pressure e of water vapor to the saturation vapor pressure e_w at the same total pressure and temperature:

$$U = \left(\frac{e}{e_w}\right)_{P,T} \cdot 100\% \tag{7.1.2}$$

For real gas conditions:

$$U = \left(\frac{e}{e'_w}\right)_{P,T} \cdot 100\% \tag{7.1.3}$$

where e'_w is the effective saturation vapor pressure with respect to water. The quotient $\frac{e'_w}{e_w} = f_w(P,T)$ is the so called f-factor and is tabulated in reference 4.

Besides the relative humidity U, there are other methods for expressing the humidity of air. It is recommended to use both the relative humidity and the following:

—the dew-point temperature
—the mass mixing ratio
—the absolute humidity

Definitions of these quantities and conversion methods are given in references 1, 2, and 3.

7.2. The role of reference materials: primary and secondary standards

A primary standard for the determination of the humidity of air is the gravimetric hygrometer (Refs. 5, 6, 7, 22). With this hygrometer one can determine the humidity of air directly with an acceptably low uncertainty. It works on the absorption principle. The mass mixing ratio of the wet air is determined by the measurement of both the mass of the water absorbed from the wet air and the mass of the dry air. The range of values for the mass mixing ratio is limited to approximately 0.0002 to 0.065.

In order to compare instruments with these primary standards, stable sources of wet air are necessary. One such source is the humidity generator.

Humidity generators are instruments for creating an atmosphere at any desired humidity and rate of air flow through the system. The four main types of generators based on thermodynamic PVT laws are: the two-pressure generator, the two-temperature generator, the divided flow generator and the low frost-point generator (Refs. 8, 9, 10, 11, 21). With the help of humidity generators, which create test atmospheres, secondary reference standards can be calibrated against a primary standard. Secondary reference standards include:

dew point hygrometer, pneumatic bridge hygrometer (Ref. 12), some electric hygrometers, and some psychrometers.

Alternatively, static fixed point generators of relative humidity can be realized by the use of hygrostatic solutions in the measurement range of

$$U = 3 \text{ to } 93\%.$$

7. Humidity

A hygrostatic solution is a saturated aqueous solution of an inorganic salt which is in thermodynamic equilibrium in a three-phase system at a given temperature. These salts can be regarded as reference materials. In the hierarchy scheme, these hygrostatic solutions are considered as secondary or working standards. The properties of these saturated solutions may be determined by comparison with secondary standards by using a humidity sensor as a comparator. However, the properties of these hygrostatic solutions may be determined by other means, virtually independent of other standards. The direct determination of the saturated vapor pressures of the hygrostatic solutions in the absence of atmospheric air may be made (Ref. 13) with a low uncertainty. The accuracy of the relative humidity standards can be improved if the corresponding f_s-factors of water vapours over hygrostatic solution in mixtures with air are known. These f-factors will not be the same as those of saturated water vapor over pure water in mixtures with air. Therefore the following relation should be used.

$$U = \frac{f_s \cdot e_s}{f_w \cdot e_w} \cdot 100\% \qquad (7.2.1)$$

However, f_s values are normally not available. The ratio f_s/f_w is normally closer to unity than either f_s or f_w (Ref. 4).

New experiments, combined with theoretical considerations, may improve the range of saturated salt solutions as reference standards for the relative humidity of air. At present, the precision of the values for hygrostatic solutions is considerably better than their accuracy.

7.3. Recommended reference materials for the measurement of relative humidity of air

Values for the water vapour pressure above hygrostatic solutions exist for more than eighty inorganic salts (Ref. 14). Data for sixteen salts are available in temperature steps of 0.5 K or even less near the points of the transition (Ref. 13). The general properties of aqueous salt systems in relation to humidity have been outlined (Ref. 15) and the experimental values have been compared with computations using free energy data.

Greenspan (Ref. 16), using the results of 21 different investigators has made an extensive evaluation of the relative humidity over hygrostatic solutions of 28 salts using the method of least squares with a regular polynomial having from two to four coefficients. In the various data compilations, a realistic estimation of the systematic and random errors of different measurements and methods is lacking. Therefore recommendations are only made for those salts of Greenspan's compilation where the reference values are based on six or more different investigations. The magnitude of the uncertainty indicates the self-consistency of the measurements. Because the values used are based on a large number of measurements, it is likely that the uncertainty includes some contribution from systematic errors. In table 7.3.1 the recommended values are summarized for the following salts: lithium chloride, magnesium chloride, magnesium nitrate, sodium chloride, potassium chloride.

Table 7.3.1. Percentage Relative Humidity of Hygrostatic Solutions from 0 to 80 °C

T/°C	Lithium Chloride	Magnesium Chloride	Magnesium Nitrate	Sodium Chloride	Potassium Chloride
0		33.66 ± 0.33	60.35 ± 0.55	75.51 ± 0.34	88.61 ± 0.53
5		33.60 ± 0.28	58.86 ± 0.43	75.65 ± 0.27	87.67 ± 0.45
10		33.47 ± 0.24	57.36 ± 0.33	75.67 ± 0.22	86.77 ± 0.39
15		33.30 ± 0.21	55.87 ± 0.27	75.61 ± 0.18	85.92 ± 0.33
20	11.31 ± 0.31	33.07 ± 0.18	54.38 ± 0.23	75.47 ± 0.14	85.11 ± 0.29
25	11.30 ± 0.27	32.78 ± 0.16	52.89 ± 0.22	75.29 ± 0.12	84.34 ± 0.26
30	11.28 ± 0.24	32.44 ± 0.14	51.40 ± 0.24	75.09 ± 0.11	83.62 ± 0.25
35	11.25 ± 0.22	32.05 ± 0.13	49.91 ± 0.29	74.87 ± 0.12	82.95 ± 0.25
40	11.21 ± 0.21	31.60 ± 0.13	48.42 ± 0.37		82.32 ± 0.25
45	11.16 ± 0.21	31.10 ± 0.13	46.93 ± 0.47		81.74 ± 0.28
50	11.10 ± 0.22	30.54 ± 0.14	45.44 ± 0.60		81.20 ± 0.31
55	11.03 ± 0.23	29.93 ± 0.16			80.70 ± 0.35
60	10.95 ± 0.26	29.26 ± 0.18			80.25 ± 0.41
65	10.86 ± 0.29	28.54 ± 0.21			79.85 ± 0.48
70	10.75 ± 0.33	27.77 ± 0.25			79.49 ± 0.57
75	10.64 ± 0.38	26.94 ± 0.29			79.17 ± 0.66
80	10.51 ± 0.44	26.05 ± 0.34			78.90 ± 0.77

7.4. Reference conditions and apparatus for reproducing a stable vapour-liquid equilibrium

A hygrostatic solution can be used for reproducing the values of relative humidity under the following reference conditions:

a) the total pressure of water vapour and air in the gaseous space above the solution is close to atmospheric pressure (101.325 kPa)

b) the solution is used at the correct temperature with an adequate thermostat to ensure the constancy the temperature with an accuracy of better than

± 0.1 K for the 0 to 50 °C range
± 0.2 K for the 50 to 80 °C range

Various devices, termed hygrostats, have been designed to ensure a stable water vapour pressure over a hygrostatic solution.

Hygrostats should be made of a non-hygroscopic material and should be easily cleaned and resistant to corrosion. A sphere is the preferred shape for the measuring vessel. The hygrostat should be equipped with suitable devices for stirring both the solution and the air above the solution surface. In the stirred hygrostats, the total surface area of the vessel walls may be as large as three times the area of the free surface of the solution. The design of salt hygrostats depends on the ultimate use of the device. Hygrostats are described in references 15, 17, 18, 19, and 20.

7.5. Guarantee of the saturation of the salt solution and purity requirements

The salts used for the preparation of the hygrostatic solutions should be of analytical-grade. The water should be either double distilled or deionised. To ensure that the aqueous salt solution is saturated, it is necessary that there is always an excess of salt. The salt must be fully covered by the solution.

REFERENCES

1. Wexler, A. *Humidity and Moisture*, Vols. 1 to 3, Reinhold: New York, **1965**.
2. Sonntag, D. *Hygrometrie*, Akademie-Verlag: Berlin, **1968**.
3. Berliner, M. A. *Izmerenija wlashnosti*, (in Russian) Energija: Moskva, 1973, (in German) Verlag Technik: Berlin, **1980**.
4. Harrison, L. P. *Fundamental Concepts and Definitions Relating to Humidity: Humidity and Moisture*, Vol. 3, Reinhold: New York, **1965**, p. 3.
5. Wexler, A.; Hyland, R. W. *National Bureau of Standards Monograph 73*, U.S. Department of Commerce: Washington, **1964**.
6. Kostyrko, K. *Izm. tech., Moskva* **1976**, *11*, 83.
7. Scholz, G. *Bulletin OIML* **1984**, *97*, 18.
8. Hasegawa, S.; Hyland, R. W.; Rhodes, S. W. *A Comparision between the NBS Two-pressure Humidity Generator and the NBS Standard-Hygrometer, Precision Measurement and Calibration-Mechanics*, National Bureau of Standards Special Publication, Vol. 8, U.S. Department of Commerce: Washington, **1972**.
9. Greenspan, L. *J. Res. Nat. Bur. Stand.*, **1973**, *77A* (5), 671.
10. Hasegawa, S.; Little, J. W. *J. Res. Nat. Bur. Stand.* **1977**, *81A*, 81.
11. Belonoshko, V. M.; Gridnjev, A. A.; Krebs, J. B.; Mandrochlebov, V. F. *Izm. tech., Moskva*, **1982**, *9*, 56.
12. Greenspan, L. *A pneumatic bridge Hygrometer for use as a working humidity standard, Precision Measurement and Calibration-Mechanics*, National Bureau of Standards Special Publication 300, Vol. 8, U.S. Department of Commerce: Washington, **1972**, p. 295.
13. Acheson, D. T. *Vapor Pressures of Saturated Aqueous Salt Solutions, in Humidity and Moisture*, Vol. 3, Reinhold: New York, **1965**.
14. O'Brien, J. *Sci. Instr.* **1948**, *25*, 73.
15. Wylie, R. G. *The properties of water-salt systems in relation to humidity*, in *Humidity and Moisture*, Vol. 3, Reinhold: New York, **1965**.

16. Greenspan, L. *J. Res. Nat. Bur. Stand.* **1977**, *81A*, 89. 53A, 19.
18. Gridnjev., A. S.; Mandrochlebov, V. F. *Izm. tech., Moskva* **1982**, *9*, 59.
19. Wyzykowska, A. *Cieploconistwo, Ogrzewnistwo Wentylecja* **1972**, *33*, 172.
20. AFNOR NF 15-014, *Enceintes et conditions d'essai; Petites enceintes de conditionnement et d'essai utilisant des solutions aqueuses.* **1973**.
21. Inamatsu, T.; Takahashi, Ch. *Bull. of NRLM* **1986**, *35*, 245.
22. Takahashi, Ch.; Inamatsu, T. *Bull. of NRLM* **1986**, *35*, 253.

7.6. Contributors

W. Künzel
Amt für Standardisierung, Messwesen
und Warenprüfung (ASMW)
Fürstenwalder Damm 388, DDR-1162 Berlin (GER. DEM. REP.)

7.7. List of Suppliers

Research and Development Centre for Standard Reference Materials
Ul. Elektoralna 2
PL 00-139 Warszawa (POLAND)

Table 7.7.1. Percentage Relative Humidity of Polish Reference Materials at 20 °C

Type	U	Mass/g
Lithium chloride $LiCl \cdot H_2O$	11.3 ± 0.3	130
Magnesium chloride $MgCl_2 \cdot 6H_2O$	33.1 ± 0.2	200
Potassium carbonate $K_2CO_3 \cdot 2H_2O$	43.2 ± 0.3	135
Sodium bromide $NaBr \cdot 2H_2O$	59.1 ± 0.4	130
Sodium chloride $NaCl$	75.5 ± 0.1	50
Ammonium sulphate $(NH_4)_2 \cdot SO_4$	81.3 ± 0.3	100
Potassium nitrate KNO_3	94.6 ± 0.7	60

The materials are available in the form of sealed glass ampoules. Their storage life in originally sealed ampoules is two years. The relative humidity U at 25 °C and 30 °C is also provided on the certificate.

8

SECTION: TEMPERATURE

COLLATORS: D. AMBROSE, L. CROVINI

CONTENTS:

8.1. Introduction
    8.1.1. Scope of the recommendation
    8.1.2. Fundamentals of temperature measurements
    8.1.3. Secondary reference points of temperature
    8.1.4. The role of reference materials

8.2. High Accuracy Fixed Points
    8.2.1. General considerations
    8.2.2. Triple points
    8.2.3. Freezing and melting points
        8.2.3.1. Purity control
    8.2.4. Liquid-vapour equilibria
        8.2.4.1. Pressure measurements

8.3. Reference Materials for High Accuracy Fixed Points
    8.3.1. Recommended points and temperature values
    8.3.2. Available reference materials
        8.3.2.1. Mercury (triple point)
        8.3.2.2. Water (ice point)
        8.3.2.3. Water (triple point)
        8.3.2.4. Phenoxybenzene (triple point)
        8.3.2.5. Gallium (triple point)

- 8.3.2.6. 1,3-Dioxolan-2-one (triple point)
- 8.3.2.7. Succinonitrile (triple point)
- 8.3.2.8. Water (steam point)
- 8.3.2.9. Indium (triple point)
- 8.3.2.10. Tin (freezing point)
- 8.3.2.11. Cadmium (freezing point)
- 8.3.2.12. Lead (freezing point)
- 8.3.2.13. Zinc (freezing point)

8.4. Lower Accuracy Fixed Points
- 8.4.1. General considerations
- 8.4.2. Fixed points for application in dynamic conditions
- 8.4.3. Recommended points and temperature values, and available reference materials
  - 8.4.3.1. 4-Nitrotoluene
  - 8.4.3.2. Naphthalene
  - 8.4.3.3. Benzil
  - 8.4.3.4. Acetanilide
  - 8.4.3.5. Benzoic acid (triple point)
  - 8.4.3.6. Benzoic acid (melting point)
  - 8.4.3.7. Diphenylacetic acid
  - 8.4.3.8. Anisic acid
  - 8.4.3.9. 2-Chloroanthraquinone
  - 8.4.3.10. Carbazole
  - 8.4.3.11. Anthraquinone
  - 8.4.3.12. Mercury (boiling point)
  - 8.4.3.13. Reference materials for differential thermal analysis

8.5. Contributors

8.6. List of suppliers

8. Temperature

8.1. Introduction

8.1.1. Scope of the recommendation

This recommendation covers reference materials for realizing fixed points of temperature, and the methods for their use, in the range from about $-50\ °C$ to $400\ °C$.

A fixed point of temperature is provided by a very pure substance undergoing a constant temperature phase transformation in nearly equilibrium conditions with all quantities influencing the transformation maintained at standard values. The stability and reproducibility that are achieved in practice depend both on the purity of the substance and on the particular technique of realization. The closer the specimen is held to the ideal thermal equilibrium during the transformation, the more reproducible is the value of the fixed point. A recommendation of a reference material for a particular application must therefore include instructions for the realization of the specified temperature.

In broad terms, the use of reference materials for the measurement of temperature may be classified as follows:

(i) The realization of the defining fixed points to establish the internationally agreed temperature scales, *i.e.*, the *International Practical Temperature Scale of 1968*, IPTS-68 (Ref. 1), and the *Provisional Temperature Scale of 1976*, EPT-76 (Ref. 2);

(ii) the realization of reference temperatures for the verification of the calibration of precision thermometers;

(iii) the *in situ* calibration of temperature scales of measuring apparatuses, such as melting-point apparatuses, DTA apparatuses, differential calorimeters, etc.

The purpose of this recommendation is not to cover applications in the first class, nor, in general, the realization of temperature scales. Those who have an interest on this matter can profitably address themselves to references 1 to 4.

All temperatures assigned to fixed points in this recommendation are on the IPTS-68.

8.1.2. Fundamentals of temperature measurements

The unit of measurement of thermodynamic temperature, symbol T, is the kelvin, symbol K, defined as the fraction 1/273.16 of the thermodynamic temperature of the triple point of water. For historical reasons connected with the way temperature scales were originally defined, it is common practice to express a temperature in terms of its difference from that of a thermal state, the ice point, 0.01 K lower than the triple point of water, thus obtaining the Celsius temperature, symbol t, with the degree Celsius, symbol $°C$, as unit. There results:

$$t/°C = T/K - 273.15 \qquad (8.1.2.1).$$

The degree Celsius is, by definition, equal in magnitude to the kelvin. A difference of temperature may be expressed in kelvins or degrees Celsius.

The direct determination of thermodynamic temperature, starting from the definition of the kelvin and proceeding in agreement with the principles of thermodynamics, is a complex process, generally unsuitable to practical applications. The IPTS-68 (Ref. 1) solves this problem, being constructed in such a way that any temperature measured on it is a close approximation to the numerically corresponding thermodynamic temperature. Moreover, it entails measurements that are easily made and are highly reproducible. The temperature of the triple point of water (273.16 K) is unique in that it is, by definition, the thermodynamic temperature and the temperature on the IPTS. The temperature values of the defining fixed points of the IPTS are assigned by the General Conference of Weight and Measures on the grounds of the best available thermodynamic temperature values at the time the scale is promulgated, and must be considered exact by convention. These values may be changed when a new IPTS is promulgated.

The scale consists of a set of defining fixed points and of specified instruments with specified equations to interpolate between the fixed points. It is worth noting that a set of fixed points alone is not sufficient to define a scale, nor do additional fixed points, besides the defining fixed points, add anything to the scale. The properties of a practical scale are fully described in reference 4.

The IPTS-68 uses both the international practical Kelvin temperature, symbol T_{68}, and the international practical Celsius temperatue, t_{68}. They are related by equation 8.1.2.1, where t_{68} and T_{68} replace t and T, respectively. It covers the range for $T_{68} \geq 13.81$ K, using three different interpolating instruments, i.e., the platinum resistance thermometer from 13.81 to 903.89 K (630.74 °C), the platinum − 10% rhodium versus platinum thermocouple from 630.74 to 1064.43 °C, and above 1064.43 °C Planck's radiation law with a standard value for the second radiation constant c_2. Table 1 of reference 1 presents the eleven defining fixed points of IPTS-68 with their assigned temperatures.

The range from 0.5 to 30 K is covered by the EPT-76 (Ref. 2), which is defined in terms of eleven fixed points, together with the differences between the temperature thereby measured, T_{76}, and the following existing scales: IPTS-68; the ^{4}He-1958 and the ^{3}He-1962 vapour pressure scales (references 5 and 6, respectively); the NBS2-20 (1965) acoustic scale (Ref. 7); various magnetic scales; NPL-75 (Ref. 8) and the NBS version of IPTS-68 which is defined by difference from NBS-55. Thus, the EPT-76 can be realized in a number of ways: either by using one of the above scales and the corresponding table of differences given in the text of EPT-76, or by using a thermodynamic interpolating thermometer, such as a gas thermometer or a magnetic thermometer, calibrated at one or more of the specified points. The EPT-76 substantially differs from IPTS-68 between 13.81 K and 25 K, but the two scales join smoothly at 27 K. Until a new IPTS, that substantially includes EPT-76, is defined, T_{76} is recommended because it is the closest approximation to thermodynamic temperature.

8. Temperature

In general, this recommendation will not include materials suitable for the realization of EPT-76.

Reference data from the literature are sometimes expressed in terms of older temperature scales as, for instance, IPTS-48 and NBS-55. Reference 9 provides numerical differences $t_{68} - t_{48}$, from -183 to 631 °C, and $T_{68} - T_{\text{NBS-55}}$ from 13 to 90 K. Numerical differences from NBS-55 and from other low temperature scales ($T < 90$ K) are given in reference 10.

In the range of temperature considered in this recommendation, IPTS-68 prescribes the platinum resistance thermometer (PRT) as the interpolating standard. Its properties and recommended techniques of application are described in references 9 and 11. Long-stem PRTs, having a resistance of about 25 Ω at the ice point (273.15 K), yield a resolution better than the equivalent of 0.2 mK when used with a suitable resistance bridge generating a measuring current not exceeding 2 mA. The long term reproducibility is best specified in terms of the ratio between the resistance at the measured temperature and that at the ice point,

$$W(T_{68}) = R(T_{68})/R(0 \ °C) \qquad (8.1.2.2).$$

It is good practice to monitor the stability of $R(0°C)$ by making measurements at the triple point of water at regular intervals, particularly after a high temperature exposure, or when cold working causes a permanent shift in $R(0 \ °C)$. By following this procedure reproducibility better than ± 1 mK from -183 to 500 °C can be achieved.

Together with the uncertainty due to the limited reproducibility of the interpolating instrument, it is also necessary to consider the combined effect of uncertainties arising from unavoidable calibration errors at the fixed points and from the non-uniqueness of the scale. This non-uniqueness results from residual differences between the values obtained with different platinum resistance thermometers for temperatures between fixed points, even when systematic errors for the calibration at the fixed points as well as other errors in measuring the resistance have, ideally, been eliminated. Estimates based on one standard deviation are reported in reference 3. In general one can say that the largest uncertainty, amounting to ± 1.5 mK, is located at about -100 °C whereas the uncertainty does not exceed ± 0.5 mK between 0 and 500 °C. The uncertainty of a complete calibration at the fixed points of the IPTS-68, as carried out at the Istituto di Metrologia G. Colonnetti is estimated in reference 14. The resulting total uncertainty, determined as one standard deviation, is the combination of the non-uniqueness uncertainty, caused by the limited reproducibility of every particular fixed point, and the uncertainty of the determination of the resistance (Ref. 9). The one-standard-deviation uncertainty is plotted in Figure 8.1.2.1. As the capabilities of the Istituto di Metrologia G. Colonnetti are very close to those of other national laboratories for providing the realization of IPTS-68, it is unlikely that any determination of a secondary point, or of any other temperature on the IPTS-68, can be carried out at the present time with an uncertainty lower than reported in reference 14.

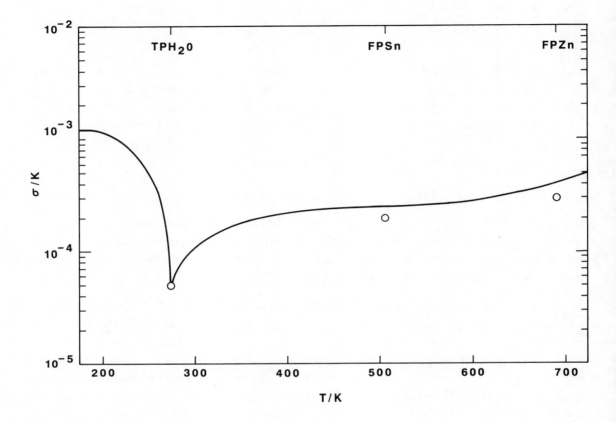

Figure 8.1.2.1: Total uncertainty (one standard deviation) of a realization of the IPTS-68 from 170 to 720 K (Ref. 14). The circles represent the reproducibility of the defining fixed points. The triple point of argon (83.798 K, not shown in the figure) is reproduced within ±0.2 mK (one-standard-deviation estimate). The uncertainty curve does not pass through tin and zinc point reproducibilities because the IPTS is defined in terms of resistance ratios with respect to the resistance at 0 °C, *i.e.* the resistance measured at the water triple point with a suitable correction. Therefore, the uncertainty of this point and that resulting from resistance changes caused by the high-temperature exposure add to the fixed point reproducibilities.

8.1.3. Secondary reference points of temperature

The IPTS-68 contains a list of Secondary Reference Points. These are given in table 6 of reference 1. Their temperatures are assigned on the grounds of the best available determinations. As a consequence, an uncertainty is associated with every temperature value, although it is not reported in IPTS-68. Hence, Secondary Reference Points have not the same rank as the Defining Fixed Points of IPTS-68. Secondary Reference Points are not substitutes for Defining Fixed Points, nor is it possible to obtain a piecewise realization of IPTS-68, by the use of sets of Secondary Reference Points.

With a view to extending the information provided by the text of IPTS-68, Working Group 2 (WG2) of the Consultative Committee for Thermometry (CCT) in its report of 1978 (Ref. 12), suggested the following possible applications of Secondary Reference Points:

(i) the realization of other units (*e.g.*, the candela through the melting point of platinum);
(ii) the internal control of a particular realization of IPTS, or the intercomparison of different realizations of IPTS;
(iii) the realization of a scale at a secondary level;
(iv) the calibration of secondary thermometers;
(v) the *in situ* calibration of instruments.

The term "secondary level scale" was introduced by the CCT to designate a scale covering partially, or totally, the range of the IPTS-68 but differing from it either in the chosen fixed points, or in the interpolating instruments, or both.

No secondary level scale, as suggested by (iii), can be regarded as a direct replacement of IPTS-68 or of part of it. The unavoidable systematic differences of a secondary level scale with respect to IPTS-68 can only be determined in national laboratories by direct comparison with a realization of IPTS-68. Hence, precise instructions for the correct use of a secondary level scale have to be made available by a national laboratory.

When Secondary Reference Points are used to calibrate thermometers, or used for the *in situ* calibration of instruments, they are not dissimilar from any comparison point, as obtained, for instance, with a standard platinum resistance thermometer and a precise temperature-controlled liquid bath. Any particular thermometer, or particular instrument, requires a minimum number of suitably spaced calibration points, depending on its principle of operation and constructional details. Secondary Reference Points can be used for one or more, or even all, calibration points. It is worth noting, however, that the use of a Secondary Reference Point, no matter how accurately it is known, will not reduce the minimum number of calibration points. Moreover, the calibration uncertainty depends on the uncertainties of all calibration points, so it would be a waste of time and effort to introduce, for instance, a very accurate Secondary Reference Point in a set of less accurate points. Lastly, the reproducibility of the thermometer to be calibrated greatly influences the final result.

A sixth application may be added to those envisaged by the WG2 of CCT, *viz*, the verification of the calibration of a thermometer, or of an instrument. For instance, the calibration of a platinum resistance thermometer may be checked at the melting point of ice (0 °C±0.002 °C) when the accuracy required is not greater than that with which the ice point can be realized. Similarly, but at a lower level of accuracy, the verification of the calibration of a melting-point apparatus is carried out with suitable reference materials (RMs) for melting temperatures. In general, when a significant deviation from the instrument calibration is detected either the instrument must be recalibrated, or the measurement procedure, including the operator's skill, must be examined and improved.

Table 6 of IPTS-68 contains only those Secondary Reference Points that can be realized with the highest reproducibility according to a well established technique of realization. Two or more independent determinations are in most cases taken into account to obtain a transition temperature value for an Secondary Reference Point. Further information on these Secondary Reference Points and on others not included in the Table 6 has been published on behalf of CCT in reference 13. This reference provides references to the original experimental works.

In accordance with the criteria proposed by WG2 of the CCT, the present recommendation reports only Secondary Reference Points for which up to date accounts of the techniques for their realization are available in the literature. In particular, it has been a condition for inclusion here that the minimum purity of the substance necessary to achieve the quoted accuracy be specified. Secondary Reference Points with more than one independent determination have been preferred.

In this monograph reference points of temperature are divided into two groups, each comprising points of comparable accuracy. They are intended to accommodate different needs. High accuracy fixed points are either those included in IPTS-68 or others of comparable accuracy and of special interest for physico-chemical measurements. Lower accuracy fixed points are those that, generally speaking, are realized with an uncertainty greater than ±20 mK. They are either melting/freezing points, or triple points, of organic substances.

8.1.4. The role of reference materials

The realization of fixed points of temperature at a high level of reproducibility can be carried out only on samples of high purity materials. This conditions is not, however, sufficient to assure the reproducibility of the point. The success of a realization depends on the purity of the substance, on the correctness of the procedure for introducing the substance into the measuring cell, and on the technique for realization of the fixed point. For instance, to realize the triple point of water it is necessary, to produce high purity distilled water, starting from natural water. Then, the water must be outgassed, transferred into a carefully cleaned cell, and sealed. The water in the sealed cell can thus be considered as a reference material. To realize the triple point, however, it is necessary to use a suitable technique to form the ice mantle and to obtain a good coupling between the thermometer and the solid-liquid interface. It is also necessary to isolate the cell from the influence of the environment, as, for instance, the effects of solar radiation, and so on. When all the necessary precautions are taken it is possible to achieve a reproducibility of ±0.1 mK, or better.

Even when a very pure substance is readily available from a qualified source, it may not be useful for generating a high accuracy reference temperature in a laboratory. For example, carbon dioxide, as bottled gas, no matter how pure, cannot be used as a RM at its triple point (−56.570 °C). The carbon dioxide only becomes a RM when it has been charged under controlled conditions into a suitable sealed cell usable for temperature experiments. Then any good experimenter, if equipped with sufficient instructions, can reproduce the transition temperature to within ±5 mK.

At a lower level of accuracy, such as that required for checking instruments like melting-point apparatus, pure substances can meaningfully be used as RMs. For achieving a reproducibility no closer than ±10 mK, several pure substances with well defined melting/freezing points are available. They can be introduced in a suitable amount into the apparatus under verification and removed once the operation has been completed. In these cases the initial purity is the prime concern.

From the examples given above we can conclude that there are three types of RMs for temperature measurements, namely:

(i) pure substances (solids, liquids and gases at room temperature) that are necessary to realize fixed points. They constitute the base material, and to obtain the final result both careful procedures and a good experimental technique must also be applied;

(ii) simple devices, *e.g.*, sealed cells, that exhibit a high reproducibility and constancy with time. They still require a careful experimental procedure, but the procedure can be well specified;

(iii) pure substances from which expendable samples may be extracted to verify the calibration of instruments with an *in situ* technique with an uncertainty to within ±20 mK in the best case.

The three volumes of Temperature, its Measurement and Control in Science and Industry (Refs. 15 to 17) are particularily useful for specific applications. Reference 18 provides useful information for the practice of precision thermometry.

REFERENCES

1. The International Practical Temperature Scale of 1968, Amended edition of 1975, Bureau International de Poids et Measures: Sèrves, **1975**. English version in *Metrologia* **1976**, *12*, 7.
2. The 1976 Provisional 0.5 K to 30 K Temperature Scale, *Metrologia* **1979**, *15*, 65.
3. Supplementary Information of the IPTS-68 and the EPT-76, Bureau International des Poids et Mesures: Sèvres, **1983**.
4. Quinn, T. J. *Temperature* Academic Press: London, **1983**.

5. Brickwedde, H.; van Dijk, H.; Durieux, M.; Clement, J. R.; Logan, J. K. *J. Res. Nat. Bur. Stand.* **1960**, *64A*, 1.
6. Sydoriak, S. G.; Roberts, T. R.; Sherman, R. H. *J. Res. Nat. Bur. Stand.* **1964**, *68A*, 559.
7. Plumb; H. H.; Cataland, G. *Metrologia* **1966**, *2*, 127.
8. Berry, K. H. *Metrologia* **1979** *15*, 89.
9. Riddle, J. L.; Furukawa, G. T.; Plumb, H. H. *Platinum Resistance Thermometry*, National Bureau of Standards Monograph 126, U.S. Department of Commerce: Washington D.C., **1973**.
10. Bedford, R. E.; Durieux, M.; Muijlwijk, R.; Barber, C. R. *Metrologia* **1969**, *5*, 47.
11. Crovini, L. *High Temperatures - High Pressures* **1979**, *11*, 151.
12. CIPM, *Comité Consultatif de Thermométrie*, 12 th Session, Bureau International des Poids et Mesures: Sèvres, T59, **1978**.
13. Crovini, L.; Bedford, R. E.; Moser, A. *Metrologia* **1977** *13*, 197
14. Crovini, L. *Standards of Temperature at IMGC: Their Realization, Dissemination, and Intercomparison*, Istituto di Metrologia G. Colonnetti Internal Report S/230, **1984**.
15. *Temperature, its Measurement and Control in Science and Industry*, Wolfe, H. C., editor, Reinhold: New York, **1954**.
16. *Temperature, its Measurement and Control in Science and Industry*, Herzfeld, H. C., Reinhold: New York, **1962**.
17. *Temperature, its Measurement and Control in Science and Industry*, Schooley, J. F., editor, American Institute of Physics: New York, **1982**.
18. Nicholas, J. V.; White, D. R. *Traceable Temperatures*, DSIR Bulletin 234, New Zealand Department of Scientific and Industrial Research: Wellington N. Z., **1982**.

8.2. High Accuracy Fixed Points

8.2.1. General considerations

High accuracy fixed points should have a reproducibility better than ±2 mK. Therefore they must be realized with substances of very high purity and stability. Methods of thermal analysis are in most cases applied to check the purity when the substance has been introduced into the device for realizing the fixed point.

The phase equilibria that can be established with such a high level of reproducibility include the triple point, the freezing/melting point and the equilibria between liquid and vapour phases and between solid and vapour phases of a pure substance. Other phase transitions, such as solid-solid transformations, eutectic points and magnetic transitions, do not appear at present to have an adequate reproducibility.

Ideally, the temperature assigned to a fixed point is that at a still interface between different phases of an ideally pure substance. In practice, the temperature sensing element cannot be placed directly on to the interface. Thus it is necessary to ensure that there is a sufficiently large isothermal zone to accommodate the thermometer without causing an appreciable heat flow between its sensing element and the surroundings. One way to reach this goal is to maintain the cell with the pure substance as far as possible isolated from the surroundings, except for a small but controlled source (or sink) of heat, by which the substance is caused to reach the transition point. When the transition temperature has been reached, the whole cell is allowed to reach equilibrium. Thus, the measuring thermometer, being thermally linked to the cell and almost totally isolated from the surroundings, reaches the temperature of the interface. This "calorimetric technique" is very often used at low temperatures with a relatively small amount of substance, usually less than 0.1 mol (Refs. 1 and 2).

Alternatively, the cell is immersed in a very uniform medium at a temperature near that of the fixed point. The substance then undergoes a slow transformation. It is necessary to allow heat exchange between the cell and the surroundings without introducing significant temperature gradients between the thermometer and the phase boundary. Hence, much larger cells than in the previous case, with much larger amounts of substance, are required in order to provide a sufficiently large isothermal zone around the thermometer. This "dynamic technique" yields the best results when applied to the realization of melting, freezing, and triple points of substances of high thermal conductivity and high thermal diffusivity. When the substance has a poor thermal conductivity, the temperature of the surrounding medium must be kept as close as possible to the fixed point, so as to minimise the heat flow.

Liquid-vapour equilibria ("boiling points") are established by promoting a continuous evaporation/condensation cycle so that a steady state is reached.

8.2.2. Triple points

At the triple point of a pure substance three phases co-exist in equilibrium. According to the phase rule this point is invariant, and the pressure and temperature are fixed. Such a useful property has attracted thermometrists since the beginning of the development of the International Temperature Scale because it is very convenient to keep a pure substance sealed in a suitable measuring cell, fully isolated from any possible sources of contamination. In general, triple points entail the following major advantages:

(i) sophisticated equipment is not required to realize triple points;

(ii) there is no possibility of external impurities contaminating the pure material sealed in the cell;

(iii) triple-point cells can easily be sent from one laboratory t another (*e.g.*, Schwab and Wichers of the National Bureau of Standards commented, "for the comparison of the scales maintained in different countries ... there is an advantage in using identical specimens of a given substance ... Sealed triple-point cells, which can easily be transported from one laboratory to another provide this advantage." (Ref. 3).

A thorough discussion of properties of triple points can be found, for instance, in reference 4. Figure 8.2.2.1 presents a cell for the realization. Following the first introduction of the triple point of water as a reference (General Conference of Weight and Measures in 1948, see references 5 and 6 for some accounts on the introduction of this fixed point), thermometrists have developed apparatus for a number of other triple points. For the sake of convenience, they can be subdivided into three categories:

(i) low temperature triple points of substances that are gaseous at room temperature;

(ii) triple points of metals;

(iii) triple points of organic substances.

All triple points of the first category except that of carbon dioxide (-56.570 °C), which has not yet been realized with a reproducibility better than 5 mK, are below the range of temperature of this recommendation. Nevertheless, it is worth while describing the technique for their realization because it is potentially applicable to other substances in other temperature ranges, such as organic substances and pure metals between -60 °C and 150 °C. Moreover, it is possible that the reproducibility of the carbon dioxide triple point may be improved in the near future and its use recommended as a high accuracy fixed point. The technique is very similar to the calorimetric procedure described by Glasgow *et al.* (Ref. 7) in which energy is added to the isolated system in pulses (whereas the points classified under (ii) and (iii) above are realized in near isothermal conditions with steady addition or abstraction of heat).

Low temperature triple points are realized in either oxygen-free high conductivity copper or stainless steel cells. The cells must be carefully outgassed and evacuated before the pure gas is introduced. There are basically two types of apparatus. The first type consists of a small measuring cell having a well to accommodate the thermometer to be calibrated and connected to a large-volume reservoir, as described, for instance, by Bonnier (Ref. 1). Alternatively, it can be connected to an external filling device, *via* thin-walled stainless steel tubing, as described by Pavese and Ferri (Ref. 2). When the cell is at room temperature, the gas is either almost totally expanded into the reservoir, preventing an excessive increase of the pressure, or it is released through the filling apparatus. In the latter case, fresh gas is introduced when the experiment is repeated.

Alternatively, a small cell can be charged with gas under pressure at room temperature. The cell volume should not ordinarily exceed 10 cm^3 and contain not more than 0.2 mole of gas. The cell body can be made of small thermal capacity but still sufficiently robust to withstand the charging pressure. After the cell has been evacuated and carefully outgassed, the pure gas is introduced and permanently sealed in. Such a device is often referred to as a "sealed cell", although this term is applicable to any practical realization of a triple point. The cell is provided with an electrical heater, usually coiled onto the external cylindrical surface. It is also equipped with one or more pockets, to accommodate one or more thermometers. A complete description of these cells can be found in reference 2.

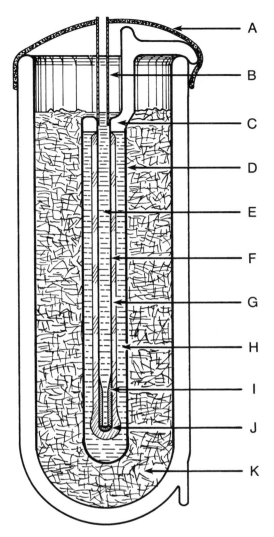

Figure 8.2.2.1 Water triple-point cell (Refs. 8, 18).
  A. Heavy black felt shield against ambient radiation.
  B. Polyethylene tube for guiding the SPRT into the thermometer well.
  C. Water vapor.
  D. Borosilicate glass cell.
  E. Water from ice bath.
  F. Thermometer well (precision bore).
  G. Ice mantle.
  H. Air-free water.
  I. Aluminum bushing with internal taper at upper end to guide the SPRT into the close-fitting inner bore.
  J. Polyurethane sponge.
  K. Finely divided ice and water.

Both types of cells are placed in a cryostat and cooled well below the triple-point temperature, so as to freeze all material contained therein. The cell is then isolated from the surroundings and pulses of heat are applied through the electrical heater. Sufficient time is allowed between two successive pulses so that the cell can return to equilibrium conditions. When the triple point has been reached each pulse causes a small fraction of the material to melt. Equal pulses melt approximately equal fractions of the substance. The melting temperature versus time curve can be recorded by taking readings of the thermometer between successive pulses. This curve spans a small range of temperature as a consequence of residual impurities in the substance and small thermal effects such as the very small residual heat flow from the surroundings which give rise to a small, but appreciable, rounding at the end of melting.

To a first approximation, the mass fraction melted is directly proportional to the heat that has been supplied through pulses since the beginning of the transformation. The fraction melted, F, is thus defined accordingly,

$$F = \frac{\text{mass fraction of specimen melted}}{\text{total mass of specimen}} = \frac{\text{heat supplied}}{\text{total heat of fusion}} \quad (8.2.2.1).$$

It is common practice in precision thermometry to plot the measured temperature versus $1/F$, fit a straight line to the plotted points, and extrapolate to $1/F = 1$ to obtain the liquidus point. It is also recommended (Ref. 8) that only the central part of the curve should be used, 15% at the beginning of melting and 15% at the end being ignored. Some accounts on the theoretical background of this technique are provided by Furukawa (Ref. 9) and by Cox and Vaughan (Ref. 10).

The temperature obtained by this method is affected by impurities. It has been suggested by Pavese (Ref. 11) that in some instances, particularly with slightly impure gases, it is advantageous to extrapolate to $1/F = 0$. It has been shown (see references 9, 11, 12, 13) that the temperature obtained in this way, if the conditions are correct, can be attributed to the ideally pure substance. Results for organic gases, such as methane, appear to support this method.

Triple points of metals are generally realized by use of a much larger amount of substance than is used in low-temperature triple-point cells. Typically, from 0.1 to 2 kg of material is used in a suitable outer container which is either permanently sealed under vacuum or capable of being evacuated. Inside this the metal may be contained in a graphite or PTFE (or similar) liner with a re-entrant tube to allow the insertion of the thermometer. The cell may be evacuated by means of a double-stage mechanical pump equipped with a nitrogen-cooled trap. Because of the relatively low pressure coefficients of freezing temperatures (ranging from -3.3×10^{-8} K Pa^{-1}, for bismuth, to 8.0×10^{-8} K Pa^{-1}, for lead) outgassing is not required. A residual pressure below 10 Pa will not affect the triple-point temperature to any appreciable extent. Oxidation, as occurs with indium, can be minimized by either applying a high vacuum or, preferably, by repeatedly filling the cell with pure argon and then pumping it away. A prolonged pumping with the cell above the melting temperature is always recommended before sealing off the apparatus. Such a technique, when applied to points below 450 °C, is sufficient to eliminate any appreciable amount of dissolved oxygen.

The technique of realization of the triple point of metals does not substantially differ from that of the corresponding freezing or melting point at standard pressure. In the temperature range of interest, with very few exceptions (for example gallium), a freezing technique is applied in order to obtain the triple-point temperature. The nucleation of a solid in the supercooled liquid phase is artificially promoted through the application of suitable techniques, generally referred to as induced techniques. They must lead to the formation of an evenly distributed thin solid mantle on the surface of the container and, particularly, on that of the thermometer well. The relatively high thermal conductivity and thermal diffusivity of metals, as compared to other substances, greatly help to meet these requirements. They are essential to ensure that the solid-liquid interface completely surrounds the thermometer well and quickly to achieve steady state conditions. Various techniques of induced nucleation are described in references 8 and 14.

Subsequently to the formation of the mantle, the cell is placed in a liquid bath, or a furnace, where a temperature slightly below that of the triple point is kept uniform and constant: if it is extremely uniform and constant it may be set very near to the triple-point temperature, so as to maintain the triple point condition inside the cell almost indefinitely by thermally isolating the cell from the colder environment. In the latter case, the sample reaches a stationary condition with a minimum amount of heat dissipated.

The realization of triple points by melting is obtained by immersing the cell with the frozen sample in an environment at a uniform temperature just above the triple point. For the gallium triple point, for instance, it is recommended that a stirred-oil bath at about 10 mK above the triple-point temperature should be used. It is also recommended that melting should be started along the outside of the sample and along the thermometer well, either by circulating oil at 40 °C, or by a suitable electrical heater (Ref. 15). These techniques are used to melt 25% of the sample so that the thermometer-sensing element is surrounded with a solid-liquid interface. With this technique the measured temperature is close to the lowest value in the melting range. No appreciable irreproducibility will result with samples of mass-fraction purity over 99.9999%.

For metals, the freezing point is generally preferred because it is initiated from the homogeneous liquid. Determination of the melting point is not recommended because of the possibility of non-equilibrium segregation of impurities during the previous freeze (Ref. 16).

To determine melting, freezing or triple-point temperatures of metals with reproducibilities of at least 1 mK, it is necessary to use samples with mass-fraction purity of 99.999% or better. Such a high degree of purity is rare for organic compounds and the triple points of only a few, such as phenoxybenzene and benzoic acid, are reproducible to better than ± 2 mK. Organic substances are generally contained in borosilicate-glass cells, with shapes and sizes similar to that of the water triple-point cell.

The procedure for setting up a triple-point, which is commonly applied to water cells and can also be used for cells containing organic liquids is as follows:

(i) the substance is heated to about 30 K above the melting point and where possible the liquid is throughly mixed to ensure homogeneity.

(ii) it is then cooled close to or just below the triple-point temperature and maintained in an environment at a sufficiently uniform temperature for the liquid to remain in the undercooled state.

(iii) strong cooling is then applied, either from outside the crucible or from inside the thermometer well, by circulation of a cool liquid, the insertion of a metal rod cooled with liquid nitrogen, or by evaporation of carbon dioxide, (see the procedure for the water triple point, Ref. 17). No matter what procedure is used, a uniform solid mantle should be generated along the cell. When this operation has been completed, and a mantle of solid has been formed extending approximately one third of the way to the outer wall of the cell, it is necessary to insert into the well a liquid, whose temperature is higher than the triple-point temperature, so that a thin sleeve of liquid melts around the well. This results in a solid-liquid interface almost totally surrounding the thermometer well, which reaches an equilibrium temperature, very close to the liquidus point of the substance, the temperature at which a very small amount of solid is in equilibrium with the liquid phase. For a detailed description of this technique, and other alternative techniques, the article of Cox and Vaughan (Ref. 10) should be consulted.

(iv) the cell should then be placed in a stirred liquid bath or some other suitable environment providing a uniform temperature very close to the triple-point temperature. The time required for equilibrium and a constant temperature in the cell to be reached will vary from substance to substance. A suitable liquid should be placed in the well to act as a heat-exchanging medium.

The depression of the equilibrium temperature between solid and liquid phases due to an impurity, $T_e - T_{ei}$ is given by equation 8.2.2.2, if it is assumed that no solid solution is formed:

$$T_e - T_{ei} = x_i / (\Delta_{fus} H R T_e^2) F \tag{8.2.2.2}$$

where x_i is the mole fraction of all liquid-soluble impurities, $\Delta_{fus}H$ is the molar enthalpy of fusion and R is the gas constant. However, when a batch of substance having impurities in the parts-per-million range is used, it is likely that the residual impurities will have physicochemical properties close to those of the host. In such cases, solid solutions may form with the host. For solid solution formation the following equation should be used:

$$T_e - T_{ei} = \frac{x_i}{A} \cdot \frac{1}{F - 1/(1 - K)} \tag{8.2.2.3},$$

where A equals $\Delta_{fus} H / R T_e^2$ and K is the distribution coefficient for the impurities between liquid and solid phases. When no solid solution forms K tends to infinity. When the concentration of the impurity is the same in both the solid and liquid phase $K = 1$. In this case the impurity has no influence on the melting temperature. When $K < 1$ the solid phase is is richer in impurities than the liquid, and, theoretically, the temperature decreases during the melting process.

When solid-soluble impurities are present, it is necessary to know the distribution coefficient K and different impurities have different K values. The introduction of a unique K, as determined by a least-squares adjustment of equation 8.2.2.3 to the experimental values, is an approximation. It is also possible that a previous freezing process may have caused an impurity distribution. McLaren (Ref. 16) has shown that this condition can occur in metals.

In general equation 8.2.2.2 is satisfactory for describing melting or freezing. With highly purified substances the experimenter needs to decide whether to use equation 8.2.2.2 or equation 8.2.2.3. When equation 8.2.2.3 is used, a value of K has to be determined.

8.2.2.1. Hydrostatic pressure correction for triple points.

The triple points of water, of metals, and of organic substances are realized in cells of sufficiently large size to require a correction for the hydrostatic pressure. Indeed, the triple-point temperature is in principle generated only at the line where the three phases meet. The measuring thermometer is usually placed at some distance below this line, and is thus subjected to a slightly different temperature corresponding to a solid-liquid equilibrium at a slightly higher pressure than that of the triple point. This overpressure is equal to the hydrostatic pressure of the column of liquid from the vapour-liquid surface to the mid-point of the thermometer sensing element. Generally, the effect of pressure on a freezing point (liquid-solid equilibrium) can be calculated from the Clausius-Clapeyron equation:

$$\frac{dT_{FP}}{dp} = \frac{T_{FP}(V_l - V_s)}{\Delta_{fus} H} \qquad (8.2.2.4),$$

where V_l and V_s are, respectively, the molar volumes of pure liquid and solid at T_{FP}, and $\Delta_{fus} H$ is the molar enthalpy of fusion of the pure substance. For practical purposes the Clausius-Clapeyron equation is linear from the standard pressure to very low pressure. Thus, equation 8.2.2.4 can also be used to estimate the hydrostatic pressure effect in a triple-point cell,

$$\frac{dT_{TP}}{dh} = \frac{T_{TP} M g}{\Delta_{fus} H} \frac{(1 - V_s)}{V_l} \qquad (8.2.2.5)$$

where M is the molar mass, g the acceleration of gravity and h is the depth of the thermometer below the surface. Values in the sixth column of table 8.2.2.1, calculated by means of equation (8.2.2.5), allow triple-point measurements to be corrected for the depth of immersion of the thermometer.

8.2.3. Freezing and melting points

Freezing and melting temperatures at atmospheric pressure are realized with techniques that are very similar to those used for triple-point determinations.

For high accuracy, air is removed from contact with the substance and replaced with either pure argon or pure nitrogen. This should always be done for high-purity metals. In principle, all melting and freezing points are affected by dissolved gases, but no appreciable change in the freezing point has ever been detected with metals when either argon or nitrogen is used as the prevailing atmosphere. Information on the effect of different dissolved gases is difficult to obtain. The effects of dissolved gases in water and phenoxybenzene are given in table 8.2.2.1. The solubilities of gases in organic compounds are reported in reference 4.

Table 8.2.2.1. Pressure dependence of triple point and freezing temperatures of some pure metals and pure compounds.

Substance	t_{TP} °C	t_{FP} °C	dT_{TP}/dp K Pa^{-1}	dT_{FP}/dp K Pa^{-1}	dT_{TP}/dh mK m^{-1}	z K
Hg[†]	−38.842	−38.836	5.4×10^{-8}		7.1	**
H$_2$O[†]	0.01	0.0*	-7.5×10^{-8}		−0.73	0.0024[‡]
Phenoxybenzene (diphenyl ether)	26.869	26.898*	0.26×10^{-6}		2.6	0.022[§]
Ga[†]	29.774	29.772	-2×10^{-8}		−1.2	**
1,3-Dioxolan-2-one	36.324		0.15×10^{-6}		2.4	
Succinonitrile	58.080		0.34×10^{-6}		3.3	
Benzoic Acid	122.370	122.383	-0.38×10^{-6}		−4.0	
In[†]	156.629$_6$	156.634		4.9×10^{-8}	3.3	**
Sn[†]		231.9681		3.3×10^{-8}	2.2	**
Cd[†]		321.108		6.2×10^{-8}	4.8	**
Pb[†]		327.502		8.0×10^{-8}	8.2	**
Zn[†]		419.58		4.3×10^{-8}	2.7	**

* Saturated with air
** In argon atmosphere
[‡] See reference 4
[†] See reference 8
[§] See also reference 18 for a detailed evaluation of the effects of various dissolved gases
t_{TP} triple-point temperature
t_{FP} freezing-point temperature
z temperature depression for dissolved air
h depth below liquid-vapour surface

Accurate pressure measurements are not required, since the pressure dependence of transition is generally small (see table 8.2.2.1). With suitable freezing/melting-point cells, the pressure above the liquid can be adjusted so as to be close to one standard atmosphere (101 325 Pa). For different pressures small corrections can be made.

8. Temperature

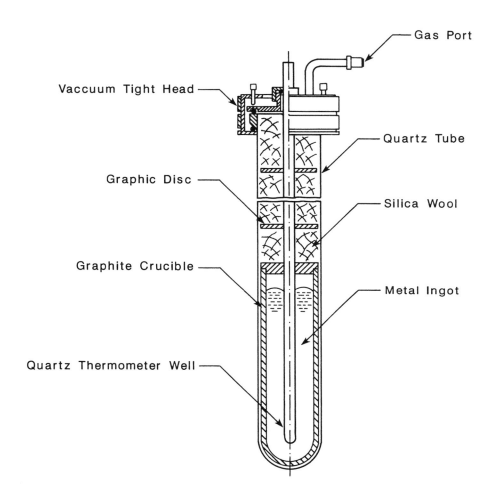

Figure 8.2.3.1: Metal freezing point apparatus for resistance and thermocouple thermometry (Refs. 8, 20).

Very accurate realizations of the freezing points of metals are described in references 8 and 16. Figure 8.2.3.1 presents a typical cell. The accurate realization of the melting point of gallium is described in reference 15. Other realizations of melting points using a calorimetric technique have been recently described by Ancsin for water and indium (Refs. 18, 19). The agreement of the freezing point of a particular sample of metal with the melting point is sometimes taken as an index of the quality of the practical realization of both points. Otherwise, the presence of incomplete solid mantles resulting from improperly induced freezes can be detected by applying temperature oscillations. Typically, the oscillation period should be about 30 minutes with the peak-to-peak amplitude about 1 K. Resulting small temperature oscillations during either a freezing or a melting plateau constitute clear evidence of the presence of an incomplete solid mantle (Ref. 20).

Another convenient secondary reference point is the melting point of ice. Its accurate realization is described in reference 8. Freezing and melting points of organic compounds are generally not realizable as high level fixed points. They are, however, useful for either lower-accuracy reference points or for purity control.

8.2.3.1. Purity control

Research methods of determining purity by the study of freezing curves depend on the change of temperature occurring as solidification progresses, while in some commercial applications the actual freezing temperature of a sample is used as the criterion of purity. The term frequently used in the latter context is the crystallizing point of the sample. For the determination of the crystallizing point, the most satisfactory reference material is a pure sample of the material being studied. The reference sample and the sample under test may then be examined in an identical way and calibration errors, from whatever cause, will tend to cancel (Ref. 21). This technique is mainly used for organic compounds.

For metals, the melting plateau provides a great deal of information on the content and distribution of impurities in the specimen. Chemical analysis, although very useful in the initial selection of a high-purity material, does not allow the calculation of the effect of impurities on the freezing temperature. This is because of lack of details in the phase diagrams at extreme dilutions as well as incomplete information on impurities present in the specimen when it has been transferred to the crucible. The melting curve, not being distorted by the effects of supercooling and nucleation, allows estimation of the total impurity on a comparative basis. Meaningful comparisons can be obtained only by melting ingots that have been previously frozen in the same way. Particular care must be taken to ensure the same type of impurity segregation occurs as in the preceding freeze. When operating under standard melting conditions (*i.e.*, same impurity segregation, furnace or liquid bath temperature, same total melting time and total melting range), melting curves are best analyzed through their "melting histogram". This histogram consists of the percentage of the total melting time spent at each constant fraction of the total melting range (for instance, at every 0.2 mK interval within the melting range) plotted against the mean temperature of the interval. As a first approximation, the narrower the histogram, the purer the material. Several examples of the application of this technique are given by McLaren (Ref. 16).

8.2.4. Liquid-vapour equilibria

A constant temperature can be realized by maintaining a pure liquid at a constant pressure in the presence of its vapour. Generally, the pressure is set to a value close to one standard atmosphere.

Liquid-vapour equilibria of liquefied gases are generally realized by the static method. In the static method one confines a sample of the substance in a vessel attached to a manometer. The pressure is measured when the sample is maintained at a constant temperature with both liquid and vapour phases present. In one realization, described by Kemp and Kemp (Ref. 22), the vessel consists of a small-volume cell within a massive, high-conductivity copper block placed inside a cryostat. The temperature is controlled by adjusting the flow of either liquid helium or nitrogen. A capillary tube connecting the cell to the pressure-measuring system is devised in such a way as to separate the vapour from the gas in the pressure-measuring device. The thermometers to be calibrated are placed in pockets surrounding the cell.

At temperatures above 0 °C dynamic methods are preferred, although static methods can be used. The dynamic method (sometimes referred to as the ebulliometric method), entails the continuous circulation of vapour and liquid inside the vessel. In the steam-point apparatus, for instance, the vapour rises and condenses on the colder parts, raising the temperature to the condensation temperature. Under steady state reflux conditions the temperature is unaffected by any change in the energy supplied. An increase in the energy applied merely increases the boil-up rate and more energy is removed by the condenser. Above a necessary minimum for satisfactory operation, the amount of energy supplied is not critical, but further increase eventually causes the apparatus to become overloaded and flooding occurs at the condenser. The essential features of the apparatus are:

(i) a heater designed to promote boiling without bumping,

(ii) adequate immersion for the thermometer. The minimum depth, for example, of a glass sheathed resistance thermometer that does not exceed 1 cm in diameter is 30 cm;

(iii) the use of shields to protect the thermometer from radiation. The higher the temperature the more important is the use of radiation shields. Two shields may be necessary at the mercury boiling point (630 K), whereas only one shield may be sufficient at the steam point (373 K). Shields prevent radiative gains from the heater and losses to the walls. Without shields, measurements at the sulphur boiling point (718 K) can be as much as 1 K low (Ref. 23),

(iv) a condenser and a means of detecting the level of the condensing vapour, such as a thermocouple in the condensing tube;

(v) even return of the refluxing liquid, usually as a steady stream down the outer walls.

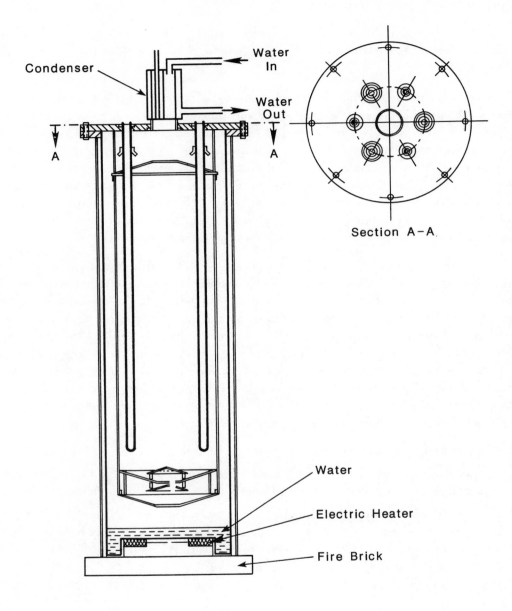

Figure 8.2.4.1: Apparatus for the realization of the boiling point of water (Ref. 8).

8. Temperature

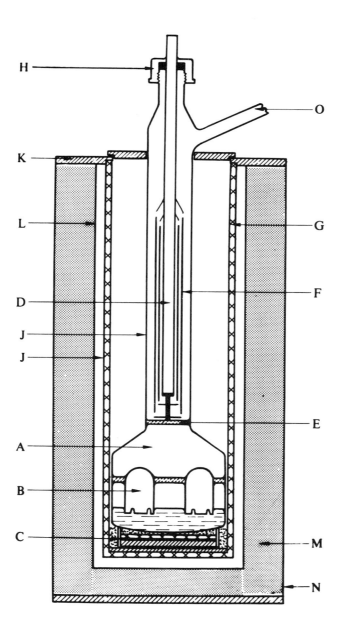

Figure 8.2.4.2: Schematic drawing of a simplified boiler (Refs. 24, 25). A, glass boiler (85 mm outside diameter); B, bubble cap; C, heating element; D, stainless steel thermometer pocket; E, glass rod to take thrust of D when apparatus is under vacuum; F, stainless steel radiation shields; G, heated jacket; H, screw-cap seal; J, positions of differential thermocouples; K, lid; L, metal canister; M, lagging; N, outer casing; O, side arm connecting with pressure main.

Further information on the realization of boiling points, particularly the steam point, is provided in reference 8. Figure 8.2.4.1, from the same reference, shows a boiler for the realization of the steam point. A simpler apparatus may be made more cheaply in glass, and a suitable design that will accept a single thermometer is shown in figure 8.2.4.2 (Refs. 24, 25).

8.2.4.1. Pressure measurements

The variation of the boiling point at atmospheric pressure with pressure varies from a minimum of 0.28 mK Pa^{-1}, for the steam point, to 0.68 mK Pa^{-1} for sulphur. Mercury has a value of 0.55 mK Pa^{-1}. Thus very accurate pressure measurements are needed to reproduce boiling points to within a few millikelvins. For example, an uncertainty of ± 1 mK in the steam point corresponds to an uncertainty of ± 3.6 Pa in a total pressure of 101 325 Pa. This corresponds to an uncertainty of $\pm 0.004\%$.

Pressures are usually measured with a mercury manometer or a dead-weight gauge. In high-accuracy measurements, and always when the static method is used, a very sensitive differential pressure transducer is interposed between the vessel and the measuring manometer. The differential pressure transducer acts as a null-detector, and the pressure of the controlling gas is measured on the manometer.

The difference in level between the point of pressure measurement and the vapour-liquid interface (or condensing interface in the dynamic method) must be determined, and the correction for the resulting hydrostatic head must be applied. This correction is particularly significant at low temperatures when the measuring cell is connected to the manometer through a long capillary tube containing gas and vapour.

Further details of methods for the accurate measurement of pressure in the determination of normal boiling points (*i.e.* at 101 325 Pa) can be found in references 8 and 26.

REFERENCES

1. Bonnier, G.; Moser, A. *Measurement* **1983**, *1*, 143.
2. Pavese, F.; Ferri, D. *Temperature, its Measurement and Control in Science and Industry* **1982**, *5*, 217.
3. Schwab, F. W.; Wichers, F. *Compte Rendus de la Quinziéme Conférence*, International Union of Pure and Applied Chemistry, 113, **1949**.
4. Vaughan, M. F. *NPL Report Chem. 86*, National Physical Laboratory: Teddington, **1978**.
5. Stimson, H. F. *Temperature, its Measurement and Control in Science and Industry* **1955**, *2*, 141.
6. Beattie, J. A.; Huang, Tzu-Ching; Benedict, M. *Proc. Am. Acad. Arts Sci.* **1938**, *72*, 137.
7. Glasgow, Jr., A. R.; Ross, G. S.; Horton, A. T.; Enagonio, D.; Dixon, H. D.; Saylor, C. P.; Furukawa, G. T.; Reilly, M. L.; Henning, J. M. *Anal. Chim. Acta* **1957**, *17*, 54.

8. Supplementary Information of the IPTS-68 and the EPT-76, Bureau International des Poids et Mesures: Sèvres, **1983**.
9. Furukawa, G. T. *Temperature, its Measurement and Control in Science and Industry* **1982**, *5*, 239.
10. Cox, J. D.; Vaughan, M. F. *Temperature, its Measurement and Control in Science and Industry* **1982**, *5*, 267.
11. Pavese, F. *Metrologia* **1979**, *15*, 47.
12. Pilcher, G. *Anal. Chim Acta* **1957**, *17*, 144.
13. Westrum, Jr., E. F.; Furukawa, G. T.; McCullogh, J. P. *Experimental Themodynamics*, McCullogh, J. P.; Scott, D. W., editors, Butterworth: London, **1968**, p. 333.
14. Furukawa, G. T.; Riddle, J. L.; Bigge, W. R.; Pfeiffer, E. R. *Standard Reference Materials: Application of some Metal SRM's as Thermometric Fixed Points*, National Bureau of Bureau of Standards Special Publication 260-77, U.S. Department of Commerce: Washington, **1982**.
15. Chattle, M. V.; Rusby, R. L.; Bonnier, G.; Moser, A.; Renoat, E.; Marcarino, P.; Bongiovanni, G.; Frassineti, G. *Temperature, its Measurement and Control in Science and Industry* **1982**, *5*, 311.
16. McLaren, E. H. *Temperature, its Measurement and Control in Science and Industry* **1962**, *3*, 185.
17. Riddle, J. L.; Furukawa, G. T.; Plumb, H. H. *Platinum Resistance Thermometry*, National Bureau of Standards Monograph 126, U.S. Department of Commerce: Washington, **1973**.
18. Ancsin, J. *Temperature, its Measurement and Control in Science and Industry* **1982**, *5*, 281.
19. Ancsin, J. *Metrologia* **1985**, *21*, 7.
20. Bongiovanni, G.; Crovini, L.; Marcarino, P. *Metrologia* **1975**, *14*, 175.
21. Enagonio, D. P.; Pearson, E. G.; Saylor, C. P. *Temperature, its Measurement and Control in Science and Industry* **1962**, *3*, 219.
22. Kemp, R. C.; Kemp, W. R. G. *Metrologia* **1978**, *14*, 9.
23. Mueller, E. F.; Burgess, H. A. *J. Am. Chem. Soc.* **1919**, *41*, 745.
24. Ambrose, D. *J. Phys. E* **1968**, *1*, 41.
25. Ambrose, D.; Sprake, C. H. S. *J. Chem. Thermodynamics* **1972**, *4*, 603.
26. *Experimental Thermodynamics Vol. II*, LeNeindre, B.; Vodar, B., editors, Butterworths: London, **1975**.

8.3. Reference Materials for High Accuracy Fixed Points

8.3.1. Recommended points and temperature values

Recommended values for high accuracy fixed points are available from the text of IPTS-68 (Ref. 1) and from an extended table published by Working Group 2 of the Comité Consultatif de Thermometrie (Ref. 2). A few more points and respective values have been described in recent publications. Table 8.3.1.1 presents the recommended high accuracy fixed points, their assigned temperatures, the estimated reproducibilities and the uncertainties of the assigned temperatures.

Table 8.3.1.1. Recommended values for High Accuracy Fixed Points

Equilibrium State	t_{68} °C	Reproducibility mK	Uncertainty mK	Purity	Reference
Triple point of mercury	−38.842	±0.2	±1	.999999	3, 4
Freezing point of mercury	−38.836	±0.2	±1	.999999	3, 4
Freezing point of water (Ice point)	0.000	±1	±2		5, 6
Triple point of water	0.01	±0.2	exact		6, 7
Triple point of phenoxybenzene (diphenyl ether)	26.869	±1.5	±2	.99998*	8, 9
Melting point of gallium	29.772	±0.2	±0.5	.999999	10, 11
Triple point of gallium	29.774	±0.2	±0.5	.999999	6, 11, 12
Triple point of 1,3-Dioxolan-2-one	36.324	±1.5	±2	.99998*	9
Triple point of succinonitrile	58.080	±0.3	±1.5	.999999	13, 14
Boiling point of water (steam point)	100.000†	±0.5	exact		6
Triple point of indium	156.629	±0.2	±1	.999999	15
Freezing point of indium	156.634	±0.2	±1	.999999	6, 15, 16
Freezing point of tin	231.9681	±0.2	exact	.999999	3, 6, 17
Freezing point of cadmium	321.108	±0.2	±1	.999999	3, 6, 17
Freezing point of lead	327.502	±0.5	±1	.999999	6, 19
Freezing point of zinc	419.58	±0.2	exact	.999999	3, 6, 20

† $t_{68}/°C = 100 + 28.0216(p/p_o - 1) - 11.642(p/p_o - 1)^2 + 7.1(p/p_o - 1)^3$
from 99.9 to 100.1 °C

Notes:

Reproducibility: estimated reproducibility of the reference temperature as provided by a single cell (three standard deviations).

Uncertainty: estimated uncertainty of the recommended temperature; the values assigned to primary fixed points of the IPTS-68 are exact by definition.

Purity: minimum required mass fraction purity for the quoted reproducibility; * refers to mole fraction purity.

Freezing point is at 101 325 Pa; melting point is at 101 325 Pa.

The boiling point is the temperature where the vapour pressure $p_o = 101\ 325$ Pa. The equations provide the equilibrium temperature t_{68} for the different pressures p.

Water used should be natural ocean water, distilled two or three times for the determination of steam and triple points. Water treated with ion-exchange resins is suitable for the ice point.

8.3.2. Available reference materials

In the following sections are collected the properties and all relevant information regarding available reference materials for high accuracy fixed points.

The accuracy of the realization of a boiling point depends critically on the measurement of pressure. When appropriate pressure-measuring equipment is available, however, accurate and reproducible boiling points can be realized by applying the procedure in Section 8.2.4 and using substances with the stated purity.

A few reference materials for freezing and triple points are not available with certification from an official supplier. They are, however, available as pure materials from qualified suppliers, and the procedure to obtain fixed points does not entail any further purification or critical measurement, and is completely described in the literature. They are therefore included as an available reference materials.

REFERENCES

1. The International Practical Temperature Scale of 1968, Amended edition of 1975, Bureau International de Poids et Measures: Sèrves, **1975**. English version in *Metrologia* **1976**, *12*, 7.
2. Crovini, L.; Bedford, R. E.; Moser, A. *Metrologia* **1977**, *13*, 197.
3. Furukawa, G. T.; Bigge, W. R.; Le point triple de mercure comme étalon thermométrique, *Comité Consultatif de Thermoétrie 11 th Session*, Annexe T14, BIPM, Sèvres, T134-T144, **1976**.
4. Furukawa, G. T.; Riddle, J. L.; Bigge, W. R.; Pfeiffer, E. R. *Standard Reference Materials: Application of some Metal SRM's as Thermometric Fixed Points*, National Bureau of Bureau of Standards Special Publication 260-77, U.S. Department of Commerce: Washington, **1982**.
5. Thomas, J. L. *J. Res. Nat. Bur. Stand.* **1934**, *12*, 323.
6. Supplementary Information of the IPTS-68 and the EPT-76, Bureau International des Poids et Mesures: Sèvres, **1983**.
7. Stimson, H. F. *Temperature, its Measurement and Control in Science and Industry* **1955**, *2*, 141.
8. Cox, J. D.; Vaughan, M. F. *Metrologia* **1980**, *16*, 105.
9. Cox, J. D.; Vaughan, M. F. *Temperature, its Measurement and Control in Science and Industry* **1982**, *5*, 267.
10. Thornton, D. D. *Clinical Chem.* **1977**, *23*, 719.
11. Chattle M. V.; Rusby, R. L.; Bonnier, G.; Moser, A.; Renoat, E.; Marcarino, P.; Bongiovanni, G.; Frassinetti, G. *Temperature, its Measurement and Control in Science and Industry* **1982**, *5*, 311.
12. Mangum, B. W. *Temperature, its Measurement and Control in Science and Industry* **1982**, *5*, 299.

13. Gliksman, M. E.; Voorkees, P. W. *Temperature, its Measurement and Control in Science and Industry* **1982**, *5*, 321.
14. Mangum, B. W. *Clinical Chem.* **1983** *29*, 1380.
15. Sawada, S. *Temperature, its Measurement and Control in Science and Industry* **1982**, *5*, 343.
16. McLaren, E. H. *Can. J. Phys.* **1958**, *36*, 1131.
17. McLaren, E. H. *Can. J. Phys.* **1957**, *35*, 1086.
18. McLaren, E. H.; Murdock, E. G. *Can. J. Phys.* **1960**, *38*, 577.
19. McLaren, E. H. *Can. J. Phys.* **1958**, *36*, 585.

8.3.2.1. Mercury

Physical property: Triple-point temperature
Units: °C, K (IPTS-68)
Recommended reference material: Mercury sealed in a borosilicate-glass vessel.
Range of variables: Triple point = −38.842 °C, 234.308 K
Contributors: J. D. Cox, M. V. Chattle

Intended usage: Calibrating resistance thermometers (especially those whose construction makes it difficult to calibrate at the defining fixed points); checking thermometers of all types.

Sources of supply and/or methods of purification: Commercially obtained metals with nominal mass fraction purity 0.999999 have proved satisfactory. Supplier (A) supplies pure mercury, certified for its triple point, as SRM 743.

Pertinent Physicochemical Data: See section 8.3.1 and table 8.2.2.1.

Technique of realization: The cell is placed in a stainless steel chamber thermostatted at about −42 °C. When the cell temperature is about −38.8 °C, the stainless steel chamber is evacuated, and freezing of the cell content is initiated by insertion of a rod, cooled in liquid nitrogen, down the re-entrant well of the cell (Ref. 1).

REFERENCES

1. Furukawa, G. T.; Riddle, J. L.; Bigge, W. R.; Pfeiffer, E. R. *Standard Reference Materials: Application of some Metal SRM's as Thermometric Fixed Points*, National Bureau of Bureau of Standards Special Publication 260-77, U.S. Department of Commerce: Washington, **1982**.

8.3.2.2. Water

Physical property: Melting point temperature (ice point)
Units: °C, K (IPTS-68)
Recommended reference material: Distilled water
Range of variables: Melting point = 0.000 °C, 273.150 K
Contributor: L. Crovini

Intended usage: Verification of the calibration of thermometers; reference temperature for thermocouples.

Sources of supply and/or methods of purification: Twice distilled natural ocean water, saturated with air; natural ocean water treated with ion-exchange resins and saturated with air. Tap water is generally a suitable substitute for natural ocean water.

Pertinent Physicochemical Data: See section 8.3.1 and table 8.2.2.1.

Technique of realization: Finely crushed or shaved ice is generally packed inside a dewar flask, with one or more thermometers immersed in it. Excess water can be drained from the bottom of the dewar using a siphon tube. The equilibrium between ice and water does not require there to be a large amount of water present; the flushing action arising from the surface of the melting ice is sufficient when measurements are required only to ±1 mK. In such cases, however, an ice point bath must contain enough water to provide good thermal contact between the ice-water interface and the thermometers. Every time the ice is melted away from a thermometer well it must be carefully repacked. Best results are achieved with thermometers immersed in an ice bath for not less than 30 cm (Refs. 1, 2).

Thermoelectrically-operated ice point devices are available for less accurate thermometry.

Useful details of the realization of an ice point are also provided in reference 3.

REFERENCES

1. Supplementary Information of the IPTS-68 and the EPT-76, Bureau International des Poids et Mesures: Sèvres, **1983**.
2. Thomas, I. L. *J. Res. Nat. Bur. Stand.*, **1934**, *12*, 323.
3. Nicholas, J. V.; White, D. R. *Traceable Temperatures*, DSIR Bulletin 234, New Zealand Department of Scientific and Industrial Research: Wellington N. Z., **1982**.

8.3.2.3. Water

Physical property: Triple-point temperature
Units: °C, K (IPTS-68)
Recommended reference material: Water sealed in a borosilicate-glass vessel.
Range of variables: Triple point = 0.01 °C, 273.16 K
Contributor: L. Crovini

Intended usage: Defining fixed point of the IPTS-68; verification of the calibration of thermometers.

Sources of supply and/or methods of purification: Supplier B supplies cells of nominal height 32 cm and 16 cm. Other cells are commercially available from qualified manufacturers such as Suppliers C and D. Methods of preparation are described in reference 1 and 2.

Pertinent Physicochemical Data: See section 8.3.1 and table 8.2.2.1.

Technique of realization: See reference 3 for a detailed description.

REFERENCES

1. Stimson, H. F. *Temperature, its Measurement and Control in Science and Industry* **1955**, *2*, 141.
2. Ambrose, D.; Collerson, R. R.; Ellender, J. H. *J. Phys. E* **1973**, *6*, 975.
3. Supplementary Information of the IPTS-68 and the EPT-76, Bureau International des Poids et Mesures: Sèvres, **1983**.

8.3.2.4. Phenoxybenzene

Physical property: Triple-point temperature
Units: °C, K (IPTS-68)
Recommended reference material: Phenoxybenzene sealed in a cylindrical cell of borosilicate glass.
Range of variables: Triple point = about 26.869 °C, 300.019 K
Actual equilibrium temperatures, when about 40% of contents are frozen, are certified for each cell.
Contributors: J. D. Cox and M. F. Vaughan.

Intended usage: Checking thermometers of all types.

Sources of supply and/or methods of purification: Supplier (B) supplies triple-point cells of nominal height 32 or 16 cm. Methods for preparing the cells have been described (Refs. 1,

2, 3). The purity of the phenoxybenzene sealed within a cell is measurable *in situ*, and is typically 0.99998 mole fraction.

Pertinent Physicochemical Data: See section 8.3.1 and table 8.2.2.1.

Technique of realization: The triple point can be established by (i) slow heating of a cell whose contents are solid, through the melting region, (ii) slow cooling of a cell whose contents are liquid through the freezing region, (iii) isothermal maintenance of a cell whose contents are partly frozen. Method (iii), the equilibrium method, is preferred, as it is the one to which the certified temperature refers (Ref. 1). A cell with molten contents is cooled in ice-water until crystallization occurs to give a mush of crystals throughout the cell. The cell is placed in an insulated enclosure within a bath or cabinet controlled at (27 ± 1) °C. A metal rod at 70 °C is placed in the re-entrant tube of the cell for 5 minutes, then the cell is removed from its enclosure and inverted; the procedure with the rod and the inversion is repeated twice more. After the cell has been allowed to remain in its enclosure for a further one hour, it is ready for use.

REFERENCES

1. Cox, J. D.; Vaughan, M. F. *Metrologia* **1980**, *16*, 105.
2. Cox, J. D.; Vaughan, M. F. *Temperature, its Measurement and Control in Science and Industry* **1982**, *5*, 267.
3. Vaughan, M. F. *NPL Report Chem. 86*, National Physical Laboratory: Teddington, **1978**.

8.3.2.5. Gallium

Physical property: Triple-point temperature
Units: °C, K (IPTS-68)
Recommended reference material: Gallium sealed in a plastic vessel.
Range of variables: Triple point = 29.774 °C, 302.924 K
Contributors: J. D. Cox, M. V. Chattle

Intended usage: Calibrating resistance thermometers (especially those whose construction makes it difficult to calibrate at the defining fixed points); checking thermometers of all types.

Sources of supply and/or methods of purification: Commercially obtained metals with nominal mass fraction purity 0.999999 have proved satisfactory. Supplier (A) supplies pure gallium in Teflon cells for the melting point as SRM 1968. When used *in vacuo* (Refs. 1, 2) they provide the triple-point temperature.

Pertinent Physicochemical Data: See section 8.3.1 and table 8.2.2.1.

Technique of realization: The cell, initially at ambient temperature, is placed in a stirred liquid bath, thermostatted at about 31 °C. A small electric heater placed in the re-entrant well of the cell is used to melt a thin layer of gallium around the well. After this heater has been switched off, a plateau temperature is reached as slow melting of the whole charge begins (Ref. 1).

REFERENCES

1. Chattle, M. V.; Rusby, R. L.; Bonnier, G.; Moser, A.; Renoat, E.; Marcarino, P.; Bongiovanni, G.; Frassinetti, G. *Temperature, its Measurement and Control in Science and Industry* **1982**, *5*, 311.
2. Mangum, B. W. *Temperature, its Measurement and Control in Science and Industry* **1982**, *5*, 299.

8.3.2.6. 1,3-Dioxolan-2-one

Physical property: Triple-point temperature
Units: °C, K (IPTS-68)
Recommended reference material: 1,3-Dioxolan-2-one sealed in a cylindrical cell of borosilicate glass.
Range of variables: Triple point = about 36.324 °C, 309.474 K
Actual equilibrium temperatures, when about 40% of the cell's contents are frozen, are certified for each cell.
Contributors: J. D. Cox; M. F. Vaughan

Intended usage: Checking thermometers of all types.

Sources of supply and/or methods of purification: Supplier (B) supplies triple-point cells of nominal height 32 or 16 cm. Methods for preparing the cells have been outlined (Ref. 1). The purity of 1,3-dioxolan-2-one sealed within a cell is measurable *in situ*, and is typically 0.99998 mole fraction.

Pertinent Physicochemical Data: See section 8.3.1. The presssure coefficient of the triple-point temperature is 2.4 mK per metre of the hydrostatic head of 1,3-dioxolan-2-one. The certified temperature may contain an allowance for this effect, on the assumption that the temperature sensor is a certain distance below the surface.

Technique of realization: The triple point can be established by (i) slow heating of a cell whose contents are solid, through the melting region, (ii) slow cooling of a cell whose contents are liquid, through the freezing region, (iii) isothermal maintenance of a cell whose contents are partly frozen. Method (iii), the equilibrium method, is preferred, as it is the one to which the certified temperature refers. A cell with molten contents is cooled in water at about 20 °C for 20 minutes; crystallization should occur when the cell is well shaken. The cell is placed in an insulated enclosure within a bath or cabinet controlled at (36 ± 1) °C. After the cell has been allowed to remain in its enclosure for one hour it is ready for use.

REFERENCES

1. Cox, J. D.; Vaughan, M. F. *Temperature, its Measurement and Control in Science and Industry* **1982**, *5*, 267.

8.3.2.7. Succinonitrile

Physical property: Triple-point temperature
Units: °C, K (IPTS-68)
Recommended reference material: Succinonitrile sealed in a borosilicate-glass vessel.
Range of variables: Triple point = 58.080 °C, 331.231 K
Contributor: L. Crovini

Intended usage: Checking thermometers of all types.

Sources of supply and/or methods of purification: Supplier (A) supplies triple-point cells of nominal height 12 cm. The estimated mass fraction purity is between 0.99999 and 0.999999. A method of purification of succinonitrile by zone-refining and of preparing the cell are described in reference 1.

Pertinent Physicochemical Data: See section 8.3.1. The pressure coefficient of the triple-point temperature is 3.3 mK per meter of hydrostatic head of succinonitrile.

Technique of realization: The triple point is easily realized either by melting or by freezing the substance (Ref. 2). Reproducible temperatures are obtained with cells immersed in a temperature-controlled, stirred liquid bath, set not more than 0.5 °C above the triple-point temperature (Ref. 2).

REFERENCES

1. Gliksman, M. E.; Voorkees, P. W. *Temperature, its Measurement and Control in Science and Industry* **1982**, *5*, 321.
2. Mangum, B. W. *Clin. Chem.* **1983**, *29*, 1380.

8.3.2.8. Water

Physical property: Boiling point (steam point)

Units: °C, K (IPTS-68)
Recommended reference material: double-distilled natural ocean water
Range of variables: Boiling point = 100.00 °C, 373.15 K, at 101 325 Pa
Contributors: L. Crovini

Intended usage: Defining fixed point of the IPTS-68; verification of the calibration of thermometers.

Sources of supply and/or methods of purification: Twice-distilled natural ocean water, saturated with air; natural ocean water treated with ion-exchange resins and saturated with air. Tap water is generally a suitable substitute for natural ocean water.

Pertinent Physicochemical Data: See section 8.3.1 and table 8.2.2.1.

Technique of realization: The general criteria for realizing a boiling point are presented in section 2.4. Figure 8.2.3 and reference 1 provide a detailed description of the boiling point apparatus and of the measurement procedure.

REFERENCES

1. Supplementary Information of the IPTS-68 and the EPT-76, Bureau International des Poids et Mesures: Sèvres, **1983**.

8.3.2.9. Indium

Physical property: Triple-point temperature
Units: °C, K (IPTS-68)
Recommended reference material: Indium sealed in a borosilicate-glass vessel.
Range of variables: Triple point = 156.629 °C, 429.779 K
Contributors: J. D. Cox and M. V. Chattle.

Intended usage: Calibrating resistance thermometers (especially those whose construction makes it difficult to calibrate at the defining fixed points); checking thermometers of all types.

Sources of supply and/or methods of purification: Commercially obtained metals of nominal mass fraction purity 0.999999 have proved satisfactory.

Pertinent Physicochemical Data: See section 8.3.1 and table 8.2.2.1.

Technique of realization: The cell is placed in an aluminium jacket in a tubular furnace. It is heated through the melting temperature, then the furnace is controlled at about 153 °C. Nucleation of the contents of the cell is achieved by inserting a rod at ambient temperature into the re-entrant well of the cell (Ref. 1).

REFERENCES

1. Sawada, S. *Temperature, its Measurement and Control in Science and Industry* **1982**, *5*, 343.

8.3.2.10. Tin

Physical property: Freezing temperature
Units: °C, K (IPTS-68)
Recommended reference material: Tin in a high-purity graphite crucible.
Range of variables: Freezing temperature = 231.9681 °C, 505.1181 K
Contributor: L. Crovini

Intended usage: Defining fixed point of the IPTS-68; verification of the calibration of thermometers.

Sources of supply and/or methods of purification: Commercially obtained metals of nominal mass fraction purity 0.999999 have proved satisfactory. It may be necessary, however, to determine the range of melting in order to estimate the purity of the sample in the conditions of use. The melting range, as defined in reference 1, must not exceed 2 mK to ensure the quoted reproducibility. Supplier (A) supplies certified samples for the freezing point as SRM 741 (primary freezing point standard, assigned temperature (231.9681 ± 0.0007)°C), and as SRM 42g (secondary freezing point standard, assigned temperature (231.967 ± 0.001)°C).

Pertinent Physicochemical Data: See section 8.3.1 and table 8.2.2.1.

Technique of realization: A crucible (high-purity graphite, spectrographic grade) is placed at the closed end of a borosilicate-glass cylinder. Alternate layers of silica wool and graphite discs thermally isolate the top of the crucible from the outside. An atmosphere of pure nitrogen, argon, or helium is put inside the tube to prevent oxidation. The tube is placed in an aluminium jacket in a tubular furnace. It is heated to about 10 K above its melting temperature to melt its contents. The furnace is set at about 229 °C and nucleation is promoted either by extracting the tube from the furnace for a minute or two, or by blowing a cool gas stream in the re-entrant well of the crucible. Further details are provided by references 1, 2, and 3.

REFERENCES

1. McLaren, E. H.; Murdock, E. G. *Can. J. Phys.* **1960**, *38*, 100.
2. Furukawa, G. T.; Riddle, J. L.; Bigge, W. R.; Pfeiffer, E. R. *Standard Reference Materials: Application of some Metal SRM's as Thermometric Fixed Points*, National Bureau of Bureau of Standards Special Publication 260-77, U.S. Department of Commerce: Washington, **1982**.

3. Supplementary Information of the IPTS-68 and the EPT-76, Bureau International des Poids et Mesures: Sèvres, **1983**.

8.3.2.11. Cadmium

Physical property: Freezing temperature
Units: °C, K (IPTS-68)
Recommended reference material: Cadmium in a high-purity graphite crucible.
Range of variables: Freezing temperature = 321.108 °C, 594.258 K
Contributor: L. Crovini

Intended usage: Verification of the calibration of thermometers; checking a particular realization of the IPTS-68.

Sources of supply and/or methods of purification: Commercially obtained metals of nominal mass fraction purity 0.99998 have proved satisfactory. It may be necessary, however, to determine the range of melting in order to estimate the purity of the sample in the conditions of use. The melting range, as defined in reference 1, must not exceed 2 mK to ensure a reproducibility better than ±0.5 mK.

Pertinent Physicochemical Data: See section 8.3.1 and table 8.2.2.1.

Technique of realization: A crucible (high-purity graphite, spectrographic grade) is placed at the closed end of a borosilicate- glass cylinder. Alternate layers of silica wool and graphite discs thermally isolate the top of the crucible from the outside. An atmosphere of pure nitrogen, argon, or helium is put inside the tube to prevent oxidation. The tube is placed in an aluminium jacket in a tubular furnace. It is heated to at least 5 K above its melting temperature to melt its contents. The furnace is set at about 316 °C; when the sample is just below the freezing point and recalescence is beginning, nucleation is promoted by inserting a cool rod, or thermometer in the re-entrant well of the crucible, and the furnace is reset at about 319 °C. Further details are provided by references 1 and 2.

REFERENCES

1. McLaren, E. H. *Can. J. Phys.* **1957**, *35*, 1086.
2. Furukawa, G. T.; Riddle, J. L.; Bigge, W. R.; Pfeiffer, E. R. *Standard Reference Materials: Application of some Metal SRM's as Thermometric Fixed Points*, National Bureau of Bureau of Standards Special Publication 260-77, U.S. Department of Commerce: Washington, **1982**.

8.3.2.12. Lead

Physical property: Freezing temperature
Units: °C, K (IPTS-68)
Recommended reference material: Lead in a high-purity graphite crucible.
Range of variables: Freezing temperature = 327.502 °C, 600.652 K
Contributor: L. Crovini

Intended usage: Verification of the calibration of thermometers; checking a particular realization of the IPTS-68.

Sources of supply and/or methods of purification: Commercially obtained metals of nominal mass fraction purity 0.99998 have proved satisfactory. It may be necessary, however, to determine the range of melting in order to estimate the purity of the sample in the conditions of use. The melting range, as defined in reference 1, must not exceed 2 mK to ensure a reproducibility better than 0.5 mK. Supplier (A) distributes a lower-purity lead as SRM 49e with a certified freezing point of (327.493 ± 0.005)°C.

Pertinent Physicochemical Data: See section 8.3.1 and table 8.2.2.1.

Technique of realization: A crucible (high-purity graphite, spectrographic grade) is placed at the closed end of a borosilicate- glass cylinder. Alternate layers of silica wool and graphite discs thermally isolate the top of the crucible from the outside. An atmosphere of pure nitrogen, argon, or helium is put inside the tube to prevent oxidation. The tube is placed in an aluminium jacket in a tubular furnace. It is heated to at least 5 K above its melting temperature to melt its contents. The furnace is set at above 322 °C; when the sample is just below the freezing point and recalescence is beginning, nucleation is promoted by inserting a cool rod, or thermometer in the re-entrant well of the crucible, and the furnace is reset at about 325 °C. Further details are provided by reference 1.

REFERENCES

1. The International Practical Temperature Scale of 1968, Amended edition of 1975, Bureau International de Poids et Measures: Sèrves, **1975**. English version in *Metrologia* **1976**, *12*, 7.

8.3.2.12. Zinc

Physical property: Freezing temperature
Units: °C, K (IPTS-68)
Recommended reference material: Zinc in a high-purity graphite crucible.

Range of variables: Freezing temperature = 419.58 °C, 692.73 K
Contributor: L. Crovini

Intended usage: Defining fixed point of the IPTS-68; verification of the calibration of thermometers.

Sources of supply and/or methods of purification: Commercially obtained metals of nominal mass fraction purity 0.999999 have proved satisfactory. It may be necessary, however, to determine the range of melting in order to estimate the purity of the sample in the conditions of use. The melting range, as defined in reference 1, must not exceed 2 mK to ensure a reproducibility better than 0.2 mK. Supplier (A) distributes high-purity zinc samples as SRM 740.

Pertinent Physicochemical Data: See section 8.3.1 and table 8.2.2.1.

Technique of realization: A crucible (high-purity graphite, spectrographic grade) is placed at the closed end of a borosilicate- glass cylinder. Alternate layers of silica wool and graphite discs thermally isolate the top of the crucible from the outside. An atmosphere of pure nitrogen, argon, or helium in the tube prevents oxidation. The tube is placed in an aluminium jacket in a tubular furnace. It is heated to 5 K above its melting temperature to melt its contents. The furnace is set at about 414 °C; when the sample is just below the freezing point and recalescence is beginning, nucleation is promoted by inserting a cool rod, or thermometer in the re-entrant well of the crucible, and the furnace is reset at about 417 °C. Further details are provided by references 2 and 3.

REFERENCES

1. McLaren, E. H.; Murdock, E. G. *Can. J. Phys.* **1960**, *38*, 577.
2. Furukawa, G. T.; Riddle, J. L.; Bigge, W. R.; Pfeiffer, E. R. *Standard Reference Materials: Application of some Metal SRM's as Thermometric Fixed Points*, National Bureau of Bureau of Standards Special Publication 260-77, U.S. Department of Commerce: Washington, **1982**.
3. McLaren, E. H. *Can. J. Phys.* **1957**, *35*, 1086.

8.4. Lower Accuracy Fixed Points

8.4.1. General considerations

For less demanding applications it is possible to use reference points with reproducibilities between 5 mK and 0.1 K. They require lower-purity materials and simplified techniques for realization so they may suitable for particular applications such as the *in situ* calibration of an instrument. Normally thermal transitions occurring at atmospheric pressure are preferred. Solid-liquid, liquid-vapour, solid-liquid-vapour, solid-vapour equilibria, and solid-solid transitions are used.

Many triple points and freezing/melting points of organic substances fall into this category because of the limited purity of commercial supplies.

8.4.2. Fixed points for application in dynamic conditions

Temperature standards are needed in many applications of thermal analysis and of analytical chemistry. For instance, melting points determined by the observation of the melting of a sample in a glass capillary tube are extensively used for characterizing organic compounds. In the measurements dealt with previously, accuracy was the prime aim. In this type of measurement the main aims are speed and convenience with, if possible, simplicity of the apparatus. Accuracy must therefore be sacrificed. In current practice, empirical methods are used in combination with standard procedures. The determination is invariably carried out with a rapidly changing temperature, and the results are affected by various factors related to the dynamic response of the capillary system, and of the associated thermometer, to the heating rate and to the mass of the specimen. Melting is not instantaneous and there will be an appreciable time interval between the onset of melting and its completion, during which the temperature of the bath rises; the temperature at which the last crystal melts is usually taken as the melting point (Ref. 1). It has been shown by a careful study (Ref. 2) that with a refined technique, at which the temperature was raised at about 0.1 K in 4 min, a reproducibility in the melting point approaching 0.03 K may be attained. Recommended methods, however, usually prescribe heating rates of 1 K min^{-1} (Refs. 1, 3, 4), or even 3 K min^{-1} (Ref. 5), and with the apparatus used the high heating rates are poorly controlled. In such conditions, discrepancies of 1 K or more can be expected, a conclusion borne out by a co-operative study sponsored by the World Health Organization (Ref. 6). In this study 15 laboratories made measurements on 13 compounds melting between 70 °C and 260 °C; the range of value observed increased from about 1.5 K for the lower-melting compounds up to 5 K for the highest.

However, even when care is taken, capillary melting points systematically differ from equilibrium melting or freezing temperatures. A comparison was carried out at National Physical Laboratory, UK (Ref. 7) at the melting temperatures of 4-nitrotoluene (52.3 °C), naphthalene (80.8 °C), benzil (95.4 °C), acetanilide (114.9 °C), benzoic acid (122.9 °C), diphenylacetic acid (148.0 °C), anisic acid (184.2 °C), 2-chloroanthraquinone (210.7 °C), carbazole (246.6 °C) and anthraquinone (285.8 °C). The meniscus temperature (*i.e.* the temperature at which a meniscus is completely formed in the capillary tube) and the liquefaction temperature (*i.e.* the temperature at which the last crystal of solid liquefies) were observed with the temperature rising at 0.2 K min^{-1}. These measurements showed that the simple rapid melting-point technique invariably gave values 0.5 K or more above the equilibrium values obtained by freezing-point determinations.

Thus it is desirable to calibrate the apparatus with several pure substances of known melting point, suitable for use in the apparatus. The amount of the reference material and some of its physical properties, such as density, heat of fusion, thermal conductivity, and heat capacity, must be close to those of the materials usually tested. Reference substances can be certified in two ways. One, developed at Istituto di Metrologia G. Colonnetti (Ref. 8), typically affords measurements with uncertainty lower than ± 0.05 K (± 0.02 K at 122.9 °C). Both melting and freezing points are measured close to equilibrium conditions, using 1 to 2 grams of substance. The other approach is to use a reference apparatus to establish the melting point of the reference material; in such an apparatus the test sample is subjected to dynamic conditions close to those of the apparatus in which the reference material is to be used, and the value determined in this way is referred to as a "dynamic melting point". The advantage of using reference materials for dynamic melting points may be illusory if the dynamic conditions change appreciably with the apparatus.

The development of reference materials for temperature is also pursued by the International Confederation for Thermal Analysis (ICTA) to satisfy the needs of differential thermal analysis (DTA), of thermogravimetry (TG), and of differential thermogravimetry (DTG). For DTA, melting points and solid-solid transitions are used; ferromagnetic-to-paramagnetic transitions are mostly preferred in the case of TG and DTG.

In general, dynamic temperature reference materials provide a temperature interval that is specified in terms of empirically defined temperatures. For melting-point apparatuses the meniscus and liquefaction temperatures are normally considered as the initial and final temperatures of the transition (Refs. 7, 8). Reference materials for melting points or solid-solid transitions used with DTA instruments are characterized by their "onset temperature", *i.e.* the temperature obtained as the intersection of the forward-extrapolated baseline with the backward-extrapolated initial side of the peak. This value results from the graphically recorded output of a DTA apparatus and the "peak temperature", as discussed in references 10, 11, and 12. Three temperatures are used to characterize the transition interval of a magnetic reference material for TG and DTG apparatus (Refs. 13, 14).

8.4.3. Recommended points and values, and available reference materials

Recommended temperatures for Lower Accuracy Fixed Points are given in table 8.4.3.1. Practically all substances are commercially available with sufficient purity for realizing the points with the given accuracy. In addition, commercially available pure metals, with mass fraction purity between 0.99998 and 0.99999 are suitable to realize any metal freeezing, or triple point of table 8.3.1.1 to within ± 0.01 K.

The reference materials presented in the following section are those available from standards laboratories.

8. Temperature

Table 8.4.3.1. Recommended Values for Lower Accuracy Fixed Points

Equilibrium State	t_{68} °C	Reproducibility mK	Uncertainty mK	Purity	References
Sublimation of CO_2	−78.477	±2	±5	.9999	15
Triple point of CO_2	−56.570	±2	±5	.9999	16, 17
Triple point of bromobenzene	−30.73	±2	±10	.99998	18
Hydrate transition of Na_2SO_4-to-$Na_2SO_4 \cdot 10H_2O$	32.374	±2	±5	.9999	19
Triple point of icosane	36.49	±2	±5	.99999*	20
Triple point of rubidium	39.27	±8	±15	.99999*	21
Hydrate transition of KF-to-KF·$2H_2O$	41.42	±2	±40	.9999	19
Hydrate transition of $Na_2HPO_4 \cdot 2H_2O$-to-$Na_2HPO_4 \cdot 7H_2O$	48.222	±2	±5	.9999	19
Melting point of 4-nitrotoluene	52.0	±40	±200	.999*	7, 25
Melting point of naphthalene	80.5	±180	±200	.999*	7, 25
Melting point of benzil	95.1	±180	±200	.999*	7, 25
Freezing point of sodium	97.819	±2	±5	.9999	22
Melting point of acetanilide	114.2	±120	±200	.999*	7, 25
Triple point of benzoic acid	122.370	±2	±3	.99998	23
Freezing point of benzoic acid	122.383	±2	±5	.99998	23, 8
Melting point of diphenylacetic acid	147.6	±120	±200	.998*	7, 25
Melting point of anisic acid	183.6	±80	±200	.999*	7, 25
Melting point of 2-chloroanthraquinone	210.4	±80	±400	.998*	7, 25
Melting point of carbazole	246.2	±80	±400	.999*	7, 25
Melting point of anthraquinone	285.2	±80	±400	.998*	7, 25
Boiling point of mercury	356.66†	±2	±10	.999999	24

† $t_{68}/°C = 356.66 + 55.552(p/p_o - 1) - 23.03(p/p_o - 1)^2 + 14.0(p/p_o - 1)^3$

Notes

Reproducibility: estimated reproducibility of the reference temperature as provided by a single cell (three standard deviations).

Purity: minimum required mass fraction purity for the quoted reproducibility; * indicates mole fraction purity.

Boiling point is the temperature where the vapour pressure p_o = 101 325 Pa; freezing point is at 101 325 Pa; melting point is at 101 325 Pa; solid-vapour equilibrium is at 101 325 Pa (sublimation point).

REFERENCES

1. Skau, E. L.; Arthur, J. C.; Wakeham, H. *Techniques of Organic Chemistry*, 3rd edition, Weissberger, A., editor, Vol. 1, Part 1, Interscience: New York, **1959**.
2. Francis, F.; Collins, F. J. E. *J. Chem. Soc.* **1936**, *137*.
3. *European Pharmacopoeia*, Part 1, Maisonneuve: Sainte-Ruffine, **1973**, p. 73.
4. *ASTM Standards*, Part 22, Method E324-69, American Society for Testing and Materials: Philadelphia, **1969**.

5. *British Pharmacopoeia 1973* **1973**, Appendix IV A, HMSO: London, **1973**.
6. Bervenmark, H.; Diding, N. A.; Örner, B. *Bull. W. Hl. O.* **1963**, *28*, 175.
7. Herington, E. F. G.; Handley, R.; Cook, A. J. *Chem and Ind. (London)*, **1956**, 292.
8. Crovini, L. and Marcarino, P. and Milazzo, G. *Anal. Chem.* **1981**, *53*, 681.
9. Cox, J. D. *NPL Report Chem. 117*, National Physical Laboratory: Teddington, **1981**.
10. Rossini, F. D.; Wagman, D. D.; Evans, W. H.; Levine, S.; Jaffe, I. *Selected Values of Chemical Thermodynaic Properties*, National Bureau of Standards Circular 500, U.S. Department of Commerce: Washington, **1952**.
11. Hedvall, J. A.; Linder, R. and Hartler, N. *Acta Chem. Scand.*, **1950**, *4* 1099.
12. *Catalog of NBS Standard Reference Material* National Bureau of Standards Special Publication 260, U.S. Department of Commerce: Washington, **1976**.
13. Norem, S. D.; O'Neill, M. J. and Gray, A. P. *Proceedings Third Toronto Symposium on Thermal Analysis*, McAdie, H. G., editor, Chem. Inst. Canada: Ottawa, **1969**, pp. 221-232.
14. McAdie, H. G. *Anal. Chem.* **1967**, *39*, 543.
15. Barber, C. R. *Brit. J. Appl. Phys.* **1966**, *17*, 391.
16. Ambrose, D. *Brit. J. Appl. Phys.* **1957**, *8*, 32.
17. Pavese, F. and Ferri, D. *Temperature, its Measurement and Control in Science and Industry* **1982**, *5*, 217.
18. Masi, J. F. and Scott, R. B. *J. Res. Natl. Bur. Stand.* **1975**, *79A*, 619.
19. Magin, R. L.; Mangum, B. W.; Staffer, J. A. and Thornton, D. D., *J. Res. Natl. Bur. Stand.* **1981**, *86A*, 191.
20. Cox, J. D.; Vaughan, M. F. *Temperature, its Measurement and Control in Science and Industry* **1982**, *5*, 267.
21. Figueroa, J. M.; Mangum, B. W. *Temperature, its Measurement and Control in Science and Industry* **1980**, *5*, 327.
22. Butkiewicz, J.; Gizmajer, W. *Comité Consultatif de Thermométrie, 13th Session*, Document CCT/80-1. BIPM: Sévres, **1980**.
23. Schwab, F. W.; Wickers, E. *J. Res. Natl. Bur. Stand.* **1945**, *34*, 333.
24. Ambrose, D.; Sprake, C. H. S. *J. Chem. Thermodynamics* **1972**, *4*, 603.
25. Herington, E. F. G. *Zone Melting of Organic Compounds*, Blackwell: Oxford, **1983**.

8.4.3.1. 4-Nitrotoluene

Physical property: (i) Initial melting temperature ("meniscus temperature")
(ii) final melting temperature ("liquefaction temperature")
measured in the presence of air at 101 kPa.
Units: °C (IPTS-68)
Recommended reference material: 4-nitrotoluene
Range of variables: Melting point of typical batches, when measured under the conditions defined below.

(i) 52.0 °C   (ii) 52.3 °C

Contributor: J. D. Cox

Intended usage: Calibrating or checking melting-point apparatus that depends on visual observation. Insofar as the measured meniscus and liquefaction temperatures can be equated with solidus and liquidus temperatures respectively, other types of melting-point apparatus which measure these latter quantities can be checked.

Sources of supply and/or methods of purification: Samples of the substance are available from Supplier B. The batches were purified by zone refining (Refs. 1, 2). Sample purity of a typical batch is about 0.999 mole fraction.

Pertinent physicochemical data: For typical sets of measurements (15 degrees of freedom) the standard deviation (reproducibility) was 0.04 °C and the overall uncertainty (2σ) was ±0.2 °C

Technique of realization: Portions of a powdered sample, kept over P_2O_5 in a desiccator, are sealed into four glass capillary tubes. These are placed in a stirred oil bath, heated to about 2 K below the initial melting temperature. The temperature of the oil-bath is then raised at 0.2 K min^{-1} and visual observation is made of the temperatures at which (i) a meniscus is first apparent, (ii) the last solid disappears. A calibrated mercury-in-glass thermometer should be used for temperature measurement. Replicate measurements are made with the tubes in other positions within the bath.

Significant difference between temperatures (i) and (ii) may reflect both impurity in the sample and the "over-run" caused by the continuing rise of bath temperature (0.2 K min^{-1}) during the melting process. It follows that the certified transition temperatures are method-dependent, and when used for calibration purposes apply only to methods that are similar to those used for certification.

The observed liquefaction temperature lies about 0.7 K above the true liquidus temperature.

REFERENCES

1. Herington, E. F. G.; Handley, R.; Cook, A. J. *Chem. and Ind. (London)*, **1956**, 292.
2. Herington, E. F. G. *Zone Melting of Organic Compounds*, Blackwell: Oxford, **1963**.

8.4.3.2. Naphthalene

Physical property: (i) Initial melting temperature ("meniscus temperature")
(ii) final melting temperature ("liquefaction temperature")
measured in the presence of air at 101 kPa.
Units: °C (IPTS-68)
Recommended reference material: naphthalene

Range of variables: Melting point of typical batches, when measured under the conditions defined below.

(i) 80.5 °C   (ii) 80.8 °C

Contributor: J. D. Cox

Intended usage: Calibrating or checking melting-point apparatus that depends on visual observation. Insofar as the measured meniscus and liquefaction temperatures can be equated with solidus and liquidus temperatures respectively, other types of melting-point apparatus which measure these latter quantities can be checked.

Sources of supply and/or methods of purification: Samples of the substance are available from Supplier B. The batches were purified by zone refining (Refs. 1, 2). Sample purity of a typical batch is > 0.999 mole fraction.

Pertinent physicochemical data: For typical sets of measurements (15 degrees of freedom) the standard deviation (reproducibility) was 0.08 °C and the overall uncertainty (2σ) was ± 0.2 °C

Technique of realization: Portions of a powdered sample, kept over P_2O_5 in a desiccator, are sealed into four glass capillary tubes. These are placed in a stirred oil bath, heated to about 2 K below the initial melting temperature. The temperature of the oil-bath is then raised at 0.2 K min^{-1} and visual observation is made of the temperatures at which (i) a meniscus is first apparent, (ii) the last solid disappears. A calibrated mercury-in-glass thermometer should be used for temperature measurement. Replicate measurements are made with the tubes in other positions within the bath.

Significant difference between temperatures (i) and (ii) may reflect both impurity in the sample and the "over-run" caused by the continuing rise of bath temperature (0.2 K min^{-1}) during the melting process. It follows that the certified transition temperatures are method-dependent, and when used for calibration purposes apply only to methods that are similar to those used for certification.

The observed liquefaction temperature lies about 0.6 K above the true liquidus temperature.

REFERENCES

1. Herington, E. F. G.; Handley, R.; Cook, A. J. *Chem. and Ind. (London)*, **1956**, 292.
2. Herington, E. F. G. *Zone Melting of Organic Compounds*, Blackwell; Oxford, **1963**.

8.4.3.3. Benzil

Physical property: (i) Initial melting temperature ("meniscus temperature")
(ii) final melting temperature ("liquefaction temperature")
measured in the presence of air at 101 kPa.
Units: °C (IPTS-68)
Recommended reference material: benzil
Range of variables: Melting point of typical batches, when measured under the conditions defined below.

(i) 95.1 °C   (ii) 95.4 °C

Contributor: J. D. Cox

Intended usage: Calibrating or checking melting-point apparatus that depends on visual observation. Insofar as the measured meniscus and liquefaction temperatures can be equated with solidus and liquidus temperatures respectively, other types of melting-point apparatus which measure these latter quantities can be checked.

Sources of supply and/or methods of purification: Samples of the substance are available from Supplier B. The batches were purified by zone refining (Refs. 1, 2). Sample purity of a typical batch is > 0.999 mole fraction.

Pertinent physicochemical data: For typical sets of measurements (15 degrees of freedom) the standard deviation (reproducibility) was 0.08 °C and the overall uncertainty (2σ) was ±0.2 °C

Technique of realization: Portions of a powdered sample, kept over P_2O_5 in a desiccator, are sealed into four glass capillary tubes. These are placed in a stirred oil bath, heated to about 2 K below the initial melting temperature. The temperature of the oil-bath is then raised at 0.2 K min^{-1} and visual observation is made of the temperatures at which (i) a meniscus is first apparent, (ii) the last solid disappears. A calibrated mercury-in-glass thermometer should be used for temperature measurement. Replicate measurements are made with the tubes in other positions within the bath.

Significant difference between temperatures (i) and (ii) may reflect both impurity in the sample and the "over-run" caused by the continuing rise of bath temperature (0.2 K min^{-1}) during the melting process. It follows that the certified transition temperatures are method-dependent, and when used for calibration purposes apply only to methods that are similar to those used for certification.

The observed liquefaction temperature lies about 0.5 K above the true liquidus temperature.

REFERENCES

1. Herington, E. F. G.; Handley, R.; Cook, A. J. *Chem. and Ind. (London)*, **1956**, 292.
2. Herington, E. F. G. *Zone Melting of Organic Compounds*, Blackwell, Oxford, **1963**.

8.4.3.4. Acetanilide

Physical property: (i) Initial melting temperature ("meniscus temperature")
(ii) final melting temperature ("liquefaction temperature")
    measured in the presence of air at 101 kPa.
Units: °C (IPTS-68)
Recommended reference material: acetanilide
Range of variables: Melting point of typical batches, when measured under the conditions defined below.
     (i) 114.2 °C   (ii) 114.9 °C
Contributor: J. D. Cox

Intended usage: Calibrating or checking melting-point apparatus that depends on visual observation. Insofar as the measured meniscus and liquefaction temperatures can be equated with solidus and liquidus temperatures respectively, other types of melting-point apparatus which measure these latter quantities can be checked.

Sources of supply and/or methods of purification: Samples of the substance are available from Supplier (B). The batches were purified by zone refining (Refs. 1 and 2). Sample purity of a typical batch is > 0.999 mole fraction.

Pertinent physicochemical data: For typical sets of measurements (15 degrees of freedom) the standard deviation (reproducibility) was 0.12 °C and the overall uncertainty (2σ) was ±0.2 °C

Technique of realization: Portions of a powdered sample, kept over P_2O_5 in a desiccator, are sealed into four glass capillary tubes. These are placed in a stirred oil bath, heated to about 2 K below the initial melting temperature. The temperature of the oil-bath is then raised at 0.2 K min^{-1} and visual observation is made of the temperatures at which (i) a meniscus is first apparent, (ii) the last solid disappears. A calibrated mercury-in-glass thermometer should be used for temperature measurement. Replicate measurements are made with the tubes in other positions within the bath.

Significant difference between temperatures (i) and (ii) may reflect both impurity in the sample and the "over-run" caused by the continuing rise of bath temperature (0.2 K min^{-1}) during the melting process. It follows that the certified transition temperatures are method-dependent, and when used for calibration purposes apply only to methods that are similar to those used for certification.

The observed liquefaction temperature lies about 0.5 K above the true liquidus temperature.

REFERENCES

1. Herington, E. F. G.; Handley, R.; Cook, A. J. *Chem. and Ind. (London)*, **1956**, 292.
2. Herington, E. F. G. *Zone Melting of Organic Compounds*, Blackwell: Oxford, **1963**.

8.4.3.5. Benzoic acid

Physical property: Triple-point temperature
Units: °C, K (IPTS-68)
Recommended reference material: Benzoic acid sealed in a cylindrical cell of borosilicate glass.
Range of variables: Triple-point temperature about 122.365 °C, 395.515 K. Actual equilibrium temperature, when about 40% of contents are frozen, is certified for each cell.
Contributor: J. D. Cox

Intended usage: Checking thermometers of all types.

Sources of supply and/or methods of purification: Supplier (B) supplies triple-point cells of nominal height 32 and 16 cm. Methods for preparing the cells have been described (Ref. 1). The purity of the benzoic acid sealed within a cell is measurable *in situ*, and is typically 0.99995 mole fraction.

Pertinent physicochemical data: The pressure coefficient of the triple-point temperature is $4._0$ mK per metre of hydrostatic head of benzoic acid. The certified temperature contains an allowance for this effect, on the assumption that the temperature sensor is at a certain distance below the surface of the benzoic acid. The uncertainty of measurement is ± 5 mK for the larger cells (height 32 cm) and ± 10 mK for the smaller cells (height 16 cm), 99% confidence level, as quoted for each certified cell from Supplier (B).

Technique of realization: The triple point can be established by (i) slow heating of a cell whose contents are solid through the melting region, (ii) slow cooling of a cell whose contents are liquid through the freezing region, (III) isothermal maintenance of a cell whose contents are partly frozen. Method (iii), the equilibrium method, is preferred as it is the one to which the certified temperature refers (Ref. 1). A cell with molten contents at 135 °C is allowed to cool slowly in an oven at 110 °C. When the cell reaches this temperature it is removed from the oven and left at ambient temperature until a dense mass of fine crystals (a mush) forms throughout the cell. It is then placed in a container at a temperature close to 122 °C. It should be noted that prolonged heating of a cell at 135 °C, or heating at a higher temperature, may cause partial dissociation of the benzoic acid to benzoic anhydride and water; these impurities will cause a depression of the triple-point temperature. Fortunately, the dissociation products can be caused to recombine by holding a cell at 100 °C overnight.

REFERENCES

1. Vaughan, M. F.; Butler, J. *NPL Report Chem. 71*, National Physical Laboratory: Teddington, **1985**.

8.4.3.6. Benzoic acid

Physical property: (i) Initial melting temperature ("meniscus temperature")
(ii) final melting temperature ("liquefaction temperature")
measured in the presence of air at 101 kPa.
Units: °C (IPTS-68)
Recommended reference material: benzoic acid
Range of variables: Melting point of typical batches, when measured under the conditions defined below.

(i) 122.5 °C   (ii) 122.9 °C

Contributor: J. D. Cox

Intended usage: Calibrating or checking melting-point apparatus that depends on visual observation. Insofar as the measured meniscus and liquefaction temperatures can be equated with solidus and liquidus temperatures respectively, other types of melting-point apparatus which measure these latter quantities can be checked.

Sources of supply and/or methods of purification: Samples of the substance are available from Supplier (B). The batches were purified by zone refining (Refs. 1, 2). Sample purity of a typical batch is > 0.999 mole fraction.

Pertinent physicochemical data: For typical sets of measurements (15 degrees of freedom) the standard deviation (reproducibility) was 0.12 °C and the overall uncertainty (2σ) was ±0.2 °C

Technique of realization: Portions of a powdered sample, kept over P_2O_5 in a desiccator, are sealed into four glass capillary tubes. These are placed in a stirred oil bath, heated to about 2 K below the initial melting temperature. The temperature of the oil-bath is then raised at 0.2 K min^{-1} and visual observation is made of the temperatures at which (i) a meniscus is first apparent, (ii) the last solid disappears. A calibrated mercury-in-glass thermometer should be used for temperature measurement. Replicate measurements are made with the tubes in other positions within the bath.

Significant difference between temperatures (i) and (ii) may reflect both impurity in the sample and the "over-run" caused by the continuing rise of bath temperature (0.2 K min^{-1}) during the melting process. It follows that the certified transition temperatures are method-dependent, and when used for calibration purposes apply only to methods that are similar to those used for certification.

The observed liquefaction temperature lies about 0.5 K above the true liquidus temperature.

REFERENCES

1. Herington, E. F. G.; Handley, R.; Cook, A. J. *Chem. and Ind. (London)*, **1956**, 292.
2. Herington, E. F. G. *Zone Melting of Organic Compounds*, Blackwell: Oxford, **1963**.

8.4.3.7. Diphenylacetic acid

Physical property: (i) Initial melting temperature ("meniscus temperature")
(ii) final melting temperature ("liquefaction temperature")
measured in the presence of air at 101 kPa.
Units: °C (IPTS-68)
Recommended reference material: diphenylacetic acid
Range of variables: Melting point of typical batches, when measured under the conditions defined below.

(i) 147.6 °C   (ii) 148.0 °C

Contributor: J. D. Cox

Intended usage: Calibrating or checking melting-point apparatus that depends on visual observation. Insofar as the measured meniscus and liquefaction temperatures can be equated with solidus and liquidus temperatures respectively, other types of melting-point apparatus which measure these latter quantities can be checked.

Sources of supply and/or methods of purification: Samples of the substance are available from Supplier (B). The batches were purified by zone refining (Refs. 1, 2). Sample purity of a typical batch is about 0.998 mole fraction.

Pertinent physicochemical data: For typical sets of measurements (15 degrees of freedom) the standard deviation (reproducibility) was 0.12 °C and the overall uncertainty (2σ) was ±0.2 °C

Technique of realization: Portions of a powdered sample, kept over P_2O_5 in a desiccator, are sealed into 4 glass capillary tubes. These are placed in a stirred oil bath, heated to about 2 K below the initial melting temperature. The temperature of the oil-bath is then raised at 0.2 K min^{-1} and visual observation is made of the temperatures at which (i) a meniscus is first apparent, (ii) the last solid disappears. A calibrated mercury-in-glass thermometer should be used for temperature measurement. Replicate measurements are made with the tubes in other positions within the bath.

Significant difference between temperatures (i) and (ii) may reflect both impurity in the sample and the "over-run" caused by the continuing rise of bath temperature (0.2 K min^{-1}) during the melting process. It follows that the certified transition temperatures are method-dependent, and when used for calibration purposes apply only to methods that are similar to those used for certification.

The observed liquefaction temperature lies about 0.7 K above the true liquidus temperature.

REFERENCES

1. Herington, E. F. G.; Handley, R.; Cook, A. J. *Chem. and Ind. (London)*, **1956**, 292.
2. Herington, E. F. G. *Zone Melting of Organic Compounds*, Blackwell: Oxford, **1963**.

8.4.3.8. Anisic acid

Physical property: (i) Initial melting temperature ("meniscus temperature")
(ii) final melting temperature ("liquefaction temperature")
measured in the presence of air at 101 kPa.
Units: °C (IPTS-68)
Recommended reference material: anisic acid
Range of variables: Melting point of typical batches, when measured under the conditions defined below.
(i) 183.6 °C   (ii) 184.2 °C

Contributor: J. D. Cox

Intended usage: Calibrating or checking melting-point apparatus that depends on visual observation. Insofar as the measured meniscus and liquefaction temperatures can be equated with solidus and liquidus temperatures respectively, other types of melting-point apparatus which measure these latter quantities can be checked.

Sources of supply and/or methods of purification: Samples of the substance are available from Supplier (B). The batches were purified by zone refining (Refs. 1, 2). Sample purity of a typical batch is > 0.999 mole fraction.

Pertinent physicochemical data: For typical sets of measurements (15 degrees of freedom) the standard deviation (reproducibility) was 0.08 °C and the overall uncertainty (2σ) was ±0.4 °C

Technique of realization: Portions of a powdered sample, kept over P_2O_5 in a desiccator, are sealed into four glass capillary tubes. These are placed in a stirred oil bath, heated to about 2 K below the initial melting temperature. The temperature of the oil-bath is then raised at 0.2 K min^{-1} and visual observation is made of the temperatures at which (i) a meniscus is first apparent, (ii) the last solid disappears. A calibrated mercury-in-glass thermometer should be used for temperature measurement. Replicate measurements are made with the tubes in other positions within the bath.

Significant difference between temperatures (i) and (ii) may reflect both impurity in the sample and the "over-run" caused by the continuing rise of bath temperature (0.2 K min^{-1})

during the melting process. It follows that the certified transition temperatures are method-dependent, and when used for calibration purposes apply only to methods that are similar to those used for certification.

The observed liquefaction temperature lies about 0.9 K above the true liquidus temperature.

REFERENCES

1. Herington, E. F. G.; Handley, R.; Cook, A. J. *Chem. and Ind. (London)*, **1956**, 292.
2. Herington, E. F. G. *Zone Melting of Organic Compounds*, Blackwell: Oxford, **1963**.

8.4.3.9. 2-Chloroanthraquinone

Physical property: (i) Initial melting temperature ("meniscus temperature")
(ii) final melting temperature ("liquefaction temperature")
measured in the presence of air at 101 kPa.
Units: °C (IPTS-68)
Recommended reference material: 2-chloroanthraquinone
Range of variables: Melting point of typical batches, when measured under the conditions defined below.
(i) 210.4 °C   (ii) 210.7 °C

Contributor: J. D. Cox

Intended usage: Calibrating or checking melting-point apparatus that depends on visual observation. Insofar as the measured meniscus and liquefaction temperatures can be equated with solidus and liquidus temperatures respectively, other types of melting-point apparatus which measure these latter quantities can be checked.

Sources of supply and/or methods of purification: Batches of the substance are available from Supplier (B). The batches were purified by zone refining (Refs. 1, 2). Sample purity of a typical batch is about 0.998 mole fraction.

Pertinent physicochemical data: For typical sets of measurements (15 degrees of freedom) the standard deviation (reproducibility) was 0.08 °C and the overall uncertainty (2σ) was ±0.4 °C

Technique of realization: Portions of a powdered sample, kept over P_2O_5 in a desiccator, are sealed into four glass capillary tubes. These are placed in a stirred oil bath, heated to about 2 K below the initial melting temperature. The temperature of the oil-bath is then raised at 0.2 K min^{-1} and visual observation is made of the temperatures at which (i) a meniscus is first apparent, (ii) the last solid disappears. A calibrated mercury-in-glass thermometer should be used for temperature measurement. Replicate measurements are made with the tubes in other positions within the bath.

Significant difference between temperatures (i) and (ii) may reflect both impurity in the sample and the "over-run" caused by the continuing rise of bath temperature (0.2 K min^{-1}) during the melting process. It follows that the certified transition temperatures are method-dependent, and when used for calibration purposes apply only to methods that are similar to those used for certification.

The observed liquefaction temperature lies about 1.0 K above the true liquidus temperature.

REFERENCES

1. Herington, E. F. G.; Handley, R.; Cook, A. J. *Chem. and Ind. (London)*, **1956**, 292.
2. Herington, E. F. G. *Zone Melting of Organic Compounds*, Blackwell: Oxford, **1963**.

8.4.3.10. Carbazole

Physical property: (i) Initial melting temperature ("meniscus temperature")
(ii) final melting temperature ("liquefaction temperature")
measured in the presence of air at 101 kPa.
Units: °C (IPTS-68)
Recommended reference material: carbazole
Range of variables: Melting point of typical batches, when measured under the conditions defined below.
(i) 246.2 °C   (ii) 246.6 °C

Contributor: J. D. Cox

Intended usage: Calibrating or checking melting-point apparatus that depends on visual observation. Insofar as the measured meniscus and liquefaction temperatures can be equated with solidus and liquidus temperatures respectively, other types of melting-point apparatus which measure these latter quantities can be checked.

Sources of supply and/or methods of purification: Samples of the substance are available from Supplier (B). The batches were purified by zone refining (Refs. 1, 2). Sample purity of a typical batch is about 0.999 mole fraction.

Pertinent physicochemical data: For typical sets of measurements (15 degrees of freedom) the standard deviation (reproducibility) was 0.08 °C and the overall uncertainty (2σ) was ±0.4 °C

Technique of realization: Portions of a powdered sample, kept over P_2O_5 in a desiccator, are sealed into four glass capillary tubes. These are placed in a stirred oil bath, heated to about 2 K below the initial melting temperature. The temperature of the oil-bath is then raised at 0.2 K min^{-1} and visual observation is made of the temperatures at which (i) a meniscus is first apparent, (ii) the last solid disappears. A calibrated mercury-in-glass thermometer should be used for temperature measurement. Replicate measurements are made with the tubes in other positions within the bath.

Significant difference between temperatures (i) and (ii) may reflect both impurity in the sample and the "over-run" caused by the continuing rise of bath temperature (0.2 K min^{-1}) during the melting process. It follows that the certified transition temperatures are method-dependent, and when used for calibration purposes apply only to methods that are similar to those used for certification.

The observed liquefaction temperature lies about 1.0 K above the true liquidus temperature.

REFERENCES

1. Herington, E. F. G.; Handley, R.; Cook, A. J. *Chem. and Ind. (London)*, **1956**, 292.
2. Herington, E. F. G. *Zone Melting of Organic Compounds*, Blackwell:, Oxford, **1963**.

8.4.3.11. Anthraquinone

Physical property: (i) Initial melting temperature ("meniscus temperature")
(ii) final melting temperature ("liquefaction temperature")
measured in the presence of air at 101 kPa.
Units: °C (IPTS-68)
Recommended reference material: anthraquinone
Range of variables: Melting point of typical batches, when measured under the conditions defined below.

(i) 285.2 °C   (ii) 285.8 °C

Contributor: J. D. Cox

Intended usage: Calibrating or checking melting-point apparatus that depends on visual observation. Insofar as the measured meniscus and liquefaction temperatures can be equated with solidus and liquidus temperatures respectively, other types of melting-point apparatus which measure these latter quantities can be checked.

Sources of supply and/or methods of purification: Samples of the substance are available from Supplier (B). The batches were purified by zone refining (Refs. 1 and 2). Sample purity of a typical batch is about 0.998 mole fraction.

Pertinent physicochemical data: For typical sets of measurements (15 degrees of freedom) the standard deviation (reproducibility) was 0.08 °C and the overall uncertainty (2σ) was ±0.4 °C

Technique of realization: Portions of a powdered sample, kept over P_2O_5 in a desiccator, are sealed into four glass capillary tubes. These are placed in a stirred oil bath, heated to about 2 K below the initial melting temperature. The temperature of the oil-bath is then raised at 0.2 K min^{-1} and visual observation is made of the temperatures at which (i) a meniscus is first apparent, (ii) the last solid disappears. A calibrated mercury-in-glass thermometer should be used for temperature measurement. Replicate measurements are made with the tubes in other positions within the bath.

Significant difference between temperatures (i) and (ii) may reflect both impurity in the sample and the "over-run" caused by the continuing rise of bath temperature (0.2 K min^{-1}) during the melting process. It follows that the certified transition temperatures are method-dependent, and when used for calibration purposes apply only to methods that are similar to those used for certification.

The observed liquefaction temperature lies about 0.7 K above the true liquidus temperature.

REFERENCES

1. Herington, E. F. G.; Handley, R.; Cook, A. J. *Chem. and Ind. (London)*, **1956**, 292.
2. Herington, E. F. G. *Zone Melting of Organic Compounds*, Blackwell: Oxford, **1963**.

8.4.3.12. Mercury

Physical property: Boiling point
Units: °C (IPTS-68)
Recommended reference material: distilled mercury
Range of variables: Boiling point = 356.66 °C, 629.81 K at 101 325 Pa.
Contributor: D. Ambrose, L. Crovini

Intended usage: calibration of thermometers of all types.

Sources of supply and/or methods of purification: Commercially obtained metals with nominal mass fraction purity of 0.999999 has proved satisfactory. SRM 743 of Supplier A can also be used. Purification methods are discussed in reference 1.

Pertinent physicochemical data: The relationship between the equilibrium temperature and the pressure is provided in table 8.4.3.1. The uncertainty of measurement is ±10 mK when the pressure is measured to within ±5 Pa.

Technique of realization: The general criteria for realizing a boiling point are presented in section 2.4. Figure 8.2.4 (see also references 2 and 3) presents a glass version of a mercury boiler. Further information on the technique of realization and the temperature to pressure relationship is provided in references 3 and 4.

REFERENCES

1. Brown, I.; Lane, J. E. *Pure and Appl. Chem.* **1976**, *45*, 1.
2. Ambrose, D. *J. Phys. E.* **1968**, 41.
3. Ambrose, D.; Sprake, C. H. S. *J. Chem. Thermodynamics* **1972**, *4*, 603.
4. Bettie, J. A.; Benedict, M.; Biaisdel, B. E. *Proc. Am. Acad. Arts Sci.* **1937**, *71*, 327.

8.4.3.13. Reference Materials for Differential Thermal Analysis

Supplier (A) distributes reference materials that have been certified by the International Confederation for Thermal Analysis (ICTA) for dynamic measurement of temperature by differential thermal analysis, differential scanning calorimetry and thermogravimetry. In the temperature range of interest of this recommendation the following reference materials are available:

GM 754
    Polystyrene; solid-solid transition at 105 °C.

GM 757
    1,2-Dichloroethane; Melting point at −32 °C.
    Cyclohexane; solid-solid transition at −83 °C.
        melting point at 7 °C.
    Diphenylether; melting point at 30 °C.
    o-terphenyl; melting point at 58 °C.

GM 758
    Potassium nitrate; solid-solid transition at 128 °C.
    Indium; melting point at 157 °C.
    Tin; melting point at 232 °C.
    Potassium perchlorate; at 300 °C.
    Silver sulfate; solid-solid transition at 430 °C.

GM 761
    Permanorm 3; solid-solid (magnetic) transition at 259 °C.
    Nickel; solid-solid (magnetic) transition at 353 °C.
    Mumetal; solid-solid (magnetic) transition at 381 °C.
    Permanorm 5; solid-solid (magnetic) transition at 454 °C.
    Trafoperm; solid-solid (magnetic) transition at 750 °C.

Information on the properties and the use of these reference materials is available from the Supplier (A) or the ICTA.

8.5. Contributors

D. Ambrose,
Department of Chemistry,
University College London,
20 Gordon Street,
London WC1H OAJ (UK)

M. V. Chattle,
Division of Quantum Metrology,
National Physical Laboratory,
Teddington, Middlesex TW11 0LW (UK)

J. D. Cox,
Division of Quantum Metrology,
National Physical Laboratory,
Teddington, Middlesex TW11 0LW (UK)

L. Crovini,
Instituto de Metrologia,
   del Consiglio Nazionale,
   delle Ricerche,
Strada delle Cacce 73,
I-10135 Torino (Italy)

M. F. Vaughan,
Division of Quantum Metrology,
National Physical Laboratory,
Teddington, Middlesex TW11 0LW (UK)

8.6. List of suppliers

A. Office of Standard Reference Materials,
U.S. Department of Commerce,
National Bureau of Standards,
Gaithersburg, MD 20899 (USA)

B. Office of Reference Materials
National Physical Laboratory,
Teddington, Middlesex, TW11 0LW (UK)

C. Jarret Instrument Co. Inc.,
2910 Lindell Court,
Wheaton, MD 20902 (USA)

D. Forshungsgemeinschaft für technisches
   Glas e. V.,
Postfach 1302,
D-6980 Wertheim 1,
(Federal Republic of Germany)

E. Yellow Springs Instrument Co.,
   Yellow Springs, OH 45387 (USA)

9

SECTION: ENTHALPY

COLLATORS: A. J. HEAD, R. SABBAH

CONTENTS:

9.1. Introduction

9.2. Reference materials for heat-capacity measurements

    In the solid state

        9.2.1. α-Aluminium oxide

        9.2.2. Platinium

        9.2.3. Copper

        9.2.4. Benzoic acid

        9.2.5. 2,2-Dimethylpropane

        9.2.6. Molybdenum

    In the solid and liquid states

        9.2.7. Naphthalene

        9.2.8. Diphenyl ether

        9.2.9. Heptane

    In the solid, liquid and real-gas states

        9.2.10. Hexafluorobenzene

    In the glass and liquid states

        9.2.11. Polystyrene (atactic)

        9.2.12. Poly(vinyl chloride)

In the liquid and real-gas states
- 9.2.13. 1,4-Dimethylbenzene
- 9.2.14. Water
- 9.2.15. Benzene
- 9.2.16. Nitrogen

In the real-gas state
- 9.2.17. Carbon dioxide

9.3. Reference materials for measurement of enthalpies of phase change

solid → solid and/or solid → liquid phase changes
- 9.3.1. 1,3-Difluorobenzene
- 9.3.2. 2,2-Dimethylpropane
- 9.3.3. Hexafluorobenzene
- 9.3.4. Diphenyl ether
- 9.3.5. Naphthalene
- 9.3.6. Benzil
- 9.3.7. Acetanilide
- 9.3.8. Benzoic acid
- 9.3.9. Diphenylacetic acid
- 9.3.10. Indium
- 9.3.11. Sodium nitrate

solid → gas phase changes
- 9.3.12. Naphthalene
- 9.3.13. Iodine

liquid → gas phase changes
- 9.3.14. Water
- 9.3.15. Benzene
- 9.3.16. 1-Propanol
- 9.3.17. Hexafluorobenzene

9. Enthalpy

9.4. Reference materials for measurement of enthalpies of reaction and related processes

solid + solid processes

9.4.1. Zirconium + barium chromate

liquid + liquid processes

9.4.2. Sulphuric acid solution + sodium hydroxide solution

9.4.3. Cyclohexane + hexane

9.4.4. 1,4-Dioxan + tetrachloromethane

9.4.5. Sucrose solution + water

9.4.6. Urea solution + water

solid + liquid processes

9.4.7. α-Quartz + hydrofluoric acid solution

9.4.8. Potassium chloride + water

9.4.9. Tris(hydroxymethyl)aminomethane + hydrochloric acid solution or sodium hydroxide solution

9.4.10. 4-Aminopyridine + perchloric acid solution

gas + gas processes

9.4.11. Hydrogen + oxygen

9.4.12. Hydrogen + chlorine

9.4.13. Methane + oxygen

9.5. Reference materials for measurement of enthalpies of combustion

Combustion of solids in oxygen

9.5.1. Benzoic acid

9.5.2. Succinic acid

9.5.3. Hippuric acid

9.5.4. Acetanilide

9.5.5. Urea

9.5.6. 4-Chlorobenzoic acid

9.5.7. 4-Fluorobenzoic acid

9.5.8. Pentafluorobenzoic acid

9.5.9. Thianthrene

9.5.10. Triphenylphosphine oxide

Combustion of liquids in oxygen

9.5.11. 2,2,4-Trimethylpentane

9.5.12. α,α,α-Trifluorotoluene

Combustion of solids in fluorine

9.5.13. Sulphur

9.5.14. Tungsten

9.6. Contributors

9.7. List of suppliers

9.1. Introduction

9.1.1. General

The general technique of calorimetry (*i.e.* the measurement of heat or energy) is applicable to a wide range of physical and chemical processes. It is used to determine heat capacities, and the enthalpies associated with changes of state, phase changes, adsorption, radiation, chemical reactions and biological processes. This technique is frequently capable of yielding results which are highly reproducible, but which may, nevertheless, be inaccurate because of faulty calibration of the measurement system.

The concept of calculating the 'water equivalent' of a calorimeter is of historical interest only, and the calibration of a modern calorimeter is achieved by measuring the signal produced when a known quantity of energy is generated within the system. The method of substitution is employed, whereby the processes of calibration and of measurement are matched as closely as possible: not only should the quantity of energy be the same, but the site and kinetics of its generation, and the temperature range (or the temperature of an isothermal system) should be as nearly as possible identical in the two series of experiments.

The energy of calibration should ideally be generated electrically but, where the nature of the calorimetric system makes this difficult to achieve, then a reference material with a well established value of the property under study may be used as a calibrant. Whichever method of calibration is employed, reference materials are needed to check the results obtained from using the system, and especially to ensure that the calorimeter, calibrated by a general method, is applicable to the particular type of material, reaction, or process for which it is intended to use it, and is not subject to systematic errors.

The need for 'test substances' in combustion calorimetry was recognized more than 50 years ago and the (then) International Union of Chemistry made important recommendations in the 1930s. During the past 30 years reference materials for calorimetry (mainly bomb, reaction and heat-capacity calorimetry) have been developed by national standardizing laboratories (such as the National Bureau of Standards in the U.S.A. and VNIIM in the U.S.S.R.) and through projects sponsored by bodies such as the Calorimetry Conference (U.S.A). Ideally, a reference material should have been studied in several specialized laboratories and its properties established by measurements involving several techniques. Some of the materials and reactions recommended in this chapter have been so studied and a few are available from national laboratories with *certified* values for their thermal properties. Others are included because, on the evidence available, they appear to be potentially suitable but more measurements are required before their use can be unequivocally recommended. They are included in the belief that they will prove useful in areas where there are no established alternatives, and in the hope that further work on them will thereby be encouraged.

It has been the intention of the present work to bring the 1974 publication (Ref. 1) up-to-date. It must be admitted, however, that the past decade has seen a decline in activity in calorimetry, especially in the more traditional areas such as combustion calorimetry: few additional reference materials have become established, although new data have been published on existing substances. There continue to be gaps in the availability of reference materials in such important fields of calorimetry as adsorption phenomena, biological processes, and the measurement of heat capacities of liquids and gases at high pressures.

In our recommendations we have indicated the temperature scale used, whenever this information was given by the authors. When the scale differed from IPTS-68 and resulted in a significant difference in the reported results, we have, whenever possible, requested the authors or the laboratory concerned to recalculate the values of the thermodynamic properties according to IPTS-68. Where this was not possible, we have quoted unchanged the values of the thermodynamic properties, but have converted the temperatures to which they refer to IPTS-68 by means of the tables and sources of information given in references 2 and 3. We gratefully aknowledge the co-operation of L. Haar (National Bureau of Standards, U.S.A.) and B. E. Gammon (Thermodynamics Research Center) in this work.

Finally, we wish to thank all those calorimetrists throughout the world who have in any way contributed to this publication.

REFERENCES

1. Cox, J. D. *Pure and Appl. Chem.* **1974**, *40*, 399.
2. Riddle, J. L.; Furukawa, G. T.; Plumb, H. H. *Platinum Resistance Thermometry*, NBS Monograph 126, National Bureau of Standards: Washington, **1973**.
3. Hust, J. G. *Cryogenics* **1969**, *9*, 443.

9.1.2. Reference materials for heat-capacity measurements

In this revision of the 1974 publication treating the same subject, we have proposed reference materials for the measurement of heat capacity in the three fundamental states of matter (solid, liquid and gas) over, wherever possible, a wide range of temperatures and pressures. We have extended the 1974 publication as follows:

(i) increased the temperature ranges for α-alumina, copper and molybdenum up to 2250, 300 and 2800 K respectively;

(ii) reduced the upper limit of the temperature range for platinum to 1500 K, since significant discrepancies are found in the results obtained at higher temperatures;

(iii) introduced three new substances, polystyrene and poly(vinyl chloride) as reference materials suitable for measurements in the glass and liquid states in the temperature range 10 to 470 K (polystyrene) and 10 to 380 K (PVC), and hexafluorobenzene as a reference material suitable for measurements in the solid, liquid, and gaseous states in the range 10 to 350 K.

(iv) added C_p data for the gaseous state of 1,4-dimethylbenzene and water, and extended those for benzene to a pressure of 202.65 kPa;

(v) extended the data for the heat capacity of nitrogen to high pressures;

(vi) replaced carbon disulphide by carbon dioxide, which is easier to obtain experimentally. Literature data, correlations and reviews are much more plentiful for carbon dioxide than for carbon disulphide.

Wherever possible, the data which appeared in the 1974 version have been replaced by more recent values or more complete tables. Nevertheless, some old data accompanying reference materials for heat capacity measurements, especially organic compounds, which appeared in that version are retained because they were determined in laboratories specialising in heat-capacity measurements. It is desirable that they should be replaced or confirmed by more recent values, obtained with appropriate techniques and modern apparatus.

In deciding to update the 1974 publication, Commission 1.4 of IUPAC wanted to increase, as far as possible, the amount of data made available on reference materials for physical property measurements. This could be achieved in two ways – by increasing the numerical data given for each substance and/or by increasing the number of reference materials proposed for each property. The results of applying these approaches are briefly outlined above. We discuss below the compounds selected as suitable for calibration and test measurements.

(i) *Solid State:* In the temperature range 0 to 300 K, we have proposed organic substances (benzoic acid, 2,2-dimethylpropane, naphthalene, diphenyl ether, hexafluorobenzene and heptane), a metal (copper) and α-alumina. The last material is non-hygroscopic, non-volatile, chemically stable, relatively cheap, available in large quantities in a very pure state, and useable from 10 to 2250 K in an oxidizing, neutral or weakly-reducing atmosphere. Being a thermal insulator it is well suited for use with apparatus intended for heat-capacity measurements on poor conductors. It is appropriate to draw the attention of thermochemists to indium which offers several advantages: it is chemically stable, can be obtained very pure (0.99999 mass fraction), has a simple crystal structure and can be used at very low temperatures (below 5 K) for the calibration of calorimeters used for heat-capacity measurements on quench-condensed thin metallic films in the temperature range 0.6 to 4.4 K. This substance is not included, however, because of some disparity in the extant data. Further measurements are needed before it can be included in a future edition of this publication.

In the temperature range 300 to 4000 K, apart from α-alumina which can be used up to 2250 K, platinum meets the requirements for use in apparatus intended for measurements on metals or alloys up to 1500 K, but it is expensive when large quantities of sample are needed. Molybdenum overcomes this disadvantage; it can be used up to 2800 K but requires a neutral or reducing atmosphere. We would have included tungsten (which has a high melting point, is chemically stable, can be obtained pure, and does not undergo any transitions up to the melting point) but the disagreement between published values of its heat capacity above 2800 K did not allow us to do so. In a recent paper (Ref. 1), Ditmars has published precise values obtained at the NBS up to 1173.15 K. According to the author, this work constitutes the first step of a study aimed at precise determination of the heat capacity of tungsten between 273 and 3600 K.

(ii) *Liquid State:* Besides water, we have proposed heptane, naphthalene, hexafluorobenzene, diphenyl ether, 1,4-dimethylbenzene and benzene. Suitable reference materials are therefore available to cover liquid heat-capacity measurements over the temperature range 183 to 647 K.

(iii) *Gas State:* As well as water, benzene, 1,4-dimethylbenzene and hexafluorobenzene, which (except for water) can be used at pressures up to 202.65 kPa, we have proposed carbon dioxide and nitrogen for which there are available heat-capacity values at high pressures.

Data for the heat capacities of liquids and gases at high pressures are too few and insufficiently well established for any materials to be included in the present proposals (Refs. 2 to 4). Nevertheless, there is an increasing need felt by industry in general and the petrochemical industry in particular for reference materials in this area and suitable data, especially on organic substances, would be greatly welcomed. For this reason we ask thermochemists to consider working in this area.

The calorimetric techniques used in the measurement of heat capacity are numerous – adiabatic, drop, dynamic (laser-flash, a.c., pulse, differential scanning) – and are chosen according to the temperature range in which measurements are made. The interested reader is invited to consult the article by Takahashi (Ref. 5) which extends the earlier work by McCullough and Scott (Ref. 6). To assist those who use the drop method, we have wherever possible given values of enthalpy as well as heat capacity.

The temperatures listed in the tables included in the recommendations refer to IPTS-68 for $T \geq 13.81$ K, either taken directly from the published values or, where necessary, converted to IPTS-68 by means of the tables given in reference 7. In the latter case we have listed both the published temperatures and the corresponding recalculated values on IPTS-68. For T < 13.81 K we have indicated the procedure used by the publishers of the selected values for the calibration of their thermometers.

The uncertainty which accompanies values of heat capacity is generally small and depends on the apparatus, the material (nature, physical form and quantity) and the temperature region in which measurements are made. For further information, we invite the reader who is interested to consult the individual entries for each substance.

REFERENCES

1. Ditmars, D. A. *High Temp.-High Pressures* **1979**, *11*, 615.
2. Schneider, G. M. *High Temp.-High Pressures* **1977**, *9*, 559.
3. Naziev, Ya. M.; Mustafaev, R. M.; Abasov, A. A. *Proc. 7th Symp. Thermophys. Prop.* ASME, NBS, Gaithersburg, MD, **1977**, 525.
4. Bury, P.; Johannin, P.; Martin, G.; Vodar, B. *Proc. 7th Symp. Thermophys. Prop.* ASME, NBS, Gaithersburg, MD, **1977**, 546.
5. Takahashi, Y. *Pure and Appl. Chem.* **1976**, *47*, 323.
6. McCullough, J. P.; Scott, D. W., editors, *Experimental Thermodynamics*, Vol. I: *Calorimetry of Non-reacting Systems*, Butterworths: London, **1968**.
7. Riddle, J. L.; Furukawa, G. T.; Plumb, H. H. *Platinum Resistance Thermometry*, NBS Monograph 126, National Bureau of Standards: Washington, **1973**.
8. Douglas, T. B. *J. Res. Nat. Bur. Stand.* **1969**, *73*, 451.

9.1.3. Reference materials for measurement of enthalpies of phase change

By phase change is understood fusion (change from the solid state to the liquid state of a substance), vaporization (change from the liquid state to the gas state of a substance), sublimation (change from the solid state to the gas state of a substance), and transition (change of phase of a substance remaining in the same state).

Compared with the 1974 version, this revision introduces three new reference materials for solid state transitions, namely 2,2-dimethylpropane, 1,3-difluorobenzene and sodium nitrate, which undergo transitions at 140.49 K (Ref. 1), 186.80 K (Ref. 2) and 549 K (Ref. 3) respectively.

With regard to fusion, the present version has been considerably improved by the inclusion of recently published data (Ref. 3). Thus eleven substances are proposed for the temperature range 204 to 580 K which will greatly assist workers who use differential scanning calorimetry, a technique which today is employed as often in university laboratories as in industry. Amongst these eleven substances, two (benzoic acid and diphenyl ether) have already featured in the list included in the 1974 version. Benzene has been omitted, partly because of the poor quality of the data which appeared in the first version and partly because of the lack of recent results for its enthalpy of fusion. For some substances, only the value obtained by Andon and Connett (Ref. 4) or by Messerly and Finke (Ref. 2) has been recommended. Although these studies appear to be very thorough, it is nevertheless desirable that their results be confirmed so that the values can be recommended with more confidence.

In general, the results obtained by adiabatic calorimetry are by far the most reproducible (uncertainty $\leq \pm 0.2$ per cent for typical references given in the texts for enthalpies of fusion) compared with those obtained from other types of calorimeters (*e.g.* Tian-Calvet, where the uncertainty is of the order of ± 2 per cent (Ref. 5)). As for the results obtained by differential scanning calorimetry, the uncertainty is always greater than ± 1 per cent and may in some cases be several per cent. The reason for this seems to us to arise principally from the quantity of substance used which, in the case of experiments carried out by adiabatic calorimetry, is of the order of several grams, whereas it is less than one gram in the experiments using a Tian-Calvet calorimeter described in reference 5, and less than 10 mg in the measurements made by differential scanning calorimetry (Ref. 6). The use of the simple and rapid technique of differential scanning calorimetry is becoming increasingly widespread. We have noted two recent studies on this subject: the first by Van Dooren and Müller (Ref. 7) treats the influence of variables on the results obtained by differential scanning calorimetry; the second by Breuer and Eysel (Ref. 3) includes an analysis of possible errors made by users of this technique. Differential scanning calorimetry has been used for the certification of new reference materials for enthalpy of fusion issued by the National Bureau of Standards (tin, SRM 2220, zinc, SRM 2221, biphenyl, SRM 2222) in 1986. The development of new reference materials for differential scanning calorimetry has been discussed by Callanan *et al.* (Ref. 15).

As already stated, the reference materials proposed for enthalpies of fusion are for the most part organic compounds and their melting points extend from 204 to 580 K. We think that it would be useful to increase this temperature range by including other materials (metals and inorganic compounds) some of which undergo transitions and would therefore be useful as reference materials for this property also. We hope that calorimetrists who share our concern will set to work in order that their results may benefit the next edition of this publication.

Regarding vaporization, we have retained the two substances proposed in the 1974 version, because of the concordance of the reported values, and have added two other compounds (1-propanol and hexafluorobenzene). For 1-propanol the agreement between published values led to its use by Majer *et al.* to test their calorimeter (Ref. 9).

We met the greatest difficulty when we came to recommend reference materials for measurements of enthalpies of sublimation. Compared with the 1974 version, we have omitted benzoic acid because recent studies (Ref. 10) have confirmed the existence of an equilibrium between monomer and dimer in the vapour phase, which had been pointed out previously by de Kruif and Oonk (Ref. 11). On the other hand, we have added iodine and retained naphthalene for which the scatter of the results remains small in comparison with all the other candidate substances in the literature that we might have proposed (benzophenone, *trans*-diphenylethylene, ferrocene, biphenyl, etc.). It is particularly important to have reference materials covering a wide range of saturation vapour pressures and temperatures. In order best to resolve discrepancies in reported values, it would be most useful for several laboratories with experience in this field to collaborate in a standardisation campaign devoted to working on representative samples taken from a single batch. The comparison of results obtained by different techniques, even from different apparatus using the same technique, would lead within a few years to the creation of a bank of high quality reference materials. Enthalpies of sublimation are generally derived from measurements of vapour pressure as a function of temperature, and are not obtained by calorimetry, which seems to be the best method to use in the case of substances having very low saturated vapour pressures.

The uncertainty of the results for enthalpies of vaporization or sublimation depends on the substance, on the saturation vapour pressure, and on the technique used (calorimetry, measurement of vapour pressure as a function of temperature, or differential scanning calorimetry). Cox and Pilcher made some quantitative estimates (Ref. 12). In general, it can be stated that at 298.15 K, enthalpies of vaporization or sublimation can be determined with an uncertainty between ± 0.1 and ± 0.5 per cent if the technique used is calorimetry, and between ± 0.1 and ± 5 per cent, according to the values of the vapour pressure of the compound, if the method employed is the measurement of vapour pressure as a function of temperature. It must be pointed out that the results obtained by differential scanning calorimetry show the lowest reproducibility and the values of the enthalpies of sublimation or vaporization obtained by this method generally differ from those obtained by other techniques (Refs. 13 to 15).

The uncertainty of the results given in Section 3 of this chapter is represented, unless otherwise indicated, by the standard deviation of the mean σ_m.

REFERENCES

1. Enokido, H. ; Shinoda, T.; Mashiko, Y. *Bull. Chem. Soc. Japan* **1969**, *42*, 84.
2. Messerly, J. F.; Finke, H. L. *J. Chem. Thermodynamics* **1970**, *2*, 867.
3. Breuer, K. H.; Eysel, W. *Thermochim. Acta* **1982**, *57*, 317.
4. Andon, R. J. L.; Connett, J. E. *Thermochim. Acta* **1980**, *42*, 241.
5. Malaspina, L.; Gigli, R.; Piacente, V. *Rev. Int. Hautes Temper. et Refract.* **1971**, *8*, 211.
6. Lowings, M. G.; McCurdy, K. G.; Hepler, L. G. *Thermochim. Acta* **1978**, *23*, 265.
7. Van Dooren, A. A.; Müller, B. W. *Thermochim. Acta* **1981**, *49*, 151, 163, 175, 185.
8. Callanan, J. E.; Sullivan, S. A.; Vecchia, D. F. *Feasibility Study for the Development of Standards using Differential Scanning Calorimetry*, NBS Special Publication 260-99, National Bureau of Standards: Gaithersburg, **1985**.
9. Majer, V.; Svoboda, V.; Hynek, V.; Pick, J. *Collect. Czech. Chem. Commun.* **1978**, *43*, 1313.
10. Murata, S.; Sakiyama, M.; Seki, S. *J. Chem. Thermodynamics* **1982**, *14*, 723.
11. de Kruif, C. G.; Oonk, H. A. *J. Chem. Ing. Techn.* **1973**, *45*, 455.

12. Cox, J. D.; Pilcher, G. *Thermochemistry of Organic and Organometallic Compounds*, Chapter 4, Academic Press: London, **1970**.
13. Beech, G.; Lintonbon, R. M. *Thermochim. Acta* **1971**, *2*, 86.
14. Mita, I.; Imai, I.; Kambe, H. *Thermochim. Acta* **1971**, *2*, 337.
15. Murray, J. P.; Cavell, K. J.; Hill, J. O. *Thermochim. Acta* **1980**, *36*, 97.

9.1.4. Reference materials for measurement of enthalpies of reaction and related processes

In the 1974 publication attention was drawn to the somewhat arbitrary distinction between reference materials for processes described as reaction, dissolution, and dilution. In the present revision, Section 4 includes reference materials for all these processes, except for reactions normally carried out in combustion bombs. The entries have been classified according to the physical states of the substances, which largely determine the type of calorimeter employed. The liquid-liquid systems, however, include processes normally studied in enthalpy-of-mixing calorimeters where the vapour spaces are small or non-existent (cyclohexane + hexane, 1,4-dioxan + tetrachloromethane), or in reaction calorimeters (sulphuric acid solution + sodium hydroxide solution), as well as dilution reactions (sucrose solution, urea solution) which are of wide applicability but often particularly useful for checking the performance of liquid flow calorimeters, including microcalorimeters.

The reaction between tris(hydroxymethyl)aminomethane (TRIS) and hydrochloric acid continues to be the most widely employed exothermic system for checking the performance of calorimeters used for studying the dissolution of a solid in a liquid, despite its susceptibility to a number of experimental variables, which have only been properly understood during the past decade. We have added the reaction between 4-aminopyridine and perchloric acid solution which seems to offer promise as an alternative and maybe simpler reaction. Both the endothermic processes for testing reaction calorimeters given in the 1974 publication have been retained. No new results have been reported for the TRIS + sodium hydroxide solution reaction, but the situation regarding potassium chloride has been greatly improved by a thorough study at the NBS, from where a *certified* reference material is now available. A suitable slow process for testing reaction calorimeters is still lacking, but the dissolution of α-quartz in hydrofluoric acid solution can be used where the calorimeter is resistant to attack by this acid.

An exothermic system is proposed for checking the performance of calorimeters for measuring the enthalpies of mixing of liquids, namely, 1,4-dioxan + tetrachloromethane. This complements the well studied endothermic mixture cyclohexane + hexane.

The dilution of urea solution has been added. Values for this system, which has advantages compared with the dilution of sucrose solution when less viscous solutions are required, are well established. Data for the dilution of solutions containing phosphate buffer (pH = 7) are also available, which are of use to those studying biochemical reactions.

All the reference materials proposed in sections V and VI of the 1974 publication have been retained, but only for those mentioned above have significant results been reported and included in revised entries.

A very large number of 'reaction' and 'solution' calorimeters have been described in the literature, ranging from the simplest, where the process takes place in a Dewar vessel, to the most elaborate adiabatic systems, capable of a precision of the order of 0.01 per cent. The accuracy attainable with the latter is often limited less by the calorimeter than by the nature of the reacting system. The extensive studies on the neutralization of TRIS by hydrochloric acid (e.g. Ref. 1) have demonstrated the importance of attention to the reactants (purity, moisture content, state of sub-division) and the surrounding atmosphere (volume of vapour space, its pressure and composition) when the highest accuracy is sought. Although published over 20 years ago, the IUPAC monograph (Ref. 2) still provides the best concise account of the principles of reaction calorimetry design.

Throughout this section, the usual practice of experimental thermochemists in presenting the uncertainty associated with numerical results by *twice* the overall standard deviation of the mean has been followed.

REFERENCES

1. Prosen, E. J.; Kilday, M. V. *J. Res. Nat. Bur. Stand.* **1973**, *77A*, 581.
2. Skinner, H. A., editor, *Experimental Thermochemistry*, Vol II, Wiley (Interscience): New York, **1962**.

9.1.5. Reference materials for measurement of enthalpies of combustion

In this section we include reference materials for reactions carried out in combustion bombs. The experimental measurements thus relate to a constant-volume system and are a determination of the energy of combustion from which the enthalpy may be readily calculated by means of the thermodynamic relation $\Delta H = \Delta U + p\Delta V$. (Reference materials for enthalpies of combustion by flame calorimetry have been dealt with in the previous section).

The last decade has seen a decline in the number of published papers on experimental combustion calorimetry, but the literature on the subject has been enriched by the publication of an authoritative treatise (Ref. 1) under the auspices of the International Union of Pure and Applied Chemistry. The chapter on 'Test and auxilary substances in combustion calorimetry' is particularly relevant to this section, and includes discussion of several candidate reference materials not yet sufficiently well established to be included in the present recommendations.

Benzoic acid continues to fulfil a unique rôle as the universal calibrant for combustion calorimeters, which are only exceptionally calibrated electrically. A further value for the energy of combustion of an NBS sample of benzoic acid (SRM 39i) 'under standard bomb conditions', determined with such an electrically-calibrated calorimeter (Ref. 2), serves to confirm the rounded value of 26434 J g^{-1}, already well established.

The present revision includes all the materials listed in the 1974 publication with the exception of tris(hydroxymethyl)aminomethane (TRIS) as a reference material for the combustion of C,H,O,N compounds, for which acetanilide has been substituted, although with only a marginally better established energy of combustion. The value for hippuric acid is known with more confidence, but the use of this substance as a reference material for nitrogen-containing compounds is complicated by the difficulty of removing water from the sample.

Of the other compounds retained from the 1974 publication, no more recent results have been reported for the energies of combustion, except for succinic acid and 4-chlorobenzoic acid, although minor changes have been made to the selected values for several compounds. More measurements are needed, in particular, on urea and pentafluorobenzoic acid to establish their energies of combustion.

We have added α,α,α-trifluorotoluene as a reference material for the combustion of liquid C,H,F,(O,N) compounds containing moderate amounts of fluorine and which need to be encapsulated before combustion; its energy of combustion is adequately known. Triphenylphosphine oxide has been included as a candidate reference material for phosphorus-containing compounds, although more determinations are needed before this substance can be accorded the status of a recommended reference material.

In the event that more combustion calorimetry is carried out in the future on bromine- and iodine-containing compounds, it is likely that the corresponding 4-halobenzoic acids, on which some measurements have been made, would be suitable as reference materials, but more work is needed on both compounds.

The precision of the results obtained using the best static- and rotating-bomb calorimeters is of the order of 0.01 per cent and results of this degree of accuracy can be obtained with highly purified compounds containing no elements other than C, H, O and N. When other elements are present, however, the accuracy is limited by the extent to which the stoichiometry of the bomb process can be controlled and determined. It is certainly true that far more inaccurate energies of combustion arise from incomplete combustion or from lack of a well-defined final state than from difficulties in the physical measurement of the energy of the bomb process. Reference materials have an important rôle to play in overcoming these problems.

Fluorine combustion calorimetry is a powerful method for studying the thermochemistry of many inorganic substances and, because the technique is difficult and the conditions of calibration (combustion of benzoic acid in oxygen) are so different from those of the measurements, the use of reference materials is particularly desirable. We have therefore included sulphur and tungsten, for which the energies of combustion in fluorine are well established, amongst our recommendations.

Throughout this section, the usual practice of combustion calorimetry in representing the uncertainty associated with numerical results by *twice* the overall standard deviation of the mean has been followed.

REFERENCES

1. Sunner, S.; Månsson, M., editors, *Combustion Calorimetry*, Pergamon: Oxford, **1979**.
2. Jochems, R.; Dekker, H.; Mosselman, C. *Rev. Sci. Inst.* **1979**, *50*, 859.

9.2. Reference materials for heat-capacity measurements

9.2.1. C_p, H, α-Aluminium oxide

Physical property: Heat capacity, enthalpy
Units: J mol^{-1} K^{-1} (molar heat capacity, C_p); J mol^{-1} (molar enthalpy, H)
J kg^{-1} K^{-1} (specific heat capacity, c_p); J kg^{-1} (specific enthalpy, h)
Recommended reference material: α-Aluminium oxide (Al_2O_3)
Range of variables: 10 to 2250 K
Physical state within the range: solid
Contributors to the first version: J. D. Cox, D. A. Ditmars, G. T. Furukawa, J. F. Martin, O. Riedel, D. R. Stull
Contributors to the revised version: S. S. Chang, J. Lielmezs, J. F. Martin, R. Sabbah

Intended usage: Highly pure α-alumina (synthetic sapphire, corundum) is recommended for testing the performance of calorimeters (Refs. 1 to 3) used for the measurement of the heat capacities of solids and liquids or for the measurement of the enthalpies of solids (Ref. 4). This material could also be used for the calibration of such calorimeters.

Sources of supply and/or methods of preparation: A highly pure grade of α-alumina (SRM 720) with a certificate giving its measured enthalpy (relative to 0 K) and heat capacity from 10 to 2250 K is available from supplier (S). Samples of α-alumina of comparable purity can be obtained (without thermal certification) from supplier (K). Sapphire discs of various thicknesses and diameters are available from supplier (A). Discs of 6 mm diameter and 0.5 mm thickness are suitable for differential scanning calorimetry. Other high purity material which might be suitable is also available from various commercial sources, *e.g.* suppliers Aldrich, Baker, Koch-Light, etc.

Pertinent physicochemical data: This material has no solid-solid transition in the quoted temperature range. Furukawa *et al.* (Ref. 5) studied a sample of α-alumina with mass fraction purity of 0.9998 to 0.9999 assessed by spectrographic analysis. They employed adiabatic calorimetry in the temperature range 13 to 380 K and a drop method with a Bunsen ice calorimeter from 273 to 1170 K. Ditmars *et al.* (Ref. 6) consolidated the results

9. Enthalpy

on a sample of comparable purity (SRM 720) in the temperature range 10 to 2250 K from C_p measured by automated adiabatic calorimetry (Ref. 7) in the temperature range 8 to 375 K and from relative enthalpy measured by drop calorimeters of different designs (Refs. 6, 8) in the temperature range 273 to 2250 K. Results obtained by all these methods agree well in the overlap ranges. Values of heat capacity and enthalpy with temperatures are given in the following table. Temperatures are reported in IPTS-68 above 13.81 K and in NBS-1965 provisional scale below 13.81 K (Ref. 7).

Table 9.2.1.1. Molar heat capacity C_p and enthalpy $H(T) - H(0\text{ K})$ of α-aluminium oxide

T_{68} K	C_p J mol^{-1} K^{-1}	$H(T)-H(0\text{ K})$ J mol^{-1}	T_{68} K	C_p J mol^{-1} K^{-1}	$H(T)-H(0\text{ K})$ J mol^{-1}	T_{68} K	C_p J mol^{-1} K^{-1}	$H(T)-H(0\text{ K})$ J mol^{-1}
10	0.0091	0.023	350	88.84	14383	1000	124.77	87986
25	0.146	0.898	400	96.08	19014	1100	126.61	100560
50	1.506	17.11	450	101.71	23965	1200	128.25	113300
75	5.685	100.32	500	106.13	29165	1400	131.08	139240
100	12.855	326.6	550	109.67	34563	1600	133.36	165700
150	31.95	1433.1	600	112.55	40121	1800	135.13	192550
200	51.12	3519.9	700	116.92	51607	2000	136.50	219720
250	67.08	6490.3	800	120.14	63468	2200	137.73	247140
300	79.41	10166	900	122.66	75612	2250	138.06	254030

The molar mass was taken as 101.9613 g mol^{-1}. The measurements made by Macleod (Ref. 9) from 400 to 1250 K agree with the values in the table, except at the highest temperatures.

Equations for C_p and $H(T) - H(0\text{ K})$ as functions of T (0 to 2200 K) have been given by Reshetnikov (Ref. 10). The values he reports for $H(T) - H(0\text{ K})$ differ from those in reference 6 by about 0.2 per cent in the range 100 to 1200 K and by 1 per cent at 2100 K. Two other papers gave results on this material in the temperature range 300 to 550 K (Ref. 11) and 80 to 400 K (Ref. 12).

REFERENCES

1. McCullough, J. P.; Scott, D. W., editors, *Experimental Thermodynamics*, Vol. I: *Calorimetry of Non-Reacting Systems*, Chapters 5 and 9. Butterworths: London, **1968**.
2. Takahashi, Y.; Yokokawa, H.; Kadokura, H.; Sekine, Y.; Mukaido, T. *J. Chem. Thermodynamics* **1979**, *11*, 379.
3. Van Dooren, A. A.; Müller, B. W. *Thermochim. Acta* **1981**, *49*, 151.
4. McCullough, J. P.; Scott, D. W., editors, *Experimental Thermodynamics*, Vol. I: *Calorimetry of Non-Reacting Systems*, Chapter 8. Butterworths: London, **1968**.
5. Furukawa, G. T.; Douglas, T. B.; McCoskey, R. E.; Ginnings, D. C. *J. Res. Nat. Bur. Stand.* **1956**, *57*, 67.

6. Ditmars, D. A.; Ishihara, S.; Chang, S. S.; Bernstein, G.; West, E. D. *J. Res. Nat. Bur. Stand.* **1982**, *87*, 159.
7. Chang, S. S. *Proc. 7th Symp. Thermophys. Prop.* ASME, NBS, Gaithersburg, MD, **1977**, 83.
8. Ditmars, D. A.; Douglas, T. B. *J. Res. Nat. Bur. Stand.* **1971**, 75A, 401.
9. Macleod, A. C. *Trans. Faraday Soc.* **1967**, *63*, 300.
10. Reshetnikov, M. A. *Zh. Fiz. Khim.* **1969**, *43*, 2238.
11. Andrews, J. T. S.; Norton, P. A.; Westrum, Jr., E. F. *J. Chem. Thermodynamics* **1978**, *10*, 949.
12. Zhicheng, T.; Lixing, Z.; Shuxia, C.; Anxue, Y.; Yi, S; Jinchun, Y.; Xiukun, W. *Scientia Sinica, Series B* **1983**, *26*, 1014.

9.2.2. C_p, H, Platinum

Physical property: Heat capacity; enthalpy
Units: J mol^{-1} K^{-1} (molar heat capacity, C_p); J mol^{-1} (molar enthalpy, H)
     J kg^{-1} K^{-1} (specific heat capacity, c_p); J kg^{-1} (specific enthalpy, h)
Recommended reference material: Platinum (Pt)
Range of variables: 298 to 1500 K
Physical state within the range: solid
Contributor to the first version: G. T. Armstrong
Contributors to the revised version: J. Lielmezs, J. F. Martin, J. Rogez, R. Sabbah

Intended usage: Because of its high chemical stability, freedom from transitions, high melting point, $T_{\text{fus}} = 2045$ K (Ref. 1), availability in high purity, and low volatility at high temperatures, platinum is recommended for checking the accuracy of apparatus (especially those which use small quantities of material such as drop calorimeters and differential scanning calorimeters) for the measurement of enthalpy and heat capacity up to high temperatures.

Sources of supply and/or methods of preparation: High purity platinum in wire form is available as SRM 680 from supplier (S). Samples of platinum of comparable purity can be obtained from supplier (K); samples of platinum of mass fraction purity greater than 0.999 are available from various sources *e.g.* (F, M), suppliers Alfa, etc.

Pertinent physicochemical data: The values of heat capacity recommended are those adopted by Hultgren *et al.* (Ref. 1) which agree with the enthalpy measurements by drop calorimetry made by Kendall *et al.* (Ref. 2) (339 to 1435 K), Jaeger *et al.* (Ref. 3) (681 to 1664 K), Jaeger and Rosenbohm (Ref. 4) (484 to 1877 K), and White (Ref. 5) (373 to 1573 K) and by laser flash calorimetry made by Yokokawa and Takahashi (Ref. 6) (80 to 1000 K). The following table gives values of the heat capacity and enthalpy (correction to IPTS-68 was made by using approximate equations given in reference 7). The molar mass was taken as 195.09 g mol^{-1}.

9. Enthalpy

Table 9.2.2.1. Molar heat capacity C_p and enthalpy $H(T) - H(298.15\text{ K})$ of platinum

$\dfrac{T_{68}}{\text{K}}$	$\dfrac{C_p}{\text{J mol}^{-1}\text{ K}^{-1}}$	$\dfrac{H(T)-H(298.15\text{ K})}{\text{J mol}^{-1}}$	$\dfrac{T_{68}}{\text{K}}$	$\dfrac{C_p}{\text{J mol}^{-1}\text{ K}^{-1}}$	$\dfrac{H(T)-H(298.15\text{ K})}{\text{J mol}^{-1}}$
298.15	25.87	0	1000	29.54	19479
400	26.47	2669	1100	30.09	22459
500	26.98	5337	1200	30.58	25490
600	27.53	8065	1300	31.16	28582
700	28.03	10843	1400	31.71	31720
800	28.57	13675	1500	32.21	34915
900	29.06	16554			

Macleod (Ref. 8) has measured the enthalpy of platinum in the temperature range 400 to 1700 K. The following equation has been fitted to his enthalpy results

$$[H(T) - H(298.15\text{ K})]/\text{J mol}^{-1} = -7658.8 + 32628.5/x + 24.529x \\ + 2.5879 \times 10^{-3}x^2, \tag{9.2.2.1}$$

where $x = T/\text{K}$ and the results fit the experimental values with a root mean square deviation of 9 J mol^{-1}. The corresponding heat capacity equation is:

$$C_p/\text{J mol}^{-1}\text{ K}^{-1} = 24.529 - 32628.5x^{-2} + 5.1758 \times 10^{-3}x. \tag{9.2.2.2}$$

Conversion of Macleod's enthalpy and heat capacity values to IPTS-68 gives the following equations:

$$[H(T) - H(298.15\text{ K})]/\text{J mol}^{-1} = -7632.7 + 25297.3/x + 24.525x \\ + 2.5565 \times 10^{-3}x^2 \tag{9.2.2.3}$$

$$C_p/\text{J mol}^{-1}\text{ K}^{-1} = 24.525 - 25297.3x^{-2} \\ + 5.113 \times 10^{-3}x. \tag{9.2.2.4}$$

These equations give values of enthalpy and heat capacity in the range 300 to 1500 K which agree with Hultgren's selected values (Ref. 1) to better than 0.2 per cent. Results obtained by Vollmer and Kohlhass (Ref. 9) and Macleod (Ref. 8) by calorimetry, by Kraftmakher (Ref. 10) and Seville (Ref. 11) using modulation methods, and by Righini and Rosso (Ref. 12) using a pulse-heating method, show an enhancement in the heat capacity of platinum above 1500 K compared with the values reported in references 1, 3, and 4.

REFERENCES

1. Hultgren, R.; Desai, P. D.; Hawkins, D. T.; Gleiser, M.; Kelley, K. K.; Wagman, D. D. *Selected Values of Thermodynamic Properties of the Elements*, Amer. Soc. Met.: Ohio, **1973**.

2. Kendall, W. B.; Orr, R. L.; Hultgren, R. *J. Chem. Engng. Data* **1962**, *7*, 516.
3. Jaeger, F. M.; Rosenbohm, E.; Bottema, J. A. *Proc. Acad. Sci. Amst.* **1932**, *35*, 763.
4. Jaeger, F. M.; Rosenbohm, E. *Proc. Acad. Sci. Amst.* **1927**, *30*, 1069.
5. White, W. P. *Phys. Rev.* **1918**, *12*, 436.
6. Yokokawa, H.; Takahashi, Y. *J. Chem. Thermodynamics* **1979**, *11*, 411.
7. Douglas, T. B. *J. Res. Nat. Bur. Stand.* **1969**, *73A*, 451.
8. Macleod, A. C. *J. Chem. Thermodynamics* **1972**, *4*, 391.
9. Vollmer, O.; Kohlhass, R. *Z. Naturforsch.* **1969**, *A24*, 1669.
10. Kraftmakher, Y. A. *High-Temp.-High Pressures* **1973**, *5*, 433.
11. Seville, A. H. *Phys. Status Solidi (a)* **1974**, *21*, 649.
12. Righini, F.; Rosso, A. *High-Temp.-High Pressure* **1980**, *12*, 335.

9.2.3. C_p, Copper

Physical property: Heat capacity
Units: J mol^{-1} K^{-1} (molar heat capacity, C_p)
J kg^{-1} K^{-1} (specific heat capacity, c_p)
Recommended reference material: Copper (Cu)
Range of variables: 1 to 300 K
Physical state within the range: solid
Contributors to the first version: J. D. Cox, J. F. Martin
Contributors to the revised version: J. W. Fisher, J. Lielmezs, D. L. Martin, J. F. Martin, R. Sabbah, R. L. Snowdon

Intended usage: Copper is recommended for testing the performance of low temperature calorimeters used for the measurement of heat capacity of solids (and possibly liquids). For use below 20 K the copper must be of very high purity and degassed because very small (ppm) amounts of transition metal and hydrogen impurities may drastically affect the heat capacity (Ref. 1). Severe cold work may affect the heat capacity significantly, especially below 20 K (Ref. 2). At higher temperatures there may be no significant differences in the heat capacities of high purity and 'workshop grade' copper. One such comparison has been made (Ref. 3) showing a difference of 0.2 per cent or less, which is comparable with the scatter of the data.

Sources of supply and/or methods of preparation: High purity degassed and annealed copper is available as Research Material RM5 from supplier (S).

Pertinent physicochemical data: Copper, the molar mass of which is taken as 63.546 g mol^{-1}, was recommended as a standard material at low temperatures by the American Calorimetry Conference. Osborne *et al.* (Ref. 4) analysed the results available at the time and recommended the following Copper Reference Equation for use in the range 1 to 25 K

$$C_p/\text{J mol}^{-1}\text{ K}^{-1} = \sum A_N (T/K)^N \qquad (9.2.3.1)$$

9. Enthalpy

where the six factors are:

$$A_1 = 6.9434 \times 10^{-4} \quad A_7 = 9.4786 \times 10^{-11}$$
$$A_3 = 4.7548 \times 10^{-5} \quad A_9 = -1.3639 \times 10^{-13}$$
$$A_5 = 1.6314 \times 10^{-9} \quad A_{11} = 5.3898 \times 10^{-17}$$

This applies to older temperature scales where the span 20 to 4.2 K was interpolated by the individual experimenter's gas thermometer and the 20 K calibration temperature and higher temperatures were on the NBS-1955 platinum thermometer scale and temperatures below 4.2 K on the 1958 ^{4}He vapour pressure scale. Many laboratories still use such scales and the above equation is recommended in the 1 to 20 K range only (data above 20 K having since been shown to be in error). The scatter of raw data about the Copper Reference Equation was about ±0.6 per cent.

The currently recommended temperature scale below 30 K is the EPT-76 scale which is very close to the Iowa State University T_x magnetic scale. The recommended reference equation for these conditions, in the 1 to 30 K range, is that of Holste et al. (Ref. 5), being equation (9.2.3.1) with the following seven factors:

$$A_1 = 6.9260 \times 10^{-4} \quad A_7 = 1.0869 \times 10^{-10} \quad A_{13} = -3.2196 \times 10^{-20}$$
$$A_3 = 4.7369 \times 10^{-5} \quad A_9 = -1.9745 \times 10^{-13}$$
$$A_5 = 1.9537 \times 10^{-9} \quad A_{11} = 1.3343 \times 10^{-16}$$

The scatter of the raw data about this equation was ±0.1 per cent. The difference between results obtained with the two recommended equations is roughly consistent with the known differences in the temperature scales. The smoothed heat capacity results from the two scales are closely similar at 20 K. It should be satisfactory to extrapolate the above equations below 1 K but sample impurity may cause divergence from these values as the temperature is reduced.

For measurements in the 30 to 300 K range on the IPTS-68 scale, the three most recent data sets (Refs. 3, 6, 7) are in reasonable agreement and the above equation with the following fourteen factors is a good fit.

$$A_0 = -0.1285753818 \times 10^1 \quad A_7 = 0.3070527023 \times 10^{-10}$$
$$A_1 = 0.3098967121 \quad A_8 = -0.1419198886 \times 10^{-12}$$
$$A_2 = -0.2924985792 \times 10^{-1} \quad A_9 = 0.4557519040 \times 10^{-15}$$
$$A_3 = 0.1418586260 \times 10^{-2} \quad A_{10} = -0.9894731263 \times 10^{-18}$$
$$A_4 = -0.3370489513 \times 10^{-4} \quad A_{11} = 0.1370529662 \times 10^{-20}$$
$$A_5 = 0.4856675621 \times 10^{-6} \quad A_{12} = -0.1074497377 \times 10^{-23}$$
$$A_6 = -0.4646773402 \times 10^{-8} \quad A_{13} = 0.3517161374 \times 10^{-27}$$

The scatter of the fitted data is generally within ±0.2 per cent above 100 K, within ±0.5 per cent in the 50 to 100 K range and within ±2 per cent in the 30 to 50 K range. The smoothed results join smoothly onto those of Holste et al. (Ref. 5), given above. It should be noted that the literature to 1967 on the heat capacity of copper over the whole temperature range has been reviewed and tables of thermodynamic properties have been published (Ref. 8).

REFERENCES

1. Cezairliyan, A.; Miller, A. P., editors, *Specific Heat of Solids*, McGraw Hill/CINDAS Data Series on Material Properties: in preparation.
2. Bevk, J. *Phil. Mag.* **1973**, *28*, 1379.
3. Martin, D. L. *Can. J. Phys.* **1960**, *38*, 17.
4. Osborne, D. W.; Flotow, H. E.; Schreiner, F. *Rev. Sci. Instrum.* **1967**, *38*, 159.
5. Holste, J. C.; Cetas, T. C.; Swenson, C. A. *Rev. Sci. Instrum.* **1972**, *43*, 670.
6. Robie, R. A.; Hemingway, B. S.; Wilson, W. H. *J. Res. U.S. Geological Survey* **1976**, *4*, 631.
7. Downie, D. B.; Martin, J. F. *J. Chem. Thermodynamics* **1980**, *12*, 779.
8. Furukawa, G. T.; Saba, W. G.; Reilly, M. L. *Critical Analysis of the Heat Capacity Data of the Literature and Evaluation of Thermodynamic Properties of Copper, Silver and Gold from 0 to 300 K*, NSRDS-NBS 18, **1968**.

9.2.4. C_{sat}, Benzoic acid

Physical property: Heat capacity
Units: J mol^{-1} K^{-1} (molar heat capacity, C_{sat})
J kg^{-1} K^{-1} (specific heat capacity, c_{sat})
Recommended reference material: Benzoic acid ($C_7H_6O_2$)
Range of variables: 10 to 350 K
Physical state within the range: solid
Contributors to the first version: J. D. Cox, J. F. Martin, O. Riedel, D. R. Stull
Contributor to the revised version: R. Sabbah

Intended usage: Benzoic acid is recommended for testing the performance of calorimeters used for the measurement of the heat capacity of solids from low temperatures to a little above ambient (Ref. 1). Use of the material in contact with base metal near to, or above, the melting point (395.52 K) is not recommended because of the material's corrosive nature.

Sources of supply and/or methods of preparation: It seems probable that samples of benzoic acid prepared as energy-of-combustion standards would also serve as heat-capacity standards (Ref. 2). These samples are available from suppliers (B, D, S), BDH.

Pertinent physicochemical data: Arvidsson *et al.* (Ref. 3) studied a sample of benzoic acid (energy-of-combustion standard SRM 39i from supplier (S)) with a mole fraction purity of 0.99997 assessed by freezing point measurements on macro-samples. Values of the variation of C_p with temperature are given in the following table. The molar mass was taken as

122.12 g mol^{-1}. Values refer to the solid in equilibrium with its vapour and were obtained with a small-sample low temperature adiabatic calorimeter. Temperatures are reported in IPTS-68 above 13.81 K. The calibration of the thermometer was extended below 13.81 K by the method of McCrackin and Chang (Ref. 4).

Table 9.2.4.1. Heat capacity C_{sat} of benzoic acid

T_{68} / K	C_{sat} / J mol^{-1} K^{-1}	T_{68} / K	C_{sat} / J mol^{-1} K^{-1}
10	2.094	180	94.65
20	11.06	200	102.88
40	31.70	220	111.42
60	45.79	240	120.26
80	55.82	260	129.31
100	64.01	280	138.44
120	71.53	300	147.64
140	79.01	320	156.87
160	86.70	350	170.70

In the work of Ardvidsson *et al.* the deviation of the experimental values from a smooth curve is less than 0.05 per cent in the range 50 to 350 K, about 0.1 per cent in the range 15 to 50 K and about 1 per cent below 15 K. In spite of the small amount of sample used and the unfavourable ratio between the heat capacity of the sample and that of the calorimeter, the agreement with previous data (Refs. 5, 6, 7) up to 300 K is good and gives strong support to the belief that the uncertainty of the measurements is better than ±0.1 per cent above 60 K, ±0.3 per cent between 15 and 60 K and about ±1 per cent in the range 8 to 15 K. A comparison of temperature scales together with an analysis of the precision of data on the variation in the heat capacity of benzoic acid with temperature is reported in reference 7.

REFERENCES

1. McCullough, J. P.; Scott, D. W., editors, *Experimental Thermodynamics*, Vol. I: *Calorimetry of Non-reacting Systems*, Chapters 5 and 6. Butterworths: London, **1968**.
2. Furukawa, G. T.; McCoskey, R. E.; King, G. J. *J. Res. Nat. Bur. Stand.* **1951**, *47*, 256.
3. Arvidsson, K.; Falk, B.; Sunner, S. *Chemica Scripta* **1976**, *10*, 193.
4. McCrackin, F. L.; Chang, S. S. *Rev. Sci. Instrum.* **1975**, *46*, 550.
5. Robie, R. A.; Hemingway, B. S. *Geological Survey Professional Paper 755*, United States Govt. Printing Office: Washington, **1972**.
6. Tatsumi, M.; Matsuo, T.; Suga, H.; Seki, S. *Bull. Chem. Soc. Japan* **1975**, *48*, 3060.
7. Rubkin, N. P.; Orlova, M. P.; Baranyuk, A. K.; Nurullaev, N. G.; Rozhnovskaya, L. N.; *Izmeritel'. Tekh.* **1974**, *7*, 29.

9.2.5. C_p, 2,2-Dimethylpropane

Physical property: Heat capacity
Units: J mol^{-1} K^{-1} (molar heat capacity, C_p), J kg^{-1} K^{-1} (specific heat capacity, c_p)
Recommended reference material: 2,2-Dimethylpropane (neopentane) (C$_5$H$_{12}$)
Range of variables: (i) 4 to 139 K (ii) 142 to 254 K
Physical state within the range: solid
Contributor to the first version: Y. Mashiko
Contributor to the revised version: R. Sabbah

Intended usage: 2,2-Dimethylpropane is recommended for testing the performance of calorimeters used for measuring the heat capacities of solids down to very low temperatures (Ref. 1). As the substance has a solid-solid transition (Ref. 2) at (140.49 ± 0.05) K, temperatures in the vicinity of the transition point (139 to 142 K) should be avoided.

Sources of supply and/or methods of preparation: A highly pure grade (mole fraction purity of 0.99997) sample of 2,2-dimethylpropane is available from supplier (C). Samples having respectively mole fraction and mass fraction purities of 0.9999 and 0.99 can be purchased from supplier (P) and Baker. The further purification of a sample of 2,2-dimethylpropane is possible by fractional distillation with treatment of the distillate by molecular sieve 5A.

Pertinent physicochemical data: Enokido et al. (Ref. 2) studied a sample of 2,2-dimethylpropane (the molar mass of which was taken as 72.151 g mol^{-1}) assessed by calorimetric study of the melting behaviour. Values of C_p at various temperatures for the solid having a mole fraction purity of 0.99997 are given in table 9.2.5.1. Temperatures were reported on IPTS-48, below 90 K on NBS-55 scale and below 10 K on ^{4}He vapour pressure scale (Ref. 3).

Table 9.2.5.1. Heat capacity C_p of 2,2-dimethylpropane

$\dfrac{T}{K}$	$\dfrac{T_{68}}{K}$	$\dfrac{C_p}{\text{J mol}^{-1}\text{ K}^{-1}}$	$\dfrac{T}{K}$	$\dfrac{T_{68}}{K}$	$\dfrac{C_p}{\text{J mol}^{-1}\text{ K}^{-1}}$	$\dfrac{T}{K}$	$\dfrac{T_{68}}{K}$	$\dfrac{C_p}{\text{J mol}^{-1}\text{ K}^{-1}}$
4.141		0.1532	81.405	81.397	60.17	161.296	161.307	107.8
6.409		0.7063	101.293	101.303	72.93	180.398	180.426	112.8
10.043		3.249	120.903	120.891	89.70	201.904	201.938	117.7
20.164	20.173	16.78	139.014	139.005	111.3	220.360	220.390	123.6
61.576	61.575	48.37				239.880	239.900	130.6
			142.424	142.417	110.0	254.044	254.055	137.5

The results for the temperature range 60 to 139 K are slightly lower and for temperatures near 250 K are much lower than those reported by Aston and Messerly (Ref. 4). Probably the less pure specimen used by the latter workers is responsible for considerable premelting.

REFERENCES

1. McCullough, J. P.; Scott, D. W., editors, *Experimental Thermodynamics*, Vol. I: *Calorimetry of Non-reacting Systems*, Chapters 5, 6 and 7. Butterworths: London, **1968**.
2. Enokido, H.; Shinoda, T.; Mashiko, Y. *Bull. Chem. Soc. Japan* **1969**, *42*, 84.
3. Shinoda, T.; Atake, T.; Chihara, H.; Mashiko, Y.; Seki, S. *Kogyo Kagaku Zasshi (J. Chem. Soc. Japan, Ind. Chem. Sect.)* **1966**, *69*, 1619.
4. Aston, J. G.; Messerly, G. H. *J. Amer. Chem. Soc.* **1936**, *58*, 2354.

9.2.6. C_p and H, Molybdenum

Physical property: Heat capacity, enthalpy
Units: J mol^{-1} K^{-1} (molar heat capacity, C_p); J mol^{-1} (molar enthalpy, H)
  J kg^{-1} K^{-1} (specific heat capacity, c_p); J kg^{-1} (specific enthalpy, h)
Recommended reference material: Molybdenum (Mo)
Range of variables: 273 to 2800 K
Physical state within the range: solid
Contributors to the first version: A. Cezairliyan, D. Ditmars
Contributors to the revised version: A. Cezairliyan, D. Ditmars, J. Lielmezs, J. Rogez, R. Sabbah

Intended usage: Pure molybdenum is recommended for testing the performance of calorimeters used for measuring the heat capacity or enthalpy of good thermally-conducting materials (solids and liquids) from 273 to 2800 K. Molybdenum has no known structural transitions within this temperature range but is oxidized rapidly in contact with air at temperatures above 773 K.

Sources of supply and/or methods of preparation: Certified molybdenum samples (mass fraction purity of 0.9995) may be obtained from supplier (S) as SRM 781. Samples of molybdenum of comparable purity can be obtained from various sources *e.g.* (F, K, M), suppliers Alfa, Baker, etc.

Pertinent physicochemical data: The relative enthalpy of NBS Standard Reference Material 781 has been measured by the drop method with two different calorimeters, a bunsen ice calorimeter and an adiabatic calorimeter, in the temperature ranges 273 to 1173 K and 1173 to 2100 K, respectively. The uncertainties in the smoothed enthalpy data derived from these measurements are believed not to exceed ±0.6 per cent at any temperature in these ranges. The heat capacity of the same material has also been measured in the temperature range 1500 to 2800 K using a millisecond-resolution pulse calorimetric technique with resistive self-heating. In the highest temperature range the smoothed heat capacity data are believed to be uncertain by no more than ±3 per cent. The details of the measurements and the results are given in reference 1. The smoothed heat capacity and enthalpy results for the

SRM 781 molybdenum are given in table 9.2.6.1. The molar mass of molybdenum was taken as 95.94 g mol^{-1}. These results are in good agreement with Hultgren's values (Ref. 2).

Table 9.2.6.1. Heat capacity C_p and enthalpy $H(T) - H(273.15 K)$ of molybdenum

T_{68} K	C_p J mol^{-1} K^{-1}	$H(T)-H(273.15\text{ K})$ J mol^{-1}	T_{68} K	C_p J mol^{-1} K^{-1}	$H(T)-H(273.15\text{ K})$ J mol^{-1}
273.15	23.56	0	1600	32.50	37376
300	23.95	637.87	1700	33.42	40671
400	25.08	3093.4	1800	34.42	44062
500	25.85	5642.2	1900	35.49	47557
600	26.46	8258.7	2000	36.65	51163
700	26.98	10931	2100	37.90	54890
800	27.44	13652	2200	39.24	58746
900	27.89	16419	2300	40.67	62740
1000	28.37	19232	2400	42.21	66884
1100	28.90	22094	2500	43.89	71188
1200	29.49	25013	2600	45.88	75673
1300	30.14	27994	2700	48.37	80381
1400	30.86	31044	2800	51.57	85371
1500	31.65	34169			

Equations for C_p and $H(T) - H(273.15\text{ K})$ or $H(T) - H(298.15\text{ K})$ as functions of T have been given in reference 1. In the range 500 to 1500 K values obtained from other sources and quoted in references 1 and 2 are often in good agreement. Betz and Frohberg (Ref. 3) determined the enthalpy of molybdenum by drop calorimetry in the range 2282 to 3383 K. Their results do not differ from the values in table 9.2.6.1 by more than 1 per cent up to 2650 K. Recently, Righini and Rosso (Ref. 4), using a pulse-heating method, determined the heat capacity of molybdenum in the range 1300 to 2500 K. Their results agree within ±0.5 per cent with the values reported in the tables over most of the temperature range.

REFERENCES

1. Ditmars, D.A.; Cezairliyan, A.; Ishihara, S.; Douglas, T. B. *Enthalpy and Heat Capacity Standard Reference Material: Molybdenum SRM 781, from 273 to 2800 K*, NBS Special Publication 260-55, National Bureau of Standards: Washington, **1977**.
2. Hultgren, R.; Desai, P. D.; Hawkins, D. T.; Gleiser, M.; Kelley K. K.; Wagman, D. D. *Selected Values of Thermodynamic Properties of the Elements*, Amer. Soc. Met.: Ohio, **1973**.
3. Betz, G.; Frohberg, M. G. *High-Temp.-High Pressures* **1980**, *12*, 169.
4. Righini, R.; Rosso, A. *Int. J. Thermophys.* **1983**, *4*, 173.

9.2.7. C_{sat}, Naphthalene

Physical property: Heat capacity
Units: J mol^{-1} K^{-1} (molar heat capacity, C_{sat})
J kg^{-1} K^{-1} (specific heat capacity, c_{sat})
Recommended reference material: Naphthalene ($C_{10}H_8$)
Range of variables: (i) 10 to 350 K (ii) 357 to 371 K
Physical states within the range: (i) solid (ii) liquid
Contributors to the first version: J. D. Cox, O. Riedel
Contributor to the revised version: R. Sabbah

Intended usage: Naphthalene is recommended for testing the performance of calorimeters for measuring the heat capacities of solids and liquids (Ref. 1).

Sources of supply and/or methods of preparation: High purity samples of naphthalene can be obtained from supplier (C). Commercial samples with a mass fraction purity ≥ 0.99 are available from suppliers Aldrich, Carlo Erba, Fluka, Koch-Light, etc. which can be used after further purification by zone refining.

Pertinent physicochemical data: McCullough et al. (Ref. 2) examined a sample of naphthalene having a mole fraction purity of 0.99985 assessed by calorimetric study of the melting behaviour. Values of C_{sat} for the solid in equilibrium with its vapour are given in table 9.2.7.1. Measurements of temperature were made with platinum resistance thermometers calibrated on IPTS-48 and, below 90 K, on the provisional scale of NBS (Ref. 3). The uncertainty of the heat capacity data was usually within ± 0.1 to ± 0.2 per cent. The 1951 International Atomic Weights and the 1951 values of the fundamental physical constants were used.

Table 9.2.7.1. Heat capacity C_{sat} of naphthalene

$\dfrac{T}{K}$	$\dfrac{T_{68}}{K}$	$\dfrac{C_{sat}}{\text{J mol}^{-1}\text{ K}^{-1}}$	$\dfrac{T}{K}$	$\dfrac{T_{68}}{K}$	$\dfrac{C_{sat}}{\text{J mol}^{-1}\text{ K}^{-1}}$	$\dfrac{T}{K}$	$\dfrac{T_{68}}{K}$	$\dfrac{C_{sat}}{\text{J mol}^{-1}\text{ K}^{-1}}$
10		1.766	120	119.99	68.50	240	240.02	129.1
20	20.00	10.94	140	139.99	77.17	260	260.01	141.1
40	40.01	30.12	160	160.01	86.48	280	280.00	153.7
60	59.99	42.65	180	180.03	96.33	300	299.99	166.9
80	79.98	51.99	200	200.03	106.7	320	319.99	181.1
100	100.01	60.25	220	220.03	117.7	340	339.99	197.2
						350	349.99	205.8

McCullough et al. (Ref. 2) also measured the heat capacity of liquid naphthalene in equilibrium with its vapour. Equation (9.2.7.1) is valid from 357 to 371 K.

$$C_{sat}(l)/\text{J mol}^{-1}\text{ K}^{-1} = 80.383 + 0.3873(T/\text{K}) \tag{9.2.7.1}$$

REFERENCES

1. McCullough, J. P.; Scott, D. W., editors, *Experimental Thermodynamics*, Vol. I: *Calorimetry of Non-reacting Systems*, Chapters 5, 6 and 9. Butterworths: London, **1968**.
2. McCullough, J. P.; Finke, H. L.; Messerly, J. F.; Todd, S. S.; Kincheloe, T. C.; Waddington, G. *J. Phys. Chem.* **1957**, *61*, 1105.
3. Hoge, H. J.; Brickwedde, F. G. *J. Res. Nat. Bur. Stand.* **1939**, *22*, 351.

9.2.8. C_{sat} and H, Diphenyl ether

Physical property: Heat capacity, enthalpy
Units: J mol^{-1} K^{-1} (molar heat capacity, C_{sat}), J mol^{-1} (molar enthalpy, H)
  J kg^{-1} K^{-1} (specific heat capacity, c_{sat}), J kg^{-1} (specific enthalpy, h)
Recommended reference material: Diphenyl ether ($C_{12}H_{10}O$)
Range of variables: (i) 10 to 300.03 K (ii) 300.03 to 570 K
Physical states within the range: (i) solid (ii) liquid
Contributors to the first version: J. D. Cox, J. F. Martin, O. Riedel, D. R. Stull
Contributor to the revised version: R. Sabbah

Intended usage: Diphenyl ether is recommended for testing the performance of calorimeters (Refs. 1, 2) used for the measurement of the heat capacities of solids and liquids. Its relatively high boiling point (532 K) permits the use of this material over a wide range of temperature and its ease of purification and the inertness of the molten substance to most materials of construction are further points in its favour.

Sources of supply and/or methods of preparation: Commercial samples with a mass fraction purity of ≥ 0.99 are available from suppliers Aldrich, Koch-Light, etc. which can be further purified by distillation and fractional crystallization. It is also necessary to remove the dissolved air and water from the sample before using it in calorimetric experiments.

Pertinent physicochemical data: Furukawa *et al.* examined a sample of diphenyl ether with mole fraction purity of 0.999987 assessed by calorimetric study of the melting behaviour. They determined the heat capacity between 18 and 360 K by adiabatic calorimetry and between 273 and 573 K by drop calorimetry (Refs. 3, 4). Table 9.2.8.1 gives values of heat capacity and enthalpy for solid and liquid diphenyl ether under their own vapour pressure. Measurements of temperature were made with platinum resistance thermometers calibrated on the IPTS-48 and, below 90 K, on the provisional scale of the NBS (Ref. 5).

9. Enthalpy

Table 9.2.8.1. Heat capacity C_{sat} and enthalpy $H(T) - H(0\text{ K})$ of diphenyl ether

$\dfrac{T}{\text{K}}$	$\dfrac{T_{68}}{\text{K}}$	$\dfrac{C_{sat}}{\text{J mol}^{-1}\text{ K}^{-1}}$	$\dfrac{H(T)-H(0\text{ K})}{\text{J mol}^{-1}}$	$\dfrac{T}{\text{K}}$	$\dfrac{T_{68}}{\text{K}}$	$\dfrac{C_{sat}}{\text{J mol}^{-1}\text{ K}^{-1}}$	$\dfrac{H(T)-H(0\text{ K})}{\text{J mol}^{-1}}$
			solid				
10		3.283	8.272	160	160.01	117.64	10823
20	20.00	18.06	109.83	180	180.03	129.98	13298
40	40.01	45.44	763.18	200	200.03	143.09	16028
60	59.99	61.92	1848.2	220	220.03	156.86	19027
80	79.98	73.77	3208.8	240	240.02	171.32	22306
100	100.01	84.35	4791.8	260	260.01	186.49	25884
120	119.99	94.82	6582.5	280	280.00	202.04	29770
140	139.99	105.88	8588.3	300	299.99	218.06	33969
				300.03	300.02	218.25	33975
			liquid				
300.03	300.02	268.42	51190	460	460.04	340.32	99872
320	319.99	277.17	56634	480	480.05	349.16	106769
340	339.99	286.25	62268	500	500.05	358.04	113845
360	360.00	295.35	68084	520	520.06	366.92	121100
380	380.00	304.39	74082	540	540.07	375.78	128534
400	400.01	313.33	80259	560	560.07	384.63	136148
420	420.02	322.36	86617	570	570.07	389.05	140022
440	440.03	331.30	93155				

The molar mass of diphenyl ether was taken as 170.20 g mol^{-1}. The variation of the heat capacity of diphenyl ether with temperature is essentially linear in the liquid region, which is a very useful property in a heat capacity reference material. The uncertainty of heat capacity values is ±0.1 to ±0.2 per cent.

REFERENCES

1. McCullough, J. P.; Scott, D. W., editors, *Experimental Thermodynamics*, Vol I: *Calorimetry of Non-reacting Systems*, Butterworths: London, **1968**.
2. Yuen, H. K.; Yosel, C. J. *Thermochim. Acta* **1979**, *33*, 281.
3. Furukawa, G. T.; Ginnings, D. C.; McCoskey, R. E.; Nelson, R. A. *J. Res. Nat. Bur. Stand.* **1951**, *46*, 195.
4. Ginnings, D. C.; Furukawa, G. T. *J. Amer. Chem. Soc.* **1953**, *75*, 522.
5. Hoge, H. J.; Brickwedde, F. G. *J. Res. Nat. Bur. Stand.* **1939**, *22*, 351.

9.2.9. C_{sat} and H, Heptane

Physical property: Heat capacity, enthalpy
Units: J mol^{-1} K^{-1} (molar heat capacity, C_{sat}), J mol^{-1} (molar enthalpy, H)
    J kg^{-1} K^{-1} (specific heat capacity, c_{sat}), J kg^{-1} (specific enthalpy, h)
Recommended reference material: Heptane (C_7H_{16})
Range of variables: (i) 10 to 182.59 K (ii) 182.59 to 400 K
Physical states within the range: (i) solid (ii) liquid
Contributors to the first version: J. D. Cox, T. B. Douglas, D. R. Douslin,
    J. F. Martin, O. Riedel, D. R. Stull
Contributor to the revised version: R. Sabbah

Intended usage: Because it can be readily purified (especially with respect to non-hydrocarbon impurities), is chemically stable up to its critical temperature (540 K), can be easily distilled into or out of a calorimeter, exhibits no solid-solid transitions and comes rapidly to thermal equilibrium, heptane is recommended for testing the performance of heat-capacity calorimeters, especially those intended for measurements over the temperature range 10 to 400 K (Ref. 1).

Sources of supply and/or methods of preparation: Samples of heptane of suitable purity (mole fraction purity > 0.999) can be obtained from supplier (C). Commercial samples with a mass fraction purity > 0.999 are available from suppliers *e.g.* B.D.H., Fluka, etc. which can be further purified by distillation. It is also necessary to remove the dissolved air and water from the sample before using it in calorimetric experiments.

Pertinent physicochemical data: Douglas *et al.* (Ref. 2) measured the heat capacity of a sample of heptane of which the mole fraction purity was ≥ 0.99997 assessed by freezing-point determination. Measurements were made by both adiabatic and drop calorimetry. Values of the heat capacity and enthalpy for solid and liquid heptane in equilibrium with their own vapour are given in table 9.2.9.1. The molar mass of heptane was taken as 100.20 g mol^{-1}. Measurements of temperature were made with platinum resistance thermometers calibrated on the IPTS-48 and, below 90 K, on the provisional scale of the NBS (Ref. 3).

The uncertainty of the listed values of heat capacity and enthalpy is about ±0.1 per cent except below 50 K where increasing tolerances must be allowed. Recently measurements were made by Schaake *et al.* (Ref. 4), by Kalinowska *et al.* (Ref. 5) and by Zhicheng *et al.* (Ref. 6) using, respectively, an automatic calorimeter in the range 83 to 287 K, a semi-automatic calorimeter in the range 185 to 300 K and an automatic calorimeter in the range 200 to 380 K. The values of Kalinowska *et al.* and Zhicheng *et al.* are in excellent agreement with the results of Douglas *et al.* in the liquid region. The deviation of the values of Schaake *et al.* from the results of Douglas *et al.* is ≤ 0.2 per cent in the range 110 to 287 K.

Table 9.2.9.1 Heat capacity C_{sat} and enthalpy $H(T) - H(0\,\text{K})$ of heptane

$\dfrac{T}{\text{K}}$	$\dfrac{T_{68}}{\text{K}}$	$\dfrac{C_{sat}}{\text{J mol}^{-1}\text{K}^{-1}}$	$\dfrac{H(T)-H(0\,\text{K})}{\text{J mol}^{-1}}$	$\dfrac{T}{\text{K}}$	$\dfrac{T_{68}}{\text{K}}$	$\dfrac{C_{sat}}{\text{J mol}^{-1}\text{K}^{-1}}$	$\dfrac{H(T)-H(0\,\text{K})}{\text{J mol}^{-1}}$
solid							
10		1.770	4.431	120	119.99	105.20	6651.2
20	20.00	11.80	65.25	140	139.99	116.83	8872.6
40	40.01	38.17	566.2	160	160.01	129.03	11328
60	59.99	60.48	1559.7	180	180.03	145.13	14054
80	79.98	78.28	2953.4	182.59	182.62	148.58	14430
100	100.01	92.77	4669.7				
liquid							
182.59	182.62	203.15	28452	300	299.99	225.44	53010
200	200.03	201.31	31973	320	319.99	233.25	57598
220	220.03	202.74	36008	340	339.99	241.67	62349
240	240.02	206.47	40097	360	360.00	250.63	67276
260	260.01	211.73	44277	380	380.00	260.10	72392
280	280.00	218.23	48574	400	400.01	270.13	77708

REFERENCES

1. McCullough, J. P.; Scott, D. W., editors, *Experimental Thermodynamics*, Vol. I: *Calorimetry of Non-reacting Systems*, Butterworths: London, **1968**.
2. Douglas, T. B.; Furukawa, G. T.; McCoskey, R. E.; Ball, A. F. *J. Res. Nat. Bur. Stand.* **1954**, *53*, 139.
3. Hoge, H. J.; Brickwedde, F. G. *J. Res. Nat. Bur. Stand.* **1939**, *22*, 351.
4. Schaake, R. C. F.; Offringa, J. C. A.; Van der Berg, G. I. K.; Van Miltenburg, J. C. *Rec. Trav. Chim. Pays-Bas* **1979**, *98*, 408.
5. Kalinowska, B.; Jedlinska, J.; Woycicki, W.; Stecki, J. *J. Chem. Thermodynamics* **1980**, *12*, 891.
6. Zhicheng, T.; Lixing, Z.; Shuxia, C.; Anxue, Y.; Yi, S.; Jinchun, Y.; Xiukun, W. *Scientia Sinica, Series B* **1983**, *26*, 1014.

9.2.10. C_{sat} and H, Hexafluorobenzene

Physical property: Heat capacity
Units: J mol^{-1} K^{-1} (molar heat capacity, C_{sat}), J mol^{-1} (molar enthalpy, H)
J kg^{-1} K^{-1} (specific heat capacity, c_{sat}), J kg^{-1} (specific enthalpy, h)
Recommended reference material: Hexafluorobenzene (C_6F_6)
Range of variables: (i) 10 to 278.30 K (ii) 278.30 to 350 K (iii) 335 to 527 K real gas at pressures up to 202.66 kPa.
Physical state within the range: (i) solid; (ii) liquid; (iii) real gas
Contributor: R. Sabbah

Intended usage: Hexafluorobenzene is recommended for testing the performance of calorimeters used for the measurements of the heat capacities of solids and liquids in equilibrium with their vapours and of vapours at pressures between zero and 202.66 kPa.

Sources of supply and/or methods of preparation: Commercial samples are available from suppliers BDH (with a mass fraction purity of 0.999), Aldrich, Fluka (with a mass fraction purity of ≥ 0.99) which can be further purified by preparative gas-liquid chromatography. It is also necessary to prevent the sample used in calorimetric experiments from being exposed to air and moisture.

Pertinent physicochemical data: (i and ii) Messerly and Finke examined a sample of hexafluorobenzene having a mole fraction purity of 0.9993 assessed by calorimetric study of the melting behaviour. They determined the heat capacity between 13 and 342 K by adiabatic calorimetry (Ref. 1). The values of heat capacity and enthalpy for solid and liquid hexafluorobenzene under their own vapour pressure are given in table 9.2.10.1. Measurements of temperature were made with platinum resistance thermometers calibrated in terms of the IPTS-48 (revised text of 1960) (Ref. 2) from 90 to 342 K and in terms of the provisional scale of the NBS below 90 K (Ref. 3).

The molar mass of hexafluorobenzene was taken as 186.057 g mol^{-1}. From 30 to 350 K, the uncertainty of the results is estimated to be less than ± 0.2 per cent. In the premelting region and in the region 200 to 250 K where anomalous thermal behaviour was observed, the uncertainty is estimated to be about ± 0.5 per cent. Below 30 K, the uncertainty of the individual measurements increases with decreasing temperature to ± 1 per cent near 12 K.

The results reported in table 9.2.10.1 for hexafluorobenzene are in overall agreement with the low-temperature thermal properties reported by Counsell *et al.* (Ref. 4).

As the difference between C_p and C_{sat} is barely significant, values in the above table for liquid hexafluorobenzene are also in agreement with values of C_p given in reference 5.

9. Enthalpy

Table 9.2.10.1. Heat capacity C_{sat} and enthalpy $H(T) - H(0\,K)$ of hexafluorobenzene

$\dfrac{T_{68}}{K}$	$\dfrac{C_{sat}}{J\,mol^{-1}\,K^{-1}}$	$\dfrac{H(T)-H(0\,K)}{J\,mol^{-1}}$	$\dfrac{T_{68}}{K}$	$\dfrac{C_{sat}}{J\,mol^{-1}\,K^{-1}}$	$\dfrac{H(T)-H(0\,K)}{J\,mol^{-1}}$
solid					
10	4.43	11.0	160	131.8	11086
20	18.44	126	180	145.2	13857
40	38.83	713	200	157.0	16883
60	55.10	1654	220	167.3	20127
80	71.04	2916	240	176.9	23568
100	86.99	4496	260	185.9	27198
120	102.5	6393	278.30	193.9	30671
140	117.4	8592			
liquid					
278.30	218.6	42254	320	226.0	51512
280	218.9	42626	340	229.9	56070
300	222.0	47033	350	231.9	58379

Table 9.2.10.2. Heat capacity C_p of hexafluorobenzene in the gas phase

	$C_p(T)/J\,mol^{-1}\,K^{-1}$							
	T_{68}/K							
p/kPa	335.15	348.15	368.15	403.15	438.15	473.15	500.15	527.15
202.66				190.53	195.74	201.81	206.32	210.55
101.32			180.24	186.70	193.49	200.21	205.03	209.61
50.660		173.24	177.47	185.01	192.39	199.38	204.49	209.14
37.996	169.55	172.27	176.81					
25.331	168.48	171.47	176.29	184.33	192.04	199.23	204.52	209.08
19.000	167.85							
12.667	167.35	170.55						
0 (ideal gas)	166.27	169.70	174.95	183.40	191.45	198.76	204.09	208.78

(iii) Experimental values obtained by Hossenlopp and Scott (Ref. 6) using vapour-flow calorimetry are reported in table 9.2.10.2 for the vapour heat capacity of hexafluorobenzene in the real and ideal gas states over large ranges of pressure and temperature (referred to the IPTS-68).

The experimental uncertainty of the reported values is less than ±0.2 per cent. The above sets of measurements agree to better than ±0.3 per cent with those of Counsell et al. (Ref. 4).

REFERENCES
1. Messerly, J. F.; Finke, H. L. *J. Chem. Thermodynamics* **1970**, *2*, 867.
2. Stimson, H. F. *J. Res. Nat. Bur. Stand.* **1961**, *65A*, 139.
3. Hoge, H. J.; Brickwedde, F. G. *J. Res. Nat. Bur. Stand.* **1939**, *22*, 351.
4. Counsell, J. F.; Green, J. H. S.; Hales, J. L.; Martin, J. F. *Trans. Faraday Soc.* **1965**, *61*, 212.
5. Gorbunova, N. I.; Grigoriev, V. A.; Simonov, V. M.; Shipova, V. A. *Int. J. Thermophysics* **1982**, *3*, 1.
6. Hossenlopp, I. A.; Scott, D. W. *J. Chem. Thermodynamics* **1981**, *13*, 405.

9.2.11. C_p, Polystyrene

Physical property: Heat capacity
Units: J mol^{-1} K^{-1} (molar heat capacity, C_p)
J kg^{-1} K^{-1} (specific heat capacity, c_p)
Recommended reference material: Polystyrene (atactic)
Range of variables: (1) 10 to 360 K (ii) 375 to 470 K
Physical state within the range: (i) glass (ii) liquid
Contributors: S. S. Chang, R. Sabbah

Intended usage: Atactic polystyrene is recommended for testing the performance of differential scanning calorimeters used for measuring the heat capacity of polymers.

Sources of supply and/or methods of preparation: Certified reference material for molecular weight determinations is available from supplier (S). Large quantities of polystyrene may be obtained from most major polymer manufacturers (*e.g.* E, I, R, V, etc.). Small quantities may be obtained from suppliers of specialized chemicals (such as Q, W, X, Aldrich, B.D.H. etc.).

Pertinent physicochemical data: Heat capacities of three Standard Reference Materials, SRM 705 for narrow molecular weight distribution (MWD) polystyrene (Ref. 1), SRM 706 for broad MWD polystyrene (Ref. 2), and SRM 1478 for narrow MWD polystyrene fraction (Ref. 3), have been determined by adiabatic calorimetry for the temperature ranges 10 to 360 K, 20 to 470 K and 6 to 380 K, respectively. The results of a large number of investigations

9. Enthalpy

were summarized by Gaur and Wunderlich (Ref. 4). The heat capacities in the glassy state below 340 K are relatively insensitive to MWD and thermal history. The heat capacity of the liquid above 370 K is highly reproducible for the individual sample. A kinetically influenced glass transition occurs around 370 K. The following values of C_p were taken from references 2, 3, 4, 5.

Table 9.2.11.1. Heat capacity C_p of polystyrene

T_{68} / K	C_p / J mol^{-1} K^{-1}	T_{68} / K	C_p / J mol^{-1} K^{-1}
10	3.34	240	101.2
20	10.76	260	110.1
40	23.6	280	119.2
60	32.9	300	128.3
80	40.7	320	137.6
100	47.7	340	146.9
120	54.6	360	156.3
140	61.4	380	194.5
160	68.7	400	201.0
180	76.2	420	207.5
200	84.2	440	214.0
220	92.5	460	220.5

The molar mass of the repeating unit, $-CH_2CH(C_6H_5)-$ was taken as 104.152 g mol^{-1}.

REFERENCES

1. Chang, S. S.; Bestul, A. B. *J. Polymer Sci.* **1968**, *A26*, 849.
2. Karasz, F. E.; Bair, H. E.; O'Reilly, J. M. *J. Phys. Chem.* **1965**, *69*, 2657.
3. Chang, S. S. *J. Polymer Sci. C, Polymer Symp.* **1984**, *71*, 59.
4. Gaur, H.; Wunderlich, B. *J. Phys. Chem. Ref. Data* **1982**, *11*, 313.
5. NBS Certificate for Standard Reference Material 705, National Bureau of Standards: Washington.

9.2.12 C_p, Poly(vinyl chloride)

Physical property: Heat capacity
Units: J mol^{-1} K^{-1} (molar heat capacity, C_p)
J kg^{-1} K^{-1} (specific heat capacity, c_p)
Recommended reference material: Poly(vinyl chloride) (PVC)
Range of variables: (i) 10 to 350 K (ii) 355 to 380 K

Physical states within the range: (i) glass (ii) liquid
Contributors: S. S. Chang, R. Sabbah

Intended usage: PVC is recommended for testing the performance of differential scanning calorimeters used for measuring the heat capacities of polymers.

Sources of supply and/or methods of preparation: Large quantities of unplasticized PVC are obtainable from most major polymer manufacturers (*e.g.* G, H, etc.). Small quantities may be obtained from suppliers of specialized chemicals (such as Q, W, X, Aldrich, Fluka, B.D.H., etc.).

Pertinent physicochemical data: Heat capacities for both bulk-polymerized and suspension-polymerized PVC have been determined by adiabatic calorimetry (Ref. 1) after being subjected to various thermal and mechanical treatments. Heat capacity differences between quenched glass and annealed glass began to be observable (> 0.1 per cent) at 270 K, and reached 1 per cent at 340 K. The influence of methods of polymerization on the heat capacity did not exceed 0.1 per cent. A glass transition occurs around 355 K. Mechanical treatment may cause some energy to be stored. Other measurements (Refs. 2, 3) are summarized in reference 1. The following table gives C_p of annealed, bulk-polymerized PVC.

Table 9.2.12.1. Heat capacity C_p of poly(vinyl chloride)

$\dfrac{T_{68}}{K}$	$\dfrac{C_p}{J\ mol^{-1}\ K^{-1}}$	$\dfrac{T_{68}}{K}$	$\dfrac{C_p}{J\ mol^{-1}\ K^{-1}}$
10	1.81	200	43.0
20	5.92	220	46.2
40	13.4	240	49.4
60	18.8	260	52.6
80	23.1	280	56.0
100	26.8	300	59.4
120	30.3	320	62.9
140	33.6	340	67.0
160	36.7	360	91.1
180	39.9	380	98.1

The molar mass of the repeating unit, $-CH_2CHCl-$, is taken as 62.499 g mol^{-1}. Heating unstabilized PVC to temperatures much higher than the glass transition temperature results in thermal degradation and is therefore not recommended.

REFERENCES

1. Chang, S. S. *J. Res. Nat. Bur. Stand.* **1977**, *82*, 9.
2. Alford, S.; Dole, M. *J. Amer. Chem. Soc.* **1955**, *77*, 4774.
3. Lebedev, B. V.; Rabinovich, I. B.; Budarina, V. A. *Vysokomol. Soedein.* **1967**, *A9*, 488.

9.2.13. C_{sat} and C_p, 1,4-Dimethylbenzene

Physical property: Heat capacity
Units: J mol^{-1} K^{-1} (molar heat capacity, C_{sat} and C_p)
   J kg^{-1} K^{-1} (specific heat capacity, c_{sat} and c_p)
Recommended reference material: 1,4-Dimethylbenzene (*p*-xylene) (C$_8$H$_{10}$)
Range of variables: (i) 286 to 570 K liquid at saturation pressure (ii) 398 to 523 K real gas at pressures up to 202.66 kPa
Physical states within the range: (i) liquid (ii) real gas
Contributors to the first version: J. D. Cox, D. R. Stull
Contributor to the revised version: R. Sabbah

Intended usage: 1,4-Dimethylbenzene is recommended for testing the performance of calorimeters (Ref. 1) used for the measurement of the heat capacities of (i) liquids in equilibrium with their vapours and (ii) vapours at pressures between 0 (ideal gas state) and 202.65 kPa.

Sources of supply and/or methods of preparation: Samples of 1,4-dimethylbenzene with mole fraction purity > 0.999 can be obtained respectively from supplier (C) and with mass fraction purity > 0.999 from Baker.

Pertinent physicochemical data: (i) Corruccini and Ginnings (Ref. 2) studied a sample of 1,4-dimethylbenzene having a mole fraction purity of 0.9993 assessed by a freezing-point method. Their experimental procedure was based on the 'drop' method and involved the use of a Bunsen ice calorimeter to receive the specimen. Values of the heat capacity of the liquid in equilibrium with its vapour at various temperatures are given in table 9.2.13.1. These values were calculated by Chao (Ref. 3) from results obtained by Corruccini and Ginnings. The temperature scale used was probably ITS-27 (Ref. 4).

(ii) Values for the vapour heat capacity of a sample of 1,4-dimethylbenzene in the real and ideal gas states over large ranges of temperature (referred to the IPTS-68) and pressure were obtained by Hossenlopp and Scott (Ref. 5) and are reported in table 9.2.13.2. The mole fraction purity of the sample was (0.9997 ± 0.0002) determined by a freezing point method. The experimental uncertainty in the values of C_p at zero pressure in the ideal gas state in table 9.2.13.2 is about ±0.2 per cent.

Table 9.2.13.1. Heat capacity C_{sat} of 1,4-dimethylbenzene in the liquid phase

$\dfrac{T}{K}$	$\dfrac{T_{68}}{K}$	$\dfrac{C_{sat}}{\text{J mol}^{-1}\text{ K}^{-1}}$	$\dfrac{T}{K}$	$\dfrac{T_{68}}{K}$	$\dfrac{C_{sat}}{\text{J mol}^{-1}\text{ K}^{-1}}$
286.39	286.38	178.30	430	430.02	230.62
290	289.99	179.29	450	450.03	238.82
310	309.99	185.37	470	470.04	247.35
330	329.99	192.16	490	490.05	256.40
350	349.99	199.43	510	510.06	266.24
370	370.00	207.00	530	530.06	277.17
390	390.01	214.75	550	550.07	289.59
410	410.02	222.62	570	570.07	303.92

Table 9.2.13.2. Heat capacity C_p of 1,4-dimethylbenzene in the gas phase

	$C_p(T)/\text{J mol}^{-1}\text{ K}^{-1}$					
	T_{68}/K					
p/kPa	398.15	423.15	448.15	473.15	498.15	523.15
202.66				199.93	206.89	214.45
152.01			190.95			
101.33		181.09	188.85	196.96	205.03	212.66
50.657	169.80	178.23	187.05	195.53	203.85	212.07
25.326	168.00	177.12	186.38	194.85	203.33	211.38
18.994	167.77	176.81				
12.662	167.35					
0 (ideal gas)	166.52	175.82	185.40	194.17	202.92	211.09

REFERENCES

1. McCullough, J. P.; Scott, D. W., editors, *Experimental Thermodynamics*, Vol I: *Calorimetry of Non-reacting Systems*, Butterworths: London, **1968**.
2. Corruccini, R. J.; Ginnings, D. C. *J. Amer. Chem. Soc.* **1947**, *69*, 2291.
3. Chao, J.; private communication.
4. Furukawa, G. T.; private communication.
5. Hossenlopp, I. A.; Scott, D. W. *J. Chem. Thermodynamics* **1981**, *13*, 423.

9.2.14. C_p and C_{sat}, Water

Physical property: Heat capacity
Units: J mol^{-1} K^{-1} (molar heat capacity, C_p and C_{sat})
J kg^{-1} K^{-1} (specific heat capacity, c_p and c_{sat})
Recommended reference material: Water (H_2O)
Range of variables: (i) 0 to 100 °C liquid at 101.325 kPa pressure (ii) 0 to 374 °C liquid at the saturated vapour pressure (iii) 361 to 487 K real gas at pressures up to 101.325 kPa
Physical State within the range: (i) and (ii) liquid (iii) real gas
Contributors to the first version: J. D. Cox, J. F. Martin, O. Riedel, D. R. Stull
Contributor to the revised version: R. Sabbah

Intended usage: Since the beginning of calorimetry, water has been used as a reference material for heat capacity measurements (Ref. 1), partly because it is universally available in high purity. Recently, it has also been used in differential scanning calorimetry (Ref. 2).

Sources of supply and/or methods of preparation: It is now possible to prepare by distillation of deionized water a sample having a high degree of purity.

Pertinent physicochemical data: (i) Values of C_p (for a pressure of 101.325 kPa) for water are known with an uncertainty of ±0.01 to ±0.02 per cent at close intervals of temperature (1 K). Values of the enthalpy of liquid water relative to the enthalpy at 0 °C, a quantity useful in experiments using the method of mixtures, are also available at 1 K intervals of temperature (Ref. 3).

Osborne *et al.* (Ref. 3) measured the heat capacity of a sample of water, using a large adiabatic calorimeter specially designed to obtain the highest possible accuracy mentioned above. Values of both C_{sat} and C_p, based on these measurements (with temperatures on IPTS-48, were presented by Ginnings and Furukawa (Ref. 4) and are given in table 9.2.14.1. A molar mass of 18.016 g mol^{-1} was used.

Table 9.2.14.1. Heat capacity C_{sat} and C_p of water in the liquid phase

t_{48} / °C	t_{68} / °C	C_{sat} / J mol^{-1} K^{-1}	C_p / J mol^{-1} K^{-1}	t_{48} / °C	t_{68} / °C	C_{sat} / J mol^{-1} K^{-1}	C_p / J mol^{-1} K^{-1}
0	0.00	75.993	75.985	50	49.99	75.320	75.318
5	5.00	75.714	75.706	55	54.99	75.350	75.348
10	10.00	75.532	75.525	60	59.99	75.385	75.385
15	14.99	75.417	75.410	65	64.99	75.428	75.428
20	19.99	75.345	75.339	70	69.99	75.476	75.478
25	24.99	75.303	75.298	75	74.99	75.532	75.536
30	29.99	75.282	75.278	80	79.99	75.594	75.601
35	34.99	75.277	75.273	85	84.99	75.667	75.675
40	39.99	75.283	75.280	90	90.00	75.746	75.757
45	44.99	75.298	75.295	95	95.00	75.835	75.850
				100	100.00	75.934	75.954

Table 9.2.14.2. Heat capacity of water C_p in the gas phase

	$C_p(T)$ / J mol^{-1} K^{-1}				
	T_{68}/K				
p/kPa	361.80	381.20	410.20	449.20	487.20
101.32		36.82	35.84	35.47	35.44
47.36	35.65	35.19	34.94	35.02	
25.09	34.76	34.65	34.64	34.79	35.08
12.33	34.38	34.32			
0[a]	33.95	34.09	34.33	34.69	35.07

[a] Values of C_p at zero pressure in ideal gas state are interpolated from the statistically calculated values listed by Wagman *et al.* (Ref. 8).

9. Enthalpy

Recently Williams *et al.* (Ref. 5) measured the specific heat of water in the range 280 to 350 K. Their data agreed with the values listed in table 9.2.14.1 to within 0.2 per cent. (Note: the corrections to $C_p^\circ(H_2O)$ to bring the published data to IPTS-68, provided by K. Harr, NBS, affect only the fourth decimal figure.)

(ii) Correlated values of C_{sat} and $H(T) - H(273.15\ K)$ are available (Ref. 6) at 1 K intervals of temperature covering the entire liquid range of water (0 to 374 °C).

(iii) Values of the heat capacity of water vapour were obtained over large ranges of pressure and temperature by McCullough *et al.* (Ref. 7). The values on IPTS-68 in table 9.2.14.2 were determined from their results, which were on IPTS-48 with 0 °C = 273.16 K. According to the authors, the uncertainty of these data should not be greater than ±0.2 per cent.

REFERENCES

1. McCullough, J. P.; Scott, D. W., editors.,*Experimental Thermodynamics*, Vol I: *Calorimetry of Non-reacting Systems*, Chapter 10, Butterworths: London, **1968**.
2. Yuen, H. K.; Yosel, C. J. *Thermochim. Acta* **1979**, *33*, 281.
3. Osborne, N. S.; Stimson, H. F.; Ginnings, D. C. *J. Res. Nat. Bur. Stand.* **1939**, *23*, 197.
4. Ginnings, D. C.; Furukawa, G. T. *J. Amer. Chem. Soc.* **1953**, *75*, 522.
5. Williams, I. S.; Street, R.; Gopal, E. S. R. *Pramana* **1978**, *11*, 519.
6. Kennan, J. H.; Keyes, F. G.; Hill, P. G.; Moore, J. G. *Steam Tables*, Wiley: New York, **1969**.
7. McCullough, J. P.; Pennington, R. E.; Waddington, G. *J. Amer. Chem. Soc.* **1952**, *74*, 4439.
8. Wagman, D. D.; Kilpatrick, J. E.; Taylor, W. J.; Pitzer, K. S.; Rossini, F. D.; *J. Res. Nat. Bur. Stand.* **1946**, *34*, 143 (Table VII).

9.2.15. C_{sat}, Benzene

Physical property: Heat capacity
Units: J mol^{-1} K^{-1} (molar heat capacity, C_{sat})
    J kg^{-1} K^{-1} (specific heat capacity, c_{sat})
Recommended reference material: Benzene (C_6H_6)
Range of variables: (i) 279 to 340 K liquid at saturation pressure (ii) 333 to 527 K real gas at pressures up to 202.65 kPa
Physical states within the range: (i) liquid (ii) real gas
Contributors to the first version: J. D. Cox, D. R. Douslin, J. F. Martin, O. Riedel, D. R. Stull
Contributors to the revised version: J. Lielmezs, R. Sabbah

Intended usage: Because benzene is not corrosive, has excellent boiling characteristics and is easy to handle experimentally, it is recommended for testing the performance of vapour flow (non-adiabatic) calorimeters (Ref. 1) for the measurement of the heat capacity of (i) liquids in equilibrium with their vapours, and (ii) vapours at pressures between 0 (ideal gas state) and 202.65 kPa.

Sources of supply and/or methods of preparation: High purity samples of benzene can be obtained from supplier (C) as well as from producers of high purity chemicals, *e.g.* B.D.H., Fluka, etc.

Pertinent physicochemical data: (i) Using an adiabatic calorimeter, Oliver *et al.* (Ref. 2) measured the heat capacity of a sample of benzene having a mole fraction purity 0.99967, as assessed by a calorimetric study of the melting behaviour. Their original data, obtained on the 1928 temperature scale were converted to the 1968 temperature scale and recomputed to give the following interpolated values for the molar heat capacity of liquid benzene at its saturation vapour pressure in the temperature range 278.67 to 340 K. The molar mass of benzene was taken as 78.113 g mol^{-1}.

Table 9.2.15.1. Heat capacity C_{sat} of benzene in the liquid phase

T_{68} / K	C_{sat} / J mol^{-1} K^{-1}	T_{68} / K	C_{sat} / J mol^{-1} K^{-1}
278.7	131.9	320	140.9
280	132.2	330	143.2
290	134.3	340	146.0
300	136.5		
310	138.5		

As the differences between C_p and C_{sat} are barely significant (Refs. 3, 4), the values in table 9.2.15.1 are in agreement with those given at 298.15, 308.15 and 318.15 K in reference 3 and do not differ above 310 K from those given in reference 4 by more than 0.5 per cent.

(ii) Experimental values obtained by Todd *et al.* (Ref. 5) are reported in table 9.2.15.2 for the vapour heat capacity of a sample of benzene (having a mole fraction purity of 0.9994±0.0002, assessed by a freezing point method) in the real and ideal gas states over large ranges of pressure and temperature (referred to IPTS-68). The earlier values of C_p° reported in reference 6 are within 0.15 per cent of a smooth curve drawn through the values given in the table 9.2.15.2. Part of the difference arises because 'curvature corrections' in the extrapolation to zero pressure were not applied in the earlier work.

Table 9.2.15.2. Heat capacity C_p of benzene in the gas phase

	$C_p(T)$/J mol^{-1} K^{-1}							
	T_{68}/K							
p/kPa	333.15	348.15	368.15	408.15	438.15	473.15	500.15	527.15
202.65				117.44	126.08	134.45	140.66	146.48
101.32			106.39	115.85	124.99	133.79	140.09	145.99
50.662		99.43	105.13	115.01	124.43	133.30	139.77	145.82
37.996	94.68	99.06	104.88					
25.331	94.23	98.71	104.59	114.69	124.20	133.09	139.60	145.70
18.998	94.00	98.53						
12.666	93.76	98.34	104.26	114.47				
0 (ideal gas)	93.32	97.99	103.98	114.29	123.93	132.94	139.47	145.59

REFERENCES

1. McCullough, J. P.; Scott, D. W., editors, *Experimental Thermodynamics*, Vol. I: *Calorimetry of Non-reacting Systems*, Chapter 10, Butterworths: London, **1968**.
2. Oliver, G. D.; Eaton, M.; Huffman, H. M. *J. Amer. Chem. Soc.* **1948**, *70*, 1502.
3. Vesely, F.; Zabransky,M.; Svoboda, V.; Pick, J. *Collect. Czech. Chem. Commun.* **1979**, *44*, 3529.
4. Gorbunova, N. I.; Grigoriev, V. A.; Simonov, V. M.; Shipova, V. A. *Int. J. Thermophysics* **1982**, *3*, 1.
5. Todd, S. S.; Hossenlopp, I. A.; Scott, D. W. *J. Chem. Thermodynamics* **1978**, *10*, 641.
6. Scott, D. W.; Waddington, G.; Smith, J. C.; Huffman, H. M. *J. Chem. Phys.* **1947**, *15*, 565.

9.2.16. C_p, Nitrogen

Physical property: Heat capacity
Units: J mol^{-1} K^{-1} (molar heat capacity, C_p)
J kg^{-1} K^{-1} (specific heat capacity, c_p)
Recommended reference material: Nitrogen (N$_2$)
Range of variables: (i) 65 to 126 K liquid at saturation pressure (ii) 100 to 1000 K real gas at pressures up to 40 MPa
Physical states within the range: (i) liquid (ii) real gas
Contributors to the first version: J. D. Cox, D. R. Stull
Contributors to the revised version: R. T. Jacobsen, R. Sabbah, L. A. Weber

Intended usage: Nitrogen is recommended for testing the performance of calorimeters used for the measurement of the heat capacities of (i) liquids and liquified gases and (ii) gases and vapours.

Sources of supply and/or methods of preparation: Nitrogen of suitable purity can be obtained commercially from various sources. Samples should be passed, at cylinder pressure, through a molecular sieve trap before being distilled into the calorimeter.

Pertinent physicochemical data: The molar mass was taken as 28.016 g mol^{-1}. (i) The measurements of Giauque and Clayton (Ref. 1) and of Weber (Ref. 2) on the gas-liquid two-phase system at its saturation vapour pressure can be expressed by the following equation:

$$C_{sat}/\text{J mol}^{-1}\text{ K}^{-1} = 88.619 - 1.3696x + 1.4443 \times 10^{-2}x^2 - 6.3700 \times 10^{-5}x^3 \\ + 1.5580x/(126.2 - x)^{1/2} \qquad (9.2.16.1).$$

where $x = T/\text{K}$. At temperatures above the normal boiling point, the data must generally be corrected for the presence of the vapour and for any PV work done (see reference 2). The data of Wiebe and Brevoort (Ref. 3) also agree well with the equation except at the higher temperatures where corrections for the presence of the vapour phase become large.

(ii) Jacobsen and Stewart (Ref. 4) report calculated values for the heat capacity of nitrogen in the range 100 to 1000 K and 0.101325 to 40 MPa (table 9.2.16.1). Values of C_p° are also given. The calculated values at 0.101325 MPa are accurate to about ± 0.1 per cent, while the higher pressure, lower temperature values are estimated to be accurate to about ± 1 per cent, based upon comparisons with the dense-gas measurements of reference 2.

Table 9.2.16.1. Heat capacity C_p of nitrogen in the gas phase

T_{68}/K	$C_p(T)$/J mol^{-1} K^{-1}							
	p/MPa							
	0	0.101325	1.0	5.0	10.0	15.0	20.0	40.0
100.00	29.10	30.03		59.62	56.30	54.22	52.75	49.36
150.00	29.11	29.37	32.13	67.00	80.11	62.49	56.07	48.46
200.00	29.11	29.23	30.38	36.68	45.54	49.94	49.85	45.33
250.00	29.11	29.18	29.82	32.80	36.41	39.22	40.88	41.44
300.00	29.12	29.17	29.58	31.39	33.46	35.16	36.44	38.28
350.00	29.16	29.20	29.48	30.71	32.09	33.26	34.19	36.09
400.00	29.25	29.27	29.48	30.37	31.37	32.23	32.94	34.64
450.00	29.39	29.40	29.56	30.24	31.00	31.66	32.23	33.69
500.00	29.58	29.59	29.72	30.25	30.85	31.37	31.83	33.09
550.00	29.82	29.84	29.94	30.36	30.84	31.27	31.65	32.74
600.00	30.11	30.12	30.20	30.55	30.94	31.30	31.62	32.57
650.00	30.42	30.43	30.50	30.79	31.12	31.42	31.69	32.52
700.00	30.76	30.76	30.82	31.06	31.34	31.60	31.83	32.56
750.00	31.09	31.10	31.15	31.35	31.59	31.81	32.01	32.66
800.00	31.43	31.44	31.48	31.66	31.86	32.05	32.23	32.80
850.00	31.76	31.77	31.81	31.96	32.14	32.30	32.46	32.97
900.00	32.09	32.09	32.13	32.26	32.41	32.56	32.70	33.15
950.00	32.40	32.40	32.43	32.55	32.69	32.81	32.93	33.34
1000.00	32.70	32.70	32.72	32.83	32.95	33.06	33.17	33.54

REFERENCES

1. Giauque, W. F.; Clayton, J. O. *J. Amer. Chem. Soc.* **1933**, *55*, 4875.
2. Weber, L. A. *J. Chem. Thermodynamics* **1981**, *13*, 389.
3. Wiebe, R.; Brevoort, M. J. *J. Amer. Chem. Soc.* **1930**, *52*, 622.
4. Angus, S. K.; de Reuck, M.; Armstrong, B., editors, *International Thermodynamic Tables of the Fluid State-6: Nitrogen.* Pergamon: Oxford, **1979**.

9.2.17. C_p, Carbon dioxide

Physical property: Heat capacity
Units: J mol^{-1} K^{-1} (molar heat capacity, C_p)
J kg^{-1} K^{-1} (specific heat capacity, c_p)
Recommended reference material: Carbon dioxide (CO_2)

Range of variables: 250 to 1000 K real gas at pressures up to 50 MPa
Physical state within the range: real gas
Contributor: R. Sabbah

Intended usage: Carbon dioxide is recommended for testing the performance of calorimeters used for the measurement of the heat capacities of gases and vapours, especially calorimeters of the vapour-flow type.

Sources of supply and/or methods of preparation: Carbon dioxide of suitable purity can be obtained commercially from various sources. As mentioned by Rivkin and Gukov (Ref. 1), small amounts of nitrogen affect the heat capacity, so it is necessary to remove nitrogen from the sample before using it in calorimetric experiments.

Pertinent physicochemical data: Altunin et al. (Ref. 2) reported calculated values for the heat capacity of carbon dioxide from which are taken the values in table 9.2.17.1 for the real gas in the range 250 to 1000 K and 0.01 to 50 MPa. Values of C_p° are also given. In two recent papers, Bottinga and Richet (Ref. 3) calculated values of C_p for the real gas in the range 400 to 2100 K and 0.1 to 500 MPa and Bender et al. (Ref. 4) gave experimental values in the range 233 to 473 K at pressures up to 1.5 MPa.

Table 9.2.17.1 Heat capacity C_p of carbon dioxide in the gas phase

T_{68}/K	$C_p(T)$/J mol^{-1} K^{-1}									
	p/MPa									
	0	0.01	0.05	0.1	0.5	1	5	10	25	50
250.00	34.83	34.9	35.1	35.4	38.2	42.8	89.3	86.0	79.9	74.4
300.00	37.21	37.2	37.3	37.5	38.7	40.3	79.0	131.0	88.1	77.3
350.00	39.39	39.4	39.5	39.6	40.2	41.2	51.8	84.9	98.6	76.6
400.00	41.33	41.3	41.4	41.4	41.9	42.5	48.2	58.4	85.6	74.7
450.00	43.06	43.1	43.1	43.1	43.5	43.9	47.6	53.1	69.2	70.9
500.00	44.61	44.6	44.6	44.7	44.9	45.2	47.9	51.5	61.6	66.9
550.00	46.03	46.0	46.3	46.1	46.3	46.5	48.5	51.0	58.1	63.6
600.00	47.32	47.3	47.3	47.4	47.5	47.7	49.2	51.1	56.3	61.1
650.00	48.49	48.5	48.5	48.5	48.6	48.8	50.0	51.5	55.4	59.5
700.00	49.56	49.6	49.6	49.6	49.7	49.8	50.8	51.9	55.0	58.4
750.00	50.54	50.5	50.6	50.6	50.6	50.7	51.5	52.5	55.0	57.8
800.00	51.43	51.4	51.4	51.5	51.5	51.6	52.3	53.1	55.1	57.5
850.00	52.25	52.3	52.3	52.3	52.3	52.4	53.0	53.6	55.3	57.4
900.00	52.99	53.0	53.0	53.0	53.1	53.1	53.6	54.2	55.6	57.4
950.00	53.67	53.7	53.7	53.7	53.7	53.8	54.2	54.7	56.0	57.5
1000.00	54.30	54.3	54.3	54.3	54.3	54.4	54.7	55.2	56.3	57.6

REFERENCES

1. Rivkin, S. L.; Gukov, V. M. *Teploenergetika* **1971**, *18*, 82.
2. Angus, S.; Armstrong, B.; de Reuck, K. M. editors, *International Thermodynamic Tables of the Fluid State-3: Carbon Dioxide*, Pergamon: Oxford, **1976**.
3. Bottinga, Y.; Richet, P. *Amer. J. Sci.* **1981**, *281*, 615.
4. Bender, R.; Bier, K.; Maurer, G. *Ber. Bunsenges. Phys. Chem.* **1981**, *85*, 778.

9.3. Reference materials for measurement of enthalpies of phase change

9.3.1. $\Delta_{fus}H$ and $\Delta_{trs}H$, 1,3-Difluorobenzene

Physical property: Enthalpy of fusion and transition
Units: J mol^{-1} or kJ mol^{-1} (molar enthalpy of fusion, $\Delta_{fus}H$ or transition, $\Delta_{trs}H$)
J kg^{-1} or J g^{-1} (specific enthalpy of fusion, $\Delta_{fus}h$ or transition $\Delta_{trs}h$)
Recommended reference material: 1,3-difluorobenzene ($C_6H_4F_2$)
Range of variables: (i) $T_{\text{triple point}} = 204.06$ K; (ii) $T_{trs} = 186.80$ K (Ref. 1) (temperatures converted to IPTS-68)
Physical state within the range: (i) solid $\rightarrow$ liquid (in equilibrium with the vapour)
(ii) solid $\rightarrow$ solid
Contributor: R. Sabbah

Intended usage: 1,3-difluorobenzene is suggested for testing calorimeters used for the measurement of the enthalpy of fusion or of transition below room temperature.

Sources of supply and/or methods of preparation: Commercial samples with a mass fraction purity > 0.99 are available from suppliers Aldrich, Fluka. It is possible to purify them further by distillation through a fractionating column. It is also necessary to prevent the sample used in calorimetric experiments from being exposed to air and moisture.

Pertinent physicochemical data: Messerly and Finke (Ref. 1) measured the enthalpy of fusion and transition of a sample having a mole fraction purity of 0.99999 assessed by calorimetric study of the melting behaviour. They found an average value of 8581 J mol^{-1} for the enthalpy of fusion at the triple point temperature and of 827.1 J mol^{-1} for the enthalpy of transition at 186.80 K. The molar mass of 1,3-difluorobenzene ws taken as 114.096 g mol^{-1}.

REFERENCE

1. Messerly, J. F.; Finke, H. L. *J. Chem. Thermodynamics* **1970**, *2*, 867.

9.3.2. $\Delta_{fus}H$ and $\Delta_{trs}H$, 2,2-Dimethylpropane

Physical property: Enthalpy of fusion, $\Delta_{fus}H$, and transition $\Delta_{trs}H$
Units: J mol^{-1} or kJ mol^{-1} (molar enthalpy of fusion, $\Delta_{fus}H$, or transition, $\Delta_{trs}H$)
 J kg^{-1} or J g^{-1} (specific enthalpy of fusion, $\Delta_{fus}h$ or transition, $\Delta_{trs}h$)
Recommended reference material: 2,2-dimethylpropane, neopentane (C$_5$H$_{12}$)
Range of variables: (i) $T_{triple\ point}$ = 256.77 K (ii) T_{trs} = (140.48 ± 0.05) K (Ref. 1, temperatures converted to IPTS-68).
Physical states within the range: (i) solid → liquid (in equilibrium with the vapour)
 (ii) solid → solid
Contributors: K. N. Roy, R. Sabbah, D. D. Sood

Intended usage: 2,2-Dimethylpropane is recommended for testing the performance of calorimeters used for the measurement of enthalpy of fusion or of transition (Ref. 2).

Sources of supply and/or methods of preparation: A highly pure grade (mole fraction purity of 0.99997) of 2,2-dimethylpropane is available from supplier (C). Samples having mole and mass fraction purities of 0.9999 and 0.99 can be purchased from supplier (P) and Baker, respectively. Further purification of a sample of 2,2-dimethylpropane is possible by fractional distillation and treatment of the distillate with molecular sieve 5A.

Pertinent physicochemical data: Enokido et al. (Ref. 1) measured the enthalpies of fusion and transition of a sample of 2,2-dimethylpropane having a mole fraction purity of 0.99997, assessed by calorimetric study of the melting behaviour. They found a value of (3096.2 ± 1.3) J mol^{-1} for the enthalpy of fusion at the triple point temperature and a value of (2630.5 ± 1.3) J mol^{-1} for the enthalpy of transition at 140.48 K. These values are reported in reference 2. The molar mass of 2,2-dimethylpropane was taken as 72.151 g mol^{-1}.

REFERENCES

1. Enokido, H.; Shinoda, T.; Mashiko, Y. *Bull. Chem. Soc. Japan* **1969**, *42*, 84.
2. Physicochemical Measurements: Catalogue of Reference Materials from National Laboratories, *Pure and Appl. Chem.* **1976**, *48*, 505.

9.3.3. $\Delta_{fus}H$, Hexafluorobenzene

Physical property: Enthalpy of fusion
Units: J mol^{-1} or kJ mol^{-1} (molar enthalpy of fusion, $\Delta_{fus}H$)
 J kg^{-1} or J g^{-1} (specific enthalpy of fusion, $\Delta_{fus}h$)
Recommended reference material: Hexafluorobenzene (C$_6$F$_6$)
Range of variables: $T_{triple\ point}$ = 278.30 K (Ref. 1) (temperature converted to IPTS-68)

Physical states within the range: solid → liquid (in equilibrium with the vapour)
Contributor: R. Sabbah

Intended usage: Hexafluorobenzene is suggested for testing the performance of calorimeters used for the measurement of enthalpy of fusion.

Sources of supply and/or methods of preparation: Commercial samples are available from BDH (with a mass fraction purity of 0.999), Aldrich, Fluka (with a mass fraction purity > 0.99) which can be further purified by preparative gas-liquid chromatography. It is also necessary to prevent the sample used in calorimetric experiments from being exposed to air and moisture.

Pertinent physicochemical data: Two sets of measurements of the enthalpy fusion of hexafluorobenzene have been made. Counsell et al. (Ref. 2) studied a sample having a mole fraction purity of 0.9997 and found (11.590 ± 0.013) kJ mol^{-1}, which is in very good agreement with the average value reported by Messerly and Finke (Ref. 1), 11.59 kJ mol^{-1}. The value found by Counsell et al. is recommended for the measurement of the enthalpy of fusion of hexafluorobenzene at the triple point temperature. The molar mass was taken as 186.057 g mol^{-1}.

REFERENCES

1. Messerly, J. F.; Finke, H. L. *J. Chem. Thermodynamics* **1970**, *2*, 867.
2. Counsell, J. F.; Green, J. H. S.; Hales, J. L.; Martin, J. F. *Trans. Faraday Soc.* **1965**, *61*, 212.

9.3.4. $\Delta_{fus}H$, Diphenyl ether

Physical property: Enthalpy of fusion
Units: J mol^{-1} or kJ mol^{-1} (molar enthalpy of fusion, $\Delta_{fus}H$)
   J kg^{-1} or J g^{-1} (specific enthalpy of fusion, $\Delta_{fus}h$)
Recommended reference material: Diphenyl ether ($C_{12}H_{10}O$)
Range of variables: $T_{\text{triple point}} = 300.02$ K (secondary fixed point on IPTS-68)
Physical states within the range: solid → liquid (in equilibrium with the vapour)
Contributors to the first version: J. D. Cox, J. F. Martin
Contributor to the revised version: R. Sabbah

Intended usage: The use of diphenyl ether as a test material for equipment used for the determination of enthalpies of fusion is suggested. Such apparatuses include classical calorimeters (adiabatic, drop, ice, radiation) and differential scanning calorimeters (Ref. 1). It has been shown (Ref. 2) that diphenyl ether can be prepared in an extremely pure state and that the molten substance is inert to most materials of construction.

Sources of supply and/or methods of preparation: Commercial samples with a mass fraction purity ≥ 0.99 are available from Aldrich, Koch-Light, etc., which can be further purified by distillation and fractional crystallization. It is also necessary to remove the dissolved air and water from the sample before using it in calorimetric experiments.

Pertinent physicochemical data: Furukawa et al. (Ref. 2) studied a sample of diphenyl ether with a mole fraction purity of 0.999987, established by calorimetric study of the melting behaviour. They reported that the enthalpy of fusion is (17216 ± 17) J mol^{-1} at the triple point temperature. The molar mass of diphenyl ether was taken as 170.20 g mol^{-1}.

REFERENCES

1. Nolan, P. S.; Lemay, Jr., H. E. *Thermochim. Acta* **1974**, *9*, 81.
2. Furukawa, G. T.; Ginnings, D. C.; McCoskey, R. E.; Nelson, R. A. *J. Res. Nat. Bur. Stand.* **1951**, *46*, 195.

9.3.5. $\Delta_{\text{fus}}H$, Naphthalene

Physical property: Enthalpy of fusion
Units: J mol^{-1} or kJ mol^{-1} (molar enthalpy of fusion, $\Delta_{\text{fus}}H$)
  J kg^{-1} or J g^{-1} (specific enthalpy of fusion, $\Delta_{\text{fus}}h$)
Recommended reference material: Naphthalene ($C_{10}H_8$)
Range of variables: $T_{\text{triple point}} = (353.39 \pm 0.02)$ K (Ref. 1)
Physical states within the range: solid $\rightarrow$ liquid (in equilibrium with the vapour)
Contributors: R. Sabbah, D. D. Sood, V. Venugopal

Intended usage: Naphthalene can be used as a calibrant for thermal analysis apparatus, especially for differential scanning calorimeters (Refs. 1 to 3).

Sources of supply and/or methods of preparation: A certified sample is available from supplier (T). A high purity sample can be obtained from supplier (C). Commercial samples with a mass fraction purity ≥ 0.99 can be purchased from Aldrich, Carlo Erba, Fluka, Koch-Light, etc. which can be used after further purification by zone refining.

Pertinent physicochemical data: From measurements using an adiabatic calorimeter and two samples having a mole fraction purity ≥ 0.9999, Andon and Connett (Ref. 1) determined two values for the enthalpy of fusion of naphthalene at the triple point temperature. The value of (19.060 ± 0.016) kJ mol^{-1} which is calculated from their results is recommended for the enthalpy of fusion of naphthalene at this temperature. The molar mass of naphthalene was taken as 128.17 g mol^{-1}. References 1, 2 and 3 review previous measurements.

REFERENCES

1. Andon, R. J. L.; Connett, J. E. *Thermochim. Acta* **1980**, *42*, 241.
2. Casellato, F.; Vecchi, C.; Girelli, A.; Casu, B. *Thermochim. Acta* **1973**, *6*, 361.
3. Wauchope, R. D.; Getzen, F. W. *J. Chem. Eng. Data* **1972**, *17*, 38.

9.3.6. $\Delta_{\text{fus}} H$, Benzil

Physical property: Enthalpy of fusion
Units: J mol^{-1} or kJ mol^{-1} (molar enthalpy of fusion, $\Delta_{\text{fus}} H$)
J kg^{-1} or J g^{-1} (specific enthalpy of fusion, $\Delta_{\text{fus}} h$)
Recommended reference material: Benzil or dibenzoyl ($C_{14}H_{10}O_2$)
Range of variables: $T_{\text{triple point}} = (368.01 \pm 0.02)$ K (Ref. 1)
Physical states within the range: solid $\to$ liquid (in equilibrium with the vapour)
Contributors: R. Sabbah, D. D. Sood, V. Venugopal

Intended usage: The use of benzil as a test material for equipment, especially differential scanning calorimeters, used for the determination of enthalpies of fusion is suggested (Ref. 1).

Sources of supply and/or methods of preparation: A certified sample is available from supplier (T). Other suitable material can be obtained from various sources, *e.g.* Aldrich, Baker, Koch-Light, Fluka, etc. and can be used after purification by zone refining.

Pertinent physicochemical data: Booss and Hauschildt (Ref. 2) measured the enthalpy of fusion of benzil by differential scanning calorimetry and reported a value of 22.7 kJ mol^{-1}. The uncertainty claimed is ± 0.8 per cent. Andon and Connett (Ref. 1) measured the enthalpy of fusion of two samples of benzil, having a mole fraction purity ≥ 0.9999, with an adiabatic calorimeter. The value of (23.545 ± 0.030) kJ mol^{-1} which is calculated from their results is recommended for the enthalpy of fusion of benzil at the triple point temperature. The molar mass of benzil was taken as 210.23 g mol^{-1}.

REFERENCES

1. Andon, R. J. L.; Connett, J. E. *Thermochim. Acta* **1980**, *42*, 241.
2. Booss, H. J.; Hauschildt, H. R. *Fresenius Z. Anal. Chem.* **1972**, *261*, 32.

9.3.7. $\Delta_{\text{fus}} H$, Acetanilide

Physical property: Enthalpy of fusion
Units: J mol^{-1} or kJ mol^{-1} (molar enthalpy of fusion, $\Delta_{\text{fus}} H$)
J kg^{-1} or J g^{-1} (specific enthalpy of fusion, $\Delta_{\text{fus}} h$)

Recommended reference material: Acetanilide (C_8H_9ON)
Range of variables: $T_{\text{triple point}} = (387.53 \pm 0.02)$ K (Ref. 1)
Physical states within the range: solid → liquid (in equilibrium with the vapour)
Contributors: R. Sabbah, D. D. Sood, V. Venugopal

Intended usage: The use of acetanilide as a test material for apparatuses using dynamic methods, such as differential scanning calorimetry, employed for the measurement of the enthalpy of fusion is suggested (Ref. 1).

Sources of supply and/or methods of preparation: A certified sample is available from supplier (T). Commercial samples with a mass fraction purity ≥ 0.99 are available from Baker, Koch-Light, etc. and can be used after purification by zone refining.

Pertinent physicochemical data: From five measurements using an adiabatic calorimeter, Andon and Connett (Ref. 1) determined a value of (21.653 ± 0.011) kJ mol^{-1} for the enthalpy of fusion of acetanilide (the sample having a mole fraction purity of 0.9997) at the triple point temperature. The molar mass of acetanilide was taken as 137.17 g mol^{-1}. Their publication mentioned that no report of any other measurements on this compound was found in the literature.

REFERENCES

1. Andon, R. J. L.; Connett, J. E. *Thermochim. Acta* **1980**, *42*, 241.

9.3.8. $\Delta_{\text{fus}}H$, Benzoic acid

Physical property: Enthalpy of fusion
Units: J mol^{-1} or kJ mol^{-1} (molar enthalpy of fusion, $\Delta_{\text{fus}}H$)
   J kg^{-1} or J g^{-1} (specific enthalpy of fusion, $\Delta_{\text{fus}}h$)
Recommended reference material: Benzoic acid ($C_7H_6O_2$)
Range of variables: $T_{\text{triple point}} = (395.52 \pm 0.02)$ K (Ref. 1)
Physical states within the range: solid → liquid (in equilibrium with the vapour)
Contributors to the first version: G. T. Armstrong, J. D. Cox
Contributors to the revised version: R. Sabbah, D. D. Sood, V. Venugopal

Intended usage: Benzoic acid is recommended for testing calorimeters used for the measurement of enthalpies of fusion, for example, adiabatic calorimeters or differential scanning calorimeters; it could also be used for the calibration of calorimeters of the latter type.

Benzoic acid attacks tin and copper at temperatures near the melting point (Ref. 2) and its use as a test (or calibration) material is therefore recommended only when the acid is in contact with inert materials.

Sources of supply and/or methods of preparation: Samples of benzoic acid prepared as reference materials for energy-of-combustion measurements could also be used as reference materials for measurements of enthalpy of fusion. These samples are available from suppliers (B, D, S) and B.D.H. A certified reference material, intended as calibrant for thermal analysis equipment, especially differential scanning calorimeters, can be purchased from supplier (T).

Pertinent physicochemical data: Using an adiabatic calorimeter, Andon and Connett (Ref. 1) determined the enthalpy of fusion at the triple point temperature of a sample of benzoic acid having a mole fraction purity ≥ 0.9999. Their result, (18.063 ± 0.021) kJ mol^{-1}, is in agreement with the value of Furukawa *et al.*, (18.00 ± 0.10) kJ mol^{-1} (Ref. 2), and is significantly higher than the value reported by Sklyankin and Strelkov, (17.80 ± 0.04) kJ mol^{-1} (Ref. 3). The molar mass of benzoic acid was taken as 122.12 g mol^{-1}.

REFERENCES

1. Andon, R. J. L.; Connett, J. E. *Thermochim. Acta* **1980**, *42*, 241.
2. Furukawa, G. T.; McCoskey, R. E.; King, G. J. *J. Res. Nat. Bur. Stand.* **1951**, *47*, 256.
3. Sklyankin, A. A.; Strelkov, P. G. *Zh. Prikl. Mekh. Tekh. Fiz.* **1960**, *2*, 100.

9.3.9. $\Delta_{fus}H$, Diphenylacetic acid

Physical property: Enthalpy of fusion
Units: J mol^{-1} or kJ mol^{-1} (molar enthalpy of fusion, $\Delta_{fus}H$)
J kg^{-1} or J g^{-1} (specific enthalpy of fusion, $\Delta_{fus}h$)
Recommended reference material: Diphenylacetic acid ($C_{14}H_{12}O_2$)
Range of variables: $T_{\text{triple point}} = (420.44 \pm 0.02)$ K (Ref. 1)
Physical states within the range: solid $\rightarrow$ liquid (in equilibrium with the vapour)
Contributors: K. N. Roy, R. Sabbah, D. D. Sood

Intended usage: The use of diphenylacetic acid as a calibrant for thermal analysis, especially differential scanning calorimeters, is recommended (Ref. 1).

Sources of supply and/or methods of preparation: A certified sample is available from supplier (T). Commercial samples with a mass fraction purity ≥ 0.99 are available from Aldrich, Koch-Light, etc. and can be used after purification by zone refining.

Pertinent physicochemical data: On the basis of five measurements using an adiabatic calorimeter, Andon and Connett (Ref. 1) determined a value of (31.271 ± 0.018) kJ mol^{-1} for the enthalpy of fusion of diphenylacetic acid (the sample having a mole fraction purity of 0.9994) at the triple point temperature. The molar mass of diphenylacetic acid was taken as 212.25 g mol^{-1}. Their publication mentioned that no earlier measurements had been reported for this substance in the literature.

REFERENCES

1. Andon, R. J. L.; Connett, J. E. *Thermochim. Acta* **1980**, *42*, 241.

9.3.10. $\Delta_{fus}H$, Indium

Physical property: Enthalpy of fusion
Units: J mol^{-1} or kJ mol^{-1} (molar enthalpy of fusion, $\Delta_{fus}H$)
J kg^{-1} or J g^{-1} (specific enthalpy of fusion, $\Delta_{fus}h$)
Recommended reference material: Indium (In)
Range of variables: $T_{\text{triple point}} = (429.78 \pm 0.01)$ K (Ref. 1)
Physical states within the range: solid $\rightarrow$ liquid (in equilibrium with the vapour)
Contributors: J. F. Martin, R. Sabbah, D. D. Sood, V. Venugopal

Intended usage: The use of indium as a test material for apparatus employed for the measurement of the enthalpy of fusion is suggested (Refs. 1, 2). Such apparatuses include adiabatic calorimeters (Refs. 1, 3) and differential scanning calorimeters (Refs. 2, 4 to 7).

Sources of supply and/or methods of preparation: Samples of high purity are available from suppliers (F, K, L, M, O, T). A certified sample is available from supplier (T). Other high purity material which might be suitable is available from various commercial sources *e.g.* Alpha, Koch-Light, etc.

Pertinent physicochemical data: Measurements by adiabatic calorimetry of the triple point temperature and the enthalpy of fusion have been made on three samples (having a mole fraction purity $\geq$ 0.9999) by Andon *et al.* (Ref. 1). They found a weighted mean value of (3259 ± 3) J mol^{-1} for the enthalpy of fusion at the triple point temperature. Their report and reference 2 review previous measurements. As the value obtained by Grønvold (Ref. 3) also appears to be reliable but is about 24 J mol^{-1} higher, we propose the value (3271 ± 12) J mol^{-1} for the enthalpy of fusion of indium. The molar mass of indium was taken as 114.82 g mol^{-1}.

REFERENCES

1. Andon, R. J. L.; Connett, J. E.; Martin, J. F. *NPL Report Chem.* 101 July, **1979**.
2. Zeeb, K. G.; Lowings, M. G.; McCurdy, K. G.; Hepler, L. G. *Thermochim. Acta* **1980**, *40*, 245.
3. Grønvold, F. *J. Therm. Anal.* **1978**, *13*, 419.
4. Lowings, M. G.; McCurdy, K. G.; Hepler, L. G. *Thermochim. Acta* **1978**, *23*, 365.
5. Richardson, M. J.; Savill, N. G. *Thermochim. Acta* **1975**, *12*, 221.
6. Van Dooren, A. A.; Muller, B. W. *Thermochim. Acta* **1981**, *49*, 151.
7. Breuer, K. H.; Eysel, W. *Thermochim. Acta* **1982**, *57*, 317.

9.3.11. $\Delta_{fus}H$ and $\Delta_{trs}H$, Sodium Nitrate

Physical property: Enthalpy of fusion
Units: J mol^{-1} or kJ mol^{-1} (molar enthalpy of fusion, $\Delta_{fus}H$, or transition, $\Delta_{trs}H$)
   J kg^{-1} or J g^{-1} (specific enthalpy of fusion, $\Delta_{fus}h$, or transition, $\Delta_{trs}h$)
Recommended reference material: Sodium nitrate (NaNO$_3$)
Range of variables: (i) $T_{fus} = 580$ K (Ref. 1) (ii) $T_{trs} = 549$ K (Ref. 2)
Physical states within the range: (i) solid $\rightarrow$ liquid (in equilibrium with the vapour)
   (ii) solid $\rightarrow$ solid
Contributors: K. N. Roy, R. Sabbah, D. D. Sood

Intended usage: The use of sodium nitrate as a test material for differential scanning calorimeters for the measurements of enthalpies of fusion is suggested (Refs. 2, 3). The substance exhibits a solid-solid transition that occurs 31 K below the melting point (Ref. 3).

Sources of supply and/or methods of preparation: A suitable sample of sodium nitrate can be prepared from analytical reagent grade material by recrystallization and/or drying procedures as described in references 3 and 4.

Pertinent physicochemical data: Measurements by differential scanning calorimetry of the enthalpy of fusion and the enthalpy of transition have been made by Zeeb et al. (Ref. 3). They found a value of (15.004 ± 0.176) kJ mol^{-1} for the enthalpy of fusion at the melting temperature and (3.410 ± 0.774) kJ mol^{-1} for the enthalpy of transition at 549 K. Their publication reviews previous measurements. The molar mass was taken as 84.999 g mol^{-1}.

REFERENCES

1. Weast, R. D., editor, *Handbook of Chemistry and Physics*, 58th ed. CRC Press: Cleveland **1977-78**.
2. Breuer, K. H.; Eysel, W. *Thermochim. Acta* **1982**, *57*, 317.
3. Zeeb, K. G.; Lowings, M. G.; McCurdy, K. G.; Hepler, L. G. *Thermochim. Acta* **1980**, *40*, 245.
4. Lowings, M. G.; McCurdy, K. G.; Hepler, L. G. *Thermochim. Acta* **1978**, *23*, 365.

9.3.12. $\Delta_{sub}H$, Naphthalene

Physical property: Enthalpy of sublimation
Units: J mol^{-1} or kJ mol^{-1} (molar enthalpy of sublimation, $\Delta_{sub}H$)
   J kg^{-1} or J g^{-1} (specific enthalpy of sublimation, $\Delta_{sub}h$)
Recommended reference material: Naphthalene (C$_{10}$H$_8$)
Range of variables: temperatures near 298 K along the saturated vapour curve
Physical states within the range: solid $\rightarrow$ real saturated vapour

Contributor to the first version: J. D. Cox
Contributor to the revised version: R. Sabbah

Intended usage: Naphthalene should be especially useful in a programme of measurement of the enthalpies of sublimation of solids having saturated vapour pressures around a pressure of 10 Pa at 298 K.

Sources of supply and/or methods of preparation: High purity samples of naphthalene can be obtained from supplier (C). Commercial samples with a mass fraction purity 0.99 are available from Aldrich, Carlo Erba, Fluka, Koch-Light, etc. which can be used after further purification by zone refining.

Pertinent physicochemical data: In a recent publication, Colomina et al. (Ref. 1) discussed literature values of the enthalpy of sublimation at 298.15 K. Most of them have been calculated from measurements of the saturated vapour pressure as a function of temperature,

Table 9.3.12.1. Enthalpy of sublimation $\Delta_{sub}H$ of naphthalene

$\dfrac{\Delta_{sub}H(298.15\ \text{K})}{\text{kJ mol}^{-1}}$	$\dfrac{T}{\text{K}}$	Method	Reference
72.68 ± 0.33	230 to 260	Knudsen gauge	2
72.05 ± 0.25	298.15	Calorimetry	3
73.00 ± 0.13	298.15	Calorimetry	4
72.5 ± 0.3	263 to 343	Diaphragm gauge	5
72.6 ± 0.6	253 to 273	Torsion-effusion	6
72.51 ± 0.07	274 to 353	Diaphragm manometer	7
72.8 ± 0.3	271 to 285	Effusion	1
72.92 ± 0.37	244 to 256	Spinning rotor friction gauge	9
72.60 ± 0.06	244 to 353	Diaphragm manometer + Spinning rotor friction gauge	9
72.52 ± 0.33	296.341	Calorimetry	8
72.47 ± 0.34			
72.70 ± 0.40			
72.6 ± 0.3	298.15	recommended value	

and some of them are extrapolated values. To check the performance of their apparatus and the efficacy of the measurement procedure, Murata et al. (Ref. 8) have recently used a calorimetric technique and reported values of $\Delta_{sub}H$ at several temperatures. As shown in table 9.3.12.1, the maximum difference between values of the enthalpy of sublimation of naphthalene is 0.9 kJ mol^{-1}. Values of $\Delta_{sub}H$ or $\Delta_{vap}H$ of naphthalene at temperatures other than 298.15 K are reported in references 1, 8, and 9.

REFERENCES

1. Colomina, M.; Jimenez, P.; Turrion, C. *J. Chem. Thermodynamics* **1982**, *14*, 779.
2. Miller, G. A. *J. Chem. Eng. Data* **1963**, *8*, 69.
3. Morawetz, E. *J. Chem. Thermodynamics* **1972**, *4*, 455.
4. Irving, R. J. *J. Chem. Thermodynamics* **1972**, *4*, 793.
5. Ambrose, D.; Lawrenson I. J. and Sprake C. H. S. *J. Chem. Thermodynamics* **1975**, *7*, 1173.
6. de Kruif, C. G. *J. Chem. Thermodynamics* **1980**, *12*, 243.
7. de Kruif, C. G.; Knipers, T.; Van Miltenburg, J. C.; Schaake, R. C. F.; Stevens, G.; *J. Chem. Thermodynamics* **1981**, *13*, 1081.
8. Murata, S.; Sakiyama, M.; Seki, S. *J. Chem. Thermodynamics* **1982**, *14*, 707.
9. Van Ekeren, P. J.; Jacobs, M. H. G.; Offringa; J. C. A.; de Kruif, C. G. *J. Chem. Thermodynamics* **1983**, *15*, 409.

9.3.13. $\Delta_{sub}H$, Iodine

Physical property: Enthalpy of sublimation
Units: J mol^{-1} or kJ mol^{-1} (molar enthalpy of sublimation, $\Delta_{sub}H$)
    J kg^{-1} or J g^{-1} (specific enthalpy of sublimation, $\Delta_{sub}h$)
Recommended reference material: Iodine (I$_2$)
Range of variables: 0 to 386.75 K along the saturated vapour curve
Physical states within the range: solid $\rightarrow$ real saturated vapour
Contributor: R. Sabbah

Intended usage: Iodine is suggested as a test material to check the performance of calorimeters used for the measurement of the enthalpies of sublimation of solids having saturated vapour pressures around 40 Pa at 298 K.

Sources of supply and/or methods of preparation: A high purity commercial sample of iodine (mass fraction 0.99999) can be obtained from supplier Alfa. Other commercial samples having a mass fraction purity $\geq$ 0.995 can be purchased from Baker, Fluka, Koch-Light, etc. Corrosion experiments show that platinum is not attacked by iodine and gold becomes slightly tarnished but the tarnish disappears on standing in air. These metals and glass can be used to contain iodine (Ref. 1).

Pertinent physicochemical data: The values for the enthalpy of sublimation listed in table 9.3.13.1 have been taken from the JANAF Thermochemical Tables (Ref. 2) and were calculated using published vapour pressure data in the temperature range 225 to 454 K, using thermodynamic functions of condensed and gaseous phases. The temperatue scale is normally IPTS-48 for measured quantities and the thermodynamic temperature scale for calculated gaseous quantities. Note that no more recent values of $\Delta_{sub}H$ for iodine have been reported in the supplements to the JANAF Tables published since 1971.

Table 9.3.13.1. Enthalpy of sublimation $\Delta_{sub}H$ of iodine at several temperatures

T_{68}/K	$\Delta_{sub}H/$ kJ mol^{-1}
0	65.52
100	65.51
200	64.07
298	62.44
300	62.41
400	44.48

REFERENCES

1. Gillespie, L. J.; Fraser, L. H. D. *J. Amer. Chem. Soc.* **1936**, *58*, 2260.
2. JANAF Thermochemical Tables, NSRDS-NBS **1971**, *37*.

9.3.14. $\Delta_{vap}H$, Water

Physical property: Enthalpy of vaporization
Units: J mol^{-1} or kJ mol^{-1} (molar enthalpy of vaporization, $\Delta_{vap}H$)
  J kg^{-1} or J g^{-1} (specific enthalpy of vaporization, $\Delta_{vap}h$)
Recommended reference material: Water (H$_2$O)
Range of variables: 0 to 374 °C along the liquid-vapour saturation curve
Physical states within the range: liquid → real saturated vapour
Contributors to the first version: J. D. Cox, J. F. Martin, I. Wadsö
Contributor to the revised version: R. Sabbah

Intended usage: In spite of its poor boiling characteristics and its corrosive and electrically conducting properties, water is recommended as a test material for checking the performance of calorimeters used for the measurement of the enthalpy of vaporization of liquids (Ref. 1).

Sources of supply and/or methods of preparation: It is now possible to prepare by distillation of deionized water a sample having a high degree of purity.

Pertinent physicochemical data: Osborne et al. (Ref. 2) studied the thermal properties of liquid and real-gas water substance in researches extending over many years (see reference 1, Chapter 11). The following data have been extracted from a summarizing paper (Ref. 2) which should be consulted for values at closer temperature intervals. The molar mass of water was taken as 18.016 g mol^{-1}. The temperature scale is not mentioned by the authors but it would be ITS-27.

Table 9.3.14.1. Enthalpy of vaporization $\Delta_{vap}H$ of water at several temperatures

t_{68} °C	$\Delta_{vap}H$ kJ mol^{-1}	t_{68} °C	$\Delta_{vap}H$ kJ mol^{-1}	t_{68} °C	$\Delta_{vap}H$ kJ mol^{-1}	t_{68} °C	$\Delta_{vap}H$ kJ mol^{-1}
0	45.054	100	40.657	200	34.962	300	25.300
25	43.990	120	39.684	220	33.468	320	22.297
40	43.350	140	38.643	240	31.809	340	18.502
60	42.482	160	37.518	260	29.930	360	12.966
80	41.585	180	36.304	280	27.795	374	2.066

The results reported in two recent papers (Refs. 3, 4) which give $\Delta_{vap}H$ for water in the range 25 to 100 °C and the independent measurements of Osborne and Ginnings (Ref. 5), Wadsö (Ref. 6), Konicek (Ref. 7) and Morawetz (Ref. 8) at 25 °C and of McCullough et al. (Ref. 9) at 65 and 100 °C are in agreement with the values in the above table.

To check the performance of their isothermal calorimeter for measuring enthalpies of vaporization at elevated temperatures and pressures, Parisod and Plattner (Ref. 10) have recently published values of $\Delta_{vap}H$ for water in the range 300 to 360 °C which differ from those in the above table by less than 1 per cent.

REFERENCES

1. McCullough, J. P.; Scott, D. W., editors, *Experimental Thermodynamics*, Vol. I: *Calorimetry of Non-reacting Systems*, Chapter 10 and 11. Butterworths: London, **1968**.
2. Osborne, N. S.; Stimson, H. F.; Ginnings, D. C. *J. Res. Nat. Bur. Stand* **1939**, *23*, 261.
3. Polak, J.; Benson, G. C. *J. Chem. Thermodynamics* **1971**, *3*, 235.
4. Majer, V.; Svoboda, V.; Hynek, V.; Pick, J. *Collect. Czech. Chem. Commun.* **1978**, *43*, 1313.

5. Osborne, N. S.; Ginnings, D. C. *J. Res. Nat. Bur. Stand.* **1947**, *39*, 453.
6. Wadsö, I. *Acta Chem. Scand.* **1966**, *20*, 536.
7. Konicek, J. *Acta. Chem. Scand.* **1973**, *27*, 1496.
8. Morawetz, E. *Chem. Scripta* **1971**, *1*, 103.
9. McCullough, J. P.; Pennington, R. E.; Waddington, G. *J. Amer. Chem. Soc.* **1952**, *74*, 4439.
10. Parisod, Ch. J.; Plattner, E. *Rev. Sci. Instrum.* **1982**, *53*, 54.

9.3.15. $\Delta_{vap}H$, Benzene

Physical property: Enthalpy of vaporization
Units: J mol^{-1} or kJ mol^{-1} (molar enthalpy of vaporization, $\Delta_{vap}H$)
    J mol kg^{-1} or J g^{-1} (specific enthalpy of vaporization, $\Delta_{vap}h$)
Recommended reference material: Benzene (C_6H_6)
Range of variables: 298.09 to 377.58 K along the liquid-vapour saturation curve
Physical states within the range: liquid $\rightarrow$ real saturated vapour
Contributors to the first version: J. D. Cox, D. R. Douslin, J. F. Martin
Contributors to the revised version: K. N. Roy, R. Sabbah, D. D. Sood.

Intended usage: Benzene is recommended for testing the performance of calorimeters used for the measurement of the enthalpy of vaporization of liquids (Ref. 1).

Sources of supply and/or methods of preparation: High purity samples of benzene can be obtained from supplier (C) as well as commercially from producers of high quality chemicals, *e.g.* B.D.H., Fluka, etc. Methods of purification of benzene are mentioned in references 2, 3, 4, 5.

Pertinent physicochemical data: Todd *et al.* (Ref. 3) measured the following values of the enthalpy of vaporization of benzene (the sample had a mole fraction purity of 0.9994 determined by a freezing temperature method) at various temperatures expressed on IPTS-68 using vapour-flow calorimetry.

The equation:
$$\ln(\Delta_{vap}H/\text{J mol}^{-1}) = A + Bx^{1/3} + Cx^{2/3} + Dx \qquad (9.3.15.1)$$

is based on the results of Todd *et al.*, where A = 13.69713, B = -8.69972, C = 8.48532, D = -3.15394, and $x = \ln[562.16/(562.16 - T/\text{K})]$, where 562.16 K is the critical temperature of benzene (Ref. 6).

This equation is suitable for interpolation in the experimental temperature range and, with reduced accuracy, for moderate extrapolation beyond that range. The results of Todd *et al.* are in good agreement with the value reported by Majer *et al.* (Ref. 2).

Table 9.3.15.1. Enthalpy of vaporiztion $\Delta_{vap}H$ of benzene at several temperatures

T_{68}/K	$\Delta_{vap}H$/J mol^{-1}	T_{68}/K	$\Delta_{vap}H$/J mol^{-1}
298.09	33843 ± 3.4	332.30	31924 ± 2.0
307.21	33330 ± 1.3	353.24	30726 ± 3.9
314.09	32950 ± 3.5	377.58	29255 ± 2.5
324.44	32369 ± 1.5		

REFERENCES

1. McCullough, J. P.; Scott, D. W., editors, *Experimental Thermodynamics*, Vol. I: *Calorimetry of Non-reacting Systems*, Chapter 10 and 11. Butterworths: London, **1962**.
2. Majer, V.; Svoboda, V.; Hynek, V.; Pick, J. *Collect. Czech. Chem. Commun.* **1978**, *43*, 1313.
3. Todd, S. S.; Hossenlopp, I. A.; Scott, D. W. *J. Chem. Thermodynamics* **1978**, *10*, 641.
4. Thorne, H. M.; Murphy, W.; Ball, J. S. *Ind. Eng. Chem. Analyt. Edn.* **1945**, *17*, 481.
5. Bugajewski, Z.; Bylicki, A.; Czerwinska, E.; Malanowski, S. *Pol. J. Chem.* **1979**, *53*, 475.
6. Ambrose, D. *J. Chem. Thermodynamics* **1981**, *13*, 1161.

9.3.16. $\Delta_{vap}H$, 1-Propanol

Physical property: Enthalpy of vaporization
Units: J mol^{-1} or kJ mol^{-1} (molar enthalpy of vaporization, $\Delta_{vap}H$)
    J kg^{-1} or J g^{-1} (specific enthalpy of vaporization, $\Delta_{vap}h$)
Recommended reference material: 1-Propanol (C$_3$H$_8$O)
Range of variables: 298.15 to 348.15 K along the liquid-vapour saturation curve
Physical states within the range: liquid → real saturated vapour
Contributors: R. Sabbah, V. Venugopal, D. D. Sood

Intended usage: 1-Propanol is suggested for testing the performance of calorimeters used for the measurement of the enthalpy of vaporization of liquids (Ref. 1).

Sources of supply and/or methods of preparation: Commercial samples having a mass fraction purity ≥ 0.995 can be purchased from B.D.H. and Fluka. Distillation can be used for further purification. It is recommended that samples be stored over molecular sieves to keep them free from water.

Pertinent physicochemical data: Majer et al. (Ref. 1) have recently determined the values of $\Delta_{vap}H$ given in table 9.3.16.1 by calorimetry in the temperature range 298.15 to 348.15 K.

Table 9.3.16.1. Enthalpy of vaporization $\Delta_{vap}H$ of 1-propanol at several temperatures

T_{68}/K	$\Delta_{vap}H/$ kJ mol^{-1}
298.15	47.37
318.15	46.09
338.15	44.50
348.15	43.63

The uncertainty of the experimental results, as indicated by the authors, was less than ±0.1 per cent. The results of Majer *et al.*, are in good agreement with the values reported previously in the literature (Refs. 2 to 7) which cover the temperature range 298 to 384 K.

REFERENCES

1. Majer, V.; Svoboda, V.; Hynek, V.; Pick, J. *Collect. Czech. Chem. Commun.* **1978**, *43*, 1313.
2. Svoboda, V.; Vesely, F.; Holub, R.; Pick, J. *Collect. Czech. Chem. Commun.* **1973**, *38*, 3539.
3. Polak, J.; Benson, G. C. *J. Chem. Thermodynamics* **1971**, *3*, 235.
4. Wadsö, I. *Acta Chem. Scand.* **1966**, *20*, 544.
5. McCurdy, K. G.; Laidler, K. J. *Can. J. Chem.* **1963**, *41*, 1867.
6. Williamson, K. D.; Harrison, R. H. *J. Chem. Phys.* **1957**, *26*, 1409.
7. Mathews, J. F.; McKetta, J. J. *J. Phys. Chem.* **1961**, *65*, 758.

9.3.17. $\Delta_{vap}H$, Hexafluorobenzene

Physical property: Enthalpy of vaporization
Units: J mol^{-1} or kJ mol^{-1} (molar enthalpy of vaporization, $\Delta_{vap}H$)
J kg^{-1} or J g^{-1} (specific enthalpy of vaporization, $\Delta_{vap}h$)
Recommended reference material: Hexafluorobenzene (C_6F_6)
Range of variables: 300.57 to 376.53 K along the liquid-vapour saturation curve
Physical states within the range: liquid → real saturated vapour
Contributor: R. Sabbah

9. Enthalpy

Intended usage: Hexafluorobenzene is suggested for testing the performance of calorimeters used for the measurement of the enthalpy of vaporization of liquids.

Sources of supply and/or methods of preparation: Commercial samples are available from suppliers B.D.H. (with mass fraction purity of 0.999), Aldrich, Fluka (with mass fraction purity of 0.99) which can be further purified by gas-liquid chromatography. It is also necessary to prevent samples used in calorimetric experiments from exposure to air and moisture.

Pertinent physicohemical data: Hossenlopp and Scott (Ref. 1) measured the following values (table 9.3.17.1) of the enthalpy of vaporization of hexafluorobenzene at various temperatures expressed on IPTS-68 using vapour-flow calorimetry. The mole fraction purity of the sample used was 0.9993 assessed by calorimetric study of the melting behaviour.

Table 9.3.17.1. Enthalpy of vaporization $\Delta_{vap}H$ of hexafluorobenzene at several temperatures

T_{68}/K	$\Delta_{vap}H/\text{J mol}^{-1}$	T_{68}/K	$\Delta_{vap}H/\text{J mol}^{-1}$
300.57	35.541 ± 0.014	353.40	31.670 ± 0.003
315.94	34.453 ± 0.010	376.53	29.833 ± 0.003
339.40	33.169 ± 0.003		

These results are in good agreement (better than 0.03 per cent) with those of Counsell *et al.* (Ref. 2) for the temperature range common to the two sets of measurements (316 to 353 K).

REFERENCES

1. Hossenlopp, I. A.; Scott, D. W. *J. Chem. Thermodynamics*, **1981**, *13*, 405.
2. Counsell, J. F.; Green, J. H. S.; Hales, J. L.; Martin, J. F. *Trans. Faraday Soc.*, **1965**, *61*, 212.

9.4. Reference materials for measurement of enthalpies of reaction and related processes

9.4.1. $\Delta_r h°$, Zirconium + Barium chromate

Physical property: Enthalpy of reaction
Units: J kg^{-1} or J g^{-1} (specific enthalpy of reaction, $\Delta_r h°$)

Recommended reference material: Zirconium + barium chromate
Range of variables: 298.15 K is the temperature normally employed
Physical states within the range: solid + solid
Contributors to the first version: G. T. Armstrong, J. D. Cox
No revision made

Intended usage: Heat-source (thermite-type, gasless) powders find use in defence and other applications. The 'heating values' of production batches of such powders are determined by calorimetry (see reference 1 for other references), and it is desirable that the calorimeters used should be calibrated under conditions similar to those of the experiments on production materials. Agglomerate mixtures of zirconium and barium chromate are used to calibrate the calorimeters (Ref. 1). Such mixtures can be readily ignited, react completely in less than one second and generate very little gas.

Sources of supply and/or methods of preparation: Zirconium plus barium chromate mixtures of three distinct nominal heating values, *viz.* -1464, -1632 and -1778 J g^{-1} are available from supplier (S).

Before use, heat-source powders of the types referred to above should be dried for two hours at 344 K and 1.3 kPa pressure in a flat metal container in an oven, which must contain no open heating coils, then cooled in a desiccator. General safety precautions in the handling of these powders are given on pages 4 and 5 of reference 1.

Pertinent physicochemical data: The heating values (*i.e.* the specific enthalpy changes for solid-state reactions, when no air, oxygen or nitrogen are in contact with the samples) of three batches of zirconium + barium chromate mixtures were determined (Ref. 1) at the National Bureau of Standards by means of an isoperibol calorimeter (Ref. 2). The samples were weighed in air.

NBS SRM 1651: Heating value $- (1460 \pm 4.8)$ J g^{-1}
NBS SRM 1652: Heating value $- (1632.3 \pm 7.3)$ J g^{-1}
NBS SRM 1653: Heating value $- (1762.0 \pm 3.0)$ J g^{-1}

REFERENCES

1. Minor, Jr., J. I.; Armstrong, G. T. *Calorimetric determination of heating values of some zirconium-barium chromate heat source powders as reference materials*, NBS Report 9928, National Bureau of Standards: Washington, **1968**.
2. Prosen, E. J.; Johnson, W. H.; Pergiel, F. Y. *J. Res. Nat. Bur. Stand* **1959**, *62*, 43.

9.4.2. $\Delta_r H°$, Sulphuric acid solution + Sodium hydroxide solution

Physical property: Enthalpy of reaction in solution
Units: J mol^{-1} or kJ mol^{-1} (molar enthalpy of reaction, $\Delta_r H°$)
J kg^{-1} or J g^{-1} (specific enthalpy of reaction, $\Delta_r h°$)
Recommended reference materials: Sulphuric acid solution + sodium hydroxide solution
Range of variables: 298.15 K is the temperature normally employed
Physical states within the range: liquid + liquid
Contributors to the first version: J. D. Cox, S. R. Gunn
No revision made

Intended usage: Calorimeters (Ref. 1) for the measurement of enthalpies of reaction in solution should be calibrated electrically. However, it is desirable to test the calorimetric procedure by measurement of the enthalpy of a reaction for which the value has been well established. For an experimental programme involving rapid exothermic reactions in solution, the reaction between sulphuric acid solution and excess sodium hydroxide solution, at defined concentrations, is recommended as a test reaction.

Sources of supply and/or methods of preparation: Instructions for preparing solutions of sulphuric acid and sodium hydroxide as calorimetric reactants have been given by Gunn (Refs. 2, 3).

Pertinent physicochemical data: Gunn (Ref. 2) has studied the enthalpy changes for the reactions between H$_2$SO$_4$(8H$_2$O) and 2.5 moles of NaOH(xH$_2$O) where x lay between 10 and 2580. By the use of a reaction calorimeter calibrated electrically Gunn (Ref. 2) found the mean value $-(150.82 \pm 0.02)$ kJ mol^{-1} at 298.15 K for the enthalpy of the reaction

$$H_2SO_4(8H_2O)(l) + 2.5[NaOH(10H_2O)](l) = [Na_2SO_4 + 0.5NaOH](35H_2O)(l) \quad (9.4.2.1.).$$

The molar masses of H$_2$SO$_4$(8H$_2$O) and NaOH(10H$_2$O) were taken as 242.200 g mol^{-1} and 220.150 g mol^{-1} respectively. Prosen and Kilday (Ref. 4) studied essentially the same process using an adiabatic, vacuum-jacketed solution calorimeter calibrated electrically, and found enthalpy values which were the same within the stated uncertainty. Gunn et al. (Ref. 3) found the mean value $-(150.80 \pm 0.02)$ kJ mol^{-1} for the same reaction at 298.15 K by the use of bomb calorimeters calibrated by the combustion of benzoic acid.

Near 298 K the temperature coefficient of the enthalpy of the reaction is given by $(d\Delta_r H°/dT) = +88$ J mol^{-1} K^{-1}.

REFERENCES

1. Skinner, H. A., editor, *Experimental Thermochemistry*, Vol. II, Chapters 9, 11 and 14. Wiley (Interscience): New York, **1962**.
2. Gunn, S. R. *J. Chem. Thermodynamics*, **1970**, *2*, 535.
3. Gunn, S. R.; Watson, J. A.; Mackle, H.; Gundry, H. A.; Head, A. J.; Månsson, M.; Sunner, S.; *J. Chem. Thermodynamics* **1970**, *2*, 549.
4. Prosen, E. J.; Kilday, M. V. *J. Res. Nat. Bur. Stand.* **1973**, *77A*, 179.

9.4.3. H^E, Cyclohexane + Hexane

Physical property: Enthalpy of mixing
Units: J mol^{-1} (molar enthalpy of mixing, $\Delta_{mix}H$,
  or excess molar enthalpy of mixing, H^E)
  J kg^{-1} or J g^{-1} (specific enthalpy of mixing, $\Delta_{mix}h$,
  or excess specific enthalpy of mixing, h^E)
Recommended reference materials: Cyclohexane + hexane ($C_6H_{12} + C_6H_{14}$)
Range of variables: 298.15 K is the reference temperature normally employed
Physical state within the range: liquid + liquid
Contributor to the first version: J. D. Cox
Contributor to the revised version: K. N. Marsh

Intended usage: Calorimeters (Ref. 1) for the measurement of the enthalpies of mixing of liquids (equal to the excess enthalpies of the corresponding mixtures, H^E, since the enthalpy of mixing in an ideal system is zero) should be calibrated electrically. However, it is good practice to prove the efficacy of the calorimetric procedure by measurement of the enthalpy of mixing of two liquids for which the value of the enthalpy of mixing has been well established. The liquid pair cyclohexane + hexane is recommended for this purpose, particularly for endothermic mixtures.

Sources of supply and/or methods of preparation: Suitable samples (Ref. 2) of cyclohexane ('Research Grade') may be obtained from supplier (U). Alternatively samples of hexane and cyclohexane may be obtained from various supply houses, for example (Ref. 2) 'AR Grade' can be obtained from BDH. The mass fraction purity of both the cyclohexane and hexane should be greater than 0.9995.

Pertinent physicochemical data: Many measurements of varying degrees of accuracy have been made on this mixture over the last 15 years. A selected set of eight series of measurements (Refs. 2 to 9), comprising over 300 measurements, made with flow, batch and dilution calorimeters at 298.15 K have been analysed by least squares. The results for $(1-x)$cyclohexane + x hexane (where x represents the mole fraction) can be expressed by the polynomial

$$H^E(298.15\ K)/\text{J mol}^{-1} = x(1-x)[864.59 + 249.92(1-2x) + 98.12(1-2x)^2$$
$$+ 31.65(1-2x)^3] \tag{9.4.3.1}$$

with a standard deviation of 0.5 J mol^{-1} or about 0.2 per cent of the maximum value of H^E. The very precise series of measurements by Marsh and Stokes (Ref. 4) and Ewing et al. (Ref. 6) are fitted by the above equation with a standard deviation of 0.12 J mol^{-1} and a maximum deviation of 0.20 J mol^{-1} or 0.09 per cent of the maximum value of H^E.

From these data the mean value of the maximum value of the enthalpy of mixing at 298.15 K is 221.0 J mol^{-1}, and this occurs at a mole fraction x of hexane equal to 0.421. Near 298 K the temperature coefficient of the enthalpy of mixing at the maximum is given (Ref. 10) by $(\text{d}H^E/\text{d}T) = -1.39$ J mol^{-1} K^{-1}.

REFERENCES

1. Skinner, H. A., editor, *Experimental Thermochemistry*, Vol. II, Chapter 15, Wiley (Interscience): New York, **1962**.
2. Watts, H.; Clarke, E. C. W.; Glew, D. N. *Can. J. Chem.* **1968**, *46*, 815.
3. McGlashan, M. L.; Stoeckli, H. F. *J. Chem. Thermodynamics* **1969**, *1*, 589.
4. Marsh, K. N.; Stokes, R. H. *J. Chem. Thermodynamics* **1969**, *1*, 223.
5. Murakami, S.; Benson, G. C. *J. Chem. Thermodynamics* **1969**, *1*, 559.
6. Ewing, M. B.; Marsh, K. N.; Stokes, R. H.; Tuxford, C. W. *J. Chem. Thermodynamics* **1970**, *2*, 751.
7. Tanaka, R.; Murakami, S.; Fujishiro, R. *Bull. Chem. Soc. Japan* **1972**, *45*, 2107.
8. Tanaka, R.; D'Arcy, P. J.; Benson, G. C. *Thermochim. Acta* **1975**, *11*, 163.
9. Arenosa, R. L.; Menduina, C.; Tardajos, G.; Diaz Pena, M. *J. Chem. Thermodynamics* **1979**, *11*, 159.
10. Ewing, M. B.; Marsh, K. N. *J. Chem. Thermodynamics* **1970**, *2*, 295.

9.4.4. H^E, 1,4-Dioxan + Tetrachloromethane

Physical property: Enthalpy of mixing
Units: J mol^{-1} (molar enthalpy of mixing, $\Delta_{\text{mix}}H$,
    or excess molar enthalpy of mixing, H^E)
J kg^{-1} or J g^{-1} (specific enthalpy of mixing, $\Delta_{\text{mix}}h$,
    or excess specific enthalpy of mixing, h^E)
Recommended reference materials: 1,4-Dioxan + tetrachloromethane
$$(C_4H_8O_2 + CCl_4)$$
Range of variables: 298.15 K is the reference temperature normally employed
Physical state within the range: liquid + liquid
Contributor: K. N. Marsh

Intended usage: Calorimeters (Ref. 1) for the measurement of the enthalpies of mixing of liquids (equal to the excess enthalpies of the corresponding mixtures, H^E, since the enthalpy of mixing in an ideal system is zero) should be calibrated electrically. However, it is good practice to prove the efficacy of the calorimetric procedure by measurement of the enthalpy of mixing of two liquids for which the value of the enthalpy of mixing has been well established. The liquid pair 1,4-dioxan + tetrachloromethane is recommended for the checking of calorimeters to be used for measuring the excess enthalpy for exothermic mixtures.

Sources of supply and/or methods of preparation: Suitable samples (Ref. 2) of 1,4-dioxan may be prepared by chemical treatment of laboratory or analytical grade material available from various suppliers. A suitable treatment is to store the sample over ferrous sulphate for one week, pass over activated alumina, store over sodium for one week, then distil from sodium and store in a dark bottle over type 4A molecular sieve. Immediately prior to use it should be distilled from $LiAlH_4$ and collected under an inert atmosphere. A suitable sample of tetrachloromethane may be obtained by distillation of analytical grade material from a column containing glass helices. Distillation from a column containing stainless steel gauze rings without a protective atmosphere can result in the formation of phosgene. Samples of tetrachloromethane should be passed over activated alumina immediately prior to use. The sample should not be exposed to light once it has passed over the alumina since under certain conditions a radical reaction with release of energy can occur, as discussed in reference 2.

Pertinent physicochemical data: A number of measurements of varying degrees of accuracy have been made on this mixture over the last 15 years. A selected set of four series of measurements (Refs. 2 to 5), comprising 158 measurements, made with flow and dilution calorimeters at 298.15 K, have been analysed by least squares. The results for x 1,4-dioxan + $(1-x)$ tetrachloromethane (where x represents the mole fraction) can be expressed by the polynomial

$$H^E(298.15\ K)/J\ mol^{-1} = x(1-x)[-1006.8 + 71.4(1-2x) + 166.9(1-2x)^2] \quad (9.4.4.1)$$

with a standard deviation of 1.0 J mol^{-1} or about 0.4 per cent of the maximum value of H^E.

From these data the mean value of the maximum value of the enthalpy of mixing at 298.15 K is -252.0 J mol^{-1}, and this occurs at a mole fraction x of 1,4-dioxan equal to 0.515.

REFERENCES

1. Skinner, H. A., editor, *Experimental Thermochemistry*, Vol. II, Chapter 15, Wiley (Interscience): New York, **1962**.
2. Costigan, M. J.; Hodges, L. J.; Marsh, K. N.; Stokes, R. H.; Tuxford, C. W. *Austral. J. Chem.* **1980**, *33*, 2103.
3. Murakami, S.; Benson, G. C. *J. Chem. Thermodynamics* **1969**, *1*, 559.
4. Becker, F.; Kiefer, M.; Koukol, H. *Z. Phys. Chem. (Frankfürt)* **1972**, *80*, 29.
5. Featherstone, J. D. B.; Dickinson, N. A. *J. Chem. Thermodynamics* **1976**, *8*, 985.

9.4.5. $\Delta_{dil}H$, Sucrose

Physical property: Enthalpy of dilution
Units: J mol^{-1} (molar enthalpy of dilution, $\Delta_{dil}H$)
Recommended reference material: Sucrose ($C_{12}H_{22}O_{11}$)
Range of variables: 298.15 K is the reference temperature normally employed
Physical state within the range: aqueous solutions
Contributor to the first version: R. N. Goldberg
Contributor to the revised version: R. N. Goldberg

Intended usage: Calorimeters (Ref. 1) for the measurement of enthalpies of mixing or dilution of aqueous solutions should be calibrated electrically. It is desirable, however, to test the calorimetric procedure by measurement of the enthalpy of mixing or dilution of solutions for which the value has been well established. The enthalpy of dilution of sucrose provides a means of detecting possible systematic errors in calibration and measurement procedures, particularly when solutions of moderate viscosity are required.

Sources of supply and/or methods of preparation: A suitable grade of sucrose is available as Standard Reference Material No. 17c from supplier (S). It is advised that freshly prepared solutions be used for the calorimetric measurements.

Pertinent physicochemical data: The best data at present available are those given by Gucker *et al.* (Ref. 2) who expressed values for the relative apparent molal enthalpy, ϕ_L, by the following equation:

$$\phi_L/\text{J mol}^{-1} = Ax - Bx^2 \qquad (9.4.5.1).$$

where $x = m/\text{mol kg}^{-1}$.

Table 9.4.5.1. Constants for equation 9.4.5.1

T/K	A	B	value of x
293.15	539.3	28.941	≤ 5.9
298.15	563.2	29.50	≤ 2.2
303.15	586.6	29.62	≤ 2.2

These workers performed their measurements by means of an adiabatic twin calorimeter and used Standard Reference Material No. 17 obtained from supplier (S).

The following solutions are deemed to be of particular interest for future interlaboratory comparisons of measurements: 0.20 mol kg^{-1} to 0.10 mol kg^{-1}; 0.20 mol kg^{-1} to 0.066 mol kg^{-1}; and 0.50 to 0.25 mol kg^{-1}.

REFERENCES

1. Skinner, H. A., editor, *Experimental Thermochemistry*, Vol. II, Chapters 9, 11, 14. Wiley (Interscience): New York, **1962**.
2. Gucker, F. T.; Pickard, H. B.; Planck, R. W. *J. Amer. Chem. Soc.* **1939**, *61*, 459. (Note that Gucker included some of the results of earlier investigations in his final equations).

9.4.6. $\Delta_{dil}H$, Urea

Physical property: Enthalpy of dilution
Units: J mol^{-1} (molar enthalpy of dilution, $\Delta_{dil}H$)
Recommended reference material: Urea (CH$_4$ON$_2$)
Range of variables: 298.15 K is the reference temperature normally employed
Physical state within the range: aqueous solutions
Contributors: M. I. Paz-Andrade, G. Pilcher

Intended usage: Calorimeters (Ref. 1) for the measurement of enthalpies of mixing or of dilution of aqueous solutions should be calibrated electrically. It is desirable, however, to test the calorimetric procedure by measurement of the enthalpy of mixing or of dilution of solutions for which the values have been well established. The enthalpy of dilution of aqueous urea solutions provides a means of detecting possible systematic errors in calibration and measurement procedures, particularly when solutions less viscous than those of sucrose are required.

Sources of supply and/or methods of preparation: A suitable grade of urea is available as 'Aristar' material from BDH. Other samples may be purified by recrystallization from water or ethanol followed by intensive drying. All solutions should be freshly prepared before use.

Pertinent physicochemical data: Gucker and Pickard (Ref. 2) measured the enthalpy of dilution of aqueous urea solutions at 298.15 K using an adiabatic twin calorimeter; their values for the relative apparent molal enthalpy, ϕ_L, were expressed by the following equation:

$$\phi_L/\text{J mol}^{-1} = -359.41x + 28.52x^2 - 1.912x^3 + 0.06157x^4 \qquad (9.4.6.1)$$

where $x = m/\text{mol kg}^{-1}$ for $x \leq 12$. Egan and Luff (Ref. 3) measured the enthalpy of solution of crystalline urea in water at 298.15 K, and from their values an expression for ϕ_L can be derived in excellent agreement with the results of Gucker and Pickard.

Paz-Andrade et al. (Ref. 4) measured enthalpies of dilution of urea solutions in a potassium phosphate buffer (pH 7.0, ionic strength 0.005 mol kg^{-1}) using a Beckman 190B twin-cell microcalorimeter. They obtained for the relative apparent molal enthalpy at 298.15 K,

$$\phi_L/\text{J mol}^{-1} = -335.5x + 15.68x^2 - 0.2988x^3 \tag{9.4.6.2}$$

where $x = m/(\text{mol kg}^{-1})$ for $x \leq 20$. The enthalpies of dilution in the phosphate buffer, which are of use to those studying biochemical reactions calorimetrically, differ only slightly from those in water.

REFERENCES

1. Skinner, H. A., editor, *Experimental Thermochemistry*, Vol. II, Chapters 9, 11, 14, Wiley (Interscience): New York, **1962**.
2. Gucker, F. T.; Pickard, H. B. *J. Amer. Chem. Soc.* **1940**, *62*, 1464.
3. Egan, E. P.; Luff, B. B. *J. Chem. Eng. Data* **1966**, *11*, 192.
4. Paz-Andrade, M. I.; Jones, M. N.; Skinner, H. A. *Eur. J. Biochem.* **1976**, *66*, 127.

9.4.7. $\Delta_r H°$, α-Silicon dioxide + Hydrofluoric acid

Physical property: Enthalpy of reaction in solution
Units: J mol^{-1} or kJ mol^{-1} (molar enthalpy of reaction, $\Delta_r H°$)
J kg^{-1} or J g^{-1} (specific enthalpy of reaction, $\Delta_r h°$)
Recommended reference material: α-Silicon dioxide (α-quartz; 'low' quartz) + hydrofluoric acid (0.244 mass-fraction)
Range of variables: 298.15 K is the reference temperature usually employed but the actual temperature of certification in this instance is 353.15 K. Information is given to permit conversion of the value to 298.15 K.
Physical states within the range: solid $\rightarrow$ liquid solution
Contributors to the first version: J. D. Cox, E. J. Prosen
Contributors to the revised version: A. J. Head, P. A. G. O'Hare

Intended usage: Calorimeters (Ref. 1) for the measurement of the enthalpies of solution of solids in hydrofluoric acid should be calibrated electrically. However, it is desirable to test the calorimetric procedure by dissolution of a solid of known enthalpy of reaction in solution. Many mineral substances dissolve rather slowly in hydrofluoric acid solution, and to increase the rate of dissolution experiments are often conducted at temperatures higher than 298 K. For an experimental programme of this type, α-quartz is recommended as a test material.

Sources of supply and/or methods of preparation: A suitable sample of α-quartz with a certified (Ref. 2) value for the enthalpy of reaction is available from supplier (S).

Pertinent physicochemical data: Kilday and Prosen (Ref. 3) have determined the enthalpy of reaction of α-quartz (NBS Standard Reference Material 1654) with a solution containing hydrofluoric acid (mass fraction 0.244). The enthalpy of reaction was found to be $-(141.93 \pm 0.07)$ kJ mol^{-1} (or $-(2362.2 \pm 1.1)$ J g^{-1}) at 353.15 K, the molar mass of α-quartz being taken as 60.0848 g mol^{-1}. These values refer to a concentration of 5 g dm^{-3} of α-quartz in hydrofluoric acid, and to a sample that passed a No. 200 sieve but was retained by a No. 400 sieve (particle size 37 to 74 μm). The equation representing the reaction and solution process is

$$SiO_2(cr, \alpha\text{-quartz}) + 155[HF(3.44H_2O)](l) = [H_2SiF_6 + 149HF](535.2H_2O)(l) \quad (9.4.7.1)$$

The certificate (Ref. 2) that accompanies the material available from supplier (S) gives instructions for the use of the material and provides references to other work. It also gives factors for calculation of the enthalpy of reaction at different concentrations of hydrofluoric acid (-7.96 kJ mol^{-1} per unit increase in mass fraction) and at lower temperatures ($d\Delta_r H^\circ/dT = -95.3$ J mol^{-1} K^{-1}). There is evidence, however, that these factors may be in error (Refs. 4, 5) and care should be exercised in using the reference material at concentrations and temperatures which differ appreciably from those relating to the certified value. There is also evidence that the enthalpy of reaction of α-quartz with hydrofluoric acid is dependent upon particle size, being more exothermic as the size decreases (Ref. 4).

REFERENCES

1. Skinner, H. A., editor, *Experimental Thermochemistry*, Vol. II, Chapters 9, 11, 14, Wiley (Interscience): New York, **1962**.
2. Cali, J. P.; *Standard Reference Material 1654, α-quartz, for hydrofluoric acid solution calorimetry*, National Bureau of Standards Certificate: Washington, **1971**.
3. Kilday, M. V.; Prosen, E. J. *J. Res. Nat. Bur. Stand.* **1973**, *77A*, 205.
4. Hemingway, B. S.; Robie, R. A. *J. Res. U. S. Geological Survey* **1977**, *5*, 413.
5. Johnson, G. K.; Flotow, H. E.; O'Hare, P. A. G.; Wise, W. S. *American Mineralogist* **1982**, *67*, 736.

9.4.8. $\Delta_{sol}H^\circ$, Potassium chloride

Physical property: Enthalpy of solution
Units: J mol^{-1} or kJ mol^{-1} (molar enthalpy of solution, $\Delta_{sol}H^\circ$)
J kg^{-1} or J g^{-1} (specific enthalpy of solution, $\Delta_{sol}h^\circ$)
Recommended reference material: Potassium chloride (KCl)
Range of variables: 298.15 K is the reference temperature normally employed
Physical states within the range: solid $\rightarrow$ liquid solution
Contributors to the first version: J. D. Cox, O. Riedel
Contributors to the revised version: E. S. Domalski, A. J. Head

Intended usage: Calorimeters (Ref. 1) for the measurement of the enthalpies of solution of solids in liquids should be calibrated electrically. However, it is desirable to test the calorimetric procedure by measurement of the enthalpy of solution of a solid in a liquid, for which the value has been well established. For an experimental programme involving endothermic dissolution, especially of a freely soluble substance dissolving in water, potassium chloride is recommended.

Sources of supply and/or methods of preparation: Suitable material, with certified value for the enthalpy of solution in water, is available from supplier (S) as SRM 1655.

Pertinent physicochemical data: The suitability of potassium chloride as a reference material for solution calorimetry has been discussed for over 50 years and many determinations of the enthalpy of solution in water have been made with results showing a wider scatter than would be expected of a reference material. The measurements were reviewed in 1977 by Montgomery *et al.* (Ref. 2) who concluded that the most reliable and consistent results had been reported for potassium chloride dried at temperatures exceeding 600 K. Other factors involving sample purity and calorimetric procedures had, however, also contributed to the scatter in published values. The most thorough study has been published in 1980 by Kilday who determined the enthalpy of solution in water of NBS SRM 1655 using both isoperibol (Ref. 3) and adiabatic (Ref. 4) calorimeters, the latter being employed to provide the certified value of (17.584 ± 0.017) kJ mol^{-1} for the standard molar enthalpy (molar mass 74.5513 g mol^{-1}) of the process

$$\text{KCl(cr)} + 500 \text{ H}_2\text{O(l)} = \text{KCl}(500 \text{ H}_2\text{O})(\text{l}) \quad (9.4.8.1).$$

This value is in very close agreement with that recently selected (Ref. 5) by the Chemical Thermodynamics Data Center of the NBS, *viz.* (17.590 ± 0.010) kJ mol^{-1}, which is based on Kilday's determinations (Refs. 3, 4) and those of Gunn (Ref. 6). The temperature coefficient of the reaction over the temperature range 296 to 303 K is given by $(d\Delta_{sol}H^\circ/dT) = -(154.8 \pm 6.4)$ J mol^{-1} K^{-1}.

REFERENCES

1. Skinner, H. A., editor, *Experimental Thermochemistry*, Vol. II, Chapters 9, 11, 14, Wiley (Interscience): New York, **1962**.
2. Montgomery, R. L.; Melaugh, R. A.; Lau, C.-C.; Meier, G. H.; Chan, H. H.; Rossini, F. D. *J. Chem. Thermodynamics* **1977**, *9*, 915.
3. Kilday, M. V. *J. Res. Nat. Bur. Stand.* **1980**, *85*, 449.
4. Kilday, M. V. *J. Res. Nat. Bur. Stand.* **1980**, *85*, 467.
5. NBS Chemical Thermodynamics Data Center (1982), private communication from V. B. Parker.
6. Gunn, S. R. *J. Phys. Chem.* **1965**, *69*, 2902.

9.4.9. $\Delta_r H°$, Tris(hydroxymethyl)aminomethane

Physical property: Enthalpy of reaction in solution
Units: J mol^{-1} or kJ mol^{-1} (molar enthalpy of reaction, $\Delta_r H°$)
  J kg^{-1} or J g^{-1} (specific enthalpy of reaction, $\Delta_r h°$)
Recommended reference material: Tris(hydroxymethyl)aminomethane
  ($C_4H_{11}O_3N$); common abbreviations: TRIS or THAM.
Range of variables: 298.15 K is the reference temperature normally employed.
Physical states within the range: solid → liquid solution
Contributors to the first version: G. T. Armstrong, J. D. Cox, A. J. Head, O. Riedel
Contributors to the revised version: A. J. Head, C. E. Vanderzee

Intended usage: Calorimeters (Ref. 1) for measuring the enthalpy of dissolution of a solid in a liquid (solution calorimeters) or the enthalpy of reaction of a solid with a relatively large volume of liquid (liquid-phase reaction calorimeters) should be calibrated electrically. It is good practice, however, to test the efficacy of the calorimetric procedure by measurement of the enthalpy of solution of a solid in a liquid, using a reaction for which the enthalpy has been determined by competent laboratories. Attested values exist for the enthalpy of solution at 298.15 K of crystalline TRIS in (a) 0.1 mol dm^{-3} hydrochloric acid, an exothermic reaction, and (b) 0.05 mol dm^{-3} sodium hydroxide solution, an endothermic reaction. TRIS is recommended as a suitable material for testing solution calorimeters and liquid-phase reaction calorimeters which are to be employed for the study of rapid dissolutions or reactions.

Sources of supply and/or methods of preparation: Suitable material, with certified values for the enthalpies of reaction with (a) 0.1 mol dm^{-3} hydrochloric acid and (b) 0.05 mol dm^{-3} sodium hydroxide solution, is available from supplier (S) as SRM 724a. Care should be taken to use the material as received and not submit it to additional drying or grinding procedures.

Although material of satisfactory purity may be obtained directly from several chemical suppliers, or by crystallization of reagent-grade material, it should be noted that complete removal of occluded solvent can only be effected by crushing and grinding, and that such mechanical action is likely to introduce stored energy into the deformed crystal lattice (40 to 90 J mol^{-1}), which persists for appreciable periods of time unless removed by thermal annealing (Ref. 2).

Pertinent physicochemical data: (a) The enthalpy of solution of TRIS in 0.1 mol dm^{-3} hydrochloric acid at 298.15 K (concentration 5 kg m^{-3}) has been extensively measured, using calorimeters capable of high precision and intended to yield results of high accuracy, and summaries of the values obtained are given in references 3 and 4. These values show an unexpectedly wide scatter and Prosen and Kilday have studied the effect of some of the experimental variables (Ref. 3). Their investigations indicate that venting of the calorimeter and the presence of carbon dioxide can significantly affect the values obtained, and that the reaction is less suitable than originally supposed. The influence of other variables has been studied by Kilday (Ref. 5). Nevertheless, provided appropriate precautions are taken, the TRIS-HCl reaction is suitable for checking the performance of calorimeters in which rapid, exothermic reactions are to be studied.

The uncertainty interval assigned to the certified value for NBS SRM 724a of $-(245.76 \pm 0.26)$ J g^{-1} for the specific enthalpy of the reaction has been considerably increased over that calculated from the precision of the measurements to allow for possible systematic errors. In the temperature range 293 to 303 K the temperature coefficient of the enthalpy of reaction is given by $(\mathrm{d}\Delta_r h°/\mathrm{d}T) = (1.435 \pm 0.023)$ J g^{-1} K^{-1}.

(b) The enthalpy of solution of TRIS in 0.05 mol dm^{-3} sodium hydroxide solution at 298.15 K (concentration 5 kg m^{-3}) was determined by Hill, Öjelund and Wadsö (Ref. 6) as (141.90 ± 0.08) J g^{-1} and by Prosen and Kilday (Ref. 3) as (141.80 ± 0.19) J g^{-1}, whose value is used for the certification of NBS SRM 724a. The uncertainty interval assigned to the result of Prosen and Kilday allows for possible errors in the analysis of sodium hydroxide solutions and for other possible systematic errors. In the temperature range 293 to 303 K the temperature coefficient of the reaction is given by $(\mathrm{d}\Delta_r h°/\mathrm{d}T) = (1.025 \pm 0.025)$ J g^{-1} K^{-1}.

REFERENCES

1. Skinner, H. A., editor, *Experimental Thermochemistry*, Vol. II, Chapters 9, 11, 14. Wiley (Interscience): New York, **1962**.
2. Vanderzee, C. E.; Waugh, D. H.; Haas, N. C. *J. Chem. Thermodynamics* **1981**, *13*, 1.
3. Prosen, E. J.; Kilday, M. V. *J. Res. Nat. Bur. Stand.* **1973**, *77A*, 581.
4. Montgomery, R. L.; Melaugh, R. A.; Lau, C.-C.; Meier, G. H.; Chan, H. H.; Rossini, F. D. *J. Chem. Thermodynamics* **1977**, *9*, 915.
5. Kilday, M. V. *J. Res. Nat. Bur. Stand.* **1980**, *85*, 449.
6. Hill, J. O.; Öjelund, G.; Wadsö, I. *J. Chem. Thermodynamics* **1969**, *1*, 111.

9.4.10. $\Delta_r H°$, 4-Aminopyridine

Physical property: Enthalpy of reaction in solution
Units: J mol^{-1} or kJ mol^{-1} (molar enthalpy of reaction, $\Delta_r H°$)
   J kg^{-1} or J g^{-1} (specific enthalpy of reaction, $\Delta_r h°$)
Recommended reference material: 4-Aminopyridine ($C_5H_6N_2$)
Range of variables: 298.15 K is the reference temperature normally employed
Physical state within the range: solid → liquid solution
Contributor: G. Pilcher

Intended usage: Calorimeters (Ref. 1) for the measurement of the enthalpy of reaction of a solid with a relatively large volume of liquid (liquid phase reaction calorimeters) should be calibrated electrically. It is desirable, however, to test the calibration and calorimetric procedure by measurement of the enthalpy of solution of a solid in a liquid for which the enthalpy of reaction has been well established. The enthalpy of solution of crystalline 4-aminopyridine in 10 per cent excess of aqueous perchloric acid at concentrations ≤ 0.05 mol dm^{-3} has been measured in three separate investigations with results in agreement. This reaction is recommended as a suitable exothermic test reaction for solution calorimeters when the experimental programme involves rapid dissolution or reactions in solution.

Sources of supply and/or methods of preparation: A suitable grade of 4-aminopyridine can be obtained by sublimation *in vacuo* of commercially available samples (*e.g.*, Aldrich Chemical Co., mass fraction purity ≥ 0.99).

Pertinent physicochemical data: Van Til and Johnson (Ref. 2) measured the enthalpy of solution of 4-aminopyridine in water and the enthalpy of neutralization of that solution by aqueous acid at 298.15 K, from which the enthalpy of dissolution of the crystalline solid in the acid can be indirectly derived. Burchfield and Hepler (Ref. 3) and Akello *et al.* (Ref. 4) measured directly the enthalpy of solution of 4-aminopyridine (cr) in excess aqueous perchloric acid. Burchfield and Hepler found no dependence of the observed enthalpy on the amount of excess acid; an excess of 10 per cent aqueous perchloric acid over the stoichiometric amount is recommended. The enthalpy of solution is a linear function of concentration of the solution produced, and the direct determinations gave:–

Burchfield and Hepler (Ref. 3):

$$\Delta_r H^\circ(298.15 \text{ K})/\text{ kJ mol}^{-1} = -(29.409 \pm 0.009) - (6.31 \pm 0.35)\ c/(\text{mol dm}^{-3}) \quad (9.4.10.1)$$

Akello *et al.* (Ref. 4):

$$\Delta_r H^\circ(298.15 \text{ K})/\text{ kJ mol}^{-1} = -(29.392 \pm 0.011) - (6.83 \pm 0.29)\ c/(\text{mol dm}^{-3}) \quad (9.4.10.2)$$

The molar mass of 4-aminopyridine was taken as 94.1158 g mol^{-1}. The results of Burchfield and Hepler (Ref. 3) and Akello *et al.* (Ref. 4) combined by least-squares analysis give the recommended equations:–

$$\Delta_r H^\circ(298.15\ K)/\text{kJ mol}^{-1} = -(29.407 \pm 0.005) - (6.41 \pm 0.17)\ c/(\text{mol dm}^{-3}) \quad (9.4.10.3)$$

$$\Delta_r h^\circ(298.15 \text{ K})/\text{J g}^{-1} = -(312.46 \pm 0.05) - (68.1 \pm 1.8)\ c/(\text{g dm}^{-3}) \quad (9.4.10.4)$$

The standard deviation of the fit was 0.025 kJ mol^{-1} and the correlation coefficient 0.969. The recommended equations are based on measurements involving concentrations ≤ 0.05 mol dm^{-3}. As this reaction is suitable over a range of concentrations, it has the advantage of permitting the testing of a particular calorimeter with a range of sample masses, hence a range of associated temperature rises.

Akello *et al.* (Ref. 4) also made measurements on this system using a Beckman 190B microcalorimeter with sample masses in the range 10 to 30 mg, and obtained results in reasonable agreement with those given above:

$$\Delta_r H^\circ(298.15 \text{ K})/\text{ kJ mol}^{-1} = -(29.354 \pm 0.032) - (6.41 \pm 1.47)\ c/(\text{mol dm}^{-3}) \quad (9.4.10.5)$$

Hence this reaction is also suitable for testing the electrical calibration of microcalorimeters.

REFERENCES

1. Skinner, H. A., editor, *Experimental Thermochemistry* Vol II, Chapters 9, 11, 14. Wiley (Interscience): New York, **1962**.
2. Van Til, A. E.; Johnson, D. C. *Thermochim. Acta* **1977**, *20*, 177.
3. Burchfield, T. E.; Hepler, L. G. *J. Chem. Thermodynamics* **1981**, *13*, 513.
4. Akello, M. J.; Paz-Andrade, M. I.; Pilcher, G. *J. Chem. Thermodynamics* **1983**, *15*, 949.

9.4.11. $\Delta_r H°$, Hydrogen + Oxygen

Physical property: Enthalpy of reaction
Units: J mol^{-1} or kJ mol^{-1} (molar enthalpy of reaction, $\Delta_r H°$)
  J kg^{-1} or J g^{-1} (specific enthalpy of reaction, $\Delta_r h°$)
Recommended reference material: Hydrogen + oxygen ($H_2 + O_2$)
Range of variables: 298.15 K is the reference temperature normally employed
Physical state within the range: gas + gas
Contributors to the first version: J. D. Cox, A. J. Head
Contributor to the revised version: A. J. Head

Intended usage: Enthalpies of gas-phase reactions are generally measured by a means of a flow calorimeter incorporating either a burner or a solid catalyst. Descriptions of such calorimeters, which may be operated isothermally, adiabatically or isoperibolically, are to be found in references 1, 2 and 3. The energy equivalent of a gas-phase reaction calorimeter may be determined by the dissipation of measured amounts of electrical energy, but in some designs of calorimeter electrical calibration may be experimentally inconvenient and then the hydrogen + oxygen reaction (either in a flame or over a catalyst) affords a convenient means for the calibration of the calorimeter. Alternatively, the hydrogen + oxygen reaction can be used to check the accuracy of an electrical calibration. Indeed the hydrogen + oxygen reaction is the only gas-gas reaction that is internationally agreed as being suitable for the calibration of reaction calorimeters.

Sources of supply and/or methods of preparation: Compressed hydrogen and oxygen of high purity are available from many manufacturers.

Pertinent physicochemical data: When a gas-reaction calorimeter is operated at a temperature close to 298 K, most of the water formed by the hydrogen + oxygen reaction is liquid and the liquid state would be the obvious reference state for water; allowance for the enthalpy of condensation of the water vapour in equilibrium with liquid water would be required. When the calorimeter is operated at a temperature above, say, 350 K it may be more convenient to adopt the gas state as the reference state for water; if the saturation vapour pressure were exceeded, allowance for the enthalpy of vaporization of liquid water would be required. The quantities necessary for these computations can be obtained from the work of the CODATA Task Group on Key Values for Thermodynamics (Ref. 4) which has selected the following values, based on experimental work by Rossini, King and Armstrong, and Keenan, Keyes, Hill and Moore:

$$H_2(g) + 0.5\ O_2(g) = H_2O(l),$$
$$\Delta_r H°(298.15\ K) = -(285.830 \pm 0.042)\ kJ\ mol^{-1} \quad (9.4.11.1)$$

$$H_2(g) + 0.5\ O_2(g) = H_2O(l),$$
$$\Delta_r H°(298.15\ K) = -(241.814 \pm 0.042)\ kJ\ mol^{-1} \quad (9.4.11.2)$$

The molar mass of water was taken as $18.0154\ g\ mol^{-1}$.

REFERENCES

1. Cox, J. D.; Pilcher, G. *Thermochemistry of Organic and Organometallic Compounds*, Academic Press: London, **1970**.
2. Rossini, F. D., editor, *Experimental Thermochemistry*, Vol. I., Wiley (Interscience): New York, **1956**.
3. Sunner, S.; Månsson, M., editors, *Combustion Calorimetry*, Pergamon: Oxford. **1979**.
4. CODATA Recommended Key Values for Thermodynamics, 1977, *J. Chem. Thermodynamics* **1978**, *10*, 903.

9.4.12. $\Delta_r H°$, Hydrogen + Chlorine

Physical property: Enthalpy of reaction
Units: $J\ mol^{-1}$ or $kJ\ mol^{-1}$ (molar enthalpy of reaction, $\Delta_r H°$)
  $J\ kg^{-1}$ or $J\ g^{-1}$ (specific enthalpy of reaction, $\Delta_r h°$)
Recommended reference material: Hydrogen + chlorine ($H_2 + Cl_2$)
Range of variables: 298.15 K is the reference temperature normally employed
Physical state within the range: gas + gas
Contributors to the first version: J. D. Cox, O. Riedel
Contributor to the revised version: A J. Head

Intended usage: Enthalpies of gas-phase reactions are generally measured by a means of a flow calorimeter incorporating either a burner or a solid catalyst. Descriptions of such calorimeters, which may be operated isothermally, adiabatically or isoperibolically, are to be found in references 1, 2 and 3. The energy equivalent of a gas-phase reaction calorimeter may be established either by use of electrical energy, or by use of the hydrogen + oxygen reaction. For checking the accuracy of a calibration, use of the hydrogen + chlorine reaction is recommended.

Sources of supply and/or methods of preparation: Compressed hydrogen and chlorine of high purity are available from many manufacturers.

Pertinent physicochemical data: In some types of gas-reaction calorimeters the hydrogen chloride formed from the reaction between hydrogen and chlorine may leave the calorimeter in the gas state. The following value, selected by the CODATA Task Group on Key Values for Thermodynamics (Ref. 4; molar mass of hydrogen chloride 36.46 g mol^{-1}), will be applicable:

$$0.5\ H_2(g) + 0.5\ Cl_2(g) = HCl(g),$$
$$\Delta_r H°(298.15\ K) = -(92.31 \pm 0.13)\ kJ\ mol^{-1} \quad (9.4.12.1)$$

In other types of gas-reaction calorimeters hydrogen chloride may be conveniently absorbed in water placed initially within the calorimeter. The following value, selected by the CODATA Task Group (Ref. 4), will then be relevant:

$$0.5\ H_2(g) + 0.5\ Cl_2(g) + \infty H_2O(l) = HCl(\infty H_2O)(l),$$
$$\Delta_r H°(298.15\ K) = -(167.080 \pm 0.088)\ kJ\ mol^{-1} \quad (9.4.12.2)$$

Data needed to correct the above value from that for infinite dilution to that for finite dilution may be taken from references 5 or 6. Unless the reactions are conducted at low pressure, corrections for non-ideality will have to be applied.

REFERENCES

1. Cox, J. D.; Pilcher, G. *Thermochemistry of Organic and Organometallic Compounds*, Academic Press: London, **1970**.
2. Rossini, F. D., editor, *Experimental Thermochemistry*, Vol. I., Wiley (Interscience): New York, **1956**.
3. Sunner, S.; Månsson, M., editors, *Combustion Calorimetry*, Pergamon: Oxford, **1979**.
4. CODATA Recommended Key Values for Thermodynamics, 1977. *J. Chem. Thermodynamics* **1978**, *10*, 903.
5. Wagman, D. D.; Evans, W. H.; Parker, V. B.; Halow, I; Bailey, S. M.; Schumm, R. H. *Selected Values of Chemical Thermodynamic Properties*, NBS Technical Note 270-3, National Bureau of Standards: Washington, **1968**.
6. Glushko, V. P. chief editor, *Termicheskie Konstanty Veshchestv*, Vol I, Table 6, Akademiya Nauk SSSR, VINITI: Moscow, **1965**.

9.4.13. $\Delta_r H^\circ$, Methane + Oxygen

Physical property: Enthalpy of reaction
Units: J mol^{-1} or kJ mol^{-1} (molar enthalpy of reaction, $\Delta_r H^\circ$)
J m^{-3} or GJ m^{-3} (volumetric enthalpy of reaction, $\Delta_r h^\circ$)
Recommended reference material: Methane + oxygen (CH$_4$ + O$_2$)
Range of variables: 298.15 K is the reference temperature normally employed but 288.71 K (60 °F) is still used in some countries
Physical state within the range: gas + gas
Contributor to the first version: G. T. Armstrong
No revision made

Intended usage: Heating values of gaseous fuels are customarily measured by means of a flow calorimeter in which gas is burned in air or oxygen. Descriptions of such calorimeters are to found in references 1 and 2. The enthalpy of combustion per unit volume is generally indicated on a chart. The calorimeter is usually run as a continuous-flow device and is most readily calibrated by occasional experiments with methane samples of known composition and heating value. Electrical calibration is essentially impossible. The calorimeters burn the gas in excess air at constant atmospheric pressure. The gas is metered by a wet test meter, hence the heating value observed is that of the gas saturated with water.

Sources of supply and/or methods of preparation: Samples in gas cylinders of methane of good quality, certified as to the enthalpy of combustion, are avaiable from supplier (J). The samples have been compared with a pure methane sample using a typical flow calorimeter. The heating value of the pure methane sample had been certified at the National Bureau of Standards, USA. The pure methane sample was obtained from supplier (N).

Pertinent physicochemical data: The certification of the pure methane sample is described by Armstrong (Ref. 3). The value for the enthalpy of combustion of the real gas is based upon the meausurements made by Rossini (Ref. 4) and by Prosen and Rossini (Ref. 5) together with the measured composition of the gas and auxiliary data on non-ideality, temperature coefficient of the reaction, and the partial pressure of water. Prosen and Rossini reported the value -890.36 kJ mol^{-1} which Armstrong (Ref. 3) recalculated to obtain the value $-(890.31 \pm 0.29)$ kJ mol^{-1}. Results in agreement with this figure have been published by Pittam and Pilcher (Ref. 6) who found the value $-(890.71 \pm 0.38)$ kJ mol^{-1}. All these values refer to a temperature of 298.15 K and to the gases in their standard states.

REFERENCES

1. Hyde, C. G.; Jones, M. W. *Gas Calorimetry, the determination of the calorific value of gaseous fuels*, 2nd edition, Benn: London, **1960**.
2. Eiseman, J. H.; Potter, E. A. *J. Res. Nat. Bur. Stand.* **1957**, *58*, 213.

3. Armstrong, G. T. *Calculation of the Heating Value of a Sample of High Purity Methane for use as a Reference Material*, NBS Technical Note 299, National Bureau of Standards: Washington, **1966**.
4. Rossini, F. D. *J. Res. Nat. Bur. Stand.* **1931**, *6*, 37; **1931**, *7*, 329.
5. Prosen, E. J.; Rossini, F. D. *J. Res. Nat. Bur. Stand.* **1945**, *34*, 263.
6. Pittam, D. A.; Pilcher, G. *J. Chem. Soc. Faraday Trans. I* **1972**, *68*, 2224.

9.5. Reference materials for measurement of enthalpies of combustion

9.5.1. $\Delta_c U°$, Benzoic acid

Physical property: Energy of combustion
Units: J mol^{-1} or kJ mol^{-1} (molar energy of combustion, $\Delta_c U°$)
J kg^{-1} or J g^{-1} (specific energy of combustion, $\Delta_c u°$)
Recommended reference material: Benzoic acid ($C_7H_6O_2$)
Range of variables: 298.15 K is the reference temperature normally employed
Physical state within the range: solid
Contributors to the first version: G. T. Armstrong, J. D. Cox, H. Feuerberg, J. Franc, B. N. Oleinik, O. Riedel
Contributors to the revised version: Yu. I. Aleksandrov, E. S. Domalski, A. J. Head, C. Mosselman, B. N. Oleinik

Intended usage: By international agreement, reached in 1934, benzoic acid is the principal reference material for measuring the energy equivalent of oxygen-bomb calorimeters. These calorimeters all have a closed chamber which contains compressed oxygen and the sample to be burnt and are provided with an electrical system for the ignition of the sample. The size, shape, material of construction, disposition of parts and usage of the bomb vary greatly (Refs. 1 to 4): some are used statically, some in a moving mode; some are used immersed in a fluid, some in contact with a metal block, and some naked; some are intended for gram samples, some for milligram samples; some are used isoperibolically, some isothermally and some adiabatically.

Benzoic acid also serves in combustion calorimetry to kindle materials which are difficult to burn, to influence the overall stoichiometry of a combustion reaction and to test analytical procedures.

Sources of supply and/or methods of preparation: Suitable grades of benzoic acid (designated as 'thermochemical' or 'calorimetric' standard) certified for the value of the energy of combustion are available from suppliers (B), (D), (S) and from BDH.

Pertinent physicochemical data: From a thermodynamic viewpoint the standard specific energy of combustion of benzoic acid at 298.15 K is the key quantity, but, since bomb-combustion reactions are conducted under conditions far from those of the thermodynamic standard state, a more practical quantity is $\Delta_c u(\text{cert})$, the specific energy of combustion certified by a standardizing laboratory as the energy evolved when one gram of benzoic acid burns under 'standard bomb conditions' (Ref. 1). Determinations of $\Delta_c u(\text{cert})$ are made using calorimeters which are calibrated by means of electrical energy. An extensive study of a single batch of benzoic acid (NBS 39i) has been made from which the weighted mean of four concordant determinations (Refs. 5 to 8) is $-(26433.6 \pm 0.9)$ J g^{-1}. This value is in good agreement with the earlier assessment by Hawtin (Ref. 9) of determinations (on various samples) made prior to 1966, viz $-(26434.4 \pm 1.2)$ J g^{-1}, the value selected by Cox and Pilcher -26434 J g^{-1} (Ref. 1), and a recently selected value of $-(26434.3 \pm 0.6)$ J g^{-1} (Ref. 10). The rounded value of -26434 J g^{-1} for the value of $\Delta_c u(\text{cert})$ for benzoic acid is thus well established. Although this value relates to essentially pure benzoic acid, it is important to note that a given batch of benzoic acid to be used for the calibration of bomb calorimeters may not be pure. Indeed, it need not be pure provided it is homogeneous and has been properly certified, but if it is not pure it cannot be used for testing analytical procedures. When benzoic acid is used under conditions remote from 'standard bomb conditions' it may be preferable to calculate the value appropriate to these conditions from the standard specific energy of combustion $\Delta_c u°$, which is 20 J g^{-1} less negative than $\Delta_c u(\text{cert})$ (Ref. 1).

REFERENCES

1. Cox, J. D.; Pilcher, G. *Thermochemistry of Organic and Organometallic Compounds*, Academic Press: London, **1970**.
2. Rossini, F. D., editor, *Experimental Thermochemistry*, Vol. I., Wiley (Interscience): New York, **1956**.
3. Skinner, H. A., editor, *Experimental Thermochemistry*, Vol. II, Wiley (Interscience): New York, **1962**.
4. Sunner, S.; Månsson, M., editors, *Combustion Calorimetry*. Pergamon: Oxford, **1979**.
5. Churney, K. L.; Armstrong, G. T. *J. Res. Nat. Bur. Stand.* **1968**, *72A*, 453.
6. Gundry, H. A., Harrop, D., Head, A. J., Lewis, G. B. *J. Chem. Thermodynamics* **1969**, *1*, 321.
7. Mosselman, C.; Dekker, H. *Rec. trav. chim.* **1969**, *88*, 161.
8. Jochems, R.; Dekker, H.; Mosselman, C. *Rev. Sci. Inst.* **1979**, *50*, 859.
9. Hawtin, P. *Nature, Lond.* **1966**, *210*, 411.
10. Aleksandrov, Yu. I.; Oleinik, B. N.; private communication (to be published).

9.5.2. $\Delta_c U°$, Succinic acid

Physical property: Energy of combustion
Units: J mol^{-1} or kJ mol^{-1} (molar energy of combustion, $\Delta_c U°$)
J kg^{-1} or J g^{-1} (specific energy of combustion, $\Delta_c u°$)
Recommended reference material: Succinic acid ($C_4H_6O_4$)
Range of variables: 298.15 K is the reference temperature normally employed

Physical state within the range: solid
Contributors to the first version: J. D. Cox, H. Feuerberg,
A. J. Head, E. F. Westrum, Jr.
Contributors to the revised version: Yu. I. Aleksandrov, A. J. Head, B. N. Oleinik

Intended usage: Energies of combustion in oxygen of most compounds containing no elements other than C, H, O, N can be accurately measured with the aid of a static bomb calorimeter (Refs. 1, 2). The energy equivalent of the calorimeter will normally be established by combustion of thermochemical-standard benzoic acid. However, many workers like to check the accuracy of the benzoic acid calibration and their experimental procedure generally, by the combustion of a test material of known energy of combustion. Succinic acid is recommended as a test material for the combustion calorimetry of C, H, O compounds, especially those that can be successfully burned without first being sealed in a capsule, although one group of workers have reported difficulty in obtaining complete combustion (Ref. 3).

Sources of supply and/or methods of preparation: A suitable sample of succinic acid can be prepared from an analytical grade specimen by four recrystallizations from distilled water followed by an effective drying procedure, either by a two-stage pelleting technique (Ref. 4) or by vacuum sublimation below 398 K (Ref. 5).

Pertinent physicochemical data: Although the energy of combustion of succinic acid has been measured many times, recent studies on the removal of water from succinic acid indicate that many of the samples used in earlier work may have contained water. Oleinik *et al.* (Ref. 5) have selected the value $-(12638.5 \pm 1.5)$ J g^{-1} based on three determinations where the purity of the succinic acid had been well established and the criteria of good calorimetry had been met (Refs. 5 to 7). Their selected value is in very close agreement with the value (12638.0 ± 1.6) J g^{-1} previously selected by Vanderzee *et al.* (Ref. 7) from a wider range of published results, after rigorous critical evaluation and re-calculation of some values. These selected values refer to the reaction (at 298.15 K):

$$C_4H_6O_4(cr) + 3.5\ O_2(g) = 4\ CO_2(g) + 3\ H_2O(l) \qquad (9.5.2.1)$$

REFERENCES

1. Sunner, S.; Månsson, M., editors, *Combustion Calorimetry*. Pergamon Press: Oxford, **1979**.
2. Rossini, F. D., editor, *Experimental Thermochemistry*, Vol. I., Wiley (Interscience): New York, **1956**.
3. Nabavian, M.; Sabbah, R.; Chastel, R.; Laffitte, M. *J. Chim. Phys.* **1977**, *74*, 115.
4. Vanderzee, C. E.; Westrum Jr., E. F. *J. Chem. Thermodynamics* **1970**, *2*, 681.
5. Oleinik, B. N.; Aleksandrov, Yu. I.; Mikina, V. D.; Hrustaleva, K. A.; Osipova, T. R. *Fourth International Conference on Thermodynamics*, Book IX, Montpellier, **1975**, p. 3.
6. Wilhoit, R. C.; Shiao, D. *J. Chem. Eng. Data* **1964**, *9*, 595.
7. Vanderzee, C. E.; Månsson, M.; Sunner, S. *J. Chem. Thermodynamics* **1972**, *4*, 533.

9.5.3. $\Delta_c U°$, Hippuric acid

Physical property: Energy of combustion
Units: J mol^{-1} or kJ mol^{-1} (molar energy of combustion, $\Delta_c U°$)
J kg^{-1} or J g^{-1} (specific energy of combustion, $\Delta_c u°$)
Recommended reference material: Hippuric acid ($C_9H_9O_3N$)
Range of variables: 298.15 K is the reference temperature normally employed
Physical state within the range: solid
Contributors to the first version: J. D. Cox, A. J. Head
Contributors to the revised version: Yu. I. Aleksandrov, A. J. Head, B. N. Oleinik

Intended usage: Energies of combustion in oxygen of most compounds containing no elements other than C, H, O, N can be accurately measured with the aid of a static- or a rotating-bomb calorimeter (Refs. 1, 2). The energy equivalent of the calorimeter will normally be established by combustion of thermochemical-standard benzoic acid. However, many workers like to check the accuracy of the benzoic acid calibration, and their experimental procedure generally, by combustion of a test material of known energy of combustion. Hippuric acid was recommended for this purpose by Huffman (Ref. 3); it is a test material for the combustion calorimetry of C, H, O, N compounds, especially those that can be burned without first being sealed in a capsule and have a relatively low nitrogen content (below 10 per cent).

Sources of supply and/or methods of preparation: Hippuric acid of sufficient purity may be obtained by crystallization of good commercial material from water (Refs. 3, 4), but there is evidence that the recrystallized material retains water (mass fraction of 6×10^{-4}) after drying *in vacuo* and storage over phosphorus pentoxide. Oleinik *et al.* (Ref. 5) have shown that the water can be completely removed by zone sublimation, provided that care is taken to avoid excessive thermal decomposition to benzoic acid. The technique successfully used by Vanderzee and Westrum (Ref. 6) for drying succinic acid does not appear to have been applied to hippuric acid and is worthy of trial.

Pertinent physicochemical data: The most reliable values for the energy of combustion of hippuric acid were obtained by Hubbard, Frow and Waddington (Ref. 4) and by Oleinik *et al.* (Ref. 5). Because of the difficulty in removing water from the sample, the former authors based their results on the mass of carbon dioxide produced rather than on the mass of sample burnt. The results of both groups of workers were compatible and lead to a recommended value of -23458 J g^{-1} for the specific standard energy of combustion at 298.15 K for the reaction

$$C_9H_9O_3N(cr) + 9.75\ O_2(g) = 9\ CO_2(g) + 4.5\ H_2O(l) + 0.5\ N_2(g) \qquad (9.5.3.1).$$

Provided that water can be removed without thermal decomposition of the sample, hippuric acid fulfils the requirements of a test substance for the combustion calorimetry of compounds with a relatively low nitrogen content.

REFERENCES

1. Sunner, S.; Månsson, M., editors, *Combustion Calorimetry*, Pergamon: Oxford, **1979**.
2. Rossini, F. D., editor, *Experimental Thermochemistry*, Vol. I, Wiley (Interscience): New York, **1956**.
3. Huffman, H. M. *J. Amer. Chem. Soc.* **1938**, *60*, 1171.
4. Hubbard, W. N.; Frow, F. R.; Waddington, G. *J. Phys. Chem.* **1961**, *65*, 1326.
5. Oleinik, B. N.; Mikina, V. D.; Aleksandrov, Yu. I.; Ushcevich, V. F.; Osipova, T. R.; Hrustaleva, K. A. *J. Chem. Thermodynamics* **1979**, *11*, 145.
6. Vanderzee, C. E.; Westrum Jr, E. F. *J. Chem. Thermodynamics* **1970**, *2*, 681.

9.5.4. $\Delta_c U°$, Acetanilide

Physical property: Energy of combustion
Units: J mol^{-1} or kJ mol^{-1} (molar energy of combustion, $\Delta_c U°$)
   J kg^{-1} or J g^{-1} (specific energy of combustion, $\Delta_c u°$)
Recommended reference material: Acetanilide (C$_8$H$_9$ON)
Range of variables: 298.15 K is the reference temperature normally employed
Physical state within the range: solid
Contributors: E. S. Domalski, A. J. Head

Intended usage: Energies of combustion of most compounds containing no elements other than C, H, O, N can be accurately measured with the aid of a static- or a rotating-bomb calorimeter (Refs. 1, 2). The energy equivalent of the calorimeter will normally be established by combustion of thermochemical-standard benzoic acid. However, many workers like to check the accuracy of the benzoic acid calibration, and their experimental procedure generally, by combustion of a test material of known energy of combustion. Acetanilide was recommended for this purpose by Johnson (Ref. 3); it is a test material for the combustion calorimetry of C, H, O, N compounds, especially those that can be burnt without first being encapsulated and without the need of auxiliary substances to effect complete combustion.

Sources of supply and/or methods of preparation: Suitable material of certified high purity (mole fraction purity 0.9999) is available from supplier (S) as a microanalytical standard, SRM 141c.

Pertinent physicochemical data: Johnson has determined the energy of combustion of acetanilide (NBS SRM 141B) using an adiabatic rotating-bomb calorimeter. The value of $-(31234 \pm 7)$ J g^{-1} obtained refers to the specific energy of combustion at 298.15 K for the reaction

$$C_8H_9ON(cr) + 9.75\ O_2(g) = 8\ CO_2(g) + 4.5\ H_2O(l) + 0.5\ N_2(g) \qquad (9.5.4.1).$$

Although this is the only published value for the energy of combustion it is supported by the value calculated from the enthalpy of the reaction between aniline and acetic anhydride and the enthalpies of formation of these two substances, viz. −31234 J g^{-1} (Ref. 4). Nevertheless, further determinations of the energy of combustion are necessary before acetanilide is established as a reference material for the combustion calorimetry of substances with relatively low nitrogen content. It offers advantages over hippuric acid in being readily available free from water.

REFERENCES

1. Sunner, S.; Månsson, M., editors, *Combustion Calorimetry*, Pergamon: Oxford, **1979**.
2. Rossini, F. D., editor, *Experimental Thermochemistry*, Vol. I, Wiley (Interscience): New York, **1956**.
3. Johnson, W. H. *J. Res. Nat. Bur. Stand.* **1975**, *79A*, 487.
4. Cox, J. D., Chapter 4 of Ref. 1.

9.5.5. $\Delta_c U°$, Urea

Physical property: Energy of combustion
Units: J mol^{-1} or kJ mol^{-1} (molar energy of combustion, $\Delta_c U°$)
J kg^{-1} or J g^{-1} (specific energy of combustion, $\Delta_c u°$)
Recommended reference material: Urea (CH$_4$ON$_2$)
Range of variables: 298.15 K is the reference temperature normally employed
Physical state within the range: solid
Contributor to the first version: G. T. Armstrong
Contributors to the revised version: Yu. I. Aleksandrov, A. J. Head, B. N. Oleinik

Intended usage: Energies of combustion in oxygen of most compounds containing no elements other than C, H, O, N can be accurately measured with the aid of a static- or a rotating-bomb calorimeter (Refs. 1, 2). The energy equivalent of the calorimeter will normally be established by combustion of thermochemical-standard benzoic acid. However, many workers like to check the accuracy of the benzoic acid calibration, and of the experimental procedure generally, by combustion of a test material of known energy of combustion. Urea is recommended as a test material for the combustion calorimetry of C, H, O, N compounds containing such a large proportion of nitrogen that complete combustion can only be obtained by the use of auxiliary materials, such as benzoic acid or paraffin oil.

Sources of supply and/or methods of preparation: Suitable material may be obtained by recrystallization of the analytical grade reagent. Urea with certified energy of combustion is obtained from supplier (S) as SRM 2152.

Pertinent physicochemical data: Johnson (Ref. 3) determined the energy of combustion of NBS reference samples of urea (SRM 912) for which the certified mass fraction purity was

9. Enthalpy

0.997 but probably approached 0.999, since he showed the moisture content to be considerably lower than the certified value. He used benzoic acid as auxiliary material and reported the value $-(10540.6 \pm 2.8)$ J g^{-1} for the energy of the reaction (at 298.15 K):

$$CH_4ON_2(cr) + 1.5\ O_2(g) = CO_2(g) + 2\ H_2O(l) + N_2(g) \qquad (9.5.5.1).$$

Johnson's value is in good agreement with that selected by Cox and Pilcher (Ref. 4) from the results of earlier studies in which the purity of the urea was not established and in which paraffin oil was used as the auxiliary substance, -10537 J g^{-1}. The agreement between determinations by different workers using different sources of urea and different auxiliary materials supports the use of urea as a reference material, and the value $-(10539 \pm 5)$ J g^{-1} (the uncertainty is estimated) is recommended for the specific energy of combustion. The use of urea has, however, been criticised (on the grounds of the alleged thermal instability and hygroscopicity) by Aleksandrov and Oleinik (Ref. 5), who suggest that 1,2,4-triazole would be a more suitable reference material for substances of high nitrogen content.

REFERENCES

1. Sunner, S.; Månsson, M., editors, *Combustion Calorimetry*, Pergamon: Oxford, **1979**.
2. Rossini, F. D., editor, *Experimental Thermochemistry*, Vol. I, Wiley (Interscience): New York, **1956**.
3. Johnson, W. H. *J. Res. Nat. Bur. Stand.* **1975**, *79A*, 487.
4. Cox, J. D.; Pilcher, G. *Thermochemistry of Organic and Organometallic Compounds*, Academic Press: London, **1970**.
5. Aleksandrov, Yu. I.; Osipova, T. R.; Ushcevich, V. F.; Murashova, S. V.; Oleinik, B. N. *Termodinam. Organ. Soedin., Gor'kii* **1979**, 65.

9.5.6. $\Delta_c U°$, 4-Chlorobenzoic acid

Physical property: Energy of combustion
Units: J mol^{-1} or kJ mol^{-1} (molar energy of combustion, $\Delta_c U°$)
J kg^{-1} or J g^{-1} (specific energy of combustion, $\Delta_c u°$)
Recommended reference material: 4-Chlorobenzoic acid (C$_7$H$_5$ClO$_2$)
Range of variables: 298.15 K is the reference temperature normally employed
Physical state within the range: solid
Contributors to the first version: J. D. Cox, A. J. Head, O. Riedel, S. Sunner
Contributors to the revised version: Yu. I. Aleksandrov, E. S. Domalski, A. J. Head, B. N. Oleinik

Intended usage: Energies of combustion in oxygen of most organic compounds containing chlorine can be accurately measured with the aid of a moving-bomb calorimeter (Refs. 1, 2, 3). Techniques involving the use of static-bomb calorimeters are less accurate. In order that the sole chlorine-containing product should be a hydrochloric acid solution, sufficient reducing agent (*e.g.* arsenious oxide solution or hydrazine hydrochloride solution) should be placed in the bomb before combustion is initiated. There is evidence that arsenious oxide solution can oxidise and hydrazine hydrochloride solution can decompose before a combustion experiment, under the conditions prevailing in some platinum-lined bombs. Experimenters should always check to see whether these reactions, which are catalysed by the platinum liner, are occurring under their experimental conditions. These side reactions can always be avoided by replacing platinum with tantalum.

The energy equivalent of a moving-bomb calorimeter will normally be established by combustion of thermochemical-standard benzoic acid. Many workers, however, like to check the accuracy of the benzoic acid calibration, and their experimental procedure generally, by the combustion of a test material of known energy of combustion. 4-Chlorobenzoic acid is recommended as a test material for the combustion calorimetry of C, H, Cl, O compounds with an atomic ratio of hydrogen to chlorine equal to or greater than unity.

Sources of supply and/or methods of preparation: Samples of 4-chlorobenzoic acid of satisfactory purity may be obtained by repeated recrystallizations of commercial material, preferably followed by vacuum sublimation.

Pertinent physicochemical data: The suitability of 4-chlorobenzoic acid as a reference substance for combustion calorimetry has been well established and the results of three groups of workers (Refs. 4 to 6), which involved the use of both platinum-lined and tantalum-lined combustion bombs and both arsenious oxide and hydrazine hydrochloride as reducing agents, have been discussed by Cox (Ref. 7). The weighted mean calculated from these values and from a more recent determination (Ref. 8) of the energy of combustion is $-(19567 \pm 2)$ J g^{-1} and applies to the reaction

$$C_7H_5O_2Cl(cr) + 7\ O_2(g) + 598\ H_2O(l) = 7\ CO_2(g) + HCl(600\ H_2O)(l) \qquad (9.5.6.1).$$

Two determinations which yielded results significantly different from the above weighted mean, *viz* $-(19582.0 \pm 7.7)$ J g^{-1} (Ref. 9) and $-(19586 \pm 2.9)$ J g^{-1} (Ref. 10) have not been included in the present selection.

Values for the energies of combustion in which the concentration of the final hydrochloric acid solution is other than HCl(600 H$_2$O) may be derived from tabulated enthalpy of formation data (Refs. 11, 12). It should be noted that the *differences* between enthalpies of formation of HCl(nH$_2$O) listed in Ref. 11 yield reliable enthalpies of dilution although the enthalpies of formation should be made more positive by 79 J mol^{-1} to be compatible with CODATA (1977) values (Ref. 13).

REFERENCES

1. Cox, J. D.; Pilcher, G. *Thermochemistry of Organic and Organometallic Compounds*, Academic Press: London, **1970**.

2. Skinner, H. A., editor, *Experimental Thermochemistry*, Vol. II, Wiley (Interscience): New York, **1962**.
3. Sunner, S.; Månsson, M., editors.,*Combustion Calorimetry*, Pergamon: Oxford, **1979**.
4. Hu, A. T.; Sinke, G. C.; Månsson, M.; Ringnér, B. *J. Chem. Thermodynamics* **1972**, *4*, 283.
5. Hajiev, S. N.; Agarunov, M. J.; Nurullaev, H. G. *J. Chem. Thermodynamics* **1974**, *6*, 713.
6. Johnson, W. H.; Prosen, E. J. *J. Res. Nat. Bur. Stand.* **1974**, *78A*, 683.
7. Cox, J. D., Chapter 4 of Ref. 3.
8. Erastov, P. A.; Kolesov, V. P.; Ushcevich, V. F.; Aleksandrov, Yu. I. *Zh. Fiz. Khim.* **1978**, *52*, 2223.
9. Kolesov, V. P.; Slavutskaya, G. M.; Alekhin, S. P.; Skuratov, S. M. *Zh. Fiz. Khim.* **1972**, *46*, 2138.
10. Platonov, V. A.; Simulin, Yu. N.; Dzhagatspanyan, R. V. *Zh. Fiz. Khim.* **1981**, *55*, 2132.
11. Wagman, D. D.; Evans, W. H.; Parker, V. B.; Halow, I.; Bailey, S. M.; Schumm, R. H. *Selected Values of Chemical Thermodynamic Properties*, NBS Technical Note 270-3, Table 10, National Bureau of Standards: Washington, **1968**.
12. Glushko, V. P., chief editor, *Termicheskie Konstanty Veshchestv*, Vol I, Table 6, Akademiya Nauk SSSR, VINITI: Moscow **1965**.
13. CODATA Recommended Key Values For Thermodynamics, 1977 *J. Chem. Thermodynamics* **1978**, *10*, 903.

9.5.7. $\Delta_c U°$, 4-Fluorobenzoic acid

Physical property: Energy of combustion
Units: J mol^{-1} or kJ mol^{-1} (molar energy of combustion, $\Delta_c U°$)
J kg^{-1} or J g^{-1} (specific energy of combustion, $\Delta_c u°$)
Recommended reference material: 4-Fluorobenzoic acid ($C_7H_5FO_2$)
Range of variables: 298.15 K is the reference temperature normally employed
Physical state within the range: solid
Contributors to the first version: J. D. Cox, A. J. Head, O. Riedel
Contributors to the revised version: Yu. I. Aleksandrov, E. S. Domalski, A. J. Head, B. N. Oleinik

Intended usage: Energies of combustion in oxygen of most organic compounds containing fluorine can be accurately measured with the aid of a platinum-lined moving-bomb calorimeter (Refs. 1 to 3). Techniques involving the use of static-bomb calorimeters are less accurate. Sufficient water should be placed in the bomb initially so that the solution of hydrofluoric acid obtained after the combustion may have a concentration of not more than 5 mol dm^{-3}. The energy equivalent of a moving-bomb calorimeter will normally be established by combustion of thermochemical-standard benzoic acid. However, many workers like to check the accuracy of the benzoic acid calibration, and the experimental procedure generally, by the combustion of a test material of known energy of combustion. 4-Fluorobenzoic acid is recommended as a test material for the combustion of C, H, F, O compounds, especially those that can be successfully burnt without first being sealed in a capsule. Combustion reactions of fluorine compounds in which the atomic ratio of hydrogen to fluorine is equal to or greater than unity differ from those of compounds in which the atomic ratio is less than unity. Thus the reactions of the former compounds yield hydrogen fluoride as the sole fluorine-containing product, whereas the reactions of the latter compounds yield hydrogen fluoride and carbon tetrafluoride. Therefore 4-fluorobenzoic acid serves as a test material only for C, H, F, O compounds with a hydrogen to fluorine ratio equal to or greater than unity.

Sources of supply and/or methods of preparation: Suitable samples may be obtained by zone-refining of commercial material (Refs. 4, 5). Johnson and Prosen (Ref. 6) have pointed out that salicylic acid is a possible contaminant of 4-fluorobenzoic acid which can remain undetected by several analytical methods, and they therefore recommended determination of hydrofluoric acid in the combustion products.

Pertinent physicochemical data: The suitability of 4-fluorobenzoic acid as a reference material in combustion calorimetry has been established by three groups of workers (Refs. 4 to 6), who obtained concordant values for the energy of the reaction

$$C_7H_5FO_2(cr) + 7\ O_2(g) + 18\ H_2O(l) = 7\ CO_2(g) + HF(20\ H_2O)(l) \qquad (9.5.7.1).$$

The published work has been assessed by Cox (Ref. 7) who recommends the selected value $-(21861 \pm 4)$ J g^{-1} for the standard energy of combustion at 298.15 K.

REFERENCES

1. Cox, J. D.; Pilcher, G. *Thermochemistry of Organic and Organometallic Compounds*, Academic Press: London, **1970**.
2. Skinner, H. A., editor, *Experimental Thermochemistry*, Vol. II, Wiley (Interscience): New York, **1962**.
3. Sunner, S.; Månsson, M., editors, *Combustion Calorimetry*. Pergamon: Oxford, **1979**.
4. Good, W. D.; Scott, D. W.; Waddington, G. *J. Phys. Chem.* **1956**, *60*, 1080.
5. Cox, J. D.; Gundry, H. A.; Head, A. J. *Trans. Faraday Soc.* **1964**, *60*, 653.
6. Johnson, W. H.; Prosen, E. J. *J. Res. Nat. Bur. Stand.* **1975**, *79A*, 481.
7. Cox, J. D. Chapter 4 of Ref. 3.

9.5.8. $\Delta_c U°$, Pentafluorobenzoic acid

Physical property: Energy of combustion
Units: J mol^{-1} or kJ mol^{-1} (molar energy of combustion, $\Delta_c U°$)
      J kg^{-1} or J g^{-1} (specific energy of combustion, $\Delta_c u°$)
Recommended reference material: Pentafluorobenzoic acid ($C_7HF_5O_2$)
Range of variables: 298.15 K is the reference temperature normally employed
Physical state within the range: solid
Contributors to the first version: J. D. Cox, A. J. Head
Contributor to the revised version: A. J. Head

Intended usage: Energies of combustion in oxygen of most organic compounds containing fluorine can be accurately measured with the aid of a platinum-lined moving-bomb calorimeter (Refs. 1 to 3). Techniques involving the use of static-bomb calorimeters are less accurate. Sufficient water should be placed in the bomb initially so that the solution of hydrofluoric acid obtained after the combustion may have a concentration of not more than 5 mol dm^{-3}. The energy equivalent of a moving-bomb calorimeter will normally be established by combustion of thermochemical-standard benzoic acid. However, many workers like to check the accuracy of the benzoic acid calibration, and the experimental procedure generally, by the combustion of a test material of known energy of combustion. Pentafluorobenzoic acid is recommended as a test material for the combustion calorimetry of C, H, F, O compounds, especially those that can be successfully burnt without prior encapsulation. Combustion reactions of fluorine compounds in which the atomic ratio of hydrogen to fluorine is greater than or equal to unity differ from those of compounds in which the atomic ratio of hydrogen to fluorine is less than unity. Thus the former compounds yield hydrogen fluoride as the sole fluorine-containing product whereas the latter compounds yield a mixture of hydrogen fluoride and carbon tetrafluoride. Therefore pentafluorobenzoic acid serves as a test material only for C, H, F, O compounds with an atomic ratio of hydrogen to fluorine less than unity.

Sources of supply and/or methods of preparation: Cox et al. (Ref. 4) showed that a suitable sample of pentafluorobenzoic acid could be obtained by zone-refining a commercial sample.

Pertinent physicochemical data: Cox, Gundry and Head (Ref. 4) suggested there was a need for a reference material for the study of the combustion of compounds with a low hydrogen to fluorine ratio, and showed that pentafluorobenzoic acid possessed the necessary properties. The value they reported, *viz.* $-(12062.4 \pm 4.8)$ J g^{-1}, relates to the reaction (at 298.15 K)

$$C_7HF_5O_2(cr) + 5\ O_2(g) + 102\ H_2O(l) = 7\ CO_2(g) + 5\ [HF(20\ H_2O)](l) \qquad (9.5.8.1).$$

It will be noted that in conformity with common practice, the equation is written as though no carbon tetrafluoride is formed. In reality, considerable amounts of carbon tetrafluoride are formed unless a hydrogen-containing auxiliary compound is burnt along with the pentafluorobenzoic acid. If enough hydrogen-containing compound is taken to ensure that the hydrogen to fluorine atomic ratio greatly exceeds unity, and if the two compounds are intimately mixed before combustion, then no carbon tetrafluoride will be formed. However, this causes the majority of the energy evolved to come from the auxiliary, which is an undesirable state of affairs. Cox, Gundry and Head chose to perform two series of experiments, in the first of which hydrocarbon oil, used as auxiliary, contributed about 40 per cent of the heat, whilst in the second benzoic acid, used as auxiliary, contributed about 60 per cent of the heat. In both series of experiments it was necessary to determine the amount of carbon tetrafluoride formed and to correct for it. The determination of the amount of carbon tetrafluoride was based on the shortfall between the number of moles of combined carbon taken, and the carbon dioxide found after the combustion. It would be desirable for further measurements of the energy of combustion of pentafluorobenzoic acid to be made.

REFERENCES

1. Cox, J. D.; Pilcher, G. *Thermochemistry of Organic and Organometallic Compounds*, Academic Press: London, **1970**.
2. Skinner, H. A., editor), *Experimental Thermochemistry*, Vol. II, Wiley (Interscience): New York, **1962**.
3. Sunner, S.; Månsson, M., editors, *Combustion Calorimetry*. Pergamon: Oxford, **1979**.
4. Cox, J. D.; Gundry, H. A.; Head, A. J. *Trans. Faraday Soc.* **1964**, *60*, 653.

9.5.9. $\Delta_c U°$, Thianthrene

Physical property: Energy of combustion
Units: J mol^{-1} or kJ mol^{-1} (molar energy of combustion, $\Delta_c U°$)
  J kg^{-1} or J g^{-1} (specific energy of combustion, $\Delta_c u°$)
Recommended reference material: Thianthrene ($C_{12}H_8S_2$)
Range of variables: 298.15 K is the reference temperature normally employed
Physical state within the range: solid
Contributors to the first version: J. D. Cox, A. J. Head, O. Riedel
Contributors to the revised version: E. S. Domalski, A. J. Head

Intended usage: Energies of combustion in oxygen of most organic compounds containing sulphur can be accurately measured with the aid of a platinum-lined moving-bomb calorimeter (Refs. 1 to 4). Techniques involving the use of static bomb calorimeters are less accurate. To ensure formation of a well defined final state, water should be placed in the bomb and sufficient gaseous nitrogen (not less than 2.5 per cent) should be present in the compressed oxygen so that the nitrogen oxides formed during the combustion process may catalyse the oxidation of all the sulphur to the S^{VI} state. It is important to ensure that the ratio between the number of moles of combined hydrogen and of combined sulphur exceeds two in the combustion. Thus for a compound with a high proportion of sulphur it may be necessary to burn an auxiliary hydrogen-containing compound to achieve the necessary hydrogen to sulphur ratio. The energy equivalent of a moving bomb calorimeter will normally be established by combustion of thermochemical-standard benzoic acid. However, many workers like to check the accuracy of the benzoic acid calibration, and of the experimental procedure generally, by combustion of a test material of known energy of combustion. Thianthrene is recommended as a test material for the combustion calorimetry of C, H, S compounds, especially those that can be successfully burned without first being enclosed in a capsule.

Sources of supply and/or methods of preparation: Purification of thianthrene to yield material suitable for use as reference material has been effected by recrystallization, sublimation and fractional freezing (Ref. 5) or by zone refining (Ref. 6). Material certified for enthalpy of combustion is issued by Supplier (S) as SRM 1656.

Pertinent physicochemical data: The suitability of thianthrene as a reference material in combustion calorimetry of sulphur compounds is well established. The results of five concordant determinations of the energy of combustion have been summarized by Johnson (Ref. 6). All five values fall within a range of 15 J g^{-1} and yield a weighted mean of $-(33468 \pm 4)$ J g^{-1}, which applies to the reaction (at 298.15 K)

$$C_{12}H_8S_2(cr) + 17\ O_2(g) + 228\ H_2O(l) = 12\ CO_2(g) + 2\ [H_2SO_4(115\ H_2O)](l) \quad (9.5.9.1).$$

Values for the energies of combustion for reactions in which the concentration of the final sulphuric acid solution is other than $H_2SO_4(115\ H_2O)$ may be deduced from tabulated enthalpy of formation data (Refs. 7, 8). It should be noted that the *differences* between enthalpies of formation of $H_2SO_4(nH_2O)$ listed in Ref. 7 yield reliable enthalpies of dilution, although the enthalpies of formation should be made more negative by 335 J mol^{-1} to be compatible with CODATA 1977 values (Ref. 9).

REFERENCES

1. Cox, J. D.; Pilcher, G. *Thermochemistry of Organic and Organometallic Compounds*, Academic Press: London, **1970**.
2. Rossini, F. D., editor, *Experimental Thermochemistry*, Vol. I, Wiley (Interscience): New York, **1956**.
3. Skinner, H. A., editor, *Experimental Thermochemistry*, Vol. II, Wiley (Interscience): New York, **1962**.
4. Sunner, S.; Månsson, M., editors, *Combustion Calorimetry*, Pergamon: Oxford, **1979**.
5. Sunner, S.; Lundin, B. *Acta Chem. Scand.* **1953**, *7*, 1112.
6. Johnson, W. H. *J. Res. Nat. Bur. Stand.* **1975**, *79A*, 561.
7. Wagman, D. D.; Evans, W. H.; Parker, V. B.; Halow, I.; Bailey, S. M.; Schumm, R. H. *Selected values of Chemical Thermodynamic Properties*, NBS Technical Note 270-3, Table 14, National Bureau of Standards: Washington, **1968**.
8. Glushko, V. P., chief editor, *Termicheskie Konstanty Veshchestv*, Vol. II, Table 17. Akademiya Nauk SSSR, VINITI: Moscow, **1966**.
9. CODATA Recommended Key Values For Thermodynamics, 1977. *J. Chem. Thermodynamics* **1978**, *10*, 903.

9.5.10. $\Delta_c U°$, Triphenylphosphine oxide

Physical property: Energy of combustion
Units: J mol^{-1} or kJ mol^{-1} (molar energy of combustion, $\Delta_c U°$)
J kg^{-1} or J g^{-1} (specific energy of combustion, $\Delta_c u°$)
Recommended reference material: Triphenylphosphine oxide ($C_{18}H_{15}OP$)
Range of variables: 298.15 K is the reference temperature normally employed
Physical state within the range: solid
Contributor: A. J. Head

Intended usage: Combustion of organophosphorus compounds has not so far attained the precision and accuracy associated with modern experimental thermochemistry, partly because complete combustion has often been difficult to achieve and partly because a mixture of oxyacids of phosphorus is produced. Harrop and Head (Ref. 1) applied the techniques of moving-bomb calorimetry (Ref. 2) to the combustion of triphenylphosphine oxide: the substance burnt completely when supported on a gold dish (platinum was attacked) and the phosphorus-containing reaction products were either a mixture of orthophosphoric acid and polyphosphoric acids (when water was used in the bomb) or orthophosphoric acid alone (when perchloric acid was used). It is suggested that these techniques are likely to be applicable to other C, H, O, P compounds and that triphenylphosphine oxide would then be a suitable reference material for checking the experimental procedure employed.

Sources of supply and/or methods of preparation: Triphenylphosphine oxide is readily purified by recrystallization followed by zone-refining.

Pertinent physicochemical data: Harrop and Head (Ref. 1) established that triphenylphosphine oxide fulfils the requirements of a reference material for combustion calorimetry and reported the value $-(35790 \pm 6)$ J g^{-1} for the specific energy of the reaction (at 298.15 K)

$$C_{18}H_{15}OP(cr) + 22.5\ O_2(g) = 18\ CO_2(g) + H_3PO_4(6\ H_2O)(l) \qquad (9.5.10.1).$$

This must be regarded as a provisional value until it is confirmed by other workers.

REFERENCES

1. Harrop, D.; Head, A. J. *J. Chem. Thermodynamics* **1977**, *9*, 1067.
2. Sunner, S.; Månsson, M., editors, *Combustion Calorimetry*, Pergamon: Oxford, **1979**.

9.5.11. $\Delta_c U°$, 2,2,4-Trimethylpentane

Physical property: Energy of combustion
Units: J mol^{-1} or kJ mol^{-1} (molar energy of combustion, $\Delta_c U°$)
J kg^{-1} or J g^{-1} (specific energy of combustion, $\Delta_c u°$)

9. Enthalpy

Recommended reference material: 2,2,4-Trimethylpentane (C_8H_{18})
Range of variables: 298.15 K is the reference temperature normally employed
Physical state within the range: liquid
Contributors to the first version: G. T. Armstrong, J. D. Cox, O. Riedel
No revision made

Intended usage: Energies of combustion in oxygen of most compounds containing no elements other than C, H, O, N can be accurately measured with the aid of a static-bomb calorimeter (Refs. 1 to 3). The energy equivalent of the calorimeter will normally be established by combustion of thermochemical-standard benzoic acid. However, many workers like to check the accuracy of the benzoic acid calibration, and of the experimental procedure generally, by the combustion of a test material of known energy of combustion. 2,2,4-Trimethylpentane is recommended as a test material for the combustion calorimetry of liquid hydrocarbons that require encapsulation (in glass or plastic (Refs. 1 to 3)) before combustion. 2,2,4-Trimethylpentane therefore serves as a test material in the establishment of the calorific values of liquid gasoline fuels.

Sources of supply and/or methods of preparation: A suitable grade of 2,2,4-trimethylpentane certified for the specific energy of combustion is available from suppliers (C) and (S).

Pertinent physicochemical data: The value quoted by Cox and Pilcher (Ref. 1) for the standard molar enthalpy of combustion of 2,2,4-trimethylpentane is $-(5461.4 \pm 1.5)$ kJ mol^{-1} which applies to the reaction (at 298.15 K)

$$C_8H_{18}(l) + 12.5\ O_2(g) = 8\ CO_2(g) + 9\ H_2O(l) \qquad (9.5.11.1).$$

The corresponding value of the standard specific energy of combustion is -47712 J g^{-1}.

An account of the determination of the energy of combustion of a sample of 2,2,4-trimethylpentane issued by the Office of Standard Reference Materials, NBS, has been given by Armstrong (Ref. 4).

REFERENCES

1. Cox, J. D.; Pilcher, G. *Thermochemistry of Organic and Organometallic Compounds*, Academic Press: London, **1970**.
2. Rossini, F. D., editor, *Experimental Thermochemistry*, Vol. I., Wiley (Interscience): New York, **1956**.
3. Sunner, S.; Månsson, M., editors, *Combustion Calorimetry*. Pergamon: Oxford, **1979**.
4. Armstrong, G. T. *Determination of the Energy of Combustion of Standard Sample 2,2,4-Trimethylpentane 217b*, National Bureau of Standards: Washington, **1963**.

9.5.12. $\Delta_c U°$, α, α, α-Trifluorotoluene

Physical property: Energy of combustion
Units: J mol^{-1} or kJ mol^{-1} (molar energy of combustion, $\Delta_c U°$)
J kg^{-1} or J g^{-1} (specific energy of combustion, $\Delta_c u°$)
Recommended reference materials: α, α, α-Trifluorotoluene (C$_7$H$_5$F$_3$)
Range of variables: 298.15 K is the reference temperature normally employed
Physical state within the range: liquid
Contributors: J. D. Cox, A. J. Head

Intended usage: Energies of combustion in oxygen of most organic compounds containing fluorine can be accurately measured with the aid of a platinum-lined moving-bomb calorimeter (Refs. 1 to 3). Techniques involving the use of static-bomb calorimeters are less accurate. Sufficient water should be placed in the bomb initially so that the solution of hydrofluoric acid obtained after the combustion may have a concentration of not more than 5 mol dm^{-3}. The energy equivalent of a moving-bomb calorimeter will normally be established by combustion of thermochemical-standard benzoic acid. However, it is desirable to check the accuracy of the benzoic acid calibration, and the experimental procedure generally, by the combustion of a test material of known energy of combustion. Combustion reactions of fluorine compounds in which the atomic ratio of hydrogen to fluorine is greater than or equal to unity differ from those in which the atomic ratio of hydrogen to fluorine is less than unity. Thus the former compounds yield hydrogen fluoride as the sole fluorine-containing product whereas the latter compounds yield a mixture of hydrogen fluoride and carbon tetrafluoride. Cox proposed α, α, α-trifluorotoluene as a suitable reference material for the combustion of liquid C, H, F, (O, N) compounds containing moderate amounts of fluorine, which yield hydrogen fluoride as the sole fluorine-containing product and which need to be encapsulated before combustion (Ref. 4).

Sources of supply and/or methods of preparation: Good, Scott and Waddington showed that very pure material could be obtained from the commercial product by fractional distillation (Ref. 5).

Pertinent physicochemical data: Two separate determinations on material with mole fraction purity of 0.99999 have been made by Good and co-workers (Refs. 5, 6) using different methods of encapsulation (quartz ampoules, necessitating the application of a thermal correction for attack by hydrofluoric acid, and polyester bags). No carbon tetrafluoride was detected in the reaction products. Later, Kolesov *et al.* made measurements on a sample of mole fraction purity 0.99984 (containing water as the only contaminant) and based the results on the mass of carbon dioxide produced (Ref. 7). They used polyamide bags to encapsulate the sample for combustion and made two series of determinations using two different moving-bomb calorimeters. Only a trace of carbon tetrafluoride was detected in the reaction products. All four values were statistically consistent and yielded a weighted mean value of $-(23052.4 \pm 2.6)$J g^{-1} for the standard energy of the reaction (at 298.15 K)

$$C_7H_5F_3(l) + 7.5\ O_2(g) + 59\ H_2O(l) = 7\ CO_2(g) + 3\ [HF(20\ H_2O)] \qquad (9.5.12.1).$$

REFERENCES

1. Sunner, S.; Månsson, M., editors, *Combustion Calorimetry*, Pergamon: Oxford, **1979**.
2. Skinner, H. A., editor, *Experimental Thermochemistry*, Vol. II, Wiley (Interscience): New York, **1962**.
3. Cox, J. D.; Pilcher, G. *Thermochemistry of Organic and Organometallic Compounds*, Academic Press: London, **1970**.
4. Cox, J. D., Chapter 4 of Ref. 1.
5. Good, W. D.; Scott, D. W.; Waddington, G. *J. Phys. Chem.* **1956**, *60*, 1080.
6. Good, W. D.; Lacina, J. L.; DePrater, B. L.; McCullough, J. P. *J. Phys. Chem.* **1964**, *68*, 579.
7. Kolesov, V. P.; Ivanov, L. S.; Alekhin, S. P.; Skuratov, S. M. *Zh. fiz. Khim.* **1970**, *44*, 2956.

9.5.13. $\Delta_c U°$, Sulphur

Physical property: Energy of combustion in fluorine
Units: J mol^{-1} or kJ mol^{-1} (molar energy of combustion, $\Delta_c U°$)
J kg^{-1} or J g^{-1} (specific energy of combustion, $\Delta_c u°$).
Recommended reference material: Sulphur (rhombic)
Range of variables: 298.15 K is the temperature normally employed
Physical state within the range: solid
Contributors: J. D. Cox, A. J. Head

Intended usage: Combustion calorimetry in fluorine is a powerful method for determining enthalpies of formation of a wide range of inorganic substances, but it is a difficult technique. The restriction on materials of construction to those not attacked by fluorine, the need to use fluorine of higher purity than is normally available from commercial sources, and the safety requirements all make extra demands on the calorimetrist (Refs. 1, 2). Fluorine-combustion calorimeters using high pressure metal bombs are usually calibrated by means of the combustion of benzoic acid in oxygen. The considerable difference between the conditions used in calibration and in measurement makes the use of reference materials for combustion in fluorine particularly desirable. Sulphur has been proposed as a reference material for the combustion of substances which ignite spontaneously in fluorine and therefore require the use of a two-compartment bomb (Refs. 1, 3).

Sources of supply and/or methods of preparation: The preparation of a highly-pure sample (mole fraction purity 0.99999) of α, rhombic sulphur from commercial roll sulphur has been described by Murphy *et al.* (Ref. 4). Much commercial material is of good quality and may only require crystallization from carbon disulphide (to ensure that only the α, rhombic form is present) to yield a sample of sufficient purity.

Pertinent physicochemical data: Three values have been reported for the standard energy of the reaction (at 298.15 K):

$$S(\alpha, \text{rhombic}) + 3\ F_2(g) = SF_6(g) \tag{9.5.13.1}$$

determined by fluorine combustion calorimetry using metal bombs: -1214.3 kJ mol^{-1} by Schröder and Sieben (Ref. 5), $-(1212.9 \pm 1.3)$ kJ mol^{-1} by Leonidov et al. (Ref. 6), and $-(1216.2 \pm 0.4)$ kJ mol^{-1} by O'Hare (Ref. 8). In addition, Hubbard et al. (Ref. 1) report an unpublished value of $-(1215.6 \pm 1.7)$ kJ mol^{-1} obtained by Gross and Hayman using glass reaction vessels. The value reported by O'Hare is based on a recalculation of a previous result (Ref. 7) and a new determination, which, however, failed to resolve the discrepancy between his results and that of Leonidov. (All the results quoted are based on the value 32.066 g mol^{-1} for the molar mass of S). We recommend the value $-(37915 \pm 90)$ J g^{-1} for the specific energy of combustion of sulphur in fluorine, where the uncertainty interval has been increased to include all published values.

REFERENCES

1. Hubbard, W. N.; Johnson, G. K.; Leonidov, V. Ya. Chapter 12 of *Combustion Calorimetry*, Sunner, S.; Månsson, M., editors, Pergamon: Oxford, **1979**.
2. Hubbard, W. N. Chapter 6 of *Experimental Thermochemistry*, Vol. II, Skinner, H. A., editor, Wiley (Interscience): New York, **1962**.
3. Cox, J. D., Chapter 4 of Ref. 1.
4. Murphy, T. J.; Clabaugh, W. S.; Gilchrist, R. *J. Res. Nat. Bur. Stand.* **1960**, *64A*, 355.
5. Schröder, J.; Sieben, F. *J. Chem. Ber.* **1970**, *103*, 76.
6. Leonidov, V. Ya.; Pervov, V. S.; Gaisinskaya, O. M.; Klyuev, L. I. *Dokl. Akad. Nauk SSSR, Fiz. Khim.* **1973**, *211*, 901.
7. O'Hare, P. A. G.; Settle, J. L.; Hubbard, W. N. *Trans. Faraday Soc.* **1966**, *62*, 558.
8. O'Hare, P. A. G. *J. Chem. Thermodynamics* **1985**, *17*, 349.

9.5.14. $\Delta_c U^\circ$, Tungsten

Physical property: Energy of combustion in fluorine
Units: J mol^{-1} or kJ mol^{-1} (molar energy of combustion, $\Delta_c U^\circ$)
J kg^{-1} or J g^{-1} (specific energy of combustion, $\Delta_c u^\circ$)
Recommended reference material: Tungsten
Range of variables: 298.15 K is the temperature normally employed.
Physical state within the range: Solid
Contributors: J. D. Cox, A. J. Head

Intended usage: Combustion calorimetry in fluorine is a powerful method for determining enthalpies of formation of a wide range of inorganic substances, but it is a difficult technique.

9. Enthalpy

The restriction on materials of construction to those not attacked by fluorine, the need to use fluorine of higher purity than is normally available from commercial sources, and the safety requirements all make extra demands on the calorimetrist (Refs. 1, 2). Fluorine-combustion calorimeters using high-pressure metal bombs are usually calibrated by means of the combustion of benzoic acid in oxygen. The considerable difference between the conditions used in calibration and in measurement makes the use of reference materials for combustion in fluorine particularly desirable. Tungsten has been proposed as a suitable substance: in the massive state it is inert to highly purified fluorine at room temperature but undergoes smooth and virtually complete conversion to tungsten hexafluoride when heated (Refs. 1, 3). Tungsten is suitable, therefore, for use in one- or two-compartment bombs, and is also useful as an auxiliary material for the combustion of other substances in fluorine.

Sources of supply and/or methods of preparation: Suitable tungsten sheet and wire of at least 0.9995 mass fraction purity can be obtained from many commercial suppliers of pure metals, *e.g.* Supplier (F). For work of the highest accuracy it is important for complete analytical data to be provided so that thermal corrections for the combustion of impurities can be calculated.

Pertinent physicochemical data: The two principal determinations, where both the tungsten and the fluorine were of high purity, gave results in close agreement for the energy of the reaction (at 298.15 K)

$$W(s) + 3 F_2(g) = WF_6(g) \tag{9.5.14.1}$$

viz. $\Delta_c U° = -(1716.8 \pm 1.7)$ kJ mol^{-1} (Ref. 4) and $-(1716.1 \pm 0.8)$ kJ mol^{-1} (Ref. 5), the molar mass of tungsten being taken as 183.85 g mol^{-1}. A third determination, where the fluorine contained one per cent of oxygen plus nitrogen (Ref. 6), yielded a result consistent with the first two, $-(1717.6 \pm 2.1)$ kJ mol^{-1}. The weighted mean value $\Delta_c u°(298.15 \text{ K}) = -(9335.9 \pm 3.8)$ J g^{-1} is calculated from these three results as the recommended value for the specific energy of combustion of tungsten in fluorine.

REFERENCES

1. Hubbard, W. N.; Johnson, G. K.; Leonidov, V. Ya. Chapter 12 of *Combustion Calorimetry*, Sunner, S.; Månsson, M., editors, Pergamon: Oxford, **1979**.
2. Hubbard, W. N. Chapter 6 of *Experimental Thermochemistry*. Vol. II, Skinner, H. A., editor, Wiley (Interscience): New York, **1962**.
3. Cox, J. D., Chapter 4 of Ref. 1.
4. O'Hare, P. A. G.; Hubbard, W. N. *J. Phys. Chem.* **1966**, *70*, 3353.
5. Schröder, J.; Sieben, F. J. *Chem. Ber.* **1970**, *103*, 76.
6. Leonidov, V. Ya.; Pervov, V. S.; Klyuev, L. I.; Gaisinskaya, O. M.; Medvedev, V. A.; Nikolaev, N. S. *Dokl. Akad. Nauk SSSR, Ser. Khim.* **1972**, *205*, 349.

9.6. Contributors

Yu. I. Aleksandrov,
VNIIM,
19 Moskovskii Prospekt,
Leningrad 198005 (USSR).

A. Cezairliyan,
National Bureau of Standards,
US Department of Commerce,
Gaithersburg, MD 20899 (USA).

S. S. Chang,
National Bureau of Standards,
US Department of Commerce,
Gaithersburg, MD 20899 (USA).

J. D. Cox,
Division of Quantum Metrology,
National Physical Laboratory,
Teddington, Middlesex TW11 0LW (UK).

D. Ditmars,
National Bureau of Standards,
US Department of Commerce,
Gaithersburg, MD 20899 (USA).

E. S. Domalski,
Chemical Thermodynamics Division,
National Bureau of Standards,
US Department of Commerce,
Gaithersburg, MD 20899 (USA).

J. W. Fisher,
National Research Council,
Division of Physics,
Ottawa K1A OR6 (Canada).

R. N. Goldberg,
Chemical Thermodynamics Division,
National Bureau of Standards,
US Department of Commerce,
Gaithersburg, MD 20899 (USA).

A. J. Head,
Division of Quantum Metrology,
National Physical Laboratory,
Teddington, Middlesex TW11 0LW (UK).

R. T. Jacobsen,
Mechanical Engineering Department,
University of Idaho,
Moscow, Idaho 83843 (USA).

J. Lielmezs,
Chemical Engineering Department,
University of British Columbia,
Vancouver, B.C. V6T 1W5 (Canada).

K. N. Marsh,
Thermodynamics Research Center,
Texas A&M University,
College Station, TX 77843 (USA).

D. L. Martin,
National Research Council,
Division of Physics,
Ottawa K1A OR6 (Canada).

J. F. Martin,
Department of Metallurgy,
University of Strathclyde,
Glasgow G1 1XN (UK).

C. Mosselman,
Department of Chemistry,
Free University,
De Boelelaan 1083,
1081 HV Amsterdam (Netherlands).

P. A. G. O'Hare,
Chemical Technology Division,
Argonne National Laboratory,
9700 South Cass Avenue,
Argonne, IL 60439 (USA).

B. N. Oleinik,
VNIIM,
19 Moskovskii Prospekt,
Leningrad 198005 (USSR).

M. I. Paz-Andrade,
Departmento de Fisica Fundamental,
Universidad de Santiago,
Santiago de Compostella (Spain)

G. Pilcher,
Department of Chemistry,
University of Manchester,
Manchester M13 9PL (UK).

J. Rogez,
Centre de Thermodynamique et de
Microcalorimétrie,
Centre National de la Recherche Scientifique,
26, rue du 141è R.I.A.,
13003 Marseille (France).

K. N. Roy,
Bhabha Atomic Research Centre,
Radiochemistry Division,
Trombay, Bombay 400085 (India).

R. Sabbah,
Centre de Thermodynamique et de
  Microcalorimétrie,
Centre National de la Recherche Scientifique,
26, rue du 141è R.I.A.,
13003 Marseille (France).

R. L. Snowdon,
National Research Council,
Division of Physics,
Ottawa K1A OR6 (Canada).

D. D. Sood,
Bhabha Atomic Research Centre,
Radiochemistry Divison,
Trombay, Bombay 400085 (India).

C. E. Vanderzee,
Department of Chemistry,
University of Nebraska,
Lincoln, NE 68588 (USA).

V. Venugopal,
Bhabha Atomic Research Centre,
Radiochemistry Divison,
Trombay, Bombay 400085 (India).

L. A. Weber,
National Bureau of Standards,
US Department of Commerce,
Gaithersburg, MD 20899 (USA).

9.7. List of suppliers

A. Agate Products Ltd,
2/4 Quintin Avenue,
Merton Park, London SW20 8LD (UK).

B. All-Union Scientific Research Institute of Metrology,
Sverdlovsk Branch,
Krasnoarmeiskaya 2-a, Sverdlovsk (USSR).

C. American Petroleum Institute,
Standard Reference Materials,
Carnegie-Mellon University,
Schenly Park,
Pittsburgh, PA 15213 (USA).

D. Bureau of Analysed Samples Ltd,
Newham Hall,
Newby, Middlesbrough,
Cleveland TS8 9EA (UK).

E. Dow Chemical,
Dow Center,
Midland, MI 48640 (USA).

F. Goodfellow Metals Ltd.,
Cambridge Science Park,
Milton Road,
Cambridge CB4 4DJ (UK).

G. B.F. Goodrich Chemical,
6100 Oak Tree Boulevard,
Independence, OH 44121 (USA).

H. Goodyear Chemical,
1144 E. Market Street,
Akron, OH 44316 (USA).

I. Hoechst A.G.,
6230 Frankfurt (M) 80 (Germany)

J. Institute of Gas Technology,
Illinois Institute of Technology,
3424 South State Street,
Chicago, IL 60616 (USA).

K. Johnson Matthey Chemical Ltd.,
74 Hatton Garden,
London EC1P 1AE (UK).

L. NV Kawacki-Billiton Metaalindustrie,
P.O. Box 38,
Arnhem (Holland).

M. Materials Research Corporation,
Route 303,
Orangeburg, NY 10962 (USA).

N. Matheson Gas Products,
932 Paterson Plank Rd.,
New Jersey 07073 (USA).

O. Metal Crystals Ltd.,
Button End,
Harston, Cambridge (UK).

P. Ministry of International Trade and Indus
National Chemical Laboratory for Industr
1-1 Honmachi Shibuya-ku,
Tokyo 151 (Japan).

Q. Monomer-Polymer Lab.,
5000 Langdon Street,
Philadelphia, PA 19124 (USA).

R. Montedison S.p.A.,
Largo G. Donegani,
20121 Milano (Italy).

S. National Bureau of Standards,
Office of Standard Reference Materials,
Gaithersburg, MD 20899 (USA).

9. Enthalpy

T. National Physical Laboratory,
Office of Reference Materials,
Division of Quantum Metrology,
Teddington, Middlesex TW11 0LW (UK).

U. Phillips Petroleum Co.,
Bartlesville, OK 74004 (USA).

V. Polysar Ltd.,
Sarnia, Ontario N7T 7M2 (Canada).

W. Polysciences, Inc.,
Warrington, PA 18976 (USA).

X. Scientific Polymer Products, Inc.,
99 Commercial Street,
Webster, NY 14580 (USA).

General suppliers

Aldrich Chemical Co.,
P.O. Box 355,
Milwaukee, WI 53201 (USA).

Alfa Technical Service,
Alfa Division of Ventron Corporation,
152 Andover Street,
Danvers, MA 01923 (USA).

J. J. Baker Chemical Co.,
222 Red School Lane,
Philipsburg, NJ 08865 (USA).

B. D. H. Chemicals Ltd.,
Broom Road,
Parkstone, Poole BH12 4NN (UK).

Carlo Erba,
Division Analytique,
Via C. Imbonati 24, 20159 Milani (Italy).

Fluka A. G.,
Chemische Fabrik,
CH-9470 Buchs (Switzerland).

Koch-Light,
37 Hollands Road,
Haverhill,
Suffolk CB9 8PU (UK).

10

SECTION: THERMAL CONDUCTIVITY

COLLATOR: B. LE NEINDRE

CONTENTS:

10.1. Introduction

10.2. Methods of measurement
- 10.2.1. Experimental methods
- 10.2.2. Convection
- 10.2.3. Radiation
- 10.2.4. Accommodation coefficient
- 10.2.5. Other corrections
- 10.2.6. Criteria for selecting reference fluids

10.3. Reference materials for the thermal conductivity of gases
- 10.3.1. Helium
- 10.3.2. Neon
- 10.3.3. Argon
- 10.3.4. Krypton
- 10.3.5. Xenon

10.4. Reference materials for the thermal conductivity of liquids
- 10.4.1. Water
- 10.4.2. Toluene
- 10.4.3. Heptane
- 10.4.4. Dimethylphthalate

10.5. Reference materials for the thermal conductivity under pressure
    10.5.1. Propane

10.6. Reference materials for the thermal conductivities of solids–high thermal conductivity materials
    10.6.1. Aluminium
    10.6.2. Copper
    10.6.3. Iron
    10.6.4. Tungsten
    10.6.5. Stainless Steel
    10.6.6. Graphite

10.7. Reference materials for the thermal conductivity of solids–low thermal conductivity materials
    10.7.1. Glass fibreboard
    10.7.2. Resin-bonded glass fibreboard
    10.7.3. Blanket

10.8. Contributors

10.9. List of suppliers

10.1. Introduction

IUPAC (Ref. 1) has recommended the symbol λ (in preference to k) for thermal conductivity. The S.I. unit for the thermal conductivity coefficient is watt per meter per Kelvin (W m^{-1} K^{-1}). Several other units have been used in the past and the relationships between these units and S.I. units is given in reference 2. The thermal conductivity coefficient is defined by Fourier's law as the time rate of transfer of energy by conduction through unit thickness across unit area for unit difference in temperature. A set of recommendations on reference materials for thermal conductivities were published in 1981 (Ref. 3).

Reference materials for the thermal conductivity of fluids should cover a wide range of temperature and pressure. The thermal conductivity of most fluids are of the same order of magnitude. Typically the thermal conductivity coefficients of gases vary from 0.020 to 0.100 W m^{-1} K^{-1} and those of liquids from 0.100 to 0.200 W m^{-1} K^{-1}. However, in the critical region, a strong enhancement of the thermal conductivity occurs.

In the experimental determination of the thermal conductivity of fluids, conditions are required under which the energy transferred by molecular interaction can be determined. The effects of other coexistent energy transport processes such as thermal radiation through the fluid layer or, in the case of partially absorbing substances, within the fluid itself, must be known and accounted for.

Likewise, natural convection, which is caused by the bulk movement of the fluid enclosed within the cell, must be minimized by choosing appropriate operating conditions. With the correct conditions, which involves relatively small temperature differences and simple geometric configurations of the solid surfaces surrounding the fluid layer, simple solutions of the Laplace equation exist which allow the absolute determination of the thermal conductivity from the measurement of the energy flow through the fluid film, the temperature difference between the bounding walls, and the linear dimensions of the apparatus.

Reliable absolute measurements of thermal conductivity require a carefully constructed apparatus and the precise and accurate measurement of the various quantities given above. Further, equally precise determinations, or proof of their absence, of possible secondary energy transfers are also necessary. In many cases it is difficult to justify the time consuming procedures required for an absolute determination when making routine or occasional measurements.

The thermal conductivity of a fluid may be obtained with a relatively simple apparatus by comparing the observed measurement with that obtained with a reference material. The linear dimensions of a cell can be compounded into a single term called the 'geometric cell constant' but when calibrating the conductivity cell with materials of known thermal conductivity, this 'geometric cell constant' is combined with the effects of energy losses or gains into another constant called the 'cell constant'. It is desirable that the chosen reference material should have a thermal conductivity similar to that of the substance to be examined, and that the rate of energy dissipation be about the same during both the calibration and the actual experiment.

The precision attainable with relative measurements is often quite adequate for technical requirements, but successful application of the relative measurement technique is frequently impeded by the lack of suitable reference materials. Further, even for absolute determinations, the occasional need arises to test some specific property of the apparatus, to check the experimental procedure employed, or to check the proper performance of the entire measuring system under the operating conditions. This can most readily be done using critically selected reference materials. It is the purpose of this report to examine a number of substances which have been proposed as reference materials for the calibration and the testing of cells to be used for the measurement of thermal conductivity and to present for these substances a set of results which have been critically evaluated.

In a thermal conductivity cell, the total heat output per unit time of the emitter Q_T is transferred through the fluid by conduction (Q_λ), free convection (Q_c) and radiation (Q_r) as well as by conduction along the solid elements (Q_s) such as centering devices in the case of steady state apparatus or the wire support in non-steady state devices.

$$Q_T = Q_\lambda + Q_c + Q_r + Q_s \tag{10.1.1}$$

This expression can be written in terms of equivalent conductivities

$$\lambda_T = \lambda_\lambda + \lambda_c + \lambda_r + \lambda_s = K Q_T/\Delta T \tag{10.1.2}$$

where K is the cell constant and ΔT the temperature difference.

Reference materials can be used to determine the cell constant K. It is often assumed that λ_c, λ_r, and λ_s are negligible. However, these contributions can be of the order of ten percent of λ_λ. The neglect of these corrections explains why differences between results from different authors are often quite large. These corrections will normally be different for each fluid under study.

To isolate the pure conductive mode requires either that the other heat transfers mechanisms be negligible or that a full mathematical treatment of the various heat transfer mechanisms be available to provide an estimation of the necessary corrections.

The theoretical evaluation of λ_c, λ_r, and λ_s is complicated and reference fluids can be used to determine the importance of these various heat transfer contributions. For example, λ_s can be determined from the thermal conductivity of noble gases (Ref. 4).

When a non-transparent fluid such as toluene is used for calibration, erroneous results will be obtained if the radiation contribution to the thermal conductivity of the test fluid is different. It is recommended that at least two reference fluids having different radiative properties be used to calibrate thermal conductivity cells.

10.2. Methods of measurement

10.2.1. Experimental methods

The majority of thermal conductivity measurements on fluids have been made by static methods using either parallel plates (Ref. 5) or coaxial cylinder (Refs. 6 to 8). In recent years nonsteady-state techniques have improved and the transient hot-wire method has been extensively used to measure the thermal conductivity of fluids. This technique has been used with both gases and liquids; however, it has been reported that the technique is actually unreliable at high temperature and in the critical region, where the thermal conductivity shows a pronounced enhancement. The working equations for the transient hot wire technique are well established and details may be found elsewhere (Refs. 9 to 13).

10.2.2. Convection

In the steady-state heat flow method, the convection is not always dissociated from the purely conductive process. The amount of heat transferred by free convection is determined from the Rayleigh number Ra which is the product of Grashof number Gr and Prandtl number Pr.

It has been established that free convection heat transfer is negligible when Ra is below 600 in a cylindrical fluid layer and below 1700 in a horizontal fluid layer heated from below. The Rayleigh number is given by:

$$Ra = Gr \times Pr = g\alpha_p \Delta T \rho^2 d^3 C_p / \eta \lambda \qquad (10.2.2.1)$$

where g is the acceleration due to gravity, $\alpha_p = -(1/\rho)(\partial \rho / \partial T)_p$ is the coefficient of thermal expansion, ΔT is the temperature gradient in the fluid layer, ρ is the density of the fluid, d is the thickness of the fluid layer, C_p is the heat capacity at constant pressure, and η is the dynamic viscosity.

The absence of convection can be checked by performing experiments at different temperature gradients. Near the critical point, where C_p diverges, convection effects can be very important. In general, the effect of convection can be rendered small by the use of small gaps ($0.2 < d < 0.4$ mm) and appropriate designs of parallel plates (Ref. 6) or coaxial cylinders (Ref. 14).

In the transient hot-wire method the onset of convection can be determined directly during the experiment. Convection is avoided by making the time scale of the measurement small enough so that the measurement are completed before the onset of convection (Refs. 9 to 12).

10.2.3. Radiation

It has been observed experimentally that within absorbing liquids and semi-transparent solids a thermal radiation heat transfer occurs which depends on the physical conditions and the bounding surfaces (Refs. 15, 16).

The contribution and effect of radiative heat transfer in thermal conductivity measurements on absorbing fluids has been subject to discussion for a long time (Refs. 17 to 20). In different instruments and different fluids the radiative contribution has been estimated to be between 2 and 40 per cent of the thermal conductivity and the correction can be of either sign (Refs. 21 to 23).

A rigorous analytical treatment of simultaneous conduction and radiation is difficult due to the coupling between the equations of energy and radiative heat transfer. Viskanta and Grosh (Refs. 18, 24) and Viskanta (Ref. 25) have studied the interaction between radiation and conduction and an extensive bibliography on the subject has been reported by Viskanta and Grosh (Ref. 26). Lick (Refs. 27, 28) has used a number of approximations such as the boundary layer type of analysis, the diffusion approximation, the picket fence model, and the linearization of the temperature gradient in order to solve the equations of simultaneous conduction and radiation in both steady-state and transient energy transfer.

Greif (Ref. 29) used the picket fence model to solve for the temperature and frequency dependent properties. The range of validity of a number of treatments has been reported by Wang and Tien (Refs. 30, 31) and Chang (Ref. 32) has used an approximate technique to obtain the radiation contribution.

Hazzak and Beck (Ref. 33) have used a differential method and Lii and Ozisik (Ref. 34) have used the normal-mode expansion technique to solve the unsteady state problem of simultaneous conduction and radiation for an absorbing, emitting isotropically scattering parallel slab. The interaction of radiation with conduction and convective transfer has been considered by Ozisik (Ref. 35). The theoretical aspects of the measurement of the thermal conductivity of semi-transparent materials, including solid materials, has been treated by Men (Refs. 36 to 40).

In most engineering applications the net heat flux in the medium is obtained by summing the conductive and radiative heat fluxes. When the fluid is not contributing to the radiation process through absorption, the radiative contribution is easy to estimate. If the fluid is completely opaque, the radiant heat loss is zero. If the fluid is completely transparent, the radiant heat loss can be calculated from

$$Q_r = 4n^2 \sigma S k(\varepsilon) \Delta T T^3 \qquad (10.2.3.1)$$

where n is the refractive index of the sample, σ is the Stefan Boltzmann constant, S is the effective heat flow area, and $k(\varepsilon)$ is a geometrical factor for the emissivities of the surfaces.

When the fluid is partially transparent, which is the general case, the process is described by a complex integro-differential equation. A number of solutions for steady-state instruments have been published (Refs. 18, 21 to 23). The solutions relate the heat transfer by radiation to the optical characteristics of the fluid (refractive index and infrared absorption spectrum) measured in the appropriate temperature ranges. Assuming that the temperature drop in the liquid layer is small and that the emitting bounding surfaces are gray, the radiative component will be a function of the wavenumber ν, the refractive index η, the optical thickness τ, the emissivity of the bounding surface $\varepsilon(\nu)$, the temperature distribution in the layer, which is generally considered to be linear, and of the spectral absorption $k(\nu)$.

To calculate the radiative component λ_r, several approximation are used. The refractive index must be known very accurately as its value is squared in the equation. The refractive index of weakly absorbing liquids is a function of wavenumber and generally a mean refractive index value is used. This value is determined by integrating over all frequencies in the infrared region. Moreover, the refractive index is a function of the density of the liquid. The emissivity is assumed to be constant, but generally it is dependent on the direction of emission and on the frequency of the radiation. The temperature gradient is assumed to be constant. In general, the spectral absorption coefficient is replaced by a mean value which depends on the thickness and on the temperature of the liquid layer.

Recently, an experimental and theoretical study of the influence of radiation on the measured thermal conductivities of liquids has been reported by Fischer *et al.* (Refs. 41 to 43).

10. Thermal Conductivity

Experiments performed in the temperature range from 253 K to 473 K using a horizontal steady-state concentric cylinder, showed that the radiation contribution to the thermal conductivity depended on the layer thickness, the temperature, and the liquid properties under study. Thermal conductivity values, free of radiation, were obtained for several liquids. The radiation free thermal conductivity values were fitted to the equation

$$\lambda/(\text{W m}^{-1} \text{ K}^{-1}) = a + bx + cx^2 \tag{10.2.3.2}$$

where $x = t/°\text{C}$ and the coefficients are given in table 10.2.3.1.

Table 10.2.3.1. Constants for equation 10.2.3.2

Substance	a	$10^4 b$	$10^7 c$	$\dfrac{\text{Range}}{°\text{C}}$
Ethanol	0.16825	−2.484	2.7996	−20 to 200
Methanol	0.20244	−2.8555	3.8831	−20 to 200
Toluene	0.13582	−2.9764	1.4995	−20 to 200
Benzene	0.14943	−3.6519	2.4241	6 to 200
Heptane	0.12738	−3.1053	2.1809	−20 to 200
R.113	0.076515	−2.0401	0.62106	−20 to 155
R.10	0.10154	−2.3791	0.65765	−20 to 155
R.11	0.090256	−2.5057	0.28750	−20 to 110
R.12	0.07281	−2.92		−20 to 65
R.22	0.09237	−4.04		−20 to 25

A complete numerical solution of the conduction-radiation equations for a transient hot-wire instrument required to determine the corrections to be applied to experimental data to yield radiation free thermal conductivities has been developed by Menashe and Wakeham (Ref. 44). The evaluation of this correction appears to be complicated and it is not practicable to apply it on a routine basis. Furthermore, many of the optical properties of the fluid and of its bounding surfaces, which are necessary to calculate the correction, have not been measured under the required conditions.

When applied to experimental data, they estimated that the uncertainty in the value of this correction arises from errors in the determination of the mean extinction coefficient and the numerical evaluation of the correction may be in error by as much as 20%. However, the correction to experimental data necessary to account for the effects of radiation absorption in the transient hot-wire method is generally less than 0.5 per cent (Ref. 45).

10.2.4. Accommodation coefficient

If the molecular mean free path of the gas in a cell approaches the thickness of the gas layer, a temperature discontinuity will occur at the bounding surfaces of the cell. This phenomenon can be accounted for by assuming that the temperature difference is that of the bounding solid surfaces and that the energy is transferred through a uniform temperature gradient. Then temperature jump distances g_1 and g_2 associated with surfaces 1 and 2 have to be added to the gap thickness:

$$Q = S\lambda\Delta T/(d + g_1 + g_2) \qquad (10.2.4.1),$$

where Q is the energy transferred over a surface S between two infinite parallel plates. The temperatures jump distance depends on the thermal accommodation coefficient α and is related to the mean free path L by

$$g = (2 - \alpha)4C\lambda L/[\alpha(\gamma + 1)\eta C_v] \qquad (10.2.4.2)$$

where γ is the ratio of the heat capacity, η is the viscosity, C_v is the heat capacity at constant volume, and C is a constant having a value of 0.49. The same equation is approximately correct for a coaxial cylinder cell. The effect of the temperature jump can be evaluated or rendered small, by working in an adequate pressure range or by using an adequate gap. This effect is most important when light gases, like helium and neon, are used as reference materials at atmospheric pressure. Further information to the corrections for the accommodation effect to be applied in coaxial cells have been reported (Refs. 46, 47). In the transient hot-wire system, it is possible to account for Knudsen effects (Ref. 9).

10.2.5. Other corrections

In a steady-state apparatus, a correction λ_s has to be made for the heat losses by the solid parts such as centering pins, electrical wires or thermocouples.

In hot-wire devices, corrections have to be made for the difference of resistance per unit length of the wires if two wires are used (Ref. 10) or for the potential lead conduction if a single wire is employed. The effect of wire tension (Ref. 11) and the effect of the distribution of measurements on the reference temperature must also be considered (Ref. 48). For electrically conducting fluids, the wire may be coated with insulating material (Refs. 49 to 51) (for which an additional heat capacity correction has to be made). Alternatively, a bare wire may be used with an alternating current.

10.2.6. Criteria for selecting reference fluids

In 1961, Ziebland (Ref. 52) and later Nieto de Castro and Wakeham (Ref. 53) suggested the following criteria for selecting reference materials for fluid thermal conductivities.

(i) The temperature interval between the freezing point and the normal boiling point *i.e.* the extent of the liquid range, must be large and include within its range some thermometric fixed point.

(ii) The liquid should be non-toxic and non-corrosive with respect to the usual engineering materials.

(iii) It should be obtainable at reasonably low coast and guaranteed high purity.

(iv) A series of liquids covering as wide a range of thermal conductivity as possible should be selected.

(v) Reliable values of liquid densities for the saturated and compressed liquids should be available in the literature, allowing reliable extrapolation of high pressure data to the saturation line.

(vi) The contribution of the radiation to the thermal conductivity measurements should be small or easy to estimate from the working equation of the instrument.

Gases

Noble gases are recommended as reference materials for gas phase measurements. They cover the required range of thermal conductivities and accurate values are available. The thermal conductivity of a pure fluid may be divided into two contributions: one arising from the transfer of energy from purely collisional or translational effects, λ', and the other from the transfer of energy through internal degrees of freedom, λ''. The first contribution, λ', can be calculated from the Chapman-Enskog equation which relates thermal conductivity to viscosity in the zero density limit. In the case of the noble gases the second contribution is zero so λ is given by

$$\lambda = \lambda' = 2.5\eta C_v f_l = (7.5/2) R \eta f_l \qquad (10.2.6.1)$$

where f_l is a small correcting factor.

There are several measurements of the thermal conductivity of noble gases on which standards could be based (Refs. 4, 54). However, as there is an exact relationship between the thermal conductivity and the viscosity of dilute monatomic gases, it is appropriate to have a set of reference thermal conductivity data which is consistent with viscosity data. Moreover, the viscosity of noble gases has been measured over a wider temperature range with greater accuracy than thermal conductivity. In the region 300 – 900 K, the viscosity data have an uncertainty which rises from ±0.2% to ±0.5%. Outside of this range, the viscosity is known with a poorer precision but still not worse than ±1.5%. The uncertainty in thermal conductivity data calculated from viscosity data would be similar. By combining viscosity data with spectroscopic data and molecular beam measurements through an intermolecular pair potential, it is possible to extend the temperature range. This is probably the best method for calculating the thermal conductivity of the noble gases. Other methods, including those based upon direct measurements, are not as accurate. The method discussed above is not applicable to molecular gases and only direct measurements can be used to obtain the thermal conductivity of these gases. Hence, the uncertainty of the data for the thermal conductivity of any molecular gas selected as a reference material will be larger. Moreover, noble gases of high purity are commercially available.

If helium or neon are used to calibrate a cell, it is better that the calibration be made at pressures between 0.1 and 3 MPa to avoid corrections due to the accommodation effect.

Liquids

In the case of liquids, the choice of standards of thermal conductivity is more difficult because radiation has a much more significant effect and because results can be obtained only from direct measurements and cannot be derived from measurements of other properties. For radiation occurring in a steady-state instrument, the solution for the simultaneous conduction–radiation equation for absorbing liquids has recently been presented by Fischer and coworkers (Refs. 41 to 43). The early work of Fritz and Poltz (Ref. 32) gave rise to controversy because the experimental observations could be attributed to either the onset of convection or to the influence of radiation. Fischer *et al.* investigated the radiation–conduction heat transfer coupling of liquids in the annular space between two concentric cylinders rotating about their horizontal axis. For a steady-state heat flux conducted across the liquid in the radial direction from the inner to the outer wall of the annulus, convection is suppressed when the direction of the density gradient coincides with the direction of centrifugal acceleration, assuming that there is a positive volume expansivity. In some of their measurements, the contribution of radiation to the measured thermal conductivity was as much as 40%. They observed that the effect of radiation decreases with decreasing distance between the plates. If the sample layer in a steady-state thermal conductivity cell is thick ($d > 0.3$ mm), a completely opaque liquid could be a useful standard, but an incompletely opaque liquid may not be, even if the absorption is strong. For this reason, it would be preferable to use a standard material which is almost transparent rather than one which is opaque. One of the materials which best satisfies this requirement is carbon disulphide, but very few thermal conductivity measurements have been made on this compound, and its boiling point is low (46.3 °C). To satisfy some of the requirements for good thermal conductivity standards which were listed before, the following liquids were selected as standards reference materials: water, toluene, heptane and dimethylphthalate.

Solids

Solid reference materials must have the following characteristics: reproducible material properties, long term stability, and suitability for the different specific measurements techniques.

Reference materials for the thermal conductivity of solids can be divided in three groups.

10. Thermal Conductivity

(i) Materials of high thermal conductivity ($100 < \lambda < 25000$ W m^{-1} K^{-1}) consist of metals and alloys. Up to now, a careful analysis of the thermal conductivity of aluminium, copper, electrolytic iron, and tungsten has been reported and it is recommended to use these metals as thermal conductivity standards. However, some other material such as graphite (Poco AXM-5Q1) or austenitic stainless steel could possibly be used as reference materials.

(ii) Reference materials for use in the low thermal conductivity range ($0.5 < \lambda < 30$ W m^{-1} K^{-1}) have not been well characterized. Measurements on cordierite ceramic (Pyroceram 9606) ($\lambda \sim 3.61$ W m^{-1} K^{-1}) to test its suitability as a reference materials for thermal conductivity measurements are underway in several laboratories.

(iii) For the very low thermal conductivity range ($0.001 < \lambda < 0.1$ W m^{-1} K^{-1}) (the thermal insulation range) extensive international programs have been developed to define thermal conductivity standards. Fibrous glass with a well defined density has been selected as a reference material.

Several reference materials have been recommended by the National Bureau of Standards.

High thermal conductivity materials

A CODATA task group on "Thermophysical Properties of Solids" has studied experimental methods and standard reference materials having high thermal conductivity ($100 < \lambda < 25000$ W m^{-1} K^{-1}). The materials which were selected by this task group were copper, electrolytic iron, tungsten, and platinum. The recommendations are based on their values (Ref. 55).

Thermal conductivity standards must represent the intrinsic thermal conductivity. There are several effects which prevent an accurate measurement of this quantity such as radiation heat losses at high temperature, chemical impurities, physical defects, size effects, magnetic field effects, and thermal or mechanical history. The imperfection concentration can often be determined by measurement of the electrical resistivity which can be related to the thermal conductivity by the Wiedeman-Franz-Lorenz law.

The thermal conductivity of a metal is generally the sum of two contributions

$$\lambda = \lambda_e + \lambda_l \qquad (10.2.6.2)$$

where λ_e is the electronic thermal conductivity component, and λ_l is the lattice thermal conductivity component.

The electronic component is related to the thermal energy of the electrons and the lattice component is related to the energy of the phonons. In pure metals, the electronic component is large and represents 80 to 95% of the thermal conductivity. Theory provides an accurate value for the low temperature dependence of the electronic component. At low temperatures, the electronic term is sensitive mainly to the scattering of electrons by phonons which leads to a resistive term R_p roughly proportional to T^2 and to the interaction of electrons with lattice defects which leads to a resistive term R_{ld} inversely proportional to the temperature.

$$\lambda = (R_p + R_{ld})^{-1} = [AT^n + (B/T)]^{-1} \qquad (10.2.6.3)$$

where n is 2, A is a constant which characterizes the metal base, and B is a constant which characterizes the concentration and type of lattice defects.

However, a plot of the experimental data as a function of temperature up to about 1.5 times the temperature at which the maximum in thermal conductivity occurs shows that A is weakly dependent of the lattice imperfections and that n is generally larger than 2 (Refs. 56, 57). At higher temperatures (> 40 K for the metals considered), the thermal conductivity at constant volume decreases slowly and approach a constant value.

In their analysis of the thermal conductivity of aluminium, copper, iron, and tungsten, Hust, White, and Laubitz (Ref. 55) derived the following equation to represent the thermal conductivity of these metals:

$$\lambda/(\text{W m}^{-1}\ \text{K}^{-1}) = (R_p + R_1 + R_{lp})^{-1} \qquad (10.2.6.4)$$

where

$$R_p = B/(T/\text{K})$$
$$R_1 = A_1 T^{A_2}/(1 + a \exp b) + R_c$$
$$a = A_1 A_3 (T/\text{K})^{(A_2 + A_4)}$$
$$b = [-A_5/(T/\text{K})]^{A_6}$$
$$R_{lp} = A_7 R_p R_1/(R_p + R_1)$$

where R_{lp} is an interaction term between R_p and R_1, R_c is a temperature dependent term which takes into account the mathematical residual deviations in R_1 and A_i and B are parameters determined by a least-squares analysis of the experimental data. There have been several compilations and reviews on the thermal conductivity of Al, Cu, Fe and W (Refs. 57 to 60). The data selected are believed to be accurate to within ± 5 to $\pm 10\%$.

The thermal conductivity of a pure metal is influenced by the imperfection concentration (chemical impurity, physical defects, etc...). These impurities also affect the electrical resistivity which can be used as a measure of the specimen purity.

Each set of data for a given specimen is characterized by a residual resistivity ratio (RRR).

$$RRR = (\rho_i/\rho_o) + 1 \qquad (10.2.6.5)$$

where ρ_o is the residual resistivity value and ρ_i is the ice point intrinsic resistivity.

Very low thermal conductivity standards

For insulation and building materials, there is a need for thermal conductivity reference materials. Unfortunately, in heat transfer processes at high temperature through porous insulating materials such as fibers, powders, and foams, thermal radiation is equally as important as thermal conduction. Moreover, in heat transfer processes through solids that are semi-transparent to infrared radiation, internal radiant heat exchange between layers at different temperatures may be of the same order of magnitude as conductive heat transfer at high temperatures. In such situations, a separate calculation of conductive and radiative heat fluxes without a consideration of the interactions between them could introduce an error in the heat transfer calculation.

Because it is difficult to separate the thermal conductivity effects from the radiation effects, it is preferable to use reference materials provided by different laboratories.

The heat transfer through insulation materials is generally measured by absolute methods such as guarded hot plates (Ref. 61) or relative ones such as a heat flow meter (Ref. 62). There are several criteria which must be considered in choosing a reference material (Refs. 63, 64). These include stability, durability, resiliency, uniformity, thermal diffusivity, isotropy, radiation properties, safety, toxicity, temperature limits, and cost.

The only materials used as reference materials are made from glass-fibre materials. Thermal conductivity values of glass-fibre specimens depend on:

(i) physical parameters of the materials including bulk density, thickness, plate emissivity, mean sample temperature and plate temperature differences,

(ii) properties of the materials such as the composition, diameter, and orientation of the fiber, and its optical scattering and absorption coefficients,

(iii) the fill gas properties including its thermal conductivity.

REFERENCES

1. Manual of symbols and terminology for physicochemical quantities and units, *Pure and Applied Chem.* **1979**, *51*, 1.
2. *Experimental thermodynamics*, Volume II. *Experimental Thermodynamics of Non-reacting fluids*, Le Neindre, B.; Vodar, B., editors, Butterworths: London, **1975**, p. 68.
3. Marsh, K. N., editor, *Pure and Appl. Chem.* **1981**, *53*, 1863.
4. Tufeu, R. *Ph.D. Thesis*, University of Paris, **1975**.
5. Le Neindre, B.; Tufeu, R. *Conductivité des liquides et des gaz*, R. 2920, Techniques de l'Ingénieur: 21 rue Cassette, 75006-PARIS.
6. Michels, A.; Sengers, J. V.; Van der Gulik, P. S. *Physica* **1962**, *28*, 1201.
7. Parsons, J. R., Jr.; Mulligan, J. C. *Rev. Sci. Instrum.* **1978**, *49*(10), 1460.
8. Friend, D. G.; Roder, H. M. *Phys. Rev.* **1985**, *A32*, 1941.
9. de Groot, J. J.; Healy, J. J.; Kestin, J. *Physica* **1976**, *82C*, 392.

10. Kestin, J.; Wakeham, W. A. *Physica* **1976**, *92A*, 102.
11. Clifford, A. A.; Kestin, J.; Wakeham, W. A. *Physica* **1980**, *100A*, 370.
12. Nieto de Castro, C. A. ; Wakeham, W. A. *Thermal conductivity*, **1978**, *15*, 236.
13. Haran E. N.; Wakeham, W. A. *J. Phys E* **1982**, *15*, 839.
14. Le Neindre, B., *Ph. D. Thesis*, University of Paris, **1969**.
15. Poltz, H.; Jugel, R. *Int. J. Heat and Mass Transfer* **1967**, *10*, 1075.
16. Schödel, G.; Grigull, U. *Proceedings of the 4th International Heat Transfer Conference*, *3*, R22, 1, Paris **1970**.
17. Jamieson, D. T.; Irving, J. B.; Tudhope, J. S. *Liquid thermal conductivity. A data survey to 1973*. H.M.S.O.: Edinburgh, **1973**.
18. Viskanta, R.; Grosh, R. J.; *Trans. ASME, J. Heat Transfer* **1963**, *84 C*, 63.
19. Saito, A.; Mani, N.; Venart, J. E. S. *Proc. 16th Nat. Heat Transfer Conf.*, *76*, CSME/CSChE -6 Canadian Society for Chemical Engineering: Ottawa, **1976**.
20. Saito, A.; Venart, J. E. S. *Proc. 6th Int. Heat Transfer Conf.* *3*, 79, (1978).
21. Leidenfrost, W. *Int. J. Heat Mass Transfer* **1964**, *7*, 447.
22. Fritz, W.; Poltz, H. *Int. J. Heat Mass Transfer* **1962**, *3*, 307.
23. Kohler, M.; *Z. Angew. Phys.* **1965**, *18*, 356.
24. Viskanta, R.; Grosh, R. J. *Intern. J. Heat Mass Transfer* **1962**, *5*, 729.
25. Viskanta, R. *J. Heat Transfer* **1965**, *87C*, 142.
26. Viskanta, R.; Grosh, R. J. *Appl. Mech. Rev.* **1964**, *17*, 91.
27. Lick, W. *Intern. J. Heat Mass Transfer* **1964**, *7*, 891.
28. Lick, W., *Proceedings of the Heat Transfer and Fluid Mechanics Institute*, Stanford University Press: Palo Alto, **1963**, pp. 14-26.
29. Greif, R. *Intern. J. Heat and Mass Transfer* **1964**, *7*, 891.
30. Wang, L. S. *Proceedings of the 3rd International Heat Transfer Conference* **1966**, *5*, 190.
31. Wang, L. S.; Tien, C. L. *J. Heat Mass Transfer* **1967**, *10*, 1327.
32. Chang, Y. P. *Amer. Inst. Aeronaut. Astronaut. J.* **1967**, *5*, 1024.
33. Hazzak, A. S.; Beck, J. V. *Intern J. Heat Mass Transfer* **1970**, *13*, 517.
34. Lii, C. C.; Ozisik, M. N. *Intern J. Heat Mass Transfer* **1972**, *15*, 1175.
35. Ozisik, M. N. *Radiative transfer and interactions with conduction and convection*, Wiley: New York **1973**.
36. Men, A. A. *High Temp.* **1973**, *11*, 252.
37. Men, A. A. *High Temp.* **1973**, *11*, 685.
38. Men, A. A. *J. Engineering Physics* **1973**, *2*, 681.
39. Men, A. A. *J. Engineering Physics* **1973**, *2*, 866.
40. Men, A. A. *Int. J. Heat Mass Transfer* **1972**, *15*, 1807.
41. Fischer, S.; Obermeier, E. *E. Proc. 9th European Thermophysical Properties Conference*, **1984**.
42. Braun, R.; Fischer, S.; Schaber, A. *Wärme und Stroffübertragung* **1983**, *17*, 121.
43. Bohne, D.; Fisher, S.; Obermeier, E. *Ber. Bunsenges, Phys. Chem.* **1984**, *88*, 739.
44. Menashe, J.; Wakeham, W.A. *Int. J. Heat and Mass Transfer* **1982**, *25*, 661.
45. Nieto de Castro, C. A.; Li, S. F. Y.; Maitland, G. C; Wakehan, W. A. *Int. J. Thermophys.* **1983**, *4*, 311.

46. Gregory, H. S. *Phil. Mag.* **1936**, *22*, 257.
47. Tufeu, R.; Le Neindre, B. *Ing. Fiz. Journal (Minsk)* **1979**, *36*, 472.
48. Calado, J. C. G.; Fareleiro J. M. N. A.; Nieto de Castro, C. A.; Wakeham, W. A. *Ref. Port. Quim.*, **1985**, (in press).
49. Hoshi, M.; Omotani, T.; Nagashima, A. *Rev. Sci. Instrum.* **1981**, *52*, 755.
50. Nagasaka, Y.; Nagashima, A. *J. Phys. E. Sci. Instrum.* **1981**, *52*, 755.
51. Allonsh, A.; Grosney, W. B.; Wakeham, W. A. *Int. J. Thermophys.* **1982**, *3*, 225.
52. Ziebland, H. *Int. J. Heat and Mass Transfer* **1961**, *2*, 273.
53. Nieto de Castro, C. A.; Wakeham, W. A. *Thermal conductivity*, *18*, **1984**, 65.
54. Kestin, J.; Paul, R.; Clifford, A. A.; Wakeham, W. A. *Physica* **1980**, *100A*, 349.
55. White, G. K.; Minges, M. L. *Thermophysical Properties of Some Key Solids*, CODATA Report 59, Pergamon: London, **1985**.
56. Cezairliyan, A.; Touloukian, Y. S. *Advances in Thermophysical Properties at Extreme Temperatures and Pressures, Third Symposium on Thermophysical Properties*, Gratch, s., editor, American Society of Mechanical Engineers: New York, **1965** p. 301.
57. Touloukian, Y. S.; Powell, R. W.; Ho, C. Y.; Klemens, P. G. *Thermophysical Properties of Matter*, Vol. 1, *Thermal Conductivity (Metallic Elements and Alloys)*, Plenum Press: New York, **1970**.
58. Ho, C. Y.; Powell, R. W.; Liley, P. E. *J. Phys. Chem. Ref. Data* **1974**, *3*, 1.
59. Childs, G. E.; Ericks, L. J.; Powell, R. L. *Thermal Conductivity of Solids at Room Temperature and Below (A review and compilation of the literature)*, National Bureau of Standards Monograph 131, U.S. Department of Commerce: Washington, **1973**.
60. Hust, J. G.; Lankford, A. B. *Thermal Conductivity of Aluminium, Copper, Iron and Tungsten for Temperatures from 1 K to the Melting Point*, National Bureau of Standards Internal Report, 84-3007, U.S. Department of Commerce: Gaithersburg, **1984**.
61. American Society for Testing and Materials, 1982 Annual Book of ASTM Standards, (Philadelphia: American Society for Testing and Materials), Part 18, ANSI/ASTM C 177-76, *Standard test method for steady-state thermal transmission properties by means of the guarded hot plate*, **1982**, p. 20.
62. American Society for Testing and Materials, 1982 Annual Book of ASTM Standards, (Philadelphia: American Society for Testing and Materials), Part 18, ANSI/ASTM C 518-76, *Standard test method for steady-state thermal transmission properties by means of the heat flow meter*, **1982**, p. 222.
63. Rennex, B. *An assessment of needs for new thermal reference materials*, National Bureau of Standards Internal Report, 85-3146, U.S. Department of Commerce: Gaithersburg, **1985**.
64. ASTM Subcommittee C16.30, *Reference Materials for Insulation Measurement Comparisons*, Thermal Transmission Measurements of Insulation, ASTM STP 660, Tye, R. P., editor, American Society for Testing and Materials, **1978**, p. 7.
65. Siu, M.C.I.; Watson, T. W.; Peavy, B. A. *NBS Certificate, Standard Reference Material 1450, Thermal Resistance–Fibrous Glass Board,*. U.S. Department of Commerce: Washington D.C. **1979**.

10.3. Reference materials for the thermal conductivity of gases

10.3.1. Helium

Physical property: Thermal conductivity, λ
Units: W m^{-1} K^{-1}
Recommended reference material: Helium (He)
Range of variables: 80 to 2000 K, 10^5 Pa
Physical state within the range: gas
Contributor: B. Le Neindre

Intended usage: Helium can be used for the calibration of thermal conductivity cells of arbitrary geometry and for the testing of the proper functioning of apparatus and its ancillary equipment within the temperature range 80 to 2000 K.

Sources of supply and/or methods of preparation: Helium of high purity is commercially available from suppliers (A), (B), (C), (D), (E), (F), and (G).

Pertinent physicochemical data: The recommended values of the thermal conductivity of helium (Ref. 1) are given in table 10.3.1.1.

REFERENCE

1. Kestin, J.; Knierim, K.; Mason, E. A.; Najafi, B.; Ro, S. T.; Waldman, M. *J. Phys. Chem. Ref. Data* **1984**, *13*, 229.

10.3.2. Neon

Physical property: Thermal conductivity, λ
Units: W m^{-1} K^{-1}
Recommended reference material: Neon (Ne)
Range of variables: 80 to 2000 K, 10^5 Pa
Physical state within the range: gas
Contributor: B. Le Neindre

Intended usage: Neon can be used for the calibration of thermal conductivity cells of arbitrary geometry and for the testing of the proper functioning of apparatus and its ancillary equipment within the temperature range 80 to 2000 K.

10. Thermal Conductivity

Table 10.3.1.1. Recommended values for the thermal conductivity of helium 80 to 2000 K

$\dfrac{T}{K}$	$\dfrac{\lambda}{mW\ m^{-1}\ K^{-1}}$	$\dfrac{T}{K}$	$\dfrac{\lambda}{mW\ m^{-1}\ K^{-1}}$	$\dfrac{T}{K}$	$\dfrac{\lambda}{mW\ m^{-1}\ K^{-1}}$
80	65.12	440	203.51	1050	374.43
100	75.54	450	206.68	1100	387.08
120	85.17	460	209.84	1150	399.58
140	94.25	480	216.10	1200	411.95
160	102.92	500	222.29	1250	424.20
180	111.26	520	228.41	1300	436.34
200	119.32	540	234.47	1350	448.36
220	127.15	550	237.48	1400	460.27
240	134.78	560	240.47	1450	472.09
260	142.23	580	246.42	1500	483.80
273	146.99	600	252.31	1550	495.43
280	149.52	650	266.86	1600	506.97
300	156.66	700	281.04	1650	518.42
320	163.68	750	295.01	1700	529.80
340	170.57	800	308.73	1750	541.10
350	173.97	850	322.24	1800	552.32
360	177.35	900	335.55	1850	563.47
380	184.02	950	348.68	1900	574.56
400	190.61	1000	361.63	2000	596.53
420	197.10				

Sources of supply and/or methods of preparation: Neon of high purity is commercially available from suppliers (A), (B), (C), (D), (E), (F), and (G).

Pertinent physicochemical data: The recommended values of the thermal conductivity of argon (Ref. 1) are given in table 10.3.2.1.

REFERENCE

1. Kestin, J.; Knierim, K.; Mason, E. A.; Najafi, B.; Ro, S. T.; Waldman, M. *J. Phys. Chem. Ref. Data* **1984**, *13*, 229.

Table 10.3.2.1. Recommended values for the thermal conductivity of neon for 80 to 2000 K

$\dfrac{T}{K}$	$\dfrac{\lambda}{mW\ m^{-1}\ K^{-1}}$	$\dfrac{T}{K}$	$\dfrac{\lambda}{mW\ m^{-1}\ K^{-1}}$	$\dfrac{T}{K}$	$\dfrac{\lambda}{mW\ m^{-1}\ K^{-1}}$
80	18.44	440	64.20	1050	114.00
100	22.26	450	65.16	1100	117.60
120	25.75	460	66.11	1150	121.15
140	28.99	480	67.98	1200	124.66
160	32.03	500	69.83	1250	128.13
180	34.90	520	71.65	1300	131.55
200	37.63	540	73.45	1350	134.94
220	40.24	550	74.34	1400	138.29
240	42.74	560	75.22	1450	141.61
260	45.16	580	76.98	1500	144.89
273	46.69	600	78.71	1550	148.15
280	47.50	650	82.96	1600	151.37
300	49.77	700	87.11	1650	154.57
320	51.97	750	91.17	1700	157.74
340	54.12	800	95.15	1750	160.88
350	55.18	850	99.04	1800	164.00
360	56.22	900	102.87	1850	167.09
380	58.28	950	106.64	1900	170.16
400	60.29	1000	110.35	2000	176.23
420	62.26				

10.3.3. Argon

Physical property: Thermal conductivity, λ
Units: W m^{-1} K^{-1}
Recommended reference material: Argon (Ar)
Range of variables: 80 to 2000 K, 10^5 Pa
Physical state within the range: gas
Contributor: B. Le Neindre

Intended usage: Argon can be used for the calibration of thermal conductivity cells of arbitrary geometry and for the testing of the proper functioning of apparatus and its ancillary equipment within the temperature range 80 to 2000 K.

10. Thermal Conductivity

Sources of supply and/or methods of preparation: Argon of high purity is commercially available from suppliers (A), (B), (C), (D), (E), (F), and (G).

Pertinent physicochemical data: The recommended values of the thermal conductivity of argon (Ref. 1) are given in table 10.3.3.1.

Table 10.3.3.1. Recommended values for the thermal conductivity of argon for 80 to 2000 K

$\dfrac{T}{K}$	$\dfrac{\lambda}{mW\ m^{-1}\ K^{-1}}$	$\dfrac{T}{K}$	$\dfrac{\lambda}{mW\ m^{-1}\ K^{-1}}$	$\dfrac{T}{K}$	$\dfrac{\lambda}{mW\ m^{-1}\ K^{-1}}$
80	5.06	440	24.26	1050	45.05
100	6.22	450	24.69	1100	46.48
120	7.43	460	25.10	1150	47.88
140	8.69	480	25.92	1200	49.26
160	9.96	500	26.73	1250	50.61
180	11.20	520	27.52	1300	51.95
200	12.41	540	28.30	1350	53.26
220	13.56	550	28.68	1400	54.55
240	14.68	560	29.07	1450	55.83
260	15.77	580	29.82	1500	57.09
273	16.45	600	30.56	1550	58.34
280	16.82	650	32.36	1600	59.57
300	17.83	700	34.11	1650	60.78
320	18.82	750	35.80	1700	61.99
340	19.79	800	37.44	1750	63.18
350	20.26	850	39.04	1800	64.36
360	20.73	900	40.59	1850	65.53
380	21.64	950	42.11	1900	66.69
400	22.53	1000	43.60	2000	68.99
420	23.41				

REFERENCE

1. Kestin, J.; Knierim, K.; Mason, E. A.; Najafi, B.; Ro, S. T.; Waldman, M. *J. Phys. Chem. Ref. Data* **1984**, *13*, 229.

10.3.4. Krypton

Physical property: Thermal conductivity, λ
Units: W m^{-1} K^{-1}
Recommended reference material: Krypton (Kr)
Range of variables: 80 to 2000 K, 10^5 Pa
Physical state within the range: gas
Contributor: B. Le Neindre

Intended usage: Krypton can be used for the calibration of thermal conductivity cells of arbitrary geometry and for the testing of the proper functioning of apparatus and its ancillary equipment within the temperature range 80 to 2000 K.

Sources of supply and/or methods of preparation: Krypton of high purity is commercially available from suppliers (A), (B), (C), (D), (E), (F), and (G).

Pertinent physicochemical data: The recommended values of the thermal conductivity of krypton (Ref. 1) are given in table 10.3.4.1.

REFERENCE

1. Kestin, J.; Knierim, K.; Mason, E. A.; Najafi, B.; Ro, S. T.; Waldman, M. *J. Phys. Chem. Ref. Data* **1984**, *13*, 229.

10.3.5. Xenon

Physical property: Thermal conductivity, λ
Units: W m^{-1} K^{-1}
Recommended reference material: Xenon (Xe)
Range of variables: 80 to 2000 K, 10^5 Pa
Physical state within the range: gas
Contributor: B. Le Neindre

Intended usage: Xenon can be used for the calibration of thermal conductivity cells of arbitrary geometry and for the testing of the proper functioning of apparatus and its ancillary equipment within the temperature range 80 to 2000 K.

Table 10.3.4.1. Recommended values for the thermal conductivity of krypton from 80 to 2000 K

$\dfrac{T}{K}$	$\dfrac{\lambda}{mW\ m^{-1}\ K^{-1}}$	$\dfrac{T}{K}$	$\dfrac{\lambda}{mW\ m^{-1}\ K^{-1}}$	$\dfrac{T}{K}$	$\dfrac{\lambda}{mW\ m^{-1}\ K^{-1}}$
80	2.72	440	13.34	1050	25.75
100	3.29	450	13.59	1100	26.59
120	3.88	460	13.84	1150	27.42
140	4.48	480	14.33	1200	28.23
160	5.09	500	14.81	1250	29.03
180	5.73	520	15.29	1300	29.82
200	6.38	540	15.75	1350	30.59
220	7.03	550	15.98	1400	31.35
240	7.68	560	16.21	1450	32.10
260	8.31	580	16.66	1500	32.84
273	8.71	600	17.10	1550	33.57
280	8.93	650	18.18	1600	34.29
300	9.52	700	19.23	1650	35.00
320	10.11	750	20.24	1700	35.70
340	10.68	800	21.21	1750	36.40
350	10.96	850	22.17	1800	37.08
360	11.23	900	23.09	1850	37.76
380	11.78	950	24.00	1900	38.43
400	12.31	1000	24.88	2000	39.76
420	12.83				

Sources of supply and/or methods of preparation: Xenon of high purity is commercially available from suppliers (A), (B), (C), (D), (E), (F), and (G).

Pertinent physicochemical data: The recommended values of the thermal conductivity of xenon (Ref. 1) are given in table 10.3.5.1.

REFERENCE

1. Kestin, J.; Knierim, K.; Mason, E. A.; Najafi, B.; Ro, S. T.; Waldman, M. *J. Phys. Chem. Ref. Data* **1984**, *13*, 229.

Table 10.3.5.1. Recommended values for the thermal conductivity of xenon from 80 to 2000 K

$\dfrac{T}{K}$	$\dfrac{\lambda}{mW\ m^{-1}\ K^{-1}}$	$\dfrac{T}{K}$	$\dfrac{\lambda}{mW\ m^{-1}\ K^{-1}}$	$\dfrac{T}{K}$	$\dfrac{\lambda}{mW\ m^{-1}\ K^{-1}}$
80	1.65	440	7.97	1050	16.20
100	1.98	450	8.14	1100	16.76
120	2.31	460	8.30	1150	17.31
140	2.64	480	8.63	1200	17.85
160	2.97	500	8.94	1250	18.38
180	3.31	520	9.26	1300	18.90
200	3.66	540	9.56	1350	19.41
220	4.02	550	9.72	1400	19.91
240	4.38	560	9.87	1450	20.41
260	4.75	580	10.16	1500	20.90
273	4.99	600	10.46	1550	21.38
280	5.12	650	11.18	1600	21.85
300	5.50	700	11.87	1650	22.32
320	5.87	750	12.54	1700	22.78
340	6.24	800	13.19	1750	23.23
350	6.42	850	13.82	1800	23.68
360	6.60	900	14.44	1850	24.13
380	6.95	950	15.04	1900	24.57
400	7.30	1000	15.63	2000	24.43
420	7.64				

10.4. Reference materials for the thermal conductivity of liquids

10.4.1. Water

Physical property: Thermal conductivity, λ
Units: $W\ m^{-1}\ K^{-1}$
Recommended reference material: Water (H_2O)
Range of variables: 280 to 370 K, 10^5 Pa
Physical state within the range: liquid
Contributor: B. Le Neindre

10. Thermal Conductivity

Intended usage: Liquid water can be used for the calibration of thermal conductivity cells of arbitrary geometry and for the testing of the proper functioning of apparatus and its ancillary equipment within the temperature range 280 to 370 K.

Sources of supply and/or methods of preparation: Distilled and degassed water is adequate.

Pertinent physicochemical data: The thermal conductivity of water is much larger than that of most other liquids. However, a careful analysis of the possible effects of electrical conduction or polarization in the liquid should be made. These effects have caused errors in many previous measurements.

In general, polar liquids like water, ammonia, or alcohol show a large thermal conductivity compared with simple liquids. A supplementary heat conduction seems to take place along hydrogen bonds. However, great care must be taken in the measurements of the thermal conductivity of these liquids or in using them as reference material. In particular, any direct contact between the substance and the power source or elements which measure the temperature gradient must be avoided as they are electrical conductors. However, this difficulty can be considerably reduced in the transient hot-wire technique by using an alternating current (Ref. 1).

The data sets used by Nieto de Castro et al. (Ref. 2) are as follows: the steady-state results of Riedel (Refs. 3, 4), Rastorguev et al. (Ref. 5), and Schmidt and Leidenfrost (Ref. 6) have an assigned accuracy of $\pm 1.5\%$. The data of Nagasaka et al. (Ref. 7) obtained with a transient hot-wire apparatus with two insulated wires have a claimed accuracy of $\pm 0.5\%$. Dietz (Refs. 1, 8) used two bare wires driven by an alternating current to measure the thermal conductivity of water under pressure. At atmospheric pressure, an accuracy of $\pm 1.2\%$ was assigned.

The recommended values of the thermal conductivity of water as a function of temperature is represented between the melting point and the boiling point by the quadratic equation (Refs. 2):

$$\lambda^* = -1.26523 + 3.70489 T^* - 1.43955 T^{*2} \qquad (10.4.1.1)$$

where $T^* = (T/K)/298.15$, $\lambda^* = \lambda/\lambda_{298.15}$, $\lambda_{298.15} = (0.6067 \pm 0.0061)$ W m^{-1} K^{-1}.

The maximum deviation of the selected data from the correlation is 1.1% with a standard deviation of ± 0.0028 W m^{-1} K^{-1}.

The agreement between the correlation for water and the International Association for the Properties of Steam (IAPS) recommendations (Ref. 9) is within $\pm 0.6\%$, which is well within the mutual uncertainties of both works, but the slopes are slightly different.

Recommended values for the thermal conductivity of water are given in table 10.4.1.1.

Table 10.4.1.1. Recommended values for the thermal conductivity of water from 280 to 370 K

$\dfrac{T}{K}$	$\dfrac{\lambda}{W\ m^{-1}\ K^{-1}}$	$\dfrac{T}{K}$	$\dfrac{\lambda}{W\ m^{-1}\ K^{-1}}$
280.00	0.5730	330.00	0.6503
290.00	0.5924	340.00	0.6598
300.00	0.6098	350.00	0.6674
310.00	0.6253	360.00	0.6731
320.00	0.6387	370.00	0.6767

REFERENCES

1. Dietz, F. J., *Ph. D. Thesis*, University of Karlsruhe, **1981**.
2. Nieto de Castro, C. A.; Li, S. F. Y.; Nagashima, A.; Trengove, R. D.; Wakeham, W. A. *J. Phys. Chem. Ref. Data*, in press.
3. Riedel, H. *Chem. Ing. Tech.* **1951**, *23*, 321.
4. Riedel, H. *Chem. Ing. Tech.* **1950**, *22*, 54.
5. Rastorguev, Yu L.; Ganiev, A.; Safronov, G. A. *Inzh. Fiz. Zh.* **1971**, *33*, 275.
6. Schmidt, E.; Leidenfrost, W. *Forsch. Ing. Wesens.* **1955**, *21*, 177.
7. Nagasaka, Y.; Okada, H.; Suzuki, J.; Nagashima, A. *Ber Bunsenges, Phys. Chem.* **1983**, *87*, 859.
8. Dietz, F. J.; De Groot, J. J.; Franck, E. U. *Ber. Bunsenges, Phys. Chem.* **1981**, *85*, 1005.
9. Sengers, J. V.; Watson, J. T. R.; Basu, R. S.; Kamgar-Parsi, B. *J. Phys. Chem. Ref. Data* **1984**, *13*, 893.

10.4.2. Toluene

Physical property: Thermal conductivity, λ
Units: $W\ m^{-1}\ K^{-1}$
Recommended reference material: Toluene (C_7H_8)
Range of variables: 230 to 360 K, 10^5 Pa
Physical state within the range: liquid
Contributor: B. Le Neindre

Intended usage: Liquid toluene can be used for the calibration of thermal conductivity cells of arbitrary geometry and for the testing of the proper functioning of apparatus and its ancillary equipment within the temperature range 230 to 360 K.

10. Thermal Conductivity

Sources of supply and/or methods of preparation: Distilled research grade toluene is suitable. This material is available from many suppliers.

Pertinent physicochemical data: Over 50 investigations have been reported in the literature. The discrepancy among the values, even near room temperature and at atmospheric pressure, is very large. Recently, measurements of the thermal conductivity of toluene have been made with the transient hot-wire technique at three different research centers (Refs. 1 to 4). The estimated accuracy for these three sets of results has been evaluated at $\pm 1\%$ (Ref. 5).

Based on these transient hot-wire data and those of Pittman (Ref. 6) at lower temperatures, two sets of equations were produced by Nieto de Castro et al. (Ref. 4). In the limited temperature range $230 < T < 360$ K, where the experimental data are of higher quality, a linear representation is adequate, but from the freezing point to 360 K, a quadratic equation is needed.

The two correlations are given in reduced form.

$$\lambda^* = 1.68182 - 0.682022 T^* \tag{10.4.2.1}$$

for $230 < T < 360$ K

$$\lambda^* = 1.45210 - 0.224229 T^* - 0.225873 T^{*2} \tag{10.4.2.2}$$

for $189 < T < 360$ K

where $T^* = T/298.15$, $\lambda^* = \lambda/\lambda_{298.15}$, $\lambda_{298.15} = (0.1311 \pm 0.0013)$ W m^{-1} K^{-1}.

The maximum deviation of the selected experimental data from equation 10.4.2.1 is $\pm 1.2\%$ with a standard deviation of 0.00067 W m^{-1} K^{-1}. The maximum deviation of the primary data from equation 10.4.2.2 is $\pm 1.5\%$ with a standard deviation of 0.00083 W m^{-1} K^{-1}.

Recommended values for the thermal conductivity of toluene are given in table 10.4.2.1.

REFERENCES

1. Nagasaka, J. Y.; Nagashima, A. *Ind. Eng. Chem. Fundam.* **1981**, *20*, 216.
2. Nieto de Castro, C. A.; Li, S. F. Y.; Maitland, G. C.; Wakeham W. A. *Int. J. Thermophys.* **1983**, *4*, 311.
3. Nieto de Castro, C. A.; Calado, J. C. G.; Wakeham, W. A., *Proc. 7th Symp., Thermophys. Prop.* 730, **1977**.
4. Nieto de Castro, C. A.; Li, S. F. Y.; Nagashima, A.; Trengove, R. D.; Wakeham, W. A. *J. Phys. Chem. Ref. Data*, in press.
5. Nieto de Castro, C. A.; Wakeham, W. A., *Thermal Conductivity*, **1984**, *18*, 65.
6. Pittman, J. F. T. *Ph. D. Thesis*, Imperial College of Science & Technology: London, **1968**.

Table 10.4.2.1. Recommended values for the thermal conductivity of toluene from 230 to 360 K at 10^5 Pa

$\dfrac{T}{K}$	$\dfrac{\lambda}{W\ m^{-1}\ K^{-1}}$	$\dfrac{T}{K}$	$\dfrac{\lambda}{W\ m^{-1}\ K^{-1}}$
230.00	0.1515	300.00	0.1305
240.00	0.1485	310.00	0.1275
250.00	0.1455	320.00	0.1245
263.00	0.1425	330.00	0.1215
270.00	0.1395	340.00	0.1185
280.00	0.1365	350.00	0.1155
290.00	0.1335	360.00	0.1125

10.4.3. Heptane

Physical property: Thermal conductivity, λ
Units: W m^{-1} K^{-1}
Recommended reference material: Heptane (C_7H_{16})
Range of variables: 190 to 370 K, 10^5 Pa
Physical state within the range: liquid
Contributor: B. Le Neindre

Intended usage: Liquid heptane can be used for the calibration of thermal conductivity cells of arbitrary geometry and for the testing of the proper functioning of apparatus and its ancillary equipment within the temperature range 190 to 370 K.

Sources of supply and/or methods of preparation: Distilled research grade heptane is suitable. Material is available from many suppliers.

Pertinent physicochemical data: After a detailed analysis of previous measurements, Nieto de Castro *et al.* (Ref. 1) selected only measurements made with the transient hot-wire technique (Refs. 2 to 4).

The reduced thermal conductivity of heptane is given as a function of reduced temperature by

$$\lambda^* = 1.78026 - 0.729932 T^* \qquad (10.4.3.1)$$

for $191 < T/K < 365$ where $T^* = T/298.15$, $\lambda^* = \lambda/\lambda_{298.15}$, and $\lambda_{298.15} = (0.1228 \pm 0.0018)$ W m^{-1} K^{-1}.

10. Thermal Conductivity

The maximum deviation of the selected data from the equation was ±1.4% with a standard deviation of 0.00067 W m^{-1} K^{-1}. However, due to a small shift in the various sets of data, an accuracy of ±1.5% was assigned to equation 10.4.3.1.

Recommended value for the thermal conductivity of heptane are given table 10.4.3.1.

Table 10.4.3.1. Recommended values for the thermal conductivity of heptane from 190 to 370 K at 10^5 Pa

$\dfrac{T}{K}$	$\dfrac{\lambda}{\text{W m}^{-1}\text{ K}^{-1}}$	$\dfrac{T}{K}$	$\dfrac{\lambda}{\text{W m}^{-1}\text{ K}^{-1}}$
190.00	0.1554	290.00	0.1253
200.00	0.1523	300.00	0.1223
210.00	0.1493	310.00	0.1193
220.00	0.1463	320.00	0.1163
230.00	0.1433	330.00	0.1135
240.00	0.1403	340.00	0.1103
250.00	0.1373	350.00	0.1073
260.00	0.1343	360.00	0.1042
270.00	0.1313	370.00	0.1012
280.00	0.1283		

REFERENCE

1. Nieto de Castro, C. A.; Li, S. F. Y.; Nagashima, A.; Trengove, R. D.; Wakeham, W. A. *J. Phys. Chem. Ref. Data*, in press.
2. Nagasaka, J. Y.; Nagashima, A. *Ind. Eng. Chem. Fundam.* **1981**, *20*, 216.
3. Nieto de Castro, C. A.; Trengove, R. D.; Wakeham, W. A., to be published.
4. Pittman, J. F. T. *Ph. D. Thesis*, Imperial College of Science & Technology: London, **1968**.

10.4.4. Dimethylphthalate

Physical property: Thermal conductivity, λ
Units: W m^{-1} K^{-1}
Recommended reference material: Dimethylphthalate ($C_{10}H_{10}O_4$)
Range of variables: 273 to 500 K, 10^5 Pa
Physical state within the range: liquid

Contributor: B. Le Neindre

Intended usage: Liquid dimethylphthalate can be used for the calibration of thermal conductivity cells of arbitrary geometry and for the testing of the proper functioning of apparatus and its ancillary equipment within the temperature range 280 to 370 K.

Sources of supply and/or methods of preparation: Material of suitable purity is available from suppliers (H) and (I). Dimethylphthalate is slightly hygroscopic and should be dried over 'Linde' type 4A molecular sieve prior to use.

Pertinent physicochemical data: Dimethylphthalate was chosen as a reference material for thermal conductivity measurements at high temperature. There are only three sets of measurements on this compound by the transient hot-wire method (Refs. 1, 2, 3). There have been a considerable number of measurements made with the steady-state method. In the steady-state method, the magnitude of the correction term for radiative energy transfer increases with the third power of the temperature; hence, a reference substance suitable for use at high temperatures should possess a high absorption coefficient in the infrared region.

The phthalic esters possess this property and moreover have good chemical stability. Samples of dimethylphthalate have been distributed amongst research groups to establish reference values for its thermal conductivity. These results were fitted by a least-square analysis and the following correlation was obtained (Ref. 4).

$$\lambda/(\text{W m}^{-1}\text{ K}^{-1}) = 0.1501 - 1.0539 \times 10^{-4}[(T/K) - 273.15]$$
$$- 2.23 \times 10^{-7}[(T/K) - 273.15]^2 \qquad (10.4.4.1)$$

for $273 < T < 493$ K. The recommended values are given in table 10.4.4.1.

Table 10.4.4.1. Recommended values for the thermal conductivity of dimethylphthalate from 273 to 500 K at 10^5 Pa

$\dfrac{T}{K}$	$\dfrac{\lambda}{\text{W m}^{-1}\text{ K}^{-1}}$	$\dfrac{T}{K}$	$\dfrac{\lambda}{\text{W m}^{-1}\text{ K}^{-1}}$	$\dfrac{T}{K}$	$\dfrac{\lambda}{\text{W m}^{-1}\text{ K}^{-1}}$
273	0.1501	350	0.1407	430	0.1277
280	0.1495	360	0.1394	440	0.1260
290	0.1483	370	0.1379	450	0.1241
300	0.1472	380	0.1364	460	0.1222
310	0.1460	390	0.1348	470	0.1203
320	0.1446	400	0.1330	480	0.1184
330	0.1434	410	0.1313	490	0.1164
340	0.1421	420	0.1296	500	0.1142

REFERENCE

1. Perkins, R. A. *Ph. D. Thesis*, Colorado School of Mines, **1983**.
2. Mani, N.; Venart, J. E. S. *Proc. 6th Symposium on Thermophysical Properties*, Liley, P. I. P., editor, American Society of Mechanical Engineers: New York, **1973**.
3. Nieto de Castro, C. A.; Calado J. C. G. and Wakeham W. A., *Seventh Symposium on Thermophysical Properties*, Cezairliyan, A., editor, American Society of Mechanical Engineers: New York, **1977** p.730.
4. Marsh, K. N., editor, *Pure and Appl. Chem.* **1981**, *53*, 1863.

10.5. Reference materials for the thermal conductivity under pressure

10.5.1. Propane

Physical property: Thermal conductivity, λ
Units: W m^{-1} K^{-1}
Recommended reference material: Propane (C_3H_8)
Range of variables: 375.5 and 380 K, 4.0 to 7.0 MPa
Physical state within the range: gas
Contributor: B. Le Neindre

Intended usage: Propane can be used for the calibration of thermal conductivity cells of any geometry working under pressure and/or for the testing of the apparatus and its ancillary equipment.

Sources of supply and/or methods of preparation: Propane of high purity is available from suppliers (A) to (G).

Pertinent physicochemical data: It is desirable that measurements be made over the complete thermal conductivity surface, which includes the physical states of the dilute gas, the dense gas, the region near critical, compressed liquid states, metastable liquid states at temperatures below critical, and pressures less than the vapor pressure. Such measurements have been made on oxygen (Ref. 1), methane (Refs. 2, 3), and ethane (Refs. 4, 5) as well as propane. The useful parameters to correlate the thermal conductivity are not pressure and temperature but density and temperature. In the regions of the phase diagram which excludes the supercritical region ($T_c < T < 1.5\, T_c$ and $0.5\, \rho_c < \rho < 1.5\, \rho_c$) and the vicinity of the saturation line, the thermal conductivity coefficient is a simple function of the density (almost independent of the temperature) and can be fitted by a simple polynomial

$$(\lambda - \lambda_o) = f(\rho) \qquad (10.5.1.1)$$

where λ_o is the thermal conductivity at atmospheric pressure.

In the critical region, the thermal conductivity coefficient shows a pronounced enhancement. Along the critical isochore, the thermal conductivity coefficient diverges on approach to T_c. This divergence can be expressed in terms of scaling laws, but there is presently no complete formulation of the phenomenon based on a theoretical understanding of the supercritical region.

To give an idea of the behavior of the thermal conductivity of a fluid in larger temperature and density ranges and eventually to use this fluid as a reference material when the thermal conductivity apparatus has to be tested under pressure, we suggest propane. For this fluid, several measurements and data compilation have been published both on the liquid and gas phases (Refs. 6 to 11). The critical temperature ($T_c = 369.85$ K) is not too high and the critical conditions are easily reproducible. The thermal conductivity of propane at a given density is typical of the thermal conductivity of a large number of fluids. For propane, a good equation of state exists from which the density can be easily calculated (Ref. 12).

As the exact formulation of the thermal conductivity of propane in the whole temperature and pressure range, including the critical region, is still subject to discussion; we report two experimental quasi isotherms above the critical temperature in tables 10.5.1.1 and 10.5.1.2. The accuracy of these measurements is better than $\pm 2\%$.

Table 10.5.1.1. Thermal conductivity of propane along the quasi isotherm 375.5 K

$\dfrac{T}{\text{K}}$	$\dfrac{p}{\text{MPa}}$	$\dfrac{\rho}{\text{kg m}^{-3}}$	$\dfrac{\lambda}{\text{W m}^{-1}\,\text{K}^{-1}}$
375.09	6.915	358.4	0.0675
375.09	6.074	341.3	0.0661
375.21	5.558	324.6	0.0647
375.17	5.051	297.0	0.0647
375.33	4.939	283.4	0.0651
375.31	4.838	267.7	0.0676
374.57	4.787	248.3	0.0684
375.67	4.747	231.1	0.0709
375.69	4.696	210.3	0.0709
375.69	4.676	202.0	0.0686
375.71	4.656	193.1	0.0660
375.79	4.646	187.4	0.0623
375.87	4.554	158.8	0.0530
375.94	4.302	122.2	0.0429
375.05	4.007	99.5	0.0387

10. Thermal Conductivity

Table 10.5.1.2. Thermal conductivity of propane along the quasi isotherm 380 K

$\dfrac{T}{K}$	$\dfrac{p}{MPa}$	$\dfrac{\rho}{kg\ m^{-3}}$	$\dfrac{\lambda}{W\ m^{-1}\ K^{-1}}$
380.34	7.523	351.2	0.0683
380.34	7.077	342.8	0.0656
380.28	6.561	330.1	0.0661
380.21	6.064	314.9	0.0651
380.08	5.517	284.6	0.0647
379.98	5.254	255.8	0.0653
379.9	5.142	236.8	0.0656
379.84	5.051	218.9	0.0660
379.80	4.975	202.0	0.0633
379.89	4.960	196.9	0.0626
379.94	4.909	184.3	0.0596
380.09	4.795	160.7	0.0528
380.20	4.534	128.0	0.0444
380.34	3.977	91.4	0.0370

REFERENCE

1. Roder, H. M. *J. Res. Nat. Bur. Stand.* **1982**, *87*, 279.
2. Prasad, R. C.; Mani, N.; Venart, J. E. S. *Int. J. Thermophys.* **1984**, *5*, 265.
3. Roder, H. M. *Int. J. Thermophys.* **1985**, *6*, 119.
4. Prasad, R. C.; Venart, J. E. S. *Int. J. Thermophys.* **1984**, *5*, 367.
5. Roder, H. M.; Nieto de Castro, C. A. *High Temp. High Press.* **1985**, 17.
6. Roder, H. M.; Nieto de Castro, C. A. *J. Chem. Eng. Data* **1982**, *27*, 12.
7. Vargaftik, N. B. *Tables on thermophysical properties of liquids and gases*, 2nd edition, Wiley: New York, **1975**.
8. Holland, P. M.; Hanley, H. J. M; Gubbins, K. E.; Haile, J. M. *J. Phys. Chem. Ref. Data* **1979**, *8*, 559.
9. Tufeu, R.; Bury, P.; Garrabos, Y.; Le Neindre, B. *Int. J. Thermophys.* **1986**, *7*, 663.
10. Aggarwal, H. C.; Springer, G. S. *J. Chem. Phys.* **1979**, *70*, 3948.
11. Prasad, R. C.; Venart, J. E. S., Presented at 9th Symposium of Thermophysical Properties, Boulder, CO, June 1985.
12. Goodwin, R. D.; Haynes, W. M. *Thermophysical Properties of Propane from 85 to 700 K at Pressures to 70 MPa*, National Bureau of Standards Monograph 170, U.S. Department of Commerce: Gaithersburg, **1982**.

10.6. Reference materials for the thermal conductivity of solids – high thermal conductivity materials

10.6.1. Aluminium

Physical property: Thermal conductivity, λ
Units: W m^{-1} K^{-1}
Recommended reference material: Aluminium (Al)
Range of variables: 1 to 900 K
Physical state within the range: solid
Class: Calibration and test material
Contributor: B. Le Neindre

Intended usage: Aluminium can be used for the calibration of thermal conductivity cells used to measure the thermal conductivity of highly conducting solids within the temperature range 1 to 900 K.

Sources of supply and/or methods of preparation: Aluminium of purity 0.9999 mass fraction can be purchased from many suppliers.

Pertinent physicochemical data: From a total of 35 publications reported in reference 1, nine data sets have been selected as primary data for annealed specimens. They cover a temperature range from 1 to 873 K and a range of RRR from 13 to 16800. The ice point intrinsic resistivity of aluminium, ρ_i, is 24.8×10^{-9} Ω m.

The following values of the parameters A_i of equation 10.2.6.4 were obtained by a non-linear least-squares fit:

$$\begin{aligned}
A_1 &= 4.716 \times 10^{18} & A_5 &= 130.9 \\
A_2 &= 2.446 & A_6 &= 2.5 \\
A_3 &= 623.6 & A_7 &= 0.8168. \\
A_4 &= -0.16
\end{aligned}$$

The residual term R_c is given by

$$\begin{aligned}
(R_c/\text{m K W}^{-1}) &= -0.0005\{\ln[z/330]\}\exp(-x^2) - 0.0013\{\ln[z/110)]\}\exp(-y^2) \\
x &= \{\ln[z/380]\}/0.6 \\
y &= \{\ln[z/94]\}/0.5
\end{aligned} \qquad (10.6.1.1).$$

where $z = T/\text{K}$

10. Thermal Conductivity

Thermal conductivity values calculated from equation 10.2.6.4 are listed in table 10.6.1.1. The maximum deviations of experimental data from the equation are of the order of ±10%.

Table 10.6.1.1. Thermal conductivity of aluminium

$\dfrac{T}{K}$	$\lambda/\text{W m}^{-1}\text{ K}^{-1}$ RRR						$\dfrac{T}{K}$	$\lambda/\text{W m}^{-1}\text{ K}^{-1}$ RRR					
	30	100	300	1000	3000	10000		30	100	300	1000	3000	10000
1	29	98	295	984	2954	9842	40	631	1172	1652	1972	2096	2143
2	57	195	589	1966	5892	19521	45	581	980	1274	1440	1497	1519
3	86	292	883	2941	8765	28499	50	526	817	997	1087	1117	1128
4	114	390	1175	3897	11475	35887	60	430	588	664	696	706	710
5	143	487	1463	4817	13885	40840	70	361	454	492	507	512	513
6	171	583	1746	5677	15853	43072	80	312	372	394	403	405	406
7	200	678	2020	6452	17272	42980	90	278	320	334	340	341	342
8	228	772	2282	7116	18109	41300	100	255	286	297	300	301	302
9	256	864	2528	7651	18406	38717	150	223	239	244	245	246	246
10	284	953	2755	8044	18260	35708	200	222	234	237	238	239	239
12	338	1122	3133	8420	17095	29474	250	224	233	235	236	237	237
14	391	1272	3398	8340	15360	23801	300	226	234	236	237	237	237
16	442	1400	3544	7960	13478	19074	400	231	237	239	239	239	239
18	489	1500	3582	7418	11658	15308	500	230	235	237	237	237	237
20	532	1572	3534	6801	9997	12366	600	226	230	231	231	231	231
25	617	1628	3178	5227	6706	7527	700	220	229	224	224	224	224
30	662	1542	2666	3843	4492	4791	800	214	217	217	218	218	218
35	664	1373	2130	2756	3036	3151	900	209	212	212	212	212	212

REFERENCE

1. Hust, J. G; Lankford, A. B. *Thermal Conductivity of Aluminium, Copper, Iron and Tungsten for Temperatures from 1 K to the Melting Point*, National Bureau of Standards Internal Report 84-3007, U.S. Department of Commerce: Gaithersburg, **1984**.

10.6.2. Copper

Physical property: Thermal conductivity, λ
Units: W m^{-1} K^{-1}
Recommended reference material: Copper (Cu)
Range of variables: 1 to 1300 K
Physical state within the range: solid

Contributor: B. Le Neindre

Intended usage: Copper can be used for the calibration of thermal conductivity cells used to measure the thermal conductivity of highly conducting solids within the temperature range 1 to 900 K.

Sources of supply and/or methods of preparation: Copper of purity 0.999 mass fraction can be purchased from many suppliers.

Pertinent physicochemical data: Twenty-two of forty-four references were selected from those listed in reference 1 as primary data sets. The data cover a temperature range from 0.2 to 1250 K and a range of RRR from 20 to 1800. Generally, the RRR of commercially pure coppers are in the range 50 to 500. The ice point intrinsic resistivity of copper, ρ_i, is 15.4×10^{-9} Ω m.

The following values of the parameter A_i of equation 10.2.6.4 were obtained by a non-linear least-square fit:

$$A_1 = 1.754 \times 10^{-8} \qquad A_5 = 70$$
$$A_2 = 2.763 \qquad A_6 = 1.765$$
$$A_3 = 1102 \qquad A_7 = 0.838/B_r^{0.1661}$$
$$A_4 = -0.165$$

with $B_r = B/0.0003$.

The residual term R_c is expressed by

$$\begin{aligned}R_c/\text{m K W}^{-1} = &-0.00012\{\ln[x/420]\}\exp[-(\{\ln[x/470]\}/0.7)^2] \\ &-0.00016\{\ln[x/73]\}\exp[-(\{\ln[x/87]\}/0.45)^2] \\ &-0.00002\{\ln[x/18]\}\exp[-(\{\ln[x/21]\}/0.5)^2]\end{aligned} \qquad (10.6.2.1).$$

where $x = T/\text{K}$. The maximum deviations of the primary data sets from equation 10.2.6.4 are of the order of $\pm 10\%$. Thermal conductivities values calculated from the equation are given in table 10.6.2.1.

REFERENCE

1. Hust, J. G; Lankford, A. B. *Thermal Conductivity of Aluminium, Copper, Iron and Tungsten for Temperatures from 1 K to the Melting Point*, National Bureau of Standards Internal Report 84-3007, U.S. Department of Commerce: Gaithersburg, **1984**.

10. Thermal Conductivity

Table 10.6.2.1. Thermal conductivity of copper

$\dfrac{T}{K}$	$\lambda/\text{W m}^{-1}\text{ K}^{-1}$ RRR					$\dfrac{T}{K}$	$\lambda/\text{W m}^{-1}\text{ K}^{-1}$ RRR				
	30	100	300	1000	3000		30	100	300	1000	3000
1	46	156	471	1574	4726	50	731	1002	1135	1196	1216
2	91	312	942	3147	9434	60	597	740	799	824	832
3	137	468	1413	4710	14044	70	513	601	634	647	651
4	183	624	1880	6243	18380	80	465	526	549	557	560
5	228	779	2343	7715	22170	90	437	485	502	508	510
6	274	933	2796	9075	25084	100	421	461	475	480	482
7	319	1085	3232	10260	26834	150	396	419	426	429	430
8	365	1235	3642	11197	27328	200	391	407	413	414	415
9	409	1380	4015	11836	26756	250	388	401	405	407	407
10	454	1520	4343	12172	25496	300	386	397	400	401	402
12	541	1778	4844	12127	22264	400	383	391	393	394	394
14	624	2002	5144	11544	19150	500	379	385	387	388	388
16	703	2186	5267	10725	16398	600	374	379	381	381	381
18	777	2324	5231	9771	13924	700	368	372	374	374	374
20	843	2408	5054	8727	11683	800	362	365	367	367	367
25	960	2381	4215	6135	7271	900	356	359	360	360	360
30	999	2119	3245	4151	4573	1000	350	352	353	353	354
35	970	1784	2436	2859	3028	1100	344	347	347	348	348
40	900	1467	1841	2047	2122	1200	339	341	342	342	342
45	814	1205	1423	1531	1568	1300	335	337	337	338	338

10.6.3. Iron

Physical property: Thermal conductivity, λ
Units: W m^{-1} K^{-1}
Recommended reference material: Iron (Fe)
Range of variables: 1 to 1000 K
Physical state within the range: solid
Contributor: B. Le Neindre

Intended usage: Iron can be used for the calibration of thermal conductivity cells used to measure the thermal conductivity of highly conducting solids within the temperature range 1 to 1000 K.

Sources of supply and/or methods of preparation: Iron of purity 0.9999 mass fraction can be purchased from many suppliers. Samples of electrolytic iron rod samples are available from supplier (H). SRM 1463 is 0.63 cm in diameter × 5.0 long and SRM is 3.17 cm diameter × 50.0 cm long. Both samples have a thermal conductivity at 293 K of 77.9 W m^{-1} K^{-1}, and they are certified from 6 to 1000 K. The ice point intrinsic resistivity of iron, ρ_i, is 87×10^{-9} Ω m.

Pertinent physicochemical data: A set of 15 references from 41 listed of references 1 were selected as primary data sets. The data cover a range of temperatures from 1.5 to 1000 K and a range of RRR from 4 to 200.

The following values for the parameters A_i of equation 10.2.6.4 were obtained by a non-linear least-squares fit:

$$A_1 = 166.9 \times 10^{-8} \qquad A_5 = 238.6$$
$$A_2 = 1.868 \qquad A_6 = 1.392$$
$$A_3 = 1.503 \times 10^{-5} \qquad A_7 = 0.0.$$
$$A_4 = -1.22$$

The residuals were represented by

$$(R_c/\text{m K W}^{-1}) = -0.004\{\ln[x/440]\}\exp[-(\{\ln[(x/650]\}/0.8)^2] \\ - 0.002[\ln(x/90)]\exp(-\{[\ln(x/90)]/0.45\}^2) \qquad (10.6.3.1).$$

where $x = T/\text{K}$. The deviations of the primary data sets from equation 10.2.6.4 are no greater than ±10%.

Thermal conductivity values calculated from the equation are given in table 10.6.3.1.

REFERENCE

1. Hust, J. G; Lankford, A. B. *Thermal Conductivity of Aluminium, Copper, Iron and Tungsten for Temperatures from 1 K to the Melting Point*, National Bureau of Standards Internal Report 84-3007, U.S. Department of Commerce: Gaithersburg, **1984**.

Table 10.6.3.1. Thermal conductivity of iron

$\dfrac{T}{K}$	$\lambda/\text{W m}^{-1}\text{ K}^{-1}$ RRR				$\dfrac{T}{K}$	$\lambda/\text{W m}^{-1}\text{ K}^{-1}$ RRR			
	10	30	100	300		10	30	100	300
1	2.5	8.1	28	84	45	91	204	336	410
2	5.1	16.3	56	168	50	94	194	292	340
3	7.6	24	83	251	60	96	170	225	247
4	10.1	32	111	333	70	94	150	183	195
5	12.6	41	138	414	80	92	133	156	164
6	15.2	49	166	492	90	89	122	138	144
7	17.7	57	192	567	100	87	114	126	130
8	20	65	218	637	150	81	94	100	102
9	23	73	244	702	200	78	88	91	92
10	25	81	269	761	250	76	82	85	85
12	30	96	315	858	300	72	77	79	79
14	35	111	357	925	400	64	67	68	69
16	40	125	393	961	500	58	60	60	61
18	45	139	422	970	600	52	53	54	54
20	49	152	445	957	700	46	47	47	47
25	61	179	471	863	800	41	41	42	42
30	71	198	462	735	900	36	37	37	37
35	79	208	429	609	1000	32	32	33	33
40	86	210	384	500					

10.6.4. Tungsten

Physical property: Thermal conductivity, λ
Units: $\text{W m}^{-1}\text{ K}^{-1}$
Recommended reference material: Tungsten (W)
Range of variables: 1 to 3000 K
Physical state within the range: solid
Class: Calibration and test material
Contributor: B. Le Neindre

Intended usage: Tungsten can be used for the calibration of thermal conductivity cells used to measure the thermal conductivity of highly conducting solids within the temperature range 1 to 3000 K. The ice point intrinsic resistivity of tungsten, ρ_i, is 48.4×10^{-9} Ω m.

Sources of supply and/or methods of preparation: Tungsten of purity 0.9999 mass fraction can be purchased from many suppliers.

Pertinent physicochemical data: Samples of tungsten rod are available from supplier (H). SRM 1465 is sintered tungsten 0.32 cm in diameter × 5.0 cm long, SRM 1466 is sintered tungsten 0.64 cm in diameter × 5.0 cm long, SRM 1467 is arc-cast tungsten 0.83 cm in diameter × 5.0 cm long, while SRM 1468 is arc-cast tungsten 1.02 cm in diameter × 5.0 cm long. All samples have a nominal thermal inductivity at 293 K of 173 W m^{-1} K^{-1}, and they are certified from 4 to 3000 K. The ice point intrinsic resistivity of tungsten, ρ_i, is 48.4×10^{-9} Ω m.

A set of 13 sets of data selected from 39 references listed in reference 1 were chosen as primary data. The primary data are in the temperature range from 2 to 3000 K and in the RRR range from 30 to 170. The following values of the parameter A_i of equation 10.2.6.4 were obtained by a non-linear least-square fit:

$$A_1 = 31.70 \times 10^{-8} \qquad A_5 = 69.94$$
$$A_2 = 2.29 \qquad A_6 = 3.557$$
$$A_3 = 541.3 \qquad A_7 = 0.0$$
$$A_4 = -0.22$$

The systematic residual is given by

$$\begin{aligned}(R_c/\text{m K W}^{-1}) = &-0.00085\{\ln[x/130]\}\exp[-(\{\ln[x/230]\}/0.7)^2]\\ &+ 0.00015\exp[-(\{\ln[x/3500]\}/0.8)^2]\\ &+ 0.0006\{\ln[x/90]\}\exp[-(\{\ln[x/80]\}/0.4)^2]\\ &+ 0.0003\{\ln[x/24]\}\exp[-(\{\ln[x/33]\}/0.5)^2]\end{aligned} \qquad (10.6.4.1).$$

where $x = T/\text{K}$. The deviations of the primary data sets from equation 10.2.6.4 are no greater than $\pm 7\%$.

Thermal conductivity values calculated from the equation are given in table 10.6.4.1.

REFERENCE

1. Hust, J. G; Lankford, A. B. *Thermal Conductivity of Aluminium, Copper, Iron and Tungsten for Temperatures from 1 K to the Melting Point*, National Bureau of Standards Internal Report 84-3007, U.S. Department of Commerce: Gaithersburg, **1984**.

Table 10.6.4.1. Thermal conductivity of tungsten

$\dfrac{T}{K}$	λ/W m^{-1} K^{-1} RRR			$\dfrac{T}{K}$	λ/W m^{-1} K^{-1} RRR			$\dfrac{T}{K}$	λ/W m^{-1} K^{-1} RRR		
	30	100	300		30	100	300		30	100	300
1	14.6	50	151	35	321	586	768	800	124	125	126
2	29	100	302	40	306	494	600	900	121	122	122
3	44	150	452	45	285	418	483	1000	118	119	119
4	59	200	602	50	262	357	398	1100	115	116	116
5	73	249	749	60	226	281	302	1200	113	114	114
6	88	299	894	70	211	250	264	1300	111	111	112
7	102	347	1033	80	204	236	246	1400	109	110	110
8	117	395	1166	90	199	225	234	1500	107	108	108
9	131	442	1291	100	195	217	224	1600	106	106	106
10	145	488	1404	150	184	197	201	1800	103	103	103
12	173	574	1595	200	180	189	191	2000	100	101	101
14	201	651	1730	250	175	182	184	2200	98	99	99
16	227	718	1802	300	169	174	176	2400	96	97	97
18	251	768	1803	400	155	158	159	2600	95	95	95
20	273	799	1734	500	143	145	146	2800	93	93	93
25	311	786	1378	600	135	136	137	3000	92	92	92
30	325	692	1020	700	129	130	130				

10.6.5. Stainless Steel

Physical property: Thermal conductivity, λ
Units: W m^{-1} K^{-1}
Recommended reference material: Stainless steel
Range of variables: 1 to 1200 K
Physical state within the range: solid
Class: Calibration and test material

Contributor: B. Le Neindre

Intended usage: Stainless steel can be used for the calibration of thermal conductivity cells used to measure the thermal conductivity of highly conducting solids within the temperature range 1 to 1200 K.

Sources of supply and/or methods of preparation: Stainless steel can be purchased from supplier H as SRM 1960 to 1962. Samples have a thermal conductivity at 293 K of W m^{-1} K^{-1} and they are certified from 1 to 1200 K.

Pertinent physicochemical data: This is a high alloy metal and the relative importance of the conduction and scattering mechanisms are different than for a pure metal. The following values for the parameters of equation 10.2.6.4 were obtained by a non-linear least-squares fit where

$$R_p/\text{m K W}^{-1} = B/(T/\text{K})^n \tag{10.6.5.1}$$

and

$$B = 15.2 \qquad A_4 = 0.592$$
$$n = 1.211 \qquad A_5 = 60$$
$$A_1 = 2.477 \times 10^{-4} \qquad A_6 = -0.1436$$
$$A_2 = 1.303 \qquad A_7 = 0.0.$$
$$A_3 = 1.918$$

Table 10.6.5.1. Thermal conductivity of stainless steel

$\dfrac{T}{\text{K}}$	$\dfrac{\lambda}{\text{W m}^{-1}\text{ K}^{-1}}$	$\dfrac{T}{\text{K}}$	$\dfrac{\lambda}{\text{W m}^{-1}\text{ K}^{-1}}$	$\dfrac{T}{\text{K}}$	$\dfrac{\lambda}{\text{W m}^{-1}\text{ K}^{-1}}$	$\dfrac{T}{\text{K}}$	$\dfrac{\lambda}{\text{W m}^{-1}\text{ K}^{-1}}$
2	0.152	14	1.588	60	6.98	400	16.16
3	0.249	16	1.858	70	7.72	500	17.78
4	0.352	18	2.132	80	8.34	600	19.23
5	0.462	20	2.407	90	8.85	700	20.54
6	0.575	25	3.092	100	9.30	800	21.75
7	0.693	30	3.763	150	10.94	900	22.86
8	0.814	35	4.404	200	12.20	1000	23.90
9	0.938	40	5.01	250	13.31	1100	24.86
10	1.064	45	5.57	300	14.32	1200	25.77
12	1.323	50	6.08				

Thermal conductivity values calculated from equation 10.2.6.4 are given in table 10.6.5.1.

10. Thermal Conductivity

REFERENCE

1. Hust, J. G; Lankford, A. B. *Update of thermal conductivity and electrical resistivity of electrolytic iron, tungsten and stainless steel*, National Bureau of Standards Special Publication 260-90, U.S. Department of Commerce: Gaithersburg, **1984**.

10.6.6. Graphite

Physical property: Thermal conductivity, λ
Units: W m^{-1} K^{-1}
Recommended reference material: Graphite
Range of variables: 5 to 2500 K
Physical state within the range: solid
Contributor: B. Le Neindre

Intended usage: Graphite can be used for the calibration of thermal conductivity cells used to measure the thermal conductivity of highly conducting solids within the temperature range 5 to 2500 K.

Sources of supply and/or methods of preparation: Graphite as AXM-5Q1 graphite can be purchased from supplier H. The samples are certified from 5 to 2500 K.

Pertinent physicochemical data: A fine grained isotropic graphite has been investigated by the National Bureau of Standards for use as an extended temperature range reference material (Ref. 1). The thermal conductivity of AMX-5Q1 graphite for a room temperature electrical resistivity of $14.5 \times 10^{-6} \Omega$ m and density of 1730 kg m^{-3} was calculated from

$$\lambda/(\text{W m}^{-1}\text{ K}^{-1}) = \lambda_b M_1 M_2 \tag{10.6.6.1}$$

where

$$\lambda_b = \frac{C_1(T/\text{K})^{C_2}}{(1 + C_3(T/\text{K})^{C_4})^{C_5}} \tag{10.6.6.2}$$

$$M_1 = [D_1 + D_2 d_o/(\text{kg m}^{-3})]/(\rho_o/\Omega \text{ m}) \tag{10.6.6.3}$$

$$M_2 = E_1 + E_2 \ln\left(\frac{T}{T_1}\right) \ln\left(\frac{T}{T_2}\right) \ln\left(\frac{T}{T_3}\right) \ln\left(\frac{T}{T_4}\right)$$

$$\ln\left(\frac{T}{T_5}\right) \ln\left(\frac{T}{T_6}\right) \tag{10.6.6.4}$$

where

$$C_1 = 0.000537 \qquad T_1 = 5.4 \text{ K}$$
$$C_2 = 2.589 \qquad T_2 = 15 \text{ K}$$
$$C_3 = 0.000202 \qquad T_3 = 58 \text{ K}$$
$$C_4 = 1.678 \qquad T_4 = 180 \text{ K}$$
$$C_5 = 2.02 \qquad T_5 = 1000 \text{ K}$$
$$D_1 = -1.851 \times 10^{-5} \qquad T_6 = 1700 \text{ K}$$
$$D_2 = 1.908 \times 10^{-8} \qquad d_o = 1730 \text{ kg m}^{-3}$$
$$E_1 = 1 \qquad \rho_o = 14.15 \times 10^{-6} \text{ } \Omega \text{ m}$$
$$E_2 = 0.0012$$

Table 10.6.6.1. Thermal conductivity of AXM-6Q1 graphite calculated for room temperature electrical resistivity $= 14.15 \times 10^{-6}$ Ω m and $\rho = 1730$ kg m^{-3}

T/K	λ/W m^{-1} K^{-1}	T/K	λ/W m^{-1} K^{-1}	T/K	λ/W m^{-1} K^{-1}	T/K	λ/W m^{-1} K^{-1}
5	0.0354	55	12.59	200	78.9	1200	49.48
6	0.0537	60	15.03	220	83.2	1300	46.68
7	0.0783	65	17.59	240	86.4	1400	44.28
8	0.1099	70	20.25	260	88.8	1500	42.22
9	0.1494	75	23.00	280	90.4	1600	40.46
10	0.1971	80	25.81	300	91.3	1700	38.95
15	0.573	85	28.65	400	90.2	1800	37.67
20	1.201	90	31.51	500	84.6	1900	36.58
25	2.095	95	34.37	600	78.0	2000	35.66
30	3.255	100	37.21	700	71.7	2100	34.89
35	4.675	120	48.16	800	65.9	2200	34.25
40	6.33	140	57.9	900	60.9	2300	33.72
45	8.23	160	66.4	1000	56.5	2400	33.29
50	10.32	180	73.3	1100	52.7	2500	32.96

Table 10.6.6.1 contains recommended value of λ for temperatures from 5 to 2500 K. The uncertainty of λ values are estimated to be $\pm 2\%$ at temperatures below 300 K and $\pm 10\%$ at 2600 K.

10. Thermal Conductivity

This graphite can be useful thermal conductivity standard with the following limitations.

(i) Only rods that show the smallest electrical resistivity versus position dependence should be selected.

(ii) Electrical resistivity and density values of the specimen at room temperature must be known.

(iii) Specimens with room temperature electrical resistivities outside the range 13.0 to 15.0 10^{-6} Ω m and densities outside the range 1700 to 1750 kg m^{-3} should be excluded.

Thermal conductivity values calculated from equation 10.6.6.1 are given in table 10.6.5.1.

REFERENCE

1. Hust, J. G. *A fined-grained isotopic graphite for use as NBS thermophysical property RM's from 5 to 2500 K*, National Bureau of Standards Special Publication 260-90, U.S. Department of Commerce: Gaithersburg, **1984**.

10.7. Reference materials for the thermal conductivity of solids – low thermal conductivity materials

10.7.1. Glass fibre board

Physical property: Thermal conductivity, λ
Units: W m^{-1} K^{-1}
Recommended reference material: Glass fibre board
Range of variables: 260 to 330 K
Physical state within the range: solid
Contributor: B. Le Neindre

Intended usage: Glass fibre board can be used for the calibration of thermal conductivity cells used for the measurement of the thermal conductivity of low thermal conductivity materials in the temperature range 100 to 330 K.

Sources of supply and/or methods of preparation: Glass fibre board (60.0 × 60.0 × 2.54 cm) is available as SRM 1450b from supplier (H).

Pertinent physicochemical data: A glass fiberboard was established as a standard reference material (S.R.M.) for thermal conductivity, by the National Bureau of Standards. It is certified over the temperature range from 100 to 330 K and the density range from 113 to 145 kg m^{-3} (Refs. 1,2). The material consists of fibrous glass made into a semirigid board with a phenolic binder. The fibers are oriented with their lengths extending primarily parallel to the face of the board. The fill glass is air or nitrogen at atmospheric pressure

(8.4×10^4 Pa at Boulder, Co). The temperature and density dependences of λ of this board are given by:

$$\lambda(T,\rho)/(\text{W m}^{-1}\text{ K}^{-1}) = a_1 + a_2\rho/(\text{kg m}^{-3}) + a_3(T/\text{K}) + a_4(T/\text{K})^3 \\ + a_5 \exp-[(T/\text{K} - 180)/75]^2 \qquad (10.7.1.1)$$

where the values of parameters, a_i depend on the specimens, ρ is the bulk density, T is the temperature and λ is the apparent thermal conductivity. For some of the lots the following a_i parameters were determined by least squares fitting as:

$$a_1 = -2.228 \times 10^{-3}$$
$$a_2 = 0.02743 \times 10^{-3}$$
$$a_3 = 0.1063 \times 10^{-3}$$
$$a_4 = 64.73 \times 10^{-12}$$
$$a_5 = 1.157 \times 10^{-3}.$$

Table 10.7.1.1. Certified values of thermal resistance of a 2.54 cm thick specimen, R_o, as a function of density and temperature

$\dfrac{T}{\text{K}}$	$R_o/(\text{m}^2\text{ K W}^{-1})$					$\dfrac{T}{\text{K}}$	$R_o/(\text{m}^2\text{ K W}^{-1})$				
	$\rho/(\text{kg m}^{-3})$						$\rho/(\text{kg m}^{-3})$				
	110	120	130	140	150		110	120	130	140	150
100	2.143	2.094	2.049	2.004	1.961	220	0.987	0.977	0.966	0.956	0.947
110	1.946	1.906	1.867	1.831	1.795	230	0.949	0.939	0.930	0.921	0.912
120	1.780	1.747	1.714	1.683	1.653	240	0.913	0.905	0.896	0.887	0.879
130	1.640	1.611	1.583	1.557	1.531	250	0.880	0.872	0.864	0.856	0.848
140	1.519	1.495	1.471	1.448	1.426	260	0.848	0.841	0.833	0.826	0.818
150	1.416	1.395	1.374	1.354	1.334	270	0.818	0.811	0.804	0.797	0.790
160	1.327	1.308	1.290	1.272	1.255	280	0.790	0.783	0.776	0.770	0.764
170	1.250	1.234	1.217	1.202	1.186	290	0.762	0.756	0.750	0.744	0.738
180	1.184	1.169	1.154	1.140	1.126	300	0.736	0.730	0.724	0.719	0.713
190	1.126	1.112	1.099	1.086	1.073	310	0.711	0.706	0.700	0.695	0.690
200	1.074	1.062	1.050	1.038	1.027	320	0.687	0.682	0.677	0.672	0.667
210	1.028	1.017	1.006	0.995	0.985	330	0.665	0.660	0.655	0.651	0.646

The deviations of equation 10.7.1.1 with respect to the experimental values are of the order of ±1.5%. The radiant heat transfer is small for this material. For certificate purposes, values of thermal resistance, R_o at a thickness of 0.0254 m are desirable.

R_o values are defined by $R_o = 0.0254/\lambda(T,\rho)$ are given in table 10.7.1.1.

The R value at differenct thickness, L, are calculated from $R = R_o L/0.0254$; however, the material is certified only for thickness close to 2.54 cm. The specimens should not be under excessive pressure between apparatus plates and compression of the specimen to a thickness less than 2.4 cm should be avoided. Experimental thermal resistance error are expected to be ±2% from 250 to 350 K and ±3% at 100 K.

The values in table 10.7.1.1 have been corrected for the thermal expansion of the measurement plates.

REFERENCE

1. Hust, J. G. *Glass fiberboard S.R.M. for thermal resistance*, National Bureau of Standards Special Publication 260-98, U.S. Department of Commerce: Gaithersburg, **1985**.
2. Siu, M. C. I.; Hust, J. G. *Standard reference material 1450 b, thermal resistance-fibrous glass board*, U.S. Department of Commerce: Gaithersburg, **1982**.

10.7.2. Resin-bonded glass fibreboard

Physical property: Thermal conductivity, λ
Units: W m^{-1} K^{-1}
Recommended reference material: Resin-bonded glass fibre board
Range of variables: 170 to 370 K
Physical state within the range: solid
Contributor: B. Le Neindre

Intended usage: Resin-bonded glass fibre board can be used for the calibration of thermal conductivity cells used to measure the thermal conductivity of low thermal conductivity materials in the temperature range 170 to 370 K.

Sources of supply and/or methods of preparation: A working group from the European Economic Community have chosen a semi-rigid resin-bonded glass fibre board with density ranging from 84 to 92 kg m^{-3}. It is available in panels $1 \times 1 \times 0.035$ m from supplier (I).

Pertinent physicochemical data: The fibres in these boards are uniformly distributed and randomly oriented in planes which are parallel to the face of the board. Consequently, the thermal conductivity perpendicular to the face of the board, which is the useful property in reference measurements, is smaller than the thermal conductivity parallel to the face of the board.

The manufacturer specifications for this material are the following:

Nominal dimensions: Thickness: 0.035 m
Density: 88 kg m^{-3}

Most frequent fibre diameter: 4.1×10^{-6} m

Resin content (mass fraction with respect to glass): (16 ± 3) per cent

Module of compression of board normal to its face at 5 per cent relative deformation: $(2.0 \pm 0.17) \times 10^5$ Pa.

The density is the ratio of the mass of the specimen by the volume. The mass is determined after drying the specimens in air at 385 K for a least 8 hours in a temperature-controlled oven and by weighing them after they have been allowed to reach temperature equilibrium with the laboratory atmosphere of 295 K at 40 to 60 per cent relative humidity.

Due to its open structure, the material can absorb moisture in proportion of the relative humidity of the environment. To avoid an ensuing effect on the thermal conductivity, the following recommendation has been made (Ref. 1).

For operation at or above ambient temperature, the specimens shall be subjected to the thermal conditioning as stated above. Unless the specimens are placed in a vapour-tight envelope during the tests, care must be taken to avoid water migration by diffusion from the surrounding thermal insulation into the specimens. This can be avoided by careful drying of the insulating material in a manner similar to that applied to the specimens.

Measurements conducted below ambient temperature require special precautions to avoid ingress of humid atmospheric air into the apparatus, causing condensation and/or freezing of water vapour within the specimens. In general, this is avoided by making such arrangements that the dew point of the atmospheric surrounding the specimens and its adjacent insulation is kept at least 5 K below the cold plate temperature.

To ensure adequately uniform contact between the specimens and the hot and cold plates of the apparatus, it had been found essential that the pressure exerted upon the specimens in the assembled apparatus should not be less than 1 kPa and not more than 2 kPa.

It is evident that if the apparatus is not equipped with a secondary temperature controlled guard around the outer edge of the specimen, a correction has to be made for lateral heat losses. Further, the thickness of the specimen under experiment must be measured accurately.

The temperature dependence of the thermal conductivity is given by

$$\lambda[\rho/(\text{kg m}^{-3}) = 88]/(\text{W m}^{-1} \text{ K}^{-1}) = 0.00141596 + 0.00010285(T/\text{K}) \qquad (10.7.2.1).$$

The thermal conductivities calculated from equation 10.7.1.2 are listed with their uncertainties in table 10.7.1.2.

10. Thermal Conductivity

Table 10.7.2.1. Thermal conductivity of resin-bonded glass fibre board at a density of 88 kg m^{-3}

$\dfrac{T}{K}$	$\dfrac{\lambda[\rho/(\text{kg m}^{-3})=88]}{\text{W m}^{-1}\text{ K}^{-1}}$	Uncertainty %	$\dfrac{T}{K}$	$\dfrac{\lambda[\rho/(\text{kg m}^{-3})=88]}{\text{W m}^{-1}\text{ K}^{-1}}$	Uncertainty %
170	0.0189_0	±2.4	280	0.0302_1	±1.5
180	0.1099_3	±2.3	290	0.0312_4	±1.4
190	0.0209_6	±2.2	300	0.0322_7	±1.4
200	0.0219_9	±2.1	310	0.0333_0	±1.3
210	0.0230_1	±1.9	320	0.0343_8	±1.3
220	0.0240_4	±1.9	330	0.0353_6	±1.3
230	0.0250_7	±1.8	340	0.0363_8	±1.2
240	0.0261_0	±1.7	350	0.0374_1	±1.2
250	0.0271_3	±1.6	360	0.0384_4	±1.2
260	0.0281_6	±1.6	370	0.0394_7	±1.2
270	0.0291_8	±1.5			

If the density of test specimens deviates from their nominal value, 88 kg m^{-3}, the thermal conductivity at any temperature between 170 K and 370 K and densities from 84 kg m^{-3} to 92 kg m^{-3} is given by

$$\lambda(\rho) = \lambda[\rho/\text{kg m}^{-3} = 88]/\{1 + 0.00130[\rho/(\text{kg m}^{-3}) - 88]\} \tag{10.7.2.2}$$

REFERENCE

1. Ziebland, H. *Certification report on a reference material for the thermal conductivity of insulating materials between 170 K and 370 K. Commission of the European Communities*, EUR 7677 EN **1982**.

10.7.3. Blanket

Physical property: Thermal conductivity, λ
Units: W m^{-1} K^{-1}
Recommended reference material: Blanket
Range of variables: 100 to 330 K
Physical state within the range: solid

Contributor: B. Le Neindre

Intended usage: Low density insulation blanket can be used for the calibration of thermal conductivity cells used to measure the thermal conductivity of highly conducting solids within the temperature range 100 to 330 K.

Sources of supply and/or methods of preparation: A transfer standard of low density glass fibre insulation is available from supplier (H).

Pertinent physicochemical data: A transfer standard has been selected by the National Bureau of Standards. The material consists of fibrous glass made into a low density blanked bonded with phenolic resin. The fibers average about 5 μm in diameter and are oriented with their lengths extending primarily parallel to the face of the blanket. The binder content is about 6% by weight.

Table 10.7.3.1. Certified values of thermal resistance of a 0.0254 m thick specimen, R_0, as a function of density and temperature

$\dfrac{T}{K}$	$R_0/(m^2\ K\ W^{-1})$				$\dfrac{T}{K}$	$R_0/(m^2\ K\ W^{-1})$			
	$\rho/(kg\ m^{-3})$					$\rho/(kg\ m^{-3})$			
	10	12	14	16		10	12	14	16
100	2.475	2.443	2.403	2.358	220	0.898	0.939	0.968	0.988
110	2.219	2.202	2.176	2.144	230	0.842	0.884	0.914	0.936
120	2.000	1.995	1.979	1.957	240	0.789	0.832	0.864	0.887
130	1.812	1.816	1.809	1.795	250	0.739	0.783	0.816	0.841
140	1.649	1.661	1.661	1.654	260	0.693	0.737	0.771	0.796
150	1.508	1.526	1.532	1.530	270	0.650	0.694	0.728	0.754
160	1.386	1.409	1.420	1.422	280	0.609	0.653	0.688	0.715
170	1.279	1.306	1.321	1.328	290	0.571	0.615	0.649	0.677
180	1.185	1.216	1.234	1.244	300	0.536	0.579	0.613	0.641
190	1.102	1.136	1.157	1.170	310	0.503	0.546	0.580	0.607
200	1.027	1.064	1.088	1.104	320	0.472	0.514	0.548	0.575
210	0.960	0.999	1.026	1.043	330	0.444	0.485	0.518	0.545

The thermal conductivity was fitted in the temperature range from 100 to 350 K and the density range from 10.5 to 16 kg m^{-3} by the equation

$$\lambda(T,\rho)/(\text{W m}^{-1}\text{ K}^{-1}) = a_1 + a_2\rho/(\text{kg m}^{-3}) + a_3 T/\text{K} + a_4(T/\text{K})^3/[\rho/(\text{kg m}^{-3})]$$
$$+ a_5 \exp -[(T/\text{K} - 180)/75]^2 \quad (10.7.3.1)$$

where

$$a_1 = -0.1059 \times 10^{-3}$$
$$a_2 = 0.1378 \times 10^{-3}$$
$$a_3 = 0.7714 \times 10^{-4}$$
$$a_4 = 0.8472 \times 10^{-8}$$
$$a_5 = 0.1339 \times 10^{-2}.$$

The deviations of this correlation with respect to experimental data are within ±2.1%.

For certification purposes, values of thermal resistance $R_o = 0.0254/\lambda(T,\rho)$ are given in table 10.7.3.1 with an accuracy of ±3% in the temperature range 250 to 300 K an increasing up to ±5% at 100 K.

The values in table 10.7.3.1 have been corrected for the thermal expansion of the measurement plates.

REFERENCE

1. Hust, J. G. *Glass Fiberblanket S.R.M for thermal resistance*, National Bureau of Standards Special Publication 260-103, U.S. Department of Commerce: Gaithersburg, **1985**.

10.8. Contributor

B. Le Neindre
L.I.M.H.P.–C.N.R.S.
Université Paris-Nord
93430-Villetaneuse (France)

10.9. List of suppliers

A. Air Liquide,
   9 avenue Descartes,
   92350 Le Plessis Robinson (France).

B. Airgaz,
   Centre d´ affaires Paris-Nord,
   Bât. Ampere 5–Boite 234,
   931153 Le Blanc Mesnil (France).

C. British Oxygen Co.,
   Dear Park Rd.,
   London SW1 (UK).

D. Messer Griesheim,
   G.M.B.H. Industriegase,
   4 Dusseldorf 2,
   Hombergerstrass E 12 (FRG).

E. Matheson Gas Products,
   P.O. Box E
   Lyndhurst, NJ 07071 (USA).

F. Union Carbide Corp.,
   Linde Division,
   270 Park Ave.,
   New York, NY (USA).

G. Airco Industrial,
   Gases Div.,
   Airco, Inc.,
   575 Mountain Ave.,
   Murray Hill, NJ (USA).

H. National Bureau of Standards,
   U.S. Department of Commerce,
   Office of Standard Reference Materials,
   Room B3, Chemistry,
   Gaithersburg, MD 20899 (USA).

I. Centre de Recherches Industrielles
   de Rantigny,
   Isover-St Gobain (France)

11

SECTION: ELECTROLYTIC CONDUCTIVITY

COLLATORS: E. JUHÁSZ, K. N. MARSH

CONTENTS:

11.1. Introduction

11.2. Methods of measurement

11.3. Reference materials for electrolytic conductivity

    11.3.1. Aqueous potassium chloride (KCl) solutions (specific concentrations)

    11.3.2. Aqueous potassium chloride (KCl) solutions (calculated from equations)

11.4. Contributors

11.5. Suppliers

11.1. Introduction

Electrolytic conductivity, κ, formerly called specific conductance, is defined by the equation

$$\kappa = j/E \qquad (11.1.1)$$

where j is the electric current density and E is the electric field strength (Ref. 1). The SI unit for electrolytic conductivity is the Siemens per metre ($S\ m^{-1}$). The molar conductivity, Λ, is defined by the equation

$$\Lambda = \kappa/c \qquad (11.1.2)$$

where c is the amount of substance concentration. The SI unit for the amount of substance concentration is mole per cubic meter.

The formula unit whose concentration is c must be specified and should be given in brackets, for example $\Lambda(\text{KCl})$, $\Lambda(\text{MgCl}_2)$, $\Lambda(1/2\ \text{MgCl}_2)$, $\Lambda(2/3\text{AlCl}_3 + 1/3\text{KCl})$. The SI unit for molar conductivity is Siemens square meter per mole (S m^2 mol^{-1}) (Ref. 1).

The recommendations given here were taken from reference 2 and recalculated by use of the conversion: international ohm = 1.000495 ohm (U.S.A.). The values were then corrected for the effect of the difference between IPTS-48 and IPTS-68 (Ref. 3) by use of the temperature dependence of the conductivity of potassium chloride solutions given by Jones and Bradshaw (Ref. 2) and Bremner and Thompson (Ref. 4). A recent absolute determination of the conductivity of one of the reference solutions at 273.15, 291.15, and 298.15 K by Saulnier and Barthel (Refs. 5, 6) has confirmed, to just within the experimental uncertainties, the values of Jones and Bradshaw. Values of the electrolytic conductivity of potassium chloride solutions at other concentrations can be obtained from empirical equations expressing the dependence of conductivity on concentration within ±0.02 per cent with reference to the various reference solutions (Refs. 7 to 11).

There is a lack of reference materials for conductance measurements in the ranges above and below that covered by potassium chloride solutions, and also for molten salts. For molten salts, the present practice is to use aqueous potassium chloride solutions to determine the cell constant at 298.15 K and then correct for the effects of thermal expansion (Ref. 12).

The following provisos apply to the information on reference materials: (a) the recommended materials have not been checked independently by the IUPAC, (b) the quality of the material may change with time, (c) the quoted sources of supply may not be the exclusive sources because no attempt has been made to seek out all possible alternative sources, and (d) the IUPAC does not guarantee any material that has been recommended.

11.2. Methods of measurement

In case of an electrolytic conductor of a uniform cross section A and length l

$$\kappa = G \frac{1}{A} \tag{11.2.1}$$

where G is the conductance of the cell containing the electrolytic conductor. The quotient $1/A$ is called the cell constant, k_{cell}. The SI unit for the cell constant is the reciprocal metre (m^{-1}).

As it is difficult and expensive to construct cells that have accurately known dimensions, it is usual to measure the electrolytic conductance of an electrolyte by use of a rigid cell for which the cell constant k_{cell} has been determined by means of a reference solution of known electrolytic conductivity. The relationship used for calibration of the cell is

$$k_{\text{cell}} = \kappa/G \tag{11.2.2}$$

11. Electrolytic Conductivity

At present the only values of conductivity known sufficiently accurately to be used for reference purposes are those of aqueous solutions of potassium chloride. Potassium chloride of high purity is available and the solutions prepared from it on a mass basis as suggested by Jones and Bradshaw (Ref. 2) are reproducible and stable.

Given a cell with constant k_{cell}, the conductivity of an electrolytic conductor is determined by measuring the resistance or the conductance (R, G) of it and those of a reference material ($R_\circ$, $G_\circ$) with known conductivity. Its value can be calculated by the equation

$$\kappa = \frac{\kappa_\circ R_\circ}{R} = \frac{\kappa_\circ G}{G_\circ} \tag{11.2.3}$$

At the electrode surfaces polarisation phenomena can occur, which influence the measured quantity. To avoid this error, different measuring methods can be applied. In each case the relationship between the conductivity and the measured quantity is established by the cell constant. Some of the methods are outlined below.

1. Two electrode measurement using AC current with a frequency high enough (usually in the kHz range) to avoid polarisation effects.

    In this case the impedance (Z) is measured;

    $$Z = R + jX \tag{11.2.4}$$

    where R is the the resistance, j is the the imaginary unit ($j = \sqrt{-1}$), X is the reactance ($X = L\omega - \frac{1}{C\omega}$), where L is the inductance, and C is the capacitance.

    Similarly the AC conductivity or admittance (Y) can be determined:

    $$Y = G + jB \tag{11.2.5}$$

    where G = the conductance, B = the susceptance (capacitive and/or inductive conductance).

    For the determination of the resistance (R) or conductance (G) in an AC circuit, the imaginary parts of the impendance (Z) or admittance (Y) must be determined. For this purpose different types of bridges e.g. Wheatstone-, Thomson-, differential-bridges and measuring cells with different electrodes are used (Ref. 3).

    The electrode effects should be eliminated by making measurements at various frequencies in properly designed conductance cells using an adequate thermostat as described by Robinson and Stokes (Ref. 14).

2. Four electrode measurements with separated current transporting and potential measuring electrodes with DC and AC current (Ref. 15).

3. Inductive or capacitive measurements by non-conductive coupling between the electrolytic conductor and the electrical measuring circuit.

REFERENCES

1. IUPAC Manual of Symbols and Terminology for Physicochemical Quantities and Units, *Pure and Appl. Chem.* **1979**, *51*, 1; and its Appendix III: Electrochemical Nomenclature, *Pure and Appl. Chem.*, **1933**, *37*, 501.
2. Jones, G.; Bradshaw, B. C. *J. Amer. Chem. Soc.* **1933**, *55*, 1780.
3. The International Practical Temperature Scale for 1968, *Metrologia*, **1969**, *5*, 35.
4. Bremner, R. W.; Thompson, T. G. *J. Amer. Chem. Soc.* **1937**, *59*, 2372.
5. Saulnier, P.; Barthel, J. *Solution Chem.* **1980**, *9*, 805.
6. Marsh, K. N. *J. Solution Chem.*, **1980**, *9*, 805.
7. Lind, J. E.; Zwolenik, J. J.; Fuoss, R. M. *J. Amer. Chem. Soc.* **1959**, *81*, 1557.
8. Chiu, Ying-Chech; Fuoss, R. M. *J. Phys. Chem.* **1968**, *72*, 4123.
9. Justice, J. C. *J. Chim. Phys.* **1968**, *65*, 353.
10. Fuoss, R. M.; Hsia, K. L. *Proc. Natl. Acad. Sci. U.S.* **1966**, *57*, 1550.
11. Sanding, R.; Feistel, R.; Grosch, A.; Einfeldt, J. quoted in Janz, G. J.; Tomkins, R. P. T. *J. Electrochem. Soc.* **1977**, *55c*, 124.
12. Janz, G. J.; Tomkins, R. P. T. *J. Electrochem. Soc.* **1977**, *55c*, 124.
13. Brakunstein, J.; Robbins, G. D. *J. Chem. Educ.* **1971**, *48*, 52.
14. Robinson, R. A.; Stokes, R. H. *Electrolyte Solutions* 2nd Edition, Ch 5, Butterworths: London, **1959**.
15. Jervis, R. E.; Muir, D. R.; Butler, S. P.; Gordon, A. R. *J. Amer. Chem. Soc.* **1953**, *75*, 2855.

11.3. Reference materials for electrolytic conductivity

11.3.1. KCl solutions (specific concentrations)

Physical property: Electrolytic Conductivity

Unit: $S\ m^{-1}$
Recommended reference material: Aqueous potassium chloride (KCl) solutions (specific concentrations)
Range of variables: 0 to 25 °C, 7×10^{-2} to 11 $S\ m^{-1}$
Physical state within the range: liquid
Contributors to the first version: E. Juhász, K. N. Marsh, T. Plebanski

Intended usage: Calibration of conductivity cells.

Sources of supply and/or methods of preparation: Potassium chloride of purity not less than 99.99 mass per cent is dissolved in water that has been distilled or passed through an ion-exchange resin to reach a conductivity of at least 1.2×10^{-4} $S\ m^{-1}$ at 298.15 K. The preparation of the solutions and corrections for the conductivity of distilled water is described by Jones and Bradshaw (Ref. 1). The conductivity of the solution is given by $\kappa(\text{soln}) = \kappa(\text{ref.}) + \kappa(\text{H}_2\text{O})$. Potassium chloride certified for the present purpose is available from supplier (A). Analytical grade potassium chloride recrystallized twice from conductivity water and dried at 770 K for 24 hours is usually suitable as a reference material. The above values has been corrected to absolute ohms and to IPTS-68 as described in the introduction.

The solutions are made as follows:

solution A: 71.1352 gram potassium chloride in 1 kilogram of aqueous solution
solution B: 7.41913 gram potassium chloride in 1 kilogram of aqueous solution
solution C: 0.745263 gram potassium chloride in 1 kilogram of aqueous solution

All values given above refer to true mass "in vacuo". When weighing in air the following densities at 293.15 K may be used: KCl (solid), 1.98×10^3 kg m^{-3}: Solution A, 1.0444×10^3 kg m^{-3}; Solution B, 1.0030×10^3 kg m^{-3}; Solution C, 0.9987×10^3 kg m^{-3}. The uncertainty in the values of the conductivity of the solutions above is $\pm 0.1\%$ at the 99% confidence level.

Pertinent physicochemical data: The values given for the three reference solutions defined above (often termed 1.0, 0.1, and 0.01 demal solutions respectively, at the three temperatures are the corrected values of Jones and Bradshaw (Ref. 1). A recent absolute determination of the conductance of solution C (defined above) by Saulnier and Barthel (Ref. 2) agrees with the Jones and Bradshaw values to within 0.01 per cent at 273.15 K and 0.07 per cent at 298.15 K (Ref. 3).

These solutions are used as primary standards in Legal Metrology (Ref. 4).

Table 11.3.1.1. Conductivity κ at various temperatures.

$\dfrac{T_{68}}{K}$	A $\dfrac{\kappa}{\text{S m}^{-1}}$	B $\dfrac{10\,\kappa}{\text{S m}^{-1}}$	C $\dfrac{10^2\,\kappa}{\text{S m}^{-1}}$
273.15	6.514	7.134	7.733
291.15	9.781	11.163	12.201
298.15	11.131	12.852	14.083

REFERENCES

1. Jones, G.; Bradshaw, B. C. *J. Amer. Chem. Soc.* **1933**, *55*, 1780.
2. Saulnier, P.; Barthel, J. *J. Solution Chem.* **1979**, *8*, 847.
3. Marsh, K. N. *J. Solution Chem.* **1980**, *9*, 805.
4. OIML International Recommendation 56: Standard solutions reproducing the conductivity of electrolytes.

11.3.2. KCl solutions (calculated from equations)

Physical property: Electrolytic Conductivity
Unit: S m^{-1}
Recommended reference material: Aqueous potassium chloride (KCl) solutions (calculated from equations)
Range of variables: 0 to 25 °C, 7×10^{-2} to 11 S m^{-1}
Physical state within the range: liquid
Contributors to the first version: E. Juhász, K. N. Marsh, T. Plebanski
Contributors to the revised version: E. Juhász, K. N. Marsh

Intended usage: Calibration of conductivity cells.

Sources of supply and/or methods of preparation: The purification and preparation is described in 11.3.1.

Pertinent physicochemical data: An alternative technique for calibrating conductance cells is to use a potassium chloride solution of known concentration together with an equation representing the molar conductance over the appropriate concentration range. The recommended equations which cover the various concentration and temperature ranges are given in terms of the molar conductivity Λ, where $\Lambda = \kappa/c$.

1. In the concentration range between $c = 10^{-4}$ to 0.04 mol dm^{-3} at 298.15 K the Justice equation (Ref. 1),

$$\Lambda/(\text{mS m}^2 \text{ mol}^{-1}) = 14.983 - 9.484 x^{1/2} + 5.861 x \log x + 22.89 x - 26.42 x^{3/2} \quad (11.3.2.1)$$

where $x = c/(\text{mol dm}^{-3})$.

2. In the concentration range between $c = 0.01$ to 0.10 mol dm^{-3} at 298.15 K the Chiu-Fuoss equation (Ref. 2),

$$\Lambda/(\text{mS m}^2 \text{ mol}^{-1}) = 14.987 - 9.485 x^{1/2} + 2.547 x \ln x + 22.0 x - 22.9 x^{3/2} \quad (11.3.2.2)$$

where $x = c/(\text{mol dm}^{-3})$.

3. In the concentration range between $c = 0.05$ and 1.0 mol dm^{-3} at 298.15 K the Rostock equation (Ref. 3),

$$\Lambda/(\text{mS m}^2 \text{ mol}^{-1}) = 14.994 - 9.925x^{1/2} + 13.575x - 12.075x^{3/2} \\ + 5.787x^2 - 1.172x^{5/2} \quad (11.3.2.3)$$

where $x = c/(\text{mol dm}^{-3})$.

4. In the concentration range $c = 10^{-4}$ to 0.05 mol dm^{-3} at 291.15, 283,15, and 273.15 K the Barthel et al. equations (Ref. 4),

291.15 K:
$$\Lambda/(\text{mS m}^2 \text{ mol}^{-1}) = 12.944 - 8.035x^{1/2} + 3.286x \log x + 15.43x \\ - 14.30x^{3/2} \quad (11.3.2.4)$$

283.15 K:
$$\Lambda/(\text{mS m}^2 \text{ mol}^{-1}) = 10.7308 - 6.495x^{1/2} + 2.706x \log x + 1.254x \\ - 11.03x^{3/2} \quad (11.3.2.5)$$

273.15 K:
$$\Lambda/(\text{mS m}^2 \text{ mol}^{-1}) = 8.1655 - 4.778x^{1/2} + 2.059x \log x + 9.38x \\ - 2.93x^{3/2} \quad (11.3.2.6)$$

where $x = c/(\text{mol dm}^{-3})$.

These equations reproduce the conductivity of the appropriate reference solutions to within ± 0.015 per cent. The experimental results on which the above equations are based have been obtained using cells whose constants have been determined using the Jones and Bradshaw reference solutions and a molar mass of potassium chloride of 74.5510 g mol^{-1}. It should be noted that if authors use the Jones and Bradshaw values when expressed in International ohms on the 1948 temperature scale then their results will be approximately on those scales irrespective of the measuring devices used. The above equations have been corrected to absolute ohms and to IPTS-68 as described in the introduction. The results of Shedlovsky (Ref. 5) in the range $c = 10^{-4}$ to 0.04 mol dm^{-3}, when corrected from the Parker and Parker standard to the Jones and Bradshaw standard, from international ohm to absolute ohm, and adjusted to the 1968 temperature scale and the 1986 value for the molecular weight of KCl are on the average 0.009 mS m^2 mol^{-1} higher than the results of Justice (Ref. 6).

REFERENCES

1. Justice, J. C. *J. Chim. Phys.* **1968**, *65*, 353.
2. Chiu, Ying-Chech; Fuoss, R. M. *J. Phys. Chem.* **1968**, *72*, 4123.
3. Sanding, R.; Reistel, R.; Grosch, A.; Einfeldt, J. quoted in Janz, G. J.; Tomkins, R. P. T. *J. Electrochem. Soc.* **1977**, *55c*, 124.
4. Barthel, J.; Feuerlein, F.; Neueder, R.; Wachter, R. *J. Solution Chem.* **1980**, *9*, 209.
5. Shedloosky, T. *J. Amer. Chem. Soc.* **1932**, *54*.
6. Hamer, W. J. Private communication, **1986**.

11.4. Contributors to the revised version

E. Juhász,
National Office of Measures,
Orszagos Meresugyi Hivatal,
H-1224 Budapest XII,
Németvölgyi út 37-39,
(HUNGARY)

K. N. Marsh,
Thermodynamics Research Center,
The Texas A&M University System,
College Station, Texas   77843,
(U.S.A.)

11.5. Suppliers

A. Research and Development Centre
   for Standard Reference Materials,
   UL. Elecktoralna 2,
   PL 00-139, Warsawa
   (POLAND)

B. National Office of Measures,
   Orszagos Meresugyi Hivatal,
   1224 Budapest XII,
   Németvölgyi út 37-39
   (HUNGARY)

12

SECTION: PERMITTIVITY

COLLATOR: K. N. MARSH

CONTENTS:

- 12.1. Introduction
- 12.2. Reference materials for permittivity in the liquid state
    - 12.2.1. Cyclohexane
    - 12.2.2. Tetrachloromethane
    - 12.2.3. Benzene
    - 12.2.4. Chlorobenzene
    - 12.2.5. 1,2-Dichloroethane
    - 12.2.6. Methanol
    - 12.2.7. Nitrobenzene
    - 12.2.8. Water
- 12.3. Reference materials for permittivity in the liquid and real-gas state
    - 12.3.1. Hydrogen
    - 12.3.2. Oxygen
- 12.4. Reference materials for permittivity in the real-gas state
    - 12.4.1. Nitrogen
    - 12.4.2. Air
- 12.5. Contributors
- 12.6. List of Suppliers

12.1. Introduction

The symbol for relative permittivity (also called dielectric constant, capacivity, or specific inductive capacity) recommended by the IUPAC is ϵ_r (Ref. 1). A primary standard is not required since the relative permittivity is defined as the dimensionless ratio of the permittivity of the dielectric to the permittivity of a vacuum. It may be determined by obtaining the ratio of the capacitance of a capacitor completely filled with the dielectric, C, to the capacitance of the capacitor when evacuated, C_o. In practice, the permittivity of the dielectric is not normally compared with vacuum but with a reference gas or air. The relative permittivity of dry, carbon dioxide free, air at 293.15 K and 101.325 kPa is $1.0005364 \pm 3 \times 10^{-7}$ (Ref. 6). This value varies with the temperature and the percentage of carbon dioxide and moisture as given in section 12.4.2. The capacitance is usually measured by a bridge or resonance method and suitable cells and bridge circuits have been described (Ref. 2). The measurement of relative permittivity can be made using either an absolute or a non-absolute technique. In the absolute method, the effective capacity of the cell in vacuum, C_o, is accurately determined from a measurement *in vacuo*. An alternative to a measurement *in vacuo* is to make a measurement using a reference gas at ambient temperature and pressure, then make the small correction necessary to obtain the value of the capacitance *in vacuo*. Most three terminal and differential cell designs allow an absolute determination to be made. Some designs of two terminal cells, where the leads capacitance can be determined without recourse to the use of a reference liquid, can be used for absolute measurements (Refs. 3, 4). The procedure is non-absolute when a liquid is used to obtain C_o. In general, it is expected that a measurement made using an absolute technique enhances the probable accuracy and such measurements have been given more weight in the evaluation. Further, the more recent precision measurements, provided the purity of the sample has been adequately established, have been given a greater weighting in the evaluation. When designing a cell, notice should be taken of the problems associated with the use of metal films on glass to form the electrode (Refs. 3, 5). The relative permittivity depends on frequency and the limiting value at zero frequency is termed the static relative permittivity. In this section the values of ϵ_r reported refer to the static relative permittivity.

Equipment for measuring relative permittivity can be either calibrated or checked with fluids of known relative permittivity, hence there is a need for selected reference materials. The criteria for selecting a reference material is that it should be chemically stable and easy to purify, its conductance should be small, and sufficient measurements of high precision should have been made in order to establish the reliability of the values (Refs. 7 to 9). Further, because of the high value of ϵ_r for water, the materials (except water) must be easy to dry.

The recommended values for the liquids are given at atmospheric pressure. The term 'atmospheric pressure' as used in this document indicates that the data refer to a sample at a nominal pressure of 10^5 Pa. The recommended values for the gases (except air) are given at 293.15 K and 101.325 kPa. Maryott and Buckley (Ref. 3) give an equation for calculating $(\epsilon_r - 1)$ for a gas in the pressure range 93 to 107 kPa and the temperature range 283 to 303 K accurate $\pm$ 0.02 per cent:

$$(\epsilon_r - 1)(T,p) = (\epsilon_r - 1)(293.15 \text{ K}, 101.325 \text{ kPa}) \left[\frac{p/\text{kPa}}{101.325[1 + 0.003411(T/\text{K} - 293.15)]} \right].$$
(12.1.1)

REFERENCES

1. *Manual of Symbols and Terminology for Physicochemical Quantities and Units*, **1979** Edition, International Union of Pure and Applied Chemistry, *Pure Appl. Chem.* **1979**, *51*, 1.
2. Vaughan, W. E.; Smyth, C. P.; Powles, J. G. *Techniques of Chemistry*, Vol. I, *Physical Methods of Chemistry*, 4th ed. Weissberger, A.; Rossiter, B. W., editors, Vol. 1, Part IV, Wiley-Interscience: New York, **1972**, p. 351.
3. Sugden, S. *J. Chem. Soc.* **1983**, 768.
4. Stokes, R. H. *J. Chem. Thermodynamics* **1973**, *5*, 379.
5. Hartshorn, L.; Parry, J. V. C.; Essen, L. *Proc. Phys. Soc. B, (London)* **1955**, *88*, 422.
6. Maryott, A. A.; Buckley, F. *Table of Dielectric Constants and Electric Dipole Moments of Substances in the Gaseous States*, National Bureau of Standards Circular 537, U.S. Department of Commerce: Washington, D.C. **1953**.
7. Amey, W. G.; Hamburger, R. *Proc. Amer. Soc. Testing Mats.* **1949**, *49*, 1079.
8. Field, R. F. *Proc. Amer. Soc. Testing Mats.* **1954**, *54*, 456.
9. Essen, L.; Froome, K. D. *Proc. Phys. Soc. (London) B64,* **1951**, 862.

12.2. Reference materials for permittivity in the liquid state

12.2.1. Cyclohexane

Physical property: Relative permittivity, ϵ_r
Units: Dimensionless
Recommended reference material: Cyclohexane, (C_6H_{12})
Range of variables: 283.15 to 348.15 K, atmospheric pressure
Physical state within the range: liquid
Contributors to the first version: H. Kienitz and K. N. Marsh.
No revision made

Intended usage: Cyclohexane can be used for the calibration and for the testing of the performance of apparatus to be used for the measurement of the relative permittivity of liquids in the temperature range 283.15 to 348.15 K.

Sources of supply and/or methods of purification: Cyclohexane is normally purified by fractional distillation of analytical grade material from sodium (Ref. 1) and the process of purification may be monitored by gas chromatography. Samples should be dried by either storing over sodium or passing through activated alumina immediately prior to use (Ref. 3). Samples of cyclohexane of purity approximately 99.95 mole per cent are available from suppliers (A), (B), and (C).

Pertinent physicochemical data: The relative permittivity ϵ_r for cyclohexane is given in table 12.2.1.1.

Table 12.2.1.1. Relative permittivity for cyclohexane

T_{68}/K	ϵ_r	T_{68}/K	ϵ_r
283.15	2.039 ± 0.0015	323.15	1.976 ± 0.001
293.15	2.024 ± 0.001	333.15	1.961 ± 0.001
298.15	2.016 ± 0.001	343.15	1.945 ± 0.001
303.15	2.008 ± 0.001	348.15	1.931 ± 0.001
313.15	1.992 ± 0.001		

The table has been compiled primarily from the absolute measurements of the relative permittivity by Stokes (Ref. 2) and Malmberg (Ref. 4). The results of these two investigations differ by 0.001 at 283 K decreasing in a regular manner to 0.0007 at 343 K. The careful measurements by Hartshorn, Parry, and Essen (Ref. 5) and by Mecke and Joeckle (Ref. 6) agree with the above recommendations at 293.15 K and 298.15 K if an uncertainty of ±0.001 is ascribed to their results. This uncertainty is five times that estimated by the authors. The compilation by Maryott and Smith (Ref. 7) in 1951 recommended the values 2.023 ± 0.002 at 293.15 K and 2.015 ± 0.002 at 298.15 K which agree with the above recommendations within the stated uncertainties.

Hartshorn, Parry, and Essen (Ref. 5) in recommending cyclohexane as a reference material noted that the relative permittivity obtained by direct distillation of the sample was the same as that obtained after 3 months drying with calcium hydride. Unfortunately the four absolute measurements referenced above show a maximum spread in the value of ϵ_r of 0.0024 at 293.15 K so that the recommended uncertainty in the relative permittivity at that temperature cannot be reduced.

Supplier (A) provides cyclohexane suitable for the calibration of cells for the determination of relative permittivity. They state that the samples do not purport to be a standard of purity and are not free from traces of water. They note that test measurements on a representative sample at 303.15 K showed an increase of 0.0007 in the relative permittivity between samples dried over 'Drierite' and those saturated with water at 293 K. For samples given a limited exposure to air up to 65 per cent relative humidity it was estimated that the change in the relative permittivity of the supplied samples would be less than 0.0002. The measured relative permittivity on representative sample bottles were: 293.15 K, 2.02280 ± 0.00004; 298.15 K, 2.01517 ± 0.00004; 303.15 K, 2.00733 ± 0.00004 with an estimated accuracy of ±0.02 per cent or better. This reference material will be discontinued when current supplies are exhausted.

REFERENCES

1. Riddick, J. A.; Bunger, W. B. *Techniques of Chemistry*, Weissberger, A., editor, Vol II, *Organic Solvents*, 3rd. ed., Wiley-Interscience, New York, **1970**, p. 592.
2. Stokes, R. H. *J. Chem. Thermodynamics* **1973**, *5*, 379.
3. Mecke, R.; Joeckle, R.; Klingenberg, G. *Ber. Bunsenges. phys. Chem.* **1962**, *66*, 239.
4. Malmberg C. G. (National Bureau of Standards), private communication.
5. Hartshorn, L.; Parry, J. V. C.; Essen, L. *Proc. Phys. Soc. B, (London)* **1955**, *68*, 422.
6. Mecke, R.; Joeckle, R. *Ber. Bunsenges. phys. Chem.* **1962**, *66*, 255.
7. Maryott, A. A.; Smith, E. R. *Table of Dielectric Constants of Pure Liquids*, National Bureau of Standards Circular 514, U.S. Department of Commerce: Washington, **1951**.

12.2.2. Tetrachloromethane

Physical property: Relative permittivity, ϵ_r
Units: Dimensionless
Recommended reference material: Tetrachloromethane, (CCl_4)
Range of variables: 273.15 K to 333.15 K, atmospheric pressure
Physical state within the range: liquid
Contributors to the first version: H. Kienitz and K. N. Marsh
No revision made

Intended usage: Tetrachloromethane can be used for the calibration and for the testing of the performance of apparatus to be used for the measurement of the relative permittivity of liquids in the temperature range 273.15 K to 333.15 K.

Sources of supply and/or methods of purification: Analytical grade tetrachloromethane that has been prepared by the direct chlorination of methane is normally purified by fractional distillation and the process of purification may be monitored by gas chromatography. Tetrachloromethane produced by the chlorination of carbon disulphide normally requires the removal of trace amounts of carbon disulphide. The usual procedure is to reflux the sample with one fifth the volume of five mass per cent aqueous sodium hydroxide solution, wash several times with water, dry with calcium chloride, then fractionally distil (Ref. 1). Before use, the sample should be passed through type 4A 'Linde' molecular sieve (Ref. 5) or distilled from phosphorus pentoxide (Ref. 3). It should be noted that tetrachloromethane gives a reduced response while carbon disulphide gives little response when using flame ionization as the detector on a gas chromatograph. Carbon disulphide and water can be determined using a thermal conductivity detector and a 'Poropak type Q' column.

Pertinent physicochemical data: The relative permittivity ϵ_r for tetrachloromethane is given in table 12.2.2.1.

Table 12.2.2.1. Relative permittivity of tetrachloromethane.

T_{68}/K	ϵ_r	T_{68}/K	ϵ_r
273.15	2.278±0.001	303.15	2.218 ± 0.001
283.15	2.258±0.001	313.15	2.198 ± 0.001
293.15	2.238±0.001	323.15	2.178 ± 0.001
298.15	2.228±0.001	333.15	2.158 ± 0.001

The table has been compiled primarily from the absolute measurements of the relative permittivity by Mopsik (Ref.2), by Stokes (Ref.3), and by Mecke and Joeckle (Ref. 5). The spread in the values of ϵ_r from these three determinations is 0.0018 at 298.15 K. The measurements by Miller (Ref. 6), using benzene as the calibrating liquid [ϵ_r = 2.2925 − 0.00198(T/K − 288.15)] agree with the values of Stokes over the temperature range 238.15 to 313.15 K with a maximum difference of 0.0007. When the value for ϵ_r (benzene, 288.15 K) of 2.2940, interpolated from the following recommendation, is used, the maximum difference is 0.0013. The measured values of Mopsik, Stokes, and Hartmann, Newmann, and Schmidt (Ref. 4) at 323.15 K show a maximum difference of 0.0016. Hartmann *et al.* used the values measured by Mecke and Joeckle to calibrate their cell at 293.15 K. The values measured by Heston and Smyth (Ref. 7) and Le Fevre (Ref. 8) differ considerably from the recommended values at higher temperatures. The compilation by Maryott and Smith (Ref. 9) in 1951 recommended the values 2.238±0.002 at 293.15 K and 2.228±0.002 at 298.15 K, which agree with the above recommendations within the stated uncertainties.

REFERENCES

1. Riddick, J. A.; Bunger, W. B. *Techniques of Chemistry*, Weissberger, A., editor, Vol II, *Organic Solvents*, 3rd. ed., Wiley-Interscience, New York, **1970**, p. 592.
2. Mopsik, F. I. *J. Chem. Phys.* **1969**, *50*, 2559.
3. Stokes, R. H. *J. Chem. Thermodynamics* **1973**, *5*, 379.
4. Hartmann, H.; Newmann, A.; Schmidt, A. P. *Ber. Bunsenges. physik. Chem.* **1968**, *72*, 877.
5. Mecke, R.; Joeckle, R. *Ber. Bunsenges. phys. Chem.* **1962**, *66*, 255.
6. Miller, J. G.; *J. Amer. Chem. Soc.* **1942**, *64*, 117.
7. Heston, W. M.; Smyth, C. P. *J. Amer. Chem. Soc.* **1950**, *72*, 99.
8. Le Fevre, R. J. W. *Trans. Faraday Soc.* **1938**, *34*, 1127.

9. Maryott, A. A.; Smith, E. R. *Table of Dielectric Constants of Pure Liquids*, National Bureau of Standards Circular 514, U.S. Department of Commerce: Washington, **1951**.

12.2.3. Benzene

Physical property: Relative permittivity, ϵ_r
Units: Dimensionless
Recommended reference material: Benzene (C_6H_6)
Range of variables: 283.15 to 333.15 K, atmospheric pressure
Physical state within the range: liquid
Contributors to the first version: H. Kienitz and K. N. Marsh
No revision made

Intended usage: Benzene can be used for the calibration and for the testing of the performance of apparatus to be used for the measurement of the relative permittivity of liquids in the temperature range 283.15 to 333.15 K.

Source of supply and/or methods of purification: Analytical grade benzene is usually purified by shaking with aliquots of one tenth its volume of concentrated sulphuric acid until no colour appears in the acid layer. After a dilute sodium carbonate wash and repeated washes with water, it is dried over calcium chloride, then sodium and distilled in an efficient column from sodium or calcium hydride (Refs. 1 to 4). Hartshorn, Parry, and Essen (Ref. 2) noted the difficulty of removal of water from benzene and considered that it was not a reliable reference material. They recommended storing benzene over calcium hydride for at least three weeks. Van der Maesen (Ref. 3) noted that, after distillation from sodium, it took six months of drying over sodium before a value constant to 10^{-4} in the relative permittivity was obtained. Mecke, Joeckle, and Klingenberg (Ref. 5) noted that it took 200 minutes of drying of a purified sample with phosphorus pentoxide before the relative permittivity became constant to 10^{-4}. The same value of the relative permittivity was obtained after drying for 10 minutes with type 4A 'Linde' molecular sieve which has been regenerated at 473 K. The recommended purification is by the standard procedure given above with a final drying over type 4A 'Linde' molecular sieve just prior to the measurement. The progress of the purification may be monitored by gas chromatography. Water may be determined using a thermal conductivity detector and a 'Poropak type Q' column. Samples of benzene of purity approximately 99.99 mole per cent are available from suppliers (B) and (C).

Pertinent physicochemical data: The relative permittivity ϵ_r for benzene is given in table 12.2.3.1.

Table 12.2.3.1. Relative permittivity for benzene.

T_{68}/K	ϵ_r	T_{68}/K	ϵ_r
283.15	2.3040 ± 0.001	313.15	2.244 ± 0.001
293.15	2.2837 ± 0.0005	323.15	2.224 ± 0.001
298.15	2.2739 ± 0.0005	333.15	2.204 ± 0.001
303.15	2.2640 ± 0.001		

This table has been compiled primarily from the absolute measurements of relative permittivity by Hartshorn, Parry, and Essen (Ref. 2), Van der Maesen (Ref. 3), Stokes (Ref. 4), and Mecke and Joeckle (Ref. 6). The spread in the values of ϵ_r for these four determinations is 0.0007 at 293.15 K. The absolute measurements of Heston and Smyth (Ref. 7) and the non-absolute measurements of Hartmann, Newmann, and Rinck (Ref. 8) agree to within 0.0012 of the recommended values over the temperature range given. Hartmann et al. calibrated their cell using a value for ϵ_r(benzene, 293.15 K) of 2.2832 (Ref. 6). The compilation by Maryott and Smith (Ref. 9) in 1951 recommended the values 2.284±0.002 at 293.15 K and 2.274±0.002 at 298.15 K which agree with the above recommendations within the stated uncertainties.

REFERENCES

1. Riddick, J. A.; Bunger, W. B. *Techniques of Chemistry*, Weissberger, A., editor, Vol II, *Organic Solvents*, 3rd. ed., Wiley-Interscience, New York, **1970**, p. 592.
2. Hartshorn, L.; Parry, J. V. C.; Essen, L. *Proc. Phys. Soc. B, (London)* **1955**, *88*, 422.
3. Van der Maesen, F. *Physica* **1949**, *15*, 481.
4. Stokes, R. H. *J. Chem. Thermodynamics* **1973**, *5*, 379.
5. Mecke, R.; Joeckle, R.; Klingenberg, G. *Ber. Bunsenges. phys. Chem.* **1962**, *66*, 239.
6. Mecke, R.; Joeckle, R. *Ber. Bunsenges. phys. Chem.* **1962**, *66*, 255.
7. Heston, W. M.; Smyth, C. P. *J. Amer. Chem. Soc.* **1950**, *72*, 99.
8. Hartmann, H.; Newmann, A.; Rinck, G. *Zeit. Physik. Chem. (Frankfurt)* **1965**, *44*, 204.
9. Maryott, A. A.; Smith, E. R. *Table of Dielectric Constants of Pure Liquids*, National Bureau of Standards Circular 514, U.S. Department of Commerce: Washington, **1951**.

12.2.4. Chlorobenzene

Physical property: Relative permittivity, ϵ_r
Units: Dimensionless

12. Permittivity

Recommended reference material: Chlorobenzene (C_6H_5Cl)
Range of variables: 293.15 to 323.15 K, atmospheric pressure
Physical state within the range: liquid
Contributors to the first version: H. Kienitz and K. N. Marsh
No revision made

Intended usage: Chlorobenzene can be used for the calibration and for the testing of the performance of apparatus to be used for the measurement of the relative permittivity of liquids in the temperature range 293.15 to 323.15 K.

Sources of supply and/or methods of purification: Analytical grade chlorobenzene is usually purified by repeated shaking with aliquots of one tenth its volume of concentrated sulphuric acid until no colour appears in the acid layer. After a potassium bicarbonate wash and repeated washes with water, it is dried over calcium chloride and fractionally distilled (Ref. 1). McAlpine and Smyth (Ref. 2) recommend a second distillation after redrying over phosphorus pentoxide. Hartmann et al. (Ref. 3) dried chlorobenzene with phosphorus pentoxide, then distilled the sample in a 25 theoretical plate column and, just prior to use, dried it with type 4A 'Linde' molecular sieve. The process of purification may be followed by gas chromatography. Chlorobenzene should be used immediately after purification since Mecke and Rosswog (Ref. 4) noted that the relative permittivity increased by 0.0055 after 72 days, presumably due to photolysis.

Pertinent physicochemical data: The relative permittivity ϵ_r for chlorobenzene is given in table 12.2.4.1

Table 12.2.4.1. Relative permittivity of chlorobenzene

T_{68}/K	ϵ_r	T_{68}/K	ϵ_r
293.15	5.70 ± 0.015	323.15	5.20 ± 0.02
298.15	5.62 ± 0.015		

This table has been compiled primarily from the absolute measurements of Mecke and Rosswog (Ref. 4), and the earlier references compiled by Maryott and Smith (Ref. 5). Hartmann et al. (Ref. 3), using the value of 5.6895 for ϵ_r (293.15 K) from reference 4, measured ϵ_r at 323.15 K. This value agrees within 0.017 with the measurement of Schornack and Eckert (Ref. 6) using a cell which was calibrated with a series of unspecified liquids. Mecke and Rosswog considered chlorobenzene to be unsuitable as a reference material because the relative permittivity does not remain constant when exposed to light.

The compilation by Maryott and Smith (Ref. 5) in 1951 recommended the values 5.708±0.006 at 293.15 K and 5.621 ± 0.006 at 298.15 K. An analysis of the measurements referenced by

Maryott and Smith and the measurement by Mecke and Rosswog indicate that the uncertainty quoted by Maryott and Smith was overly optimistic. Additional measurements on this compound seem desirable.

REFERENCES

1. Riddick, J. A.; Bunger, W. B. *Techniques of Chemistry*, Weissberger, A., editor, Vol II, *Organic Solvents*, 3rd. ed., Wiley-Interscience, New York, **1970**, p. 592.
2. McAlpine, K. B.; Smyth, C. P. *J. Phys. Chem.* **1935**, *55*, 55.
3. Hartmann, H.; Newmann, A.; Rinck, G. *Zeit. Physik. Chem. (Frankfurt)* **1965**, *44*, 204.
4. Mecke, R.; Rosswog, K. *Ber. Bunsenges. physik. Chem.* **1956**, *60*, 47.
5. Maryott, A. A.; Smith, E. R. *Table of Dielectric Constants of Pure Liquids*, National Bureau of Standards Circular 514, U.S. Department of Commerce: Washington, **1951**.
6. Schornack, L. G.; Eckert, C. A. *J. Phys. Chem.* **1970**, *74*, 3014.

12.2.5. 1,2–Dichloroethane

Physical property: Relative permittivity, ϵ_r
Units: Dimensionless
Recommended reference material: 1,2–Dichloroethane ($C_2H_4Cl_2$)
Range of variables: 293.15 to 298.15 K, atmospheric pressure
Physical state within the range: liquid
Contributors to the first version: H. Kienitz and K. N. Marsh
No revision made

Intended usage: 1,2–Dichloroethane can be used for the calibration and for the testing of the performance of apparatus to be used for the measurement of the relative permittivity of liquids in the temperature range 293.15 to 298.15 K.

Sources of supply and/or methods of purification: Analytical grade 1,2–dichloroethane is purified for relative permittivity measurements by washing with dilute potassium hydroxide, then water, drying over calcium chloride or phosphorus pentoxide followed by fractional distillation (Refs. 1,2). The process of purification may be monitored by gas chromatography. Hartshorn *et al.* (Ref. 3) noted that a purified sample stored for several months showed the presence of a considerable concentration of ions. On removal of these ions (by the application of 300 V D.C. across the electrodes) they found that the relative permittivity decreased from 10.85 to 10.66 over a period of 24 hours. Samples of 1,2–dichloroethane are available from Supplier (A).

Pertinent physicochemical data: The relative permittivity ϵ_r for 1,2–dichloroethane is given in table 12.2.5.1.

12. Permittivity

Table 12.2.5.1. Relative permittivity of 1,2-dichloroethane

T_{68}/K	ϵ_r	T_{68}/K	ϵ_r
293.15	10.65 ± 0.01	298.15	10.37 ± 0.01

This table has been compiled primarily from the absolute measurements by Hartshorn *et al.* (Ref. 3), Vernon *et al.* (Ref. 4), Sugden (Ref. 2), and Davies (Ref. 5). Determinations at temperatures other than those given above show no consistency. The compilation of Maryott and Smith (Ref. 6) in 1951 recommended the value 10.65 ± 0.01 at 293.15 K and 10.36 ± 0.01 at 298.15 K in agreement with the above recommendations. Their evaluation was primarily based on the results given in references 2, 4, and 5.

Supplier (A) provides 1,2-dichloroethane for the calibration of cells for the determination of relative permittivity. They state that the sample does not purport to be a standard for purity and is not free from traces of water. The measured relative permittivity on representative bottled samples were: 293.15 K, 10.6493 ± 0.0008; 298.15 K, 10.3551 ± 0.0008; 303.15 K, 10.075 ± 0.0011. They state that the relationship

$$\epsilon_r(t/^\circ\text{C}) = 11.9480 - 7.03068 \times 10^{-2}(t/^\circ\text{C}) + 2.7548 \times 10^{-4}(t/^\circ\text{C})^2 - 4.22 \times 10^{-7}(t/^\circ\text{C})^3$$
(12.2.5.1)

can be used to calculate the measured relative permittivity from 283.15 to 313.15 K without significant error and with an estimated accuracy of ±0.05 per cent or better. This reference material will be discontinued when current supplies are exhausted.

REFERENCES

1. Riddick, J. A.; Bunger, W. B. *Techniques of Chemistry*, Weissberger, A., editor, Vol II, *Organic Solvents*, 3rd. ed., Wiley-Interscience, New York, **1970**, p. 592.
2. Sugden, S. *J. Chem. Soc.* **1933**, 768.
3. Hartshorn, L.; Parry, J. V. C.; Essen, L. *Proc. Phys. Soc. B, (London)* **1955**, *88*, 422.
4. Vernon, A. A.; Wyman, J.; Avery, R. A. *J. Amer. Chem. Soc.* i**1945**, *67*, 1477.
5. Davies, R. M. *Phil. Mag.* **1936**, *21*, 1008.
6. Maryott, A. A.; Smith, E. R. *Table of Dielectric Constants of Pure Liquids*, National Bureau of Standards Circular 514, U.S. Department of Commerce: Washington, **1951**.

12.2.6. Methanol

Physical property: Relative permittivity, ϵ_r
Units: Dimensionless
Recommended reference material: Methanol (CH_3OH)
Range of variables: 283.15 to 313.15 K, atmospheric pressure
Physical state within the range: liquid
Contributors to the first version: H. Kienitz and K. N. Marsh.
No revision made

Intended usage: Methanol can be used for the calibration and for the testing of the performance of apparatus to be used for the measurement of the relative permittivity of liquids in the temperature range 283.15 to 313.15 K.

Sources of supply and/or methods of purification: Analytical grade methanol is usually purified by distilling in an efficient column from magnesium activated with iodine (Ref. 1). Maryott purified methanol by distilling from magnesium ribbon (Ref. 2) while Srinivasan and Kay (Ref. 3) purified methanol by passage through a mixed bed ion exchange column from which all water had been leached, followed by fractional distillation under nitrogen. Methanol should be stored under an inert atmosphere prior to use. The process of purification can be followed by gas chromatography. The sample can be analysed for its water content by the use of a 'Poropak type Q' column and a thermal conductivity detector.

Pertinent physicochemical data: The relative permittivity ϵ_r for methanol is given in table 12.2.6.1

Table 12.2.6.1. Relative permittivity of methanol

T_{68}/K	ϵ_r	T_{68}/K	ϵ_r
283.15	35.70 ± 0.03	298.15	32.66 ± 0.02
293.15	33.66 ± 0.03	313.15	29.86 ± 0.05

This table has been compiled primarily from the absolute measurements of Srinivasan and Kay (Ref. 3). The above recommended value at 298.15 K agrees, within the uncertainties given, with the non-absolute measurements by Albright and Gosting (Ref. 4), Hartmann *et al.* (Ref. 5) and Coleman (Ref. 6). At 313.15 K the measurements of Srinivasan and Kay (Ref. 3) agree to ±0.03 with the measurements of Albright and Gosting (Ref. 4) but the non-absolute measurements of Hartmann *et al.* (Ref. 5) and Dannhauser and Bake (Ref. 7) are lower by 0.1. The compilation by Maryott and Smith (Ref. 8) in 1951 recommended the values 33.62 ± 0.07 and 32.63 ± 0.07 at 293.15 and 298.15 K respectively, which agree with the above recommendations within the stated uncertainties.

REFERENCES

1. Riddick, J. A.; Bunger, W. B. *Techniques of Chemistry*, Weissberger, A., editor, Vol II, *Organic Solvents*, 3rd. ed., Wiley-Interscience, New York, **1970**, p. 592.
2. Maryott, A. A. *J. Amer. Chem. Soc.* **1941**, *63*, 3079.
3. Srinivasan, K. R.; Kay, R. L. *J. Solution Chem.* **1975**, *4*, 299.
4. Albright, P. S.; Gosting, L. J. *J. Amer. Chem. Soc.* **1946**, *68*, 1061.
5. Hartmann, H.; Newmann, A.; Schmidt, A. P. *Ber. Bunsenges. physik. Chem.* **1968**. *72*, 877.
6. Coleman, C. F. *J. Phys. Chem.* **1968**, *72*, 365.
7. Dannhauser, W.; Bake, L. W. *J. Chem. Phys.* **1964**, *140*, 3058.
8. Maryott, A. A.; Smith, E. R. *Table of Dielectric Constants of Pure Liquids*, National Bureau of Standards Circular 514, U.S. Department of Commerce: Washington, **1951**.

12.2.7. Nitrobenzene

Physical property: Relative permittivity, ϵ_r
Units: Dimensionless
Recommended reference material: Nitrobenzene ($C_6H_5NO_2$)
Range of variables: 293.15 to 298.15 K, atmospheric pressure
Physical state within the range: liquid
Contributors to the first version: H. Kienitz and K. N. Marsh
No revision made

Intended usage: Nitrobenzene can be used for the calibration and for the testing of the performance of apparatus to be used for the measurement of the relative permittivity of liquids in the temperature range 293.15 to 298.15 K.

Sources of supply and/or methods of purification: Analytical grade nitrobenzene is usually purified by repeated fractional crystallization followed by distillation under reduced pressure (Ref. 1). Hartshorn *et al.* (Ref. 2) repeatedly distilled (at atmospheric pressure) an analytical grade sample, the final distillation being from finely powdered alumina. No change was observed after allowing the liquid to stand over powdered calcium hydride for four weeks. Samples of nitrobenzene are available from supplier (A).

Pertinent physicochemical data: The relative permittivity ϵ_r for nitrobenzene is given in table 12.2.7.1.

Table 12.2.7.1 Relative permittivity for nitrobenzene.

$T_{68}/$K	ϵ_r	$T_{68}/$K	ϵ_r
293.15	35.72 ± 0.04	298.15	34.78 ± 0.04

This table has been compiled primary from the absolute measurements of Hartshorn *et al.* (Ref. 2) and Sugden (Ref. 3). There have been a few non- absolute measurements at other temperatures but they shown considerable disagreement. The compilation by Maryott and Smith (Ref. 4) in 1951 recommended the values 35.74 ± 0.07 at 293.15 K and 34.82 ± 0.07 at 298.15 K which agree with the above recommendations within the stated uncertainties. Further measurements on this liquid would be desirable.

Supplier (A) provides nitrobenzene suitable for the calibration of cells used for the determination of relative permittivity. They state that the samples do not purport to be a standard of purity and are not free from traces of water. The relative permittivity on representative bottled samples were: 293.15 K, 35.7037 ± 0.001; 298.15 K, 34.7416 ± 0.001; 303.15 K, 33.8134 ± 0.003. They state that the relationship

$$\epsilon_r(t/^\circ\text{C}) = 39.9278 - 0.226899(t/^\circ\text{C}) + 8.0801 \times 10^{-4}(t/^\circ\text{C})^2 - 1.267 \times 10^{-6}(t/^\circ\text{C})^3 \quad (12.2.7.1)$$

can be used to calculate the measured relative permittivity from 283.15 to 313.15 K without significant error with an estimated accuracy of ±0.04 per cent. The reference material will be discontinued when current supplies are exhausted.

REFERENCES

1. Riddick, J. A.; Bunger, W. B. *Techniques of Chemistry*, Weissberger, A., editor, Vol II, *Organic Solvents*, 3rd. ed., Wiley-Interscience, New York, **1970**, p. 592.
5. Hartshorn, L.; Parry, J. V. C.; Essen, L. *Proc. Phys. Soc. B, (London)* **1955**, *88*, 422.
3. Sugden, S. *J. Chem. Soc.* **1933**, 768.
4. Maryott, A. A.; Smith, E. R. *Table of Dielectric Constants of Pure Liquids*, National Bureau of Standards Circular 514, U.S. Department of Commerce: Washington, **1951**.

12.2.8. Water

Physical property: Relative permittivity, ϵ_r
Units: Dimensionless
Recommended reference material: Water (H_2O)
Range of variables: 273.15 to 373.15 K, atmospheric pressure
Physical state within the range: liquid
Contributors to the first version: H. Kienitz and K. N. Marsh
Contributor to the first version: K. N. Marsh

Intended usage: Water can be used for the calibration and for the testing of the performance of apparatus to be used for the measurement of the relative permittivity of liquids in the temperature range 273.15 to 373.15 K.

Sources of supply and/or methods of purification: Either singly distilled or deionized water is suitable for relative permittivity measurements.

12. Permittivity

Pertinent physicochemical data: The relative permittivity ϵ_r for water is given in table 12.2.8.1.

Table 12.2.8.1. Relative permittivity for water.

T_{68}/K	ϵ_r	T_{68}/K	ϵ_r
273.15	87.85 ± 0.07	323.15	69.87 ± 0.05
283.15	83.95 ± 0.07	333.15	66.73 ± 0.06
293.15	80.20 ± 0.05	343.15	63.72 ± 0.09
298.15	78.36 ± 0.05	353.15	60.84 ± 0.12
303.15	76.60 ± 0.05	363.15	58.09 ± 0.15
313.15	73.16 ± 0.05	373.15	55.46 ± 0.20

There have been a number of critical evaluations of the relative permittivity of water over a wide temperature and pressure range (Refs. 1 to 5). The evaluation by Uematsu and Frank (Ref. 1) forms the basis of the values given in the International Association of the Property of Steam Tables. The other major evaluation is that of Bradley and Pitzer (Ref. 2).

At temperatures below 323.15 K, the most reliable values are from the absolute measurements by Malmberg and Maryott (Ref. 6), Owen *et al.* (Ref. 7), Kay and coworkers (Refs. 8 to 11), and Dunn and Stokes (Ref. 11). The values given in references 7 to 11 agree to within 0.05 in the relative permittivity from 273.15 to 313.15 K. The results of Malmberg and Maryott and Dunn and Stokes are lower by about 0.15 from the values in references 7 to 11 at 273.15 K but decrease to agreement within ± 0.03 at 303.15 K. The results of Dunn and Stokes (Ref. 12) can be considered as absolute measurements since they used a conductance technique to determine the capacitance of the cell when evacuated. The non-absolute measurements of Lees (Ref. 13) agree to within 0.04 with those of Owen *et al.* in the range 273.15 to 323.15 K. Thus there are two sets of measurements which are inconsistent below 303 K, hence the estimated uncertainty is greater at lower temperatures. The absolute measurements by Rusche and Good (Ref. 14) are higher by about 0.20 units from the recommended values. The compilation by Maryott and Smith (Ref. 15) in 1951 recommended the values 80.37 ± 0.08 and 78.54 ± 0.08 at 293.15 and 298.15 K respectively. These two estimates are considerably higher than the recent determinations referenced above. At temperatures above 323.15 K, the measurements of Malmberg and Maryott (Ref. 6) increasingly deviate from the measurements of Wyman (Ref. 16), Åkerlöf (Ref. 17) and Heger (Ref. 18). Further, the analysis by both Uematsu and Frank (Ref. 1) and Bradley and Pitzer (Ref. 2), taking into account other measurements in the range 370 to 600 K, suggest that the higher temperature results of Malmberg and Maryott are in error. Below 323 K the differences between the previous recommendations and the values calculated from the Bradley and Pitzer equation were considerably less than the estimated uncertainties.

The recommended values from 273 to 373 K given in table 12.2.8.1 were calculated from the equation given by Bradley and Pitzer (Ref. 2). Because of the considerable discrepancy between the values of Malmberg and Maryott and Bradley and Pitzer (0.26 at 373.15 K), the estimated uncertainties in the relative permittivity of water above 323 K have been increased significantly compared with the previous recommendation.

The differences between the values calculated by the Bradley and Pitzer equation and the Uematsu and Frank equation in the range 273 to 373 K are considerably less than the estimated uncertainties given in table 12.2.8.1.

REFERENCES

1. Uematsu, M.; Frank, E. U. *J. Phys. Chem. Ref. Data* **1980**, *9*, 1291. See also Harr, L.; Gallagher, J. S.; Kell, G. S. *NBS/NRC Steam Tables*, Hemisphere: New York, **1985**.
2. Bradley, D. J.; Pitzer, K. S. *J. Phys. Chem.* **1979**, *83*, 1500.
3. Kodakovsky, I. L.; Dorofeyeva, V. A. *Geokhimiya*, **1981**, *8*, 1174.
4. Helgeson, H. C.; Kirkham, D. H. *Amer. J. Sci.* **1974**, *274*, 1199.
5. Beyer, R. P.; Staples, B. R. *J. Soln. Chem.* **1986**, *15*, 749.
6. Malmberg, C. C.; Maryott, A. A. *J. Res. Nat. Bur. Stand.* **1956**, *56*, 1.
7. Owen, B. B.; Miller, R. C.; Milner, C. E.; Cogan, H. L. *J. Phys. Chem.* **1961**, *65*, 2065.
8. Srinivasan, K. R.; Kay, R. L. *J. Chem. Phys.* **1974**, *60*, 3645.
9. Kay, R. L.; Vidulich, G. A.; Pribadi, K. S. *J. Phys. Chem.* **1969**, *73*, 445.
10. Vidulich, G. A,; Evans, D. F.; Kay, R. L. *J. Phys. Chem.* **1967**, *71*, 656.
11. Vidulich, G. A.; Kay, R. L. *J. Phys. Chem.* **1962**, *66*, 383.
12. Dunn, L. A.; Stokes, R. H. *Trans. Faraday Soc.* **1969**, *65*, 2906.
13. Lees, W. L. *Ph.D. Thesis*, Harvard University, **1949**.
14. Rusche, E. W.; Good, W. B. *J. Chem. Phys.* **1966**, *45*, 4667.
15. Maryott, A. A.; Smith, E. R. *Table of Dielectric Constants of Pure Liquids*, National Bureau of Standards Circular 514, U.S. Department of Commerce: Washington, **1951**.
16. Wyman, J. *Phys. Rev.* **1930**, *35*, 623.
17. Åkerlöf, G. *J. Amer. Chem. Soc.* **1932**, *54*, 4125.
18. Heger, K. Dissertation, Univ. Karlsruhe, Karlsruhe, Germany, **1969**.

12.3. Reference materials for permittivity in the liquid and real gas state.

12.3.1. Hydrogen

Physical property: Relative permittivity, ϵ_r
Units: Dimensionless
Recommended reference material: Hydrogen (H_2O)
Range of variables: 14 to 20 K, saturated vapour line and at 293.15 K, 101.325 kPa

12. Permittivity

Physical state within the range: liquid and gas
Contributors to the first version: H. Kienitz and K. N. Marsh
No revision made

Intended usage: Liquid hydrogen can be used for the calibration and testing of the performance of apparatus to be used for the measurement of relative permittivity at low temperatures. Hydrogen gas at atmospheric pressure can be used for a similar purpose at 293.15 K.

Sources of supply and/or methods of purification: For measurements on the liquid a normal hydrogen sample should be prepared electrolytically. It should not have been liquified previously. For measurements at 293.15 K research grade hydrogen gas dried prior to use is satisfactory.

Pertinent physicochemical data: A. Liquid State - The relative permittivity ϵ_r for hydrogen is given in table 12.3.1.1.

Table 12.3.1.1. Relative permittivity for hydrogen in the liquid state.

T_{68}/K	ϵ_r	T_{68}/K	ϵ_r
14.035	1.2534	18.34	1.2399
14.745	1.2514	18.90	1.2351
15.38	1.2494	19.555	1.2351
16.295	1.2465	19.92	1.2338
17.29	1.2432	20.375	1.2320

This table has been prepared primarily from absolute measurements by Kogan, Milenko, and Grigorova (Ref. 1). They claim an uncertainty in ϵ_r of $\pm 5 \times 10^{-5}$. The earlier measurements by Werner and Keesom (Ref. 2) are lower than the above results by an average of about 0.0007 but between 14 and 15 K the difference increases to 0.002. Werner and Keesom estimate their uncertainty as $\pm 5 \times 10^{-4}$. The measurement of Van Itterbeck and Spaepen (Ref. 3) at 20.35 K is lower than that of Kogan *et al.* by 0.006. The estimated uncertainty in the values given in the table are ± 0.0008. The compilation of Maryott and Smith (Ref. 4) in 1951 recommended the value of 1.228 at 20.4 K which is lower than the above recommendation.

B. Gas State - There have been no precise measurements on the relative permittivity of hydrogen gas at 293.15 K since the compilation by Maryott and Buckley (Ref. 5) so their recommended value remains.

$$(\epsilon_r - 1)10^6[H_2(g, 293.15 \text{ K}, 101.325 \text{ kPa})] = 253.8 \pm 0.3 \qquad (12.3.1.1)$$

REFERENCES

1. Kogan, V. S.; Milenko, Yu. Ya.; Grigorova, T. K. *Physica* **1971**, *53*, 125.
2. Werner, W.; Keesom, W. H. *Commun. Phys. Lab. Leiden 16*, **1926**, No.178a.
3. Van Itterbeck, A.; Spaepen, J. *Physica* **1942**, *9*, 339.
4. Maryott, A. A.; Smith, E. R. *Table of Dielectric Constants of Pure Liquids*, National Bureau of Standards Circular 514, U.S. Department of Commerce: Washington, **1951**.
5. Maryott, A. A.; Buckley, F. *Table of Dielectric Constants and Electric Dipole Moments of Substances in the Gaseous States*, National Bureau of Standards Circular 537, U.S. Department of Commerce: Washington, D.C. **1953**.

12.3.2. Oxygen

Physical property: Relative permittivity, ϵ_r
Units: Dimensionless
Recommended reference material: Oxygen (O_2)
Range of variable: 54.5 K to 84 K, saturated vapour curve and at 293.15 K, 101.325 kPa
Physical state within the range: liquid and gas
Contributors to the first version: H. Kienitz and K. N. Marsh
No revision made

Intended usage: Liquid oxygen can be used for the calibration and for the testing of the performance of apparatus to be used for the measurement of relative permittivity at low temperatures. Oxygen gas at atmospheric pressure can be used for a similar purpose at 293.15 K.

Sources of supply and/or methods of purification: Laboratory grade oxygen purified by passing through a silica gel trap at 77 K before liquefaction is recommended (Ref. 1). For measurements at 293.15 K research grade oxygen dried prior to uswe is satisfactory (Ref. 1).

Pertinent physicochemical data: A. Liquid State - The relative permittivity ϵ_r for oxygen is given in table 12.3.2.1.

Table 12.3.2.1. Relative permittivity for oxygen in the liquid state.

T_{68}/K	ϵ_r	T_{68}/K	ϵ_r
54.478	1.5685	68.000	1.5384
55.000	1.5674	72.000	1.5294
56.000	1.5651	76.000	1.5203
62.000	1.5518	80.000	1.5111
64.000	1.5473	84.000	1.5018

12. Permittivity

This table has been prepared primarily from absolute measurements by Younglove (Ref. 1) who claims an uncertainty of ±0.0005 in ϵ_r. The interppolated value at 81.95 K (1.5066±0.0005) is in excellent agreement with the measurements of Lewis and Smyth (Ref. 2), (1.505±0.001). The compilation by Maryott and Smith (Ref. 3) in 1951 recommended the value of 1.507 at 80 K which differs from the value recommended in the table, 1.511. The difference occurs because of the difficulty in assessing the accuracy of the earlier measurements (Refs. 4, 5).

B. Gas State - The only recent precise measurements on oxygen gas are those by Dunn (Ref. 6). He obtained

$$(\epsilon_r - 1)10^6[O_2(g, 293.15 \text{ K}, 101.325 \text{ kPa})] = 494.3 \pm 0.2 \qquad (12.3.2.1)$$

which agrees within the uncertainty limits with the value of 494.7±0.2 recommended by Maryott and Buckley (Ref. 7). In light of the result of Dunn (Ref. 6) the recommendation becomes

$$(\epsilon_r - 1)10^6[O_2(g, 293.15 \text{ K}, 101.325 \text{ kPa})] = 494.6 \pm 0.2 \qquad (12.3.2.2)$$

REFERENCES

1. Younglove, B. A. *J. Res. Nat. Bur. Stand.* **1972**, *76A*, 37.
2. Lewis, G. L.; Smyth, C. P. *J. Amer. Chem. Soc.* 1939, *61*, 3063.
3. Maryott, A. A.; Smith, E. R. *Table of Dielectric Constants of Pure Liquids*, National Bureau of Standards Circular 514, U.S. Department of Commerce: Washington, **1951**.
4. Kanda, E. *Bull. Chem. Soc. Japan* **1937**, *12*, 473.
5. Werner, W.; Keesom, W. H. *Proc. Koninkl. Nederland. Akad. Wetenschap* **1926**, *29*, 306.
6. Dunn, A. F. *Canad. J. Physics* **1964**, *42*, 1489.
7. Maryott, A. A.; Buckley, F. *Table of Dielectric Constants and Electric Dipole Moments of Substances in the Gaseous States*, National Bureau of Standards Circular 537, U.S. Department of Commerce: Washington, D.C. **1953**.

12.4. Reference materials for permittivity in the real gas state

12.4.1. Nitrogen

Physical property: Relative permittivity, ϵ_r
Units: Dimensionless
Recommended reference material: Oxygen (N_2)
Range of variables: 293.15 K, 101.325 kPa
Physical state within the range: gas
Contributors to the first version: H. Kienitz and K. N. Marsh
No revision made

Intended usage: Nitrogen gas at atmospheric pressure can be used for the calibration and the testing of the performance of apparatus to be used for the measurement of relative permittivity.

Sources of supply and/or methods of purification: Research grade nitrogen dried with P_2O_5 prior to use is satisfactory (Ref. 1).

Pertinent physicochemical data: Dunn (Ref. 1) made very precise measurements on the relative permittivity of nitrogen gas and obtained

$$(\epsilon_r - 1)10^6[N_2(g, 293.15\ K, 101.325\ kPa)] = 547 \pm 0.2 \qquad (12.4.1.1)$$

which agrees within the uncertainty limits with the value of 548.0 ± 0.5 recommended by Maryott and Buckley (Ref. 2). The new recommendation is

$$(\epsilon_r - 1)10^6[N_2(g, 293.15\ K, 101.325\ kPa)] = 547.7 \pm 0.3 \qquad (12.4.1.2)$$

REFERENCES

1. Dunn, A. F. *Canad. J. Physics.* **1964**, *42*, 1489.
2. Maryott, A. A.; Buckley, F. *Table of Dielectric Constants and Electric Dipole Moments of Substances in the Gaseous States*, National Bureau of Standards Circular 537, U.S. Department of Commerce: Washington, D.C. **1953**.

12.4.2. Air

Physical property: Relative permittivity, ϵ_r
Units: Dimensionless
Recommended reference material: Air (dry, carbon dioxide free)
Range of variables: 293.15 K, 101.325 kPa
Physical state within the range: gas
Contributors to the first version: H. Kienitz and K. N. Marsh
No revision made

Intended usage: Air (dry, carbon dioxide free) at atmospheric pressure can be used for the calibration and the testing of the performance of apparatus to be used for the measurement of relative permittivity.

Sources of supply and/or methods of purification: Air passed over sodium hydroxide to remove carbon dioxide then dried over phosphorus pentoxide is suitable (Ref. 1).

Pertinent physicochemical data: There have been no significant precise measurements since the compilation by Maryott and Buckley (Ref. 2). The recommended value for dry carbon dioxide free air remains at

$$(\epsilon_r - 1)10^6[\text{Air}(293.15\ K, 101.325\ kPa)] = 536.4 \pm 0.3 \qquad (12.4.2.1)$$

Essen and Froome (Ref. 1) have given an equation for the dielectric constant of air which contains carbon dioxide and moisture

$$(\epsilon_r - 1)10^6(T/K, p/Pa) = -0.4 + \frac{1.5519(p_1/Pa)}{T/K} + \frac{2.6612(p_2/Pa)}{T/K} + 1.2940(p_3/Pa)\left(1 + \frac{5746}{T/K}\right)/(T/K) \qquad (12.4.2.2)$$

where p_1, p_2, and p_3 are the partial pressures of air, carbon dioxide, and water vapour respectively. This equation has been adjusted to agree with the recommended value of dry, carbon dioxide free, air given above.

REFERENCES

1. Essen, L.; Froome, K. D. *Proc. Phys. Soc. B* **1951** *64*, 862.
2. Maryott, A. A.; Buckley, F. *Table of Dielectric Constants and Electric Dipole Moments of Substances in the Gaseous States*, National Bureau of Standards Circular 537, U.S. Department of Commerce: Washington, D.C. **1953**.

12.5. Contributors to the revised version

K. N. Marsh,
Thermodynamics Research Center,
Texas A&M University,
College Station, TX 77843,
(U.S.A.)

12.6. List of Suppliers

A. Office of Standard Reference Materials,
National Bureau of Standards,
Gaithersburg, MD 20899
(U.S.A.)

B. Philips Petroleum,
Special Chemical Branch,
Bartlesville, OK 74004
(U.S.A.)

C. A.P.I. Samples,
Carnegie-Mellon University,
Schenley Park,
Pittsburg, PA 15213
(U.S.A.)

13

SECTION: POTENTIOMETRIC ION ACTIVITIES

COLLATOR: E. JUHÁSZ

CONTENTS:

13.1. Introduction

13.2. Definitions
- 13.2.1. Notional definition
- 13.2.2. Operational definition
- 13.2.3. Bracketting procedure

13.3. Standard reference solutions for pH
- 13.3.1. Reference value pH standard
- 13.3.2. Primary pH standards
- 13.3.3. Working (operational) pH standards

13.4. Standard reference solutions for pD

13.5. Standard reference solutions for ionic activity

13.6. Certified reference materials for pH
- 13.6.1. Reference value pH reference material
- 13.6.2. Primary pH reference materials

13.7. Certified reference materials for pD

13.8. Certified reference materials for ionic activity

13.9. Contributors

13.10. List of Suppliers

13.1. Introduction

This recommendation is based on two IUPAC Recommendations:

a. IUPAC Subcommittee on calibration and test materials (I.4): Recommended reference materials for realization of physicochemical properties, Section: Potentiometric ion activities (Ref. 1) and

b. IUPAC Commission on Electroanalytical Chemistry (V. 5) and Commission on Electrochemistry (I. 3): Definition of pH scales, standard reference values, measurement of pH and related terminology (Ref. 2).

The concept of pH (Ref. 3) is unique amongst the physicochemical quantities. In the IUPAC Manual of Symbols and Terminology for Physicochemical Quantities and Units (Ref. 4), its (notional) definition is

$$\mathrm{pH} = -\log_{10} a_{\mathrm{H}} \qquad (13.1.1).$$

This definition involves a single ion activity and is thus immeasurable. It is therefore defined operationally in terms of the operation or method used to measure it, that is, by means of the cell or variants of it:

$$\text{reference electrode} \,|\, \mathrm{KCl}(\geq 3.5\ \mathrm{mol\ kg^{-1}}) \,\|\, \text{Solution} \,|\, \mathrm{H_2} \,|\, \mathrm{Pt}$$

called the operational cell (see Section 2.2). The cell is standardized by solution(s) of assigned pH value. Such standard reference solutions are buffer solutions whose pH values are assigned from measurements on cells without liquid junction. With the use of several standard reference solutions, this constitutes the multi-primary standard approach pioneered by work at the U.S. National Bureau of Standards (NBS) from 1943 to 1969 (Ref. 5). Most national pH scale recommendations and OIML recommendations have been based on this multi-standard system. The situation has recently been reviewed (Ref. 1) by IUPAC Commission I.4 and this revealed differences in the number of primary reference standards recommended and pH values assigned. The British Standard pH scale (Ref. 6), in contrast, recommends only one primary standard (potassium hydrogen phthalate) determined by the cell without liquid junction and a number of secondary standards determined by the operational cell. The basis of this different approach is the argument that to define the pH scale (i.e. to define the line of operational cell electromotive force vs. pH) only one point, or primary reference standard, is required, the other parameter being the theoretical slope (59.159 mV/pH at 298.15 K). It must be emphasized that the definition of the pH scale is quite distinct from the measurement of pH with glass-reference electrode-pH meter assemblies where several standards are used in order to take into account possible deficiencies in the electrode and meter performance.

In permitting the validity of either approach, with the proviso that results are properly reported, indicating which system and standards have been used, it is hoped that there will be

13. Potentiometric Ion Activities

a greater realization of the problems of pH measurement by the scientific community. Further research is needed to establish the respective merits of the two approaches and only then can a thermodynamically significant and metrologically sound pH scale be recommended.

Consequently this recommendation is a compromise, which recognizes that pH experts are divided and IUPAC is unable to recommend, unreservedly, on scientific grounds one or other of the two approaches to pH scale definition.

13.2. Definitions

13.2.1. Notional definition

The concept of pH was first introduced (Ref. 3) as

$$\mathrm{pH} = -\log_{10} c_\mathrm{H} \tag{13.2.1.1}$$

where c_H is the hydrogen ion concentration (in mol dm^{-3}). It was subsequently modified to

$$\mathrm{pH} = \log_{10} a_\mathrm{H} \tag{13.2.1.2}$$

where a_H is the (relative) hydrogen ion activity. This can only be a notional definition, for pH involves a single ion activity which is immeasurable. Equation (13.2.1.2) is often written as

$$\mathrm{pH} = -\log_{10} m_\mathrm{H}\gamma_\mathrm{H} \quad \text{or} \quad \mathrm{pH} = -\log_{10} c_\mathrm{H} y_\mathrm{H} \tag{13.2.1.3}$$

where m_H is the molality of hydrogen ions, c_H is the amount-of-substance concentration of hydrogen ion and γ_H and y_H are single ion activity coefficients of hydrogen ion on the two scales respectively. However, pH is a dimensionaless quantity, thus it is not correct to write, in isolation, the logarithm of a quantity other than a dimensionless number, and the full forms of Equation (13.2.1.3) are either

$$\mathrm{pH} = -\log_{10}\left(m_\mathrm{H}\gamma_\mathrm{H}/m^\ominus\right) \tag{13.2.1.4}$$

or

$$\mathrm{pH} = -\log_{10}\left(c_\mathrm{H} y_\mathrm{H}/c^\ominus\right) \tag{13.2.1.5}$$

where $m^\ominus$ and $c^\ominus$ are arbitrary constants, representing the standard state condition, numerically equal to either 1 mol kg^{-1} or 1 mol dm^{-3} respectively. For most purposes the difference between the two scales for dilute aqueous solutions can be ignored; it depends on the density of water and for the solutions considered in this section the difference reaches a maximum of about 0.001 at 298.15 K rising to 0.02 at 393.15 K.

13.2.2. Operational definition

It is universally agreed that the definition of pH difference is an operational one. The electromotive force $E(X)$ of the cell:

$$\text{reference electrode} \mid \text{KCl(aq, concentrated)} \mid\mid \text{solution X} \mid H_2 \mid Pt \qquad (I)$$

is measured and likewise the electromotive force (e.m.f.) $E(S)$ of the cell:

$$\text{reference electrode} \mid \text{KCl(aq, concentrated)} \mid\mid \text{solution S} \mid H_2 \mid Pt \qquad (II)$$

is measured with both cells being at the same temperature throughout and the hydrogen gas pressures and the reference electrode being identical in the two cells. The two bridge solutions may be any molality of potassium chloride not less than 3.5 mol kg^{-1}, provided they are the same (Ref. 7). The pH of solution X, denoted by pH(X), is then related to the pH of the standard reference solution S, denoted by pH(S), by the definition:

$$\text{pH}(X) = \text{pH}(S) + \frac{E(S) - E(X)}{(RT/F)\ln 10} \qquad (13.2.2.1)$$

where R denotes the gas constant, T the thermodynamic temperature, and F the Faraday constant.

To a good approximation, the hydrogen electrodes in both cells may be replaced by other hydrogen-ion-responsive electrodes, e.g. quinhydrone; in particular, in most measurements a single glass electrode, transferred between the cells, replaces the two hydrogen electrodes.

13.2.3. Bracketting procedure

Alternatively, a bracketting procedure may be adopted as follows: the electromotive force $E(X)$ is measured, and likewise the electromotive forces $E(S_1)$ and $E(S_2)$ of two cells II with standard solutions S_1 and S_2 such that the $E(S_1)$ and $E(S_2)$ values are on either side of, and as near as possible to, $E(X)$. The pH of solution X is then obtained by assuming linearity between pH and E, that is to say:

$$\frac{\text{pH}(X) - \text{pH}(S_1)}{\text{pH}(S_2) - \text{pH}(S_1)} = \frac{E(X) - E(S_1)}{E(S_2) - E(S_1)} \qquad (13.2.3.1).$$

Using this procedure it will not be possible to select both standard reference buffers to match in pH, composition and ionic strength. The purpose of the bracketting procedure is to compensate for deficiencies in the electrodes and measuring system.

13.3. Standard reference solutions for pH

The difference between the pH of two solutions having been defined as above, the definition of pH must be completed by assigning a value of pH at each temperature to one or more chosen solutions designated as standard solutions.

13.3.1. Reference value pH reference material

The qualities of potassium hydrogen phthalate as the most studied (Refs. 8 to 12) of all the primary pH reference materials are recognized and so it is designated as the reference value pH reference material (RVS) at specified molality of 0.05 mol kg^{-1}. This solutions is called a reference value pH standard in reference 2 and is designated by pH (RVS). The terms used here conform to those recommended by the ISO international vocabulary of basic and general terms in metrology. The reference value method is the cell without liquid junction (Cell III).

$$\text{Pt(Pd)} \,|\, \text{H}_2(\text{g}, \, p = 101\,325 \text{ Pa} = 1 \text{ atm}) \,|\, \text{RVS}, \text{Cl}^- \,|\, \text{AgCl} \,|\, \text{Ag} \qquad \text{(III)}$$

The e.m.f. of cell (III) is given by

$$E = E^\ominus - \text{k} \log_{10} [m_\text{H} m_\text{Cl} \gamma_\text{H} \gamma_\text{Cl}/(m^\ominus)^2] \qquad (13.3.1.1)$$

where m signifies molality and γ is the activity coefficient of the subscripted species, $E^\ominus$ is the standard e.m.f., $m^\ominus = 1$ mol kg^{-1} and k $= RT(\ln 10)/F$.

Equation (13.3.1.1) can be rearranged to

$$-\log_{10}(m_\text{H} \gamma_\text{H} \gamma_\text{Cl}/m^\ominus) = (E - E^\ominus)/\text{k} + \log_{10}(m_\text{Cl}/m^\ominus) \qquad (13.3.1.2).$$

The standard e.m.f is derived from measurements on the cell

$$\text{Pt} \,|\, \text{H}_2(\text{g}, \, p = 101\,325 \text{ Pa} = 1 \text{ atm}) \,|\, \text{HCl}(0.01 \text{ mol kg}^{-1}) \,|\, \text{AgCl} \,|\, \text{Ag} \qquad \text{(IV)}$$

and calculated from Equation (13.3.1.1) with $\gamma_\text{H}\gamma_\text{Cl} = \gamma_\pm^2$, where $\gamma_\pm$ is the mean ionic activity coefficient of HCl at 0.01 mol kg^{-1}. The quantity $-\log_{10}(m_\text{H}\gamma_\text{H}\gamma_\text{Cl}/m^\ominus)$ is calculated from measured E values for each of several molalities m_Cl of chloride ion, plotted against m_Cl and extrapolated to $m_\text{Cl} = 0$. Then, pH(RVS) is calculated from

$$\text{pH(RVS)} = [-\log_{10}(m_\text{H}\gamma_\text{H}\gamma_\text{Cl}/m^\ominus)]_{m_\text{Cl} \to 0} + \log_{10} \gamma_\text{Cl} \qquad (13.3.1.3)$$

where γ_Cl is obtained from the Bates-Guggenheim convention

$$\log_{10} \gamma_\text{Cl} = \text{A}(I/m^\ominus)^{1/2}/[1 + 1.5(I/m^\ominus)^{1/2}]; \, (I \leq 0.1 \text{ mol kg}^{-1}) \qquad (13.3.1.4)$$

and A is the Debye-Hückel (temperature dependent) limiting slope and $I = \frac{1}{2}\sum_i m_i z_i^2$ the ionic strength. Values of A, $\gamma_\pm$ at 0.01 mol kg^{-1}, and $E^\ominus$ are available (Refs. 13, 14).

13.3.2. Primary pH Reference materials

Certain substances which meet the criteria of:

(1) preparation in highly pure state reproducibly, and availability as certified reference material;

(2) stability of solution over a reasonable period of time;

(3) having a low value of the residual liquid junction potential;

can be designated as primary reference materials in aqueous solution of specified concentration. These solutions are called primary pH standards in reference 2 and are designated by pH (PS). Their number shall be reviewed from time to time. At present OIML (Ref. 15) recommends 9, while IUPAC (Ref. 2) recommends 7 solutions, including the reference value pH standard. The pH(PS) values assigned to these primary standards are derived from measurements on cells without liquid junction

$$\text{Pt} \,|\, \text{H}_2(\text{g}, \, p = 101\,325 \text{ Pa} = 1 \text{ atm}) \,|\, \text{PS}, \text{Cl}^- \,|\, \text{AgCl} \,|\, \text{Ag} \qquad (V)$$

Values of pH(PS) have been assigned by the same method as that used for the reference value solution.

If the definition of pH given above is adhered to, then the pH(X) of a solution using cells I and II (see 2.2) may be slightly dependent on which standard solution is used. These deviations are due to (i) the Bates-Guggenheim convention (see Equation (13.3.1.3)) adopted for the single ion activity coefficient of the chloride ion (in order to obtain, from the analysis of measurements on the cell without liquid junction, pH(PS) values) being applied to all standard reference solutions; and (ii) variations in the liquid junctions resulting from the different ionic compositions and mobilities of the several standards and from differences in the geometry of the liquid-liquid boundary. In fact, such variations in measured pH(X) are usually at the 0.02 level and are too small to be of practical significance for most measurements.

13.3.3. Secondary pH reference materials

Certain substances which meet the criteria of:

(1) preparation in highly pure state reproducibly;

(2) stability of solution over a reasonable period of time

can be designated as secondary standards in aqueous solution of specified concentration. These solutions are called operational pH standards in (Ref. 2) and are designated by pH(OS). Their number is in principle unlimited but values are available now for 15 solutions (Ref. 2). The values of pH(OS) are assigned by comparison with the pH(RVS) in

13. Potentiometric Ion Activities

cells with liquid junction, the operational cells, where the liquid junctions are formed within vertical 1 mm bore capillary tubes.

$$\text{Pt(Pd)} \mid \text{H}_2 \mid \text{RVS} \| \text{KCl}(\geq 3.5 \text{ mol dm}^{-3}) \| \text{OS} \mid \text{H}_2 \mid \text{Pt} \qquad \text{(VI)}$$

13.4. Standard reference solutions for pD

The National Bureau of Standards has adopted (Refs. 7, 13, 16) a pD scale based on deuterium ion activity. The pD(S) values listed correspond to values of $\log_{10}(1/a_D)$, where a_D is the conventional activity of the deuterium ion referred to the standard state on the scale of molality. These reference materials have been certified at NBS using a procedure strictly analogous to the assignment of pH(S) values for ordinary water except a deuterium gas electrode was used as the deuterium ion indicator electrode (Ref. 17). There are four steps in the assignment of the pD(S) values to reference materials (a) determination of the acidity function $p(a_D\gamma_{Cl})$ (Ref. 18); (b) evaluation of the limiting acidity function $p(a_D\gamma_{Cl})°$; (c) computation of pa_D from $p(a_D\gamma_{Cl})°$; and (d) identification of pa_D with pD(S). Emf measurements are made on a cell without transference consisting of the deuterium gas electrode and silver/silver chloride electrode (Ref. 19). Emf measurements on the buffer solutions are made on cells without liquid junction at various temperatures and concentrations of chloride ion and the pD(S) values are derived by a method of calculation analogous to that described for the assignment of pH(S) values (Ref. 19). These solutions are recommended for the calibration of pH meters to be used for the measurement of pD in deuterium oxide.

13.5. Standard reference solutions for ionic activity

The National Bureau of Standards certifies samples of potassium chloride, sodium chloride and potassium fluoride ionic activity reference materials for K^+, Na^+, Cl^-, and F^-. These three materials are intended for use in the calibration of ion-selective electrodes.

The NBS has adopted the convention (Refs. 20, 21) for single ion activities based on the hydration theory proposed by Stokes and Robinson (Ref. 22). The details of this activity scale have been reviewed (Ref. 23). Briefly, the single ion activity coefficients are calculated from experimental values of the mean molal activity coefficient (γ_{MX}), the osmotic coefficient (ϕ), and an assigned hydration number (h). For univalent electrolytes the defining equations are:

$$\log \gamma_M = \log \gamma_{MX} + 0.00782(h_M - h_X)m\phi \qquad (13.5.1)$$
$$\log \gamma_X = \log \gamma_{MX} + 0.00782(h_X - h_M)m\phi \qquad (13.5.2).$$

As in the case of the pH scale, it is necessary to make certain extra-thermodynamic assumptions. The 'conventional' step in the treatment is the assignment of hydration numbers to the individual ionic species. The hydration numbers are calculated from smoothed values of γ_{MX} and ϕ using the Stokes-Robinson hydration theory. The hydration number for the electrolyte is then 'split' between the cation and anion by assigning a hydration value to one of the ions. For the present scale, the hydration number for the chloride ion was taken to be zero, and the other ionic hydration numbers are referred to this convention (Ref. 22). Ionic activity for unassociated electrolytes derived on the basis of this hydration convention have been found to be consistent with the observed responses of ion-selective electrodes.

This treatment has been used to compute single ion activity coefficients for the ions K^+, Na^+, Cl^- and F^- for three reference solutions. The pK, pNa, pCl and pF values listed correspond to $\log(1/a_i)$ where $i = K^+$, Na^+, Cl^- or F^- and where a_i is the conventional activity of the ions referred to the standard state on the scale of molality.

It is recommended that the reference solutions be used at a concentration similar to that of the sample to minimize liquid junction potential errors. Use of a bracketting technique (two reference solutions which bracket the sample solution concentration) will minimize errors due to non-Nernstian response of the electrodes and will increase the reliability of measurements.

REFERENCES

1. Durst, R. A.; Cali, J. P. *Pure Appl. Chem.* **1978**, *50*, 1485.
2. Definition of pH scales, standard reference values, measurement of pH and related terminology. *Pure Appl. Chem.* **1985**, *57*, 531.
3. Sorensen, S. P. L. *Biochem. Z.* **1909**, *21*, 131; *C. R. Trav. Lab. Carlsberg* **1909**, *8*, 1.
4. *Manual of Symbols and Terminology for Physicochemical Quantities and Units*, 2nd rev., *Pure Appl. Chem.* **1979**, *51*, 1.
5. Durst, R. A. *Standard Reference Materials; Standardization of pH Measurements*, NBS Special Publication, 260-53, U.S. Department of Commerce: Washington, **1975**.
6. British Standards Institution, Specification for pH Scale, BS1647, **1984**.
7. Bates, R. G.; Guggenheim, E. A. *Pure Appl. Chem.* **1960**, *1*, 163.
8. Hamer, W. J.; Acree, S. F. *J. Res. Nat. Bur. Stand.* **1944**, *32*, 215.
9. Hamer, W. J.; Pinching G. D.; Acree, S. F. *J. Res. Nat. Bur. Stand.* **1946**, *36*, 47.
10. Hetzer, H. B.; Durst, R. A.; Robinson, R. A.; Bates, R. G. *J. Res. Nat. Bur. Stand.* **1977**, *81A*, 21.
11. Etz, E. unpublished work (1973) quoted in ref. 12.
12. Bütikofer, H. P.; Covington, A. K. *Anal. Chim. Acta*, **1979**, *108*, 179.
13. Bates, R. G. *Determination of pH. Theory and Practice*, 2nd ed., Wiley: New York, **1973**.
14. Bates, R. G.; Robinson, R. A. *J. Soln. Chem.* **1980**, *9*, 455.
15. OIML International Recommendation 54: pH scale for aqueous solutions, **1981**.
16. Bates, R. G., editor, *National Bureau of Standards Technical Note* **1968**, *453*, 15.
17. Bates, R. G. *Anal. Chem.* **1968**, *40*(6), 28A.
18. Bates, R. G.; Bower, V. E. *J. Res. Nat. Bur. Std.* **1954**, *53*, 283.
19. Bates, R. G. *J. Res. Nat. Bur. Stds.* **1962**, *66A*, 179.

20. Bates, R. G.; Staples, B. R.; Robinson, R. A. *Anal. Chem.* **1970**, *42*, 867.
21. Durst, R. A., editor, *National Bureau of Standards Technical Note* **1970**, *543*, 21.
22. Stokes, R. H.; Robinson, R. A. *J. Amer. Chem. Soc.* **1948**, *70*, 1870.
23. Bates, R. G. *Pure Appl. Chem.* **1973**, *36*, 407.

13.6. Certified Reference Materials for pH

13.6.1. Reference value pH reference material (RVS)

Physical property: pH
Unit: Dimensionless
Recommended reference material: Potassium hydrogen phthalate ($KHC_8H_4O_4$)
Range of variables: 0 to 95 °C
Physical state within the range: aqueous solution, $m = 0.05$ mol kg^{-1}

Sources of supply and/or methods of purification: Certified samples are available from suppliers (A), (B), (C), (D), (E), (F), (G), (H), and (I).

Table 13.6.1. Potassium hydrogen phthalate ($KHC_8H_4O_4$); aqueous solution, $m = 0.05$ mol kg^{-1}

$t/°C$	pH(RVS)	$t/°C$	pH(RVS)	$t/°C$	pH(RVS)
0	4.000	35	4.018	65	4.097
5	3.998	37	4.022	70	4.116
10	3.997	40	4.027	75	4.137
15	3.998	45	4.038	80	4.159
20	4.001	50	4.050	85	4.183
25	4.005	55	4.064	90	4.21
30	4.011	60	4.080	95	4.24

The values are from reference 1.

REFERENCE

1. Definition of pH scales, standard reference values, measurement of pH and related terminology. *Pure Appl. Chem.* **1985**, *57*, 531.

13.6.2. Primary pH Reference Materials

Physical property: pH
Unit: Dimensionless
Recommended reference materials: see below

Sources of supply and/or methods of purification: Certified samples are available from suppliers (A) to (I). Some suppliers may not have all the reference materials.

Table 13.6.2.1. Potassium hydrogen tartrate ($KHC_4H_4O_6$), aqueous solution saturated at 25 °C

$t/°C$	pH(S)	$t/°C$	pH(S)	$t/°C$	pH(S)
25	3.557	45	3.547	70	3.580
30	3.552	50	3.549	80	3.609
35	3.549	55	3.554	90	3.650
38	3.548	60	3.560	95	3.674
40	3.547				

The values are from references 1, 2, 3, and 4. The uncertainty is ±0.005 from 0 to 60 °C and ±0.01 from 70 to 95 °C.

Table 13.6.2.2. Potassium dihydrogen citrate ($KH_2C_6H_5O_7$); aqueous solution, $m = 0.05$ mol kg^{-1}

$t/°C$	pH(S)	$t/°C$	pH(S)	$t/°C$	pH(S)
0	3.863	20	3.788	37	3.756
5	3.840	25	3.776	40	3.754
10	3.820	30	3.766	50	3.749
15	3.802	35	3.759		

The values are from references 1, 2, 3, and 4. The uncertainty is ±0.005 from 0 to 50 °C.

Table 13.6.2.3. Disodium hydrogen phosphate (Na_2HPO_4); Potassium dihydrogen phosphate (KH_2PO_4), aqueous solution $\begin{cases} m(Na_2HPO_4) = 0.025 \text{ mol kg}^{-1} \\ m(KH_2PO_4) = 0.025 \text{ mol kg}^{-1} \end{cases}$

$t/°C$	pH(S)	$t/°C$	pH(S)	$t/°C$	pH(S)	$t/°C$	pH(S)
0	6.982	20	6.878	37	6.839	60	6.836
5	6.949	25	6.863	40	6.836	70	6.845
10	6.921	30	6.851	45	6.832	80	6.859
15	6.898	35	6.842	50	6.831	90	6.876

The values are from references 1, 2, 3, and 4. The uncertainty is ±0.005 from 0 to 50 °C and ±0.01 from 60 to 90 °C.

Table 13.6.2.4. Disodium hydrogen phosphate (Na_2HPO_4); Potassium dihydrogen phosphate (KH_2PO_4),

aqueous solution $\begin{cases} m(Na_2HPO_4) = 0.03043 \text{ mol kg}^{-1} \\ m(KH_2PO_4) = 0.008695 \text{ mol kg}^{-1} \end{cases}$

$t/°C$	pH(S)	$t/°C$	pH(S)	$t/°C$	pH(S)
0	7.534	20	7.429	37	7.386
5	7.500	25	7.413	40	7.380
10	7.472	30	7.400	45	7.373
15	7.448	35	7.389	50	7.367

The values are from references 1 and 4. The uncertainty is ±0.005 from 0 to 50 °C.

Table 13.6.2.5. Sodium tetraborate ($Na_2B_4O_7 \cdot 10H_2O$); aqueous solution, $m = 0.01$ mol kg^{-1}

$t/°C$	pH(S)	$t/°C$	pH(S)	$t/°C$	pH(S)	$t/°C$	pH(S)
0	9.464	20	9.225	37	9.088	70	8.921
5	9.395	25	9.180	40	9.068	80	8.884
10	9.332	30	9.139	50	9.011	90	8.850
15	9.276	35	9.102	60	8.962	95	8.833

The values are from references 1, 3, and 4. The uncertainty is ±0.005 from 0 to 50 °C and ±0.01 from 60 to 95 °C.

Table 13.6.2.6. Sodium hydrogen carbonate (NaHCO$_3$); Sodium carbonate (Na$_2$CO$_3$), aqueous solution $\begin{cases} m(\text{NaHCO}_3) = 0.025 \text{ mol kg}^{-1} \\ m(\text{Na}_2\text{CO}_3) = 0.025 \text{ mol kg}^{-1} \end{cases}$

$t/^\circ$C	pH(S)	$t/^\circ$C	pH(S)	$t/^\circ$C	pH(S)
0	10.317	20	10.062	37	9.910
5	10.245	25	10.012	40	9.889
10	10.179	30	9.966	45	9.856
15	10.118	35	9.926	50	9.828

The values are from references 1, 3, and 4. The uncertainty is ±0.005 from 0 to 50 °C.

Table 13.6.2.7. Potassium tetraoxalate (KH$_3$(C$_2$O$_4$)$_2$ · H$_2$O);* aqueous solution, $m = 0.05$ mol kg^{-1}

$t/^\circ$C	pH(S)	$t/^\circ$C	pH(S)	$t/^\circ$C	pH(S)
0	1.666	30	1.683	55	1.715
5	1.668	35	1.688	60	1.723
10	1.670	38	1.691	70	1.743
15	1.672	40	1.694	80	1.766
20	1.675	45	1.700	90	1.792
25	1.679	50	1.707	95	1.806

The values are from references 2, 3, and 4. The uncertainty is ±0.005 from 0 to 60 °C and ±0.01 from 70 to 95 °C.

Table 13.6.2.8. Calcium hydroxide (Ca(OH)$_2$),* aqueous solution saturated at 25 °C

$t/^\circ$C	pH(S)	$t/^\circ$C	pH(S)	$t/^\circ$C	pH(S)
0	13.423	25	12.454	45	11.841
5	13.207	30	12.289	50	11.705
10	13.003	35	12.133	55	11.574
15	12.810	38	12.043	60	11.449
20	12.627	40	11.984		

The values are from references 3 and 4. The uncertainty is ±0.005 from 0 to 60 °C.

Table 13.6.2.9. Tris(hydroxymethyl)aminomethane;* Tris(hydroxymethyl)aminomethane hydochloride, aqueous solution $\begin{cases} m(\text{Tris}) &= 0.0167 \text{ mol kg}^{-1} \\ m(\text{Tris} \cdot \text{HCl}) &= 0.0500 \text{ mol kg}^{-1} \end{cases}$

$t/°C$	pH(S)	$t/°C$	pH(S)	$t/°C$	pH(S)
0	8.471	20	7.840	37	7.382
5	8.303	25	7.699	40	7.307
10	8.142	30	7.563	45	7.186
15	7.988	35	7.433	50	7.070

The values are from references 2, 5, and 6. The uncertainty is ±0.005 from 0 to 50 °C.

* These studies are referred to by the NBS as secondary pH standards because of the greater variability of the liquid junction potentials.

The primary pH reference materials listed in Section 13.6.2 can be certified as secondary pH reference materials by comparison with the reference value reference material. In this case the certified pH values slightly differ from those in the tables due to liquid junction potentials especially at pH < 3 and > 10. See Table 2 of reference 2.

REFERENCES

1. Definition of pH scales, standard reference values, measurement of pH and related terminology. *Pure Appl. Chem.* **1985**, *57*, 531.
2. Potentiometric ion activities. *Pure Appl. Chem.* **1978**, *50*, 1485.
3. Bates, R. G. *Determination of pH. Theory and Practice*, 2nd ed., Wiley: New York, **1973**.
4. OIML International Recommendation 54: pH scale of aqueous solutions, **1981**.
5. Durst, R. A., editor, *National Bureau of Standards Technical Note* **1970**, *543*, 21.
6. Durst, R. A.; Staples, B. R. *Clin. Chem.*, **1972**, *18*, 206.

13.7. Certified Reference Materials for pD

Physical property: pD
Unit: Dimensionless **Recommended reference materials:** see below

Sources of supply and/or methods of purification: Certified samples are available from supplier (E).

Table 13.7.1. Potassium dihydrogen phosphate (KH_2PO_4); Disodium hydrogen phosphate (Na_2HPO_4), deuterium oxide solution $\begin{cases} m(KH_2PO_4) = 0.025 \text{ mol kg}^{-1} \\ m(Na_2HPO_4) = 0.025 \text{ mol kg}^{-1} \end{cases}$

$t/°C$	pD(S)	$t/°C$	pD(S)	$t/°C$	pD(S)
5	7.539	25	7.428	40	7.387
10	7.504	30	7.411	45	7.381
15	7.475	35	7.397	50	7.377
20	7.449				

The values are from references 1 and 2 to 7. The uncertainty is ±0.01 from 0 to 50 °C.

Table 13.7.2. Sodium bicarbonate ($NaHCO_3$); Sodium carbonate (Na_2CO_3), deuterium oxide solution $\begin{cases} m(NaHCO_3) = 0.025 \text{ mol kg}^{-1} \\ m(Na_2CO_3) = 0.025 \text{ mol kg}^{-1} \end{cases}$

$t/°C$	pD(S)	$t/°C$	pD(S)	$t/°C$	pD(S)
5	10.998	25	10.736	40	10.597
10	10.923	30	10.685	45	10.559
15	10.855	35	10.638	50	10.527
20	10.793				

The values are from references 1 and 2 to 7. The uncertainty is ±0.01 from 0 to 50 °C.

REFERENCES

1. Potentiometric ion activities. *Pure Appl. Chem.* **1978**, *50*, 1485.
2. Bates, R. G., editor, *National Bureau of Standards Technical Note* **1968**, *453*, 15.
3. Bates, R. G.; Guggenheim, E. A. *Pure Appl. Chem.* **1960**, *1*, 163.
4. Bates, R. G. *Determination of pH. Theory and Practice*, 2nd ed., Wiley: New York, **1973**.
5. Bates, R. G. *Anal. Chem.* **1968**, *40*(6), 28A.
6. Bates, R. G.; Bower, V. E. *J. Res. Nat. Bur. Std.* **1954**, *53*, 283.
7. Bates, R. G. *J. Res. Nat. Bur. Stds.* **1962**, *66A*, 179.

13.8. Certified Reference Materials for Ionic Activity

Physical property: pX
Unit: Dimensionless
Recommended reference materials: see below

13. Potentiometric Ion Activities

Sources of supply and/or methods of purification: Certified samples are available from suppliers (E) and (I).

Table 13.8.1. Potassium chloride (KCl); aqueous solution, $m = (0.001 \text{ to } 2.0)$ mol kg^{-1} at 25 °C

Molality mol kg^{-1}	Amount concentration mol dm^{-3}	γ_{K^+}	γ_{Cl^-}	pK$^+$	pCl$^-$
0.001	0.000997	0.965	0.965	3.016	3.016
0.01	0.00997	0.901	0.901	2.045	2.045
0.1	0.0994	0.772	0.768	1.112	1.115
0.2	0.1983	0.723	0.713	0.841	0.846
0.3	0.2967	0.693	0.680	0.682	0.690
0.5	0.4916	0.659	0.639	0.482	0.495
1.0	0.9692	0.623	0.586	0.206	0.232
1.5	1.4329	0.611	0.558	0.0376	0.078
2.0	1.8827	0.609	0.539		

The values are from references 1 and 2 to 5. The uncertainty is ±0.01.

Table 13.8.2. Sodium chloride (NaCl); aqueous solution, $m = (0.001 \text{ to } 2.0)$ mol kg^{-1} at 25 °C

Molality mol kg^{-1}	Amount concentration mol dm^{-3}	γ_{Na^+}	γ_{Cl^-}	pNa$^+$	pCl$^-$
0.001	0.000997	0.965	0.965	3.015	3.015
0.01	0.00997	0.903	0.902	2.044	2.045
0.1	0.0995	0.783	0.773	1.106	1.112
0.2	0.1987	0.744	0.727	0.828	0.838
0.3	0.2975	0.721	0.697	0.664	0.680
0.5	0.4941	0.701	0.662	0.455	0.480
1.0	0.9789	0.696	0.620	0.157	0.208
1.5	1.4543	0.718	0.602		
2.0	1.9200	0.752	0.593		

The values are from references 1 and 2 to 5. The uncertainty is ±0.01.

Table 13.8.3. Potassium fluoride (KF); aqueous solution, $m = (0.0001 \text{ to } 2.0)$ mol kg^{-1} at 25 °C

Molality mol kg^{-1}	Amount concentration mol dm^{-3}	Activity coefficient ($\gamma_{F^-} = \gamma_{K^+} = \gamma_\pm$)	pF$^-$
0.0001	0.0000997	0.988	4.00$_5$
0.0005	0.000499	0.975	3.31$_2$
0.001	0.000997	0.965	3.01$_6$
0.005	0.00498	0.927	2.33$_4$
0.01	0.00997	0.902	2.04$_5$
0.05	0.0498	0.818	1.38$_8$
0.1	0.0996	0.773	1.11$_2$
0.2	0.1990	0.726	0.83$_8$
0.3	0.2982	0.699	0.67$_8$
0.5	0.4961	0.670	0.47$_5$
0.75	0.7424	0.652	0.31$_1$
1.0	0.9873	0.645	0.19$_0$
1.5	1.4729	0.646	+0.01$_4$
2.0	1.9523	0.658	−0.11$_9$

The values are from references 1 and 2 to 5. The uncertainty is ±0.01.

REFERENCES

1. Durst, R. A.; Staples, B. R. *Pure Appl. Chem.* **1978**, *50*, 1485.
2. Bates, R. G.; Staples, B. R.; Robinson, R. A. *Anal. Chem.* **1970**, *42*, 867.
3. Durst, R. A., editor, *National Bureau of Standards Technical Note* **1970**, *543*, 21.
4. Stokes, R. H.; Robinson, R. A. *J. Amer. Chem. Soc.* **1948**, *70*, 1870.
5. Bates, R. G. *Pure Appl. Chem.* **1973**, *36*, 407.

13.9. Contributors to the revised version

E. Juhász,
National Office of Measurements,
Némétölgyi út 37,
H-1124, Budapest XII
(Hungary)

R. A. Durst
National Bureau of Standards,
US Department of Commerce,
Gaithersburg, MD 20899
(U.S.A.)

13.10. List of Suppliers

A. BDH Chemicals, Ltd.
   Poole, Dorset
   (U.K.)

B. Kanto Chemical Co., Inc.
   3-7 Hon-machi
   Nihonbashi
   Chou-ku, Tokyo
   (Japan)

C. E. Merck
   D-6100 Darmstadt
   (Germany)

D. National Office of Measures
   Névetvölgyi út 37
   Budapest XII
   (Hungary)

E. Office of Standard Reference Materials
   US Department of Commerce
   National Bureau of Standards
   Gaithersburg, MD 20899
   (U.S.A.)

F. Polish Committee for Standardization
   and Measures
   Division of Physicochemical Metrology
   UL. Elektoralna 2,
   PL-00139 Warsaw
   (Poland)

G. Ridel-De Haen AG.
   D-3016 Seelze
   (Germany)

H. Wako Pure Chemical Industries, Ltd.
   3-10 Dosho-machi
   Higashi-ku
   Osaka
   (Japan)

I. Institute of Metrology
   Sos. Vitan-Birzesti nr. 11
   Sector 5,
   Bucharest
   (Rumania)

14

SECTION: OPTICAL ROTATION

COLLATOR: B. COXON

CONTENTS:

14.1. Introduction

14.2. Reference materials for the measurement of optical rotation

    14.2.1. Sucrose

    14.2.2. Anhydrous dextrose

    14.2.3. Quartz control plates

14.3. Contributors

14.4. List of Suppliers

14.1. Introduction

The techniques of polarimetry and saccharimetry were formulated by Biot early in the nineteenth century and the classical methods and apparatus used have been described in detail by Bates et al. (Ref. 1), Heller (Ref. 2) and Asmus (Ref. 3). Saccharimetry is extensively used in the sugar industry, and polarimetry has long been used for the physical characterization and for the study of the reaction kinetics of optically active organic compounds, including the steroids, carbohydrates, peptides, and proteins. Optical rotatory dispersion techniques have proved to be particularly applicable to proteins. The calibration materials which have been employed are sucrose solutions, dextrose solutions and quartz plates. These three materials are proposed as suitable calibration and test materials.

Optical rotations of liquids, solutions and quartz plates may be measured by use of either visual or photoelectric (automatic) polarimeters (Refs. 4 to 6). The half-shade principle has been extensively used in the design of visual polarimeters (Ref. 1). In this technique, a collimated, usually monochromatic, beam of light is plane polarized by a special prism. The plane of polarization of half the cross-sectional area of the light beam is then rotated by a second prism so as to differ from the plane of the remainder of the light by a small angle (approx 2°), known as the half shade angle. The planes of polarization of the entire beam of light are rotated equally by the sample, resolved by an analyser prism, focussed by a telescope, and viewed by the eye. By rotation of the analyser prism, a null point is detected when both halves of the field appear equally dark. The null point is determined in both the presence and absence of the sample, and the difference between the corresponding analyser angles measured on a circular scale gives the optical rotation of the sample.

In photoelectric polarimeters, the plane of polarization of the light beam is usually modulated, either mechanically by oscillation of the polarizer, or electromagnetically by the Faraday effect produced by application of an alternating current to a coil surrounding the material of the Faraday cell (Refs. 4 to 6). After passing through the sample and analyser, the modulated light beam is converted to an electrical signal by a photomultiplier or photodiode, and detection of the signal component at the modulation frequency is then used to indicate the null point of the analyser, the angular position of which may be measured either electrically or mechanically.

Recent improvements to polarimetric technique include the introduction of laser light sources and the use of highly precise optical encoders for the measurement of angles with a sensitivity of 3 μrad (Refs. 4, 6, 7). Laser light sources are both spatially coherent and monochromatic and as a result of their small beam divergence and diameter, 1 to 2 mW lasers can provide as much as 1000 times more useful light to a polarimeter than can a conventional atomic discharge lamp (Ref. 4). For visual polarimeters based on the Lippich half-shade method, a typical polarimetric sensitivity of 0.04 mrad has been quoted (Ref. 4) for both lamp and laser light sources. For photoelectric polarimeters using polarization modulation, a typical sensitivity of 0.04 mrad has been indicated for lamp sources (0.006 mrad with signal averaging) and 0.003 mrad for laser sources (Ref. 4).

Highly accurate measurements of optical rotation require that many different parameters be considered, including source wavelength, cell window birefringence, sample temperature, ambient magnetic field, light collimation, optical alignment, and sample concentration (Ref. 6).

The conventional measures of optical rotation have been the specific optical rotatory power or specific optical rotation. That is the rotation in circular degrees given by a 10 cm thickness of liquid or solution containing 1 g of optically active material per millilitre. The customary symbol has been $[\alpha]_\lambda^t$ where t is the Celsius temperature and λ is the wavelength of light used. A temperature of 20 °C and a wavelength in air of 546.1 or 589.3 nm are usually used. The latter wavelength is generated by the 'sodium D line' and the conventional symbol $[\alpha]_D^t$ is in widespread use for the characterization of organic materials. The use of the 'sodium D line' is thought to be less than ideal because of the presence of a closely spaced doublet of spectral lines of slightly variable relative intensities. The use of the helium-neon laser (632.8

14. Optical Rotation

nm) as an alternative has been discussed (Ref. 7).

The conventional specific rotation for liquids is given by $[\alpha] = \alpha/l\rho$, and for solids in solution, $[\alpha] = 100\alpha/lc$ where α is the rotation angle in circular degrees, l is the cell path length in dm, ρ is the density of the liquid in g cm^{-3}, and c is the concentration of dissolved material in g/100 ml. For solids, degrees per millimetre have also been used.

From the above relations it can be seen that the conventional unit of specific rotation is deg ml g^{-1} dm^{-1}. In the sugar industry, measurements are made on the "sugar scale" (°S) where 100 °S is the rotation given by a 2 dm tube filled with "normal" sucrose solution containing an apparent mass in air of 26.000 g of sugar per 100.000 cm^3 of solution at 20 °C and at the vacuum wavelength ($\lambda = 546.2271$ nm) of the green line of the mercury isotope ^{198}Hg. The quantity 100 °S corresponds to an angle, α, of 40.765°, an angle which is also given by a 'normal' quartz plate (1.5934 mm thick). (Note:- the weight in air given here is for weighing in air with brass weights, at 101.3 kPa and 50% relative humidity. The corresponding mass is 26.016 g).

On the basis of a more recent extensive set of measurements at the PTB and Braunschweig Sugar Institute (Ref. 8), it has been proposed that the 100°S point be redefined by the angle $\alpha = 40.7772°$. The general magnitude of this proposed change has been confirmed by measurements at the U.S. National Bureau of Standards ($\alpha = 40.782°$, 1977) and in Braunschweig ($\alpha = 40.7777°$, 1977 and $\alpha = 40.7768°$, 1979). A new value ($\alpha = 40.777 \pm 0.001°$) was officially adopted by the International Commission for Uniform Methods of Sugar Analysis (ICUMSA) in May 1986, and is scheduled to take effect on July 1, 1988.

The SI unit of angle is the radian, and McGlashan (Ref. 9) has proposed the symbol α_m for the specific optical rotatory power with the unit rad m^2 kg^{-1}. This specific optical rotatory power is the rotation in radians given by a 1 m thickness of liquid or solution containing 1 kg of optically active solute per m^3 of solution. Thus

$$\begin{aligned}\alpha_m/(\text{rad m}^2 \text{ kg}^{-1}) &= [\alpha] \times \pi/180 \times 10^{-2}/(\text{deg ml g}^{-1} \text{ dm}^{-1}) \\ &= [\alpha] \times 1.745329 \times 10^{-4}/(\text{deg ml g}^{-1} \text{ dm}^{-1})\end{aligned} \quad (14.1.1)$$

For sucrose, the value of $[\alpha]_{546.1 \text{ nm}}^{20}/(\text{deg ml g}^{-1} \text{ dm}^{-1})$ is 78.342, which corresponds to the value 1.3674×10^{-2} for $\alpha_m/(\text{rad m}^2 \text{ kg}^{-1})$. The SI unit, rad m^2 kg^{-1} is inconveniently large and it is proposed that the unit mrad m^2 kg^{-1} be adopted thus:

$$\alpha_m/(\text{mrad m}^2 \text{ kg}^{-1}) = [\alpha] \times 0.1745329/(\text{deg ml g}^{-1}\text{dm}^{-1}) \quad (14.1.2).$$

It should be noted that Heller (Ref. 2) has suggested a unit, the Biot, which is 10.00028 times smaller than the proposed unit, mrad m^2 kg^{-1}.

It would be desirable to have α_m values for calibration and test materials quoted for 25 °C as well as for 20 °C.

It has been suggested that national laboratories who issue samples of sucrose, dextrose or quartz plates as calibration and test materials should give concentrations in kg m^{-3} and specific rotatory power in mrad m^2 kg^{-1}.

REFERENCES

1. Bates, F. J. and associates, *U.S. National Bureau of Standards Circular C440*, U.S. Department of Commerce: Washington, **1942**.
2. Heller, W. *Techniques of Organic Chemistry*, 3rd edition, Weissberger, A., editor, Vol. I, Part 3, Interscience: New York, **1960**, p. 2147.
3. Asmus, E. *Methoden der Organischen Chemie [Houben-Weyl]*, Band III, Teil 2, Georg Thieme Verlag: Stuttgart, **1955**, p. 425.
4. Cummings, A. L.; Layer, H. P.; Hocken, R. J. *Lasers in Chemical Analysis*, Hieftje, G. M.; Travis, J. C.; Lytle, F. E., editors, The Humana Press: Clifton, NY, **1981**, p. 291.
5. Olson, W. B. *Opt. Eng.* **1973**, *12*(3), 102.
6. Cummings, A. L.; Hocken, R. J. *Precision Engineering* **1982**, *4*, 33.
7. Gates, J. W. *Chem. and Ind.* **1958**, 190.
8. *Proc. 16th. Session International Commission for Uniform Methods of Sugar Analysis*, P.O. Box 35, Wharf Road, Peterborough, England PE2 9PU, **1974**.
9. McGlashan, M. L. *Physicochemical Quantities and Units*, Monograph 15, Royal Institute of Chemistry: London, **1968**.

14.2. Reference materials for the measurement of optical rotation

14.2.1. Sucrose

Physical property: Specific optical rotatory power, α_m
Unit: mrad m^2 kg^{-1}
Recommended reference material: Sucrose ($C_{12}H_{22}O_{11}$)
Range of variables: 20 °C; wavelengths 546.2271 nm, 589.4400 nm, 632.9914 nm
(*in vacuo*)
Physical state within the range: aqueous solution
Contributors to the first version: I. Brown, J. P. Cali, J. Franc, J. E. Lane, T. Plebanski
Contributor to the revised version: B. Coxon

Intended usage: Sucrose can be used for calibration and for checking the performance of manually operated and automatically recording light polarimeters and saccharimeters employed for determination of the optical rotation of liquids or dissolved solids. Details of the apparatus and of the methods are available (Refs. 1, 2, 3)

Sources of supply and/or methods of preparation: Samples of sucrose are available from suppliers (B) and (C). As sucrose must be kept dry during storage to avoid the possibility

14. Optical Rotation

of inversion, samples should ideally be stored in sealed ampoules under vacuum or in dry inert gas.

Pertinent physicochemical data: The following values of specific rotation were calculated from data given on National Bureau of Standards certificates for a sample of sucrose labelled SRM 17d. The values refer to a solution of mass concentration 260.16 kg m^{-3}.

$$\alpha_{m,\,546.2271\,\text{nm}}^{20\,°\text{C}}/(\text{mrad m}^2\text{ kg}^{-1}) = 13.677 \pm 0.003 \quad (14.2.1.1)$$

$$\alpha_{m,\,589.4400\,\text{nm}}^{20\,°\text{C}}/(\text{mrad m}^2\text{ kg}^{-1}) = 11.613 \quad (14.2.1.2)$$

(calculated from an interpolation function)

$$\alpha_{m,\,632.9914\,\text{nm}}^{20\,°\text{C}}/(\text{mrad m}^2\text{ kg}^{-1}) = 9.9779 \pm 0.0033 \quad (14.2.1.3)$$

(the uncertainty given includes a minor contribution for sample inhomogeneity and is approximately two standard deviations of the certified value for wavelength *in vacuo*.)

REFERENCES

1. Bates, F. J. and associates, *U.S. National Bureau of Standards Circular C440*, U.S. Department of Commerce: Washington, **1942**.
2. Heller, W. *Techniques of Organic Chemistry*, 3rd edition, Weissberger, A., editor, Vol. I, Part 3, Interscience: New York, **1960**, p. 2147.
3. Asmus, E. *Methoden der Organischen Chemie [Houben-Weyl]*, Band III, Teil 2, Georg Thieme Verlag: Stuttgart, **1955** p. 425.

14.2.2. Anhydrous dextrose

Physical property: Specific optical rotatory power, α_m
Unit: mrad m^2 kg^{-1}
Recommended reference material: Anhydrous dextrose (D-glucose) (C$_6$H$_{12}$O$_6$)
Range of variables: 20 °C; wavelengths 546.2271 nm, 589.4400 nm, 632.9913 nm
(*in vacuo*)
Physical state within the range: aqueous solution
Contributors to the first version: I. Brown, J. P. Cali, J. Franc, J. E. Lane
Contributor to the revised version: B. Coxon

Intended usage: D-Glucose can be used for calibration and for checking the performance of manually operated and automatically recording light polarimeters and saccharimeters employed for the determination of the optical rotation of liquids and dissolved solids. Details of the apparatus and of the methods are available (Refs. 1 to 3).

Sources of supply and/or methods of preparation: Samples of D-glucose are available from supplier (B). Samples of D-glucose absorb moisture on the surface and this water can be removed by drying the specimen in vacuum at temperatures between 60 and 70 °C.

Pertinent physicochemical data: The following equation (Ref. 1) expresses the specific rotation of dextrose solutions as a function of dextrose mass concentration, c, for the range 60 to 320 kg m^{-3},

$$\alpha_{m,546.1\text{ nm}}^{20\ ^\circ\text{C}}/(\text{mrad m}^2\text{ kg}^{-1}) = 10.827 + 0.000\ 7430c/(\text{kg m}^{-3}) \qquad (14.2.2.1).$$

The following value of specific rotation for a wavelength of 589.25 nm which refers to a mass concentration of 40.0 kg m^{-3} was derived from the same source (Ref. 1):

$$\alpha_{m,589.25\text{ nm}}^{20\ ^\circ\text{C}}/(\text{mrad m}^2\text{ kg}^{-1}) = 9.20 \qquad (14.2.2.2).$$

More recently, the values:

$$\alpha_{m,546\text{ nm}}^{20\ ^\circ\text{C}}/(\text{mrad m}^2\text{ kg}^{-1}) = 10.9 \qquad (14.2.2.3)$$

and

$$\alpha_{m,589\text{ nm}}^{20\ ^\circ\text{C}}/(\text{mrad m}^2\text{ kg}^{-1}) = 9.25 \qquad (14.2.2.4)$$

have been measured for U.S. National Bureau of Standards dextrose SRM 41b at a mass concentration of 200 kg m^{-3}. This material has been renewed (SRM 41c, 1986), and measurements of it at a mass concentration of 200 kg m^{-3} have given

$$\alpha_{m,546\text{ nm}}^{20\ ^\circ\text{C}}/(\text{mrad m}^2\text{ kg}^{-1}) = 11.01 \qquad (14.2.2.5)$$

$$\alpha_{m,589\text{ nm}}^{20\ ^\circ\text{C}}/(\text{mrad m}^2\text{ kg}^{-1}) = 9.318 \qquad (14.2.2.6)$$

$$\alpha_{m,633\text{ nm}}^{20\ ^\circ\text{C}}/(\text{mrad m}^2\text{ kg}^{-1}) = 7.987 \qquad (14.2.2.7)$$

Unlike previous certifications of SRM 41, 41a, and 41b, the measurements for SRM 41c were performed without spiking of the solution with ammonia, a procedure that has been commonly employed to achieve rapid equilibration of tautomers.

REFERENCES

1. Bates, F. J. and associates, *U.S. National Bureau of Standards Circular C440*, U.S. Department of Commerce: Washington, **1942**.
2. Heller, W. *Techniques of Organic Chemistry*, 3rd edition, Weissberger, A., editor, Vol. I, Part 3, Interscience: New York, **1960**, p. 2147.
3. Asmus, E. *Methoden der Organischen Chemie [Houben-Weyl]*, Band III, Teil 2, Georg Thieme Verlag: Stuttgart, **1955**, p. 425.

14.2.3. Quartz control plates

Physical property: Optical rotation, α
Unit: mrad
Recommended reference materials: Quartz control plates
Range of variables: 20 °C; wavelengths 546.2271 nm, 589.4400 nm, 632.9914 nm (*in vacuo*) (Ref. 1)
Physical state within the range: solid
Contributors to the first version: I. Brown, J. Dyson, J. E. Lane
Contributor to the revised version: B. Coxon

Intended Usage: Quartz control plates can be used for calibration and for checking the performance of saccharimeters employed for the determination of sucrose in solutions. These plates are normally calibrated on the sugar scale. They could be calibrated in milliradians for use in the calibration of polarimeters. Details of the plates and an account of their uses are given in reference 2.

Sources of supply and/or methods of preparation: Quartz control plates are available from suppliers (A), (D), (E), and (F). Certificates giving the rotation on the sugar scale for individual plates from these suppliers are available from the National Physical Laboratory, Teddington, Middlesex (U.K.), from the Center for Analytical Chemistry, National Bureau of Standards, Gaithersburg, MD 20899 (U.S.A.), and from Supplier (F).

Pertinent physicochemical data: Values of the optical rotation for individual plates are given in certificates issued with the plates.

REFERENCES

1. *Proc. 18th Session International Commission for Uniform Methods of Sugar Analysis*, P.O. Box 35, Wharf Road, Peterborough, England PE2 9PU, **1982**.
2. Bates, F. J. and associates, *U.S. National Bureau of Standards Circular C440*, U.S. Department of Commerce: Washington, **1942**.

14.3. Contributor to the revised version

B. COXON,
National Bureau of Standards,
U.S. Department of Commerce,
Gaithersburg, MD 20899
(U.S.A)

14.4. List of Suppliers

(A) Messrs. Bellingham and Stanley Ltd,
    Polyfract Works,
    Longfield Road,
    Tunbridge Wells, Kent
    (U.K.)

(B) Office of Standard Reference Materials,
    National Bureau of Standards,
    Gaithersburg, MD, 20899
    (U.S.A.)

(C) Polish Committee of Standardization,
    and Measures,
    Division of Physicochemical Metrology,
    Ul. Elektoralna 2, PL 00-139, Warsaw
    (Poland)

(D) Optical Activity Ltd.,
    Bury Road,
    Ramsey, Huntingdon
    Cambridgeshire, PE17 1NA
    (U.K.)

(E) Schmidt and Haensch,
    Naumannstrasse 33,
    1 Berline 62
    (W. Germany)

(F) Rudolf Instruments,
    40 Pier Lane,
    Fairfield, NJ 07006,
    (U.S.A.)

15

SECTION: OPTICAL REFRACTION

COLLATORS: A. FELDMAN, I. H. MALITSON

CONTENTS:

- 15.1. Introduction
- 15.2. Reference materials for the measurement of optical refraction (refractive index)
  - 15.2.1. Water
  - 15.2.2. 2,2,4-Trimethylpentane
  - 15.2.3. Hexadecane
  - 15.2.4. *trans*-Bicyclo[4,4,0]decane
  - 15.2.5. 1-Methylnaphthalene
  - 15.2.6. Toluene
  - 15.2.7. Methylcyclohexane
  - 15.2.8. Silicone liquids
  - 15.2.9. Borosilicate glass
  - 15.2.10. Soda-lime glass
- 15.3. Contributors
- 15.4. List of Suppliers

15.1. Introduction

The symbol for refractive index recommended by the IUPAC (Ref. 1) is n. A primary standard is not required for refractive index measurements because the refractive index of any transparent material can be determined accurately on a goniometer with monochromatic illumination by the use of the minimum deviation method described by Tilton and Taylor (Ref. 2). When applied to a liquid the method requires a carefully made hollow prism with the temperature precisely controlled by a thermostat. Such apparatus is not used for routine measurements. Commercial refractometers require reference materials for calibration. Most commercial refractometers are either critical angle instruments or are hollow prism instruments. Flow differential refractometers based on the Fresnel principle and hollow split prisms are available. An automatic refractometer that measures the refractive index of liquids in a matter of seconds and displays the results on a digital readout is available.

One of the major problems in the accurate measurement of the refractive index of liquids is the attainment of adequate control of the temperature of the sample. In some commercial instruments there are difficulties with enclosure of the sample and such designs may allow evaporation of the liquid to occur which lowers the sample temperature and which may cause changes in the concentration of mixtures. Inadequate closure may also cause the contamination of the sample by air or by water vapour. The most satisfactory methods of measurement are those based on the use of a hollow prism operated in a constant temperature enclosure. A good general description of the apparatus and methods has been given by Bauer and Lewin (Ref. 3). Additional useful information is given by Tilton (Ref. 4), Tilton and Taylor (Ref. 5) and Fishter (Ref. 6).

If a laboratory is to do accurate measurements over a wide range of indices, it should have a series of certified reference materials for which the refractive indices are known over an appropriate range of wavelengths and temperatures. Suitable reference materials for this purpose are distilled water, selected hydrocarbons, other certified liquids, and optical glasses. A survey of selected laboratories listed in the ISO Directory of Certified Reference Materials (Ref. 7) has shown that a limited number of certified reference materials can be purchased directly. Certain National Laboratories that do not issue certified standards provide a measurement service on specimens submitted for calibration. The National Bureau of Standards (U.S.A.) issues certified liquid and glass standards. The National Physical Laboratory (Gt. Britain) certifies submitted solid and liquid specimens (Ref. 8) with an accuracy of 10^{-5}. The Institut d'Optique Theorique et Appliquee (France) measures, on request, refractive index values accurate to 10^{-5}. The Physikalisch-Technische Bundesanstalt (W. Germany) has the capability of index measurements accurate to 10^{-6}. The American Petroleum Institute (U.S.A.) sells a series of hydrocarbons certified as to refractive index. A commercial laboratory (Ref. 9), not listed in the ISO directory, can supply a variety of refractive index liquids that cover the index range from n_D from 1.30 to 2.31 at 25 °C.

Water is recommended to be the first choice as a calibration and test material for refractive index measurements of liquids. The refractive index of air-free water relative to that of dry

15. Optical Refraction

air at temperatures from 0 to 60 °C for light of 13 wavelengths has been measured with great care by Tilton and Taylor (Ref. 2) with an uncertainty of less than 1×10^{-6}. The differences between these values and those of another sample of purified water due to likely isotope variations are of the order of a few parts in 10^7. Selected values of refractive index are given (Ref. 10) for air-saturated water over the temperature range 0 to 60 °C. These values are approximately three parts in 10^7 higher than those reported by Tilton and Taylor (Ref. 2).

In the past the values for the refractive index of water given in the International Critical Tables have been widely used but these are based on the earlier measurements made by Flatow (Ref. 11) and are in marked disagreement with the more recent values reported by Tilton and Taylor (Ref. 2).

The temperature scale used by Tilton and Taylor in 1938 was the 'International Temperature Scale' and in the range 0 to 60 °C it is identical with the International Temperature Scale of 1948. This, however, differs from the present recommended scale, the International Practical Temperature Scale of 1968, by as much as 0.01 K in this temperature range. The temperature coefficient of the refractive index of water is approximately 1.7×10^{-4} K^{-1}. The values given by Tilton and Taylor have been recalculated for the wavelength 589.26 nm to give the table of values recommended by this IUPAC Commission. For example, for 50 °C (t_{48}) Tilton and Taylor give for n_D the value 1.3290369 but the value for 50 °C (t_{68}) which equals 50.01 °C (t_{48}) is 1.3290352. When the accuracy of refractive index values required for calibration purposes is not greater than one in the fifth decimal place, the values for water quoted by Tilton and Taylor (Ref. 2) tabulated for temperatures relative to the International Practical Temperature Scale 1948 (IPTS 1948) may be used without alteration for temperatures relative to IPTS 1968. When a greater accuracy is required for temperatures expressed on IPTS 1968, the table of values provided here with the data sheets can be used for the wavelength 589.26 nm at temperatures in the range 0 to 60 °C. Values on the IPTS 1968 can be obtained for other wavelengths by first converting the temperature to the IPTS 1948 and using the values in the equation (3) provided by Tilton and Taylor (Ref. 2).

Various hydrocarbons with certified values for refractive index are available for refractometer calibration. Samples of 2,2,4-trimethylpentane and toluene are available from Supplier (A) and Supplier (B). Samples of hexadecane, *trans*-decahydronaphthalene, 1-methylnaphthalene and methylcyclohexane are available from supplier (A). Two silicone liquids are available from supplier (B).

For those observers who prefer solids to check the performance of their instruments, optical quality materials are commercially available. Optical glasses might be thought to be the most suitable materials for use as calibration materials because they are stable and have very low temperature coefficients of refractive index. However, although they are suitable for the calibration and testing of certain critical angle instruments, they are not always suitable for use with hollow prism or flow instruments for which the use of a liquid may be essential. High quality optical glasses are available (Ref. 12) that cover the index range from $n_D = 1.435$ to 1.7, the limit of most commercial refractometers. Samples of optical glasses can be certified as to refractive index values by National Laboratories. The National Bureau of Standards (U.S.A.) issues two certified glass standards designed for calibration of refractometers ($n_D = 1.488$ and $n_D = 1.518$). For low refractive index settings single crystal and polycrystalline LiF ($n = 1.392$), CaF_2 ($n = 1.434$) and BaF_2 ($n = 1.476$) and synthetic fused SiO_2 ($n = 1.458$) are available from a number of companies (Ref. 13). Samples of these materials would have to be certified by National Laboratories for their use as reference materials. It is recommended that solid standards for use on a refractometer be prepared as illustrated in figure 15.1.1.

In summary, six hydrocarbons, water, two silicone liquids and two glasses are recommended as suitable calibration and test materials. Together they form a group of calibration and test materials with refractive index values spaced at suitable intervals and covering a range of temperatures and wavelengths. The following data sheets give details of the properties of these materials.

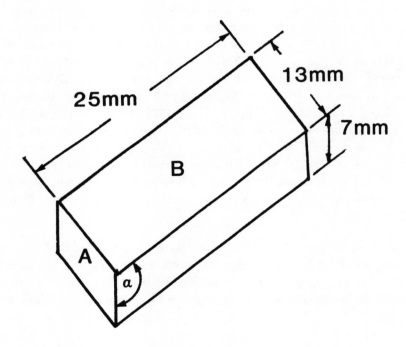

Figure 15.1.1. Recommended form for solid refractive index standard. Polish face A flat to within one half wavelength 632.8 nm. Polish face B to within one wavelength 632.8 nm. All other faces may be ground. Angle α between faces A and B to be $90° \pm 3$ min with no chips or bevels on edge. All other edges may be beveled. Opposite pairs of long faces should be parallel to within 10 min.

REFERENCES

1. *IUPAC Manual of Symbols and Terminology for Physicochemical Quantities and Units*, Butterworths: London, **1969**.
2. Tilton, L. W.; Taylor, J. K. *J. Res. Nat. Bur. Stand.* **1938**, *20*, 419.
3. Bauer, N.; Lewin, S. Z. *Techniques of Organic Chemistry*, Vol. I, *Physical Methods*, Part II, Third Edition, Weissberger, A., editor, Interscience: New York, **1960**, p. 1211.
4. Tilton, L. W. *J. Opt. Soc. Amer.* **1942**, *32*, 371.
5. Tilton, L. W.; Taylor, J. K. *Physical Methods in Chemical Analysis*, Vol. I, Second Revised edition, Berl, W. G., editor, Academic Press: New York, **1960**, p. 441.
6. Fishter, G. E. *Applied Optical Engineering*, Vol. IV, Chapter 10, Kingslake, R., editor, Academic Press: New York **1967**.
7. *ISO (International Organization of Standardization) Directory of Certified Reference Materials*, First Edition, 1 rue de Varembe, Case postale 56, CH-1211 Geneve 20, Switzerland, **1982**.
8. Debenham, M.; Dew, G. D. *J. Phys. E: Sci. Instrum*, **1981**, *14*, 544.
9. R. P. Cargille Laboratories, Inc., 55 Commerce Road, Cedar Grove, New Jersey 07009, U.S.A.
10. *TRC Thermodynamic Tables – Non-Hydrocarbons*, Thermodynamic Research Center, Texas A & M University, College Station, Texas. Loose-leaf data extant, **1983**.
11. Flatow, E. *Ann. Phys.*, **1903**, *Lpz. 12*, 95.
12. Manufacturers of high-quality optical glass are:
    a) Chance Pilkington Ltd., St. Asaph (U.K.)
    b) Jenaer Glaswerk Schott & Gen., Mainz (W. Germany)
    c) Schott Optical Glass Inc., Duryea, PA (U.S.A)
    d) Corning France and Corning Glass Works (U.S.A)
    e) Saint-Gobain Industries (France)
    f) Hoya Corp., Tokyo (Japan) and Hoya Optics U.S.A., Inc.
    g) Ohara Optical Glass Mfg. Co., Ltd., Tokyo (Japan)
13. A good source of locating international suppliers of optical materials is The Optical Industry and Systems Purchasing Directory, published by The Optical Publishing Co., Inc., P.O. Box 1146, Berkshire Commons, Pittsfield, MA 01201 (U.S.A).

15.2. Reference materials for the measurement of optical refraction

15.2.1. Water

Physical property: Optical Refraction (refractive index), n
Unit: Dimensionless
Recommended reference material: Water (H_2O)
Range of variables: 0 to 60 °C, wavelength 589.26 nm

Physical state within the range: liquid
Contributors to the first version: I. Brown, H. Feuerberg, J. Franc, E. F. G. Herington, J. F. Lane, T. Plebanski.
Contributors to the revised version: A. Feldman, I. H. Malitson.

Intended usage: Water can be used for the calibration of refractometers employed for measurements on solids and liquids. Details of methods and apparatus are given by Tilton (Ref. 1).

Sources of supply and/or methods of preparation: A suitable sample of purified, degased water can be prepared by the distillation from a well steamed glass apparatus with the sample collected hot. Details are given by Tilton and Taylor (Ref. 2); the method recommended by Harkins and Brown (Ref. 3) is also suitable.

Pertinent physicochemical data: Values for the refractive index of air-free water at thirteen wavelengths and at temperatures from 0 to 60 °C (IPTS 1948) are given by Tilton and Taylor (Ref. 2). Table 15.2.1.1 gives the refractive index for the sodium D line (589.26 nm) at one degree intervals over the temperature range 0 to 60 °C where the temperatures refer to IPTS 1968. These values were calculated in the following way.

Temperatures on the IPTS 1948 corresponding to the temperatures at one degree intervals on the IPTS 1968 were calculated by the use of equation (15.2.1.1) which is a modification of equation (12) given by Bedford and Kirby (Ref. 4).

$$t_{48} = t_{68} - 0.045(t_{68}/100)[(t_{68}/100) - 1][(t_{68}/419.58) - 1][(t_{68}/630.74) - 1] \\ - t_{68}(t_{68} - 100)(\delta_{68} - \delta_{48})/[\delta_{68}(100 - 2t_{68}) + 10^4] \quad (15.2.1.1)$$

where $(\delta_{68} - \delta_{48}) = 0.0049768$ and $\delta_{68} = 1.496334$. Then the required refractive index values were calculated for a wavelength of 589.26 nm by equation (15.2.1.2) given by Tilton and Taylor (Ref. 2).

$$n_D^t = n_D^{20} - [B/(\Delta x)^3 + A(\Delta x)^2 + C\ \Delta x]/[10^7(x + D)] \quad (15.2.1.2)$$

where $x = t_{48}/°C$, $A = 2352.12$, $B = 6.3649$, $C = 76087.9$, and $D = 65.7081$; $n_D^{20} = 1.3329877$, $\Delta x = (t_{48}/°C - 20)$. These values of the constants are given by Tilton and Taylor (Ref. 2).

15. Optical Refraction

Table 15.2.1.1. Refractive index of air-free water at temperatures from 0 to 60 °C (IPTS 1968) and at wavelength 589.26 nm[a]

$t_{68}/°C$	n_D	$t_{68}/°C$	n_D	$t_{68}/°C$	n_D
0	1.3339493	20	1.3329870	40	1.3306084
1	9474	21	8965	41	4614
2	9397	22	8027	42	3121
3	9264	23	7055	43	1604
4	9077	24	6052	44	0064
5	8837	25	5017	45	1.3298501
6	8546	26	3951	46	6916
7	8205	27	2855	47	5308
8	7812	28	1730	48	3677
9	7382	29	0575	49	2025
10	6902	30	1.3319392	50	0352
11	6378	31	8180	51	1.3288656
12	5812	32	6941	52	6940
13	5204	33	5675	53	5203
14	4556	34	4382	54	3444
15	3868	35	3062	55	1666
16	3142	36	1717	56	1.3279867
17	2378	37	0346	57	8047
18	1577	38	1.3308950	58	6208
19	0741	39	7529	59	4349
-	-	-	-	60	2470

[a] Intensity-weighted mean of doublet, Sodium D_1, D_2

REFERENCES

1. Tilton, L. W. *J. Opt. Soc. Am.* **1942**, *32*, 371.
2. Tilton, L. W.; Taylor, J. K. *J. Res. Nat. Bur. Stand.* **1938**, *20*, 419.
3. Harkins, W. D.; Brown, F. E. *J. Amer. Chem. Soc.* **1919**, *41*, 499.
4. Bedford, R. E.; Kirby, C. G. *Metrologia* **1969**, *5*, 83.

15.2.2. 2,2,4-Trimethylpentane

Physical property: Optical Refraction (refractive index), n
Unit: Dimensionless

Recommended reference material: 2,2,4-Trimethylpentane (C_8H_{18})
Range of variables: 20 to 30°C, 7 wavelength 435.83 to 667.81 nm
Physical state within the range: liquid
Contributors to the first version: I. Brown, H. Feuerberg, J. Franc, E. F. G. Herington, J. F. Lane, T. Plebanski.
Contributors to the revised version: A. Feldman, I. H. Malitson.

Intended usage: Samples of 2,2,4-trimethylpentane can be used to calibrate refractometers employed for the measurements on solids and liquids. Details of methods and apparatus are given by Tilton (Ref. 1).

Sources of supply and/or methods of purification: Samples with certified values for the refractive index are available from suppliers (A) and (B); see also reference 2.

Pertinent physicochemical data: Supplier (A) – the following values of refractive index, which apply to an air-saturated sample available were determined by the former Optical Instrument Section of the National Bureau of Standards, U.S.A. The uncertainty in these values is estimated to be ±0.00003.

Table 15.2.2.1. Supplier (A), Sample 217X, (99.968 ±0.006) mole per cent pure

Wavelength/nm	20 °C	25 °C	30 °C
667.81	1.38916	1.38670	1.38424
656.28	1.38945	1.38698	1.38452
589.26[a]	1.39145	1.38898	1.38650
546.07	1.39316	1.39068	1.38820
501.57	1.39544	1.39294	1.39044
486.13	1.39639	1.39389	1.39138
435.83	1.40029	1.39776	1.39523

[a] Intensity-weighted mean of doublet, Sodium D_1, D_2

Supplier (B) – The refractive index of four samples was determined at seven wavelengths and at temperatures near 20, 25, and 30 °C. The measurements were made on a precision spectrometer by means of a minimum-deviation method. Emission lamps of Hg, He, H_2, and Na were used as wavelength sources. Each sample was contained in a prismatic cell with plane-parallel windows during the measurement process. Water from the temperature-controlled bath was circulated through the cell housing to maintain a constant temperature during the test. The temperature coefficient of refractive index was determined for each sample at each wavelength. An average temperature coefficient for the four samples at each wavelength was used to determine the refractive index value for the wavelength at 20, 25, and 30 °C. The refractive index value given for each temperature is an average of the values determined for the four samples at that wavelength and temperature. These average values are considered to be representative refractive index values accurate to 2×10^{-4} for this lot of material.

Table 15.2.2.2. Supplier (B), Sample SRM 217c

Wavelength/nm	20 °C	25 °C	30 °C
667.8	1.3894	1.3868	1.3844
656.3	1.3897	1.3871	1.3847
589.3[a]	1.3917	1.3891	1.3866
546.1	1.3934	1.3908	1.3883
501.6	1.3957	1.3931	1.3906
486.1	1.3967	1.3940	1.3915
435.8	1.4006	1.3979	1.3954

[a]Intensity-weighted mean of doublet, Sodium D_1, D_2

REFERENCES

1. Tilton, L. W. *J. Opt. Soc. Am.* **1942**, *32*, 371.
2. *IUPAC Catalogue of Physicochemical Standard Substances*, Butterworths: London, **1972**.

15.2.3. Hexadecane

Physical property: Optical Refraction (refractive index), n
Unit: Dimensionless
Recommended reference material: Hexadecane ($C_{16}H_{34}$)
Range of variables: 20 to 100 °C, 7 wavelengths 435.83 to 667.81 nm
Physical state within the range: liquid
Contributors to the first version: I. Brown, H. Feuerberg, J. Franc, E. F. G. Herington, J. E. Lane, T. Plebanski
Contributors to the revised version: A. Feldman, I. H. Malitson

Intended usage: Samples of hexadecane can be used to calibrate refractometers employed for making measurements on solids and liquids. Details of methods and apparatus are given by Tilton (Ref. 1).

Sources of supply and/or methods of preparation: Samples with certified values for the refractive index are available from supplier (A).

Pertinent physicochemical data: The following values of refractive index, which apply to an air-saturated sample available from supplier (A), were determined for temperatures of 20, 25, and 30 °C by the American Petroleum Institute Research Project 6 at the Carnegie-Mellon University, and for 80 and 100 °C by the Research Laboratory of the Sun Oil Company. The uncertainty in the values is estimated to be ±0.00008 for temperatures of 20, 25, and 30 °C and ±0.0002 for 80 and 100 °C. Values for other temperatures may be obtained by linear interpolation between the tabulated values.

Table 15.2.3.1. Hexadecane available from supplier (A)
Supplier (A), Sample 568-X (99.96 ±0.04) mole per cent pure

Wavelength/nm	20 °C	25 °C	30 °C	80 °C	100 °C
667.81	1.43204	1.43001	1.42798	-	-
656.28	1.43235	1.43032	1.42829	1.4078	1.3998
589.26[a]	1.43453	1.43250	1.43047	1.4098	1.4017
546.07	1.43640	1.43436	1.43232	1.4117	1.4034
501.57	1.43888	1.43684	1.43480	-	-
486.13	1.43993	1.43788	1.43583	1.4150	1.4069
435.83	1.44419	1.44213	1.44007	1.4191	1.4108

[a]Intensity-weighted mean of doublet, Sodium D_1, D_2

REFERENCE

1. Tilton, L. W. *J. Opt. Soc. Am.* **1942**, *32*, 371.

15.2.4. *trans*-Bicyclo[4,4,0]decane

Physical property: Optical Refraction (refractive index), n
Unit: Dimensionless
Recommended reference material: *trans*-Bicyclo[4,4,0]decane ($C_{10}H_{18}$)
  trans-Decahydronaphthalene
Range of variables: 20 to 100 °C, 7 wavelengths 435.83 to 667.81 nm
Physical state within the range: liquid
Contributors to the first version: I. Brown, H. Feuerberg, J. Franc, E. F. G. Herington, J. E. Lane, T. Plebanski
Contributors to the revised version: A. Feldman, I. H. Malitson

Intended usage: Samples of *trans*-bicyclo[4,4,0]decane can be used to calibrate refractometers employed for making measurements on solids and liquids. Details of methods and apparatus are given by Tilton (Ref. 1).

Sources of supply and/or methods of preparation: Samples with certified values for the refractive index are available from supplier (A).

Pertinent physicochemical data: The following values of refractive index, which apply to an air-saturated sample available from supplier (A), were determined for temperatures of 20, 25, and 30 °C by the American Petroleum Institute Research Project 6 at the Carnegie-Mellon University, and for 80 and 100 °C by the Research Laboratory of the Sun Oil Company. The uncertainty in the values is estimated to be ±0.00008 for temperatures of 20, 25, and 30 °C and ±0.0002 for 80 and 100 °C. Values for other temperatures may be obtained by linear interpolation between the tabulated values.

Table 15.2.4.1. *trans*-Bicyclo[4,4,0]decane available from supplier (A)
Supplier (A), Sample 561-X   (99.98 ±0.02) mole per cent pure

Wavelength/nm	20 °C	25 °C	30 °C	80 °C	100 °C
667.81	1.46654	1.46438	1.46222	-	-
656.28	1.46688	1.46472	1.46256	1.4411	1.4324
589.26[a]	1.46932	1.46715	1.46498	1.4434	1.4347
546.07	1.47141	1.46923	1.46705	1.4453	1.4367
501.57	1.47420	1.47200	1.46980	-	-
486.13	1.47535	1.47315	1.47095	1.4492	1.4409
435.83	1.48011	1.47789	1.47567	1.4536	1.4448

[a]Intensity-weighted mean of doublet, Sodium D_1, D_2

REFERENCE
1. Tilton, L. W. *J. Opt. Soc. Am.* **1942**, *32*, 371.

15.2.5. 1-Methylnaphthalene

Physical property: Optical Refraction (refractive index), n
Unit: Dimensionless
Recommended reference material: 1-Methylnaphthalene ($C_{11}H_{10}$)
Range of variables: 20 to 100 °C, 7 wavelengths 435.83 to 667.81 nm
Physical state within the range: liquid
Contributors to the first version: I. Brown, H. Feuerberg, J. Franc, E. F. G. Herington, J. E. Lane, T. Plebanski
Contributors to the revised version: A. Feldman, I. H. Malitson

Intended usage: Samples of 1-methylnaphthalene can be used to calibrate refractometers employed for making measurements on solids and liquids. Details of methods and apparatus are given by Tilton (Ref. 1).

Sources of supply and/or methods of preparation: Samples with certified values for the refractive index are available from supplier (A).

Pertinent physicochemical data: The following values of refractive index, which apply to an air-saturated sample available from supplier (A), were determined for temperatures of 20, 25, and 30 °C by the American Petroleum Institute Research Project 6 at the Carnegie-Mellon University, and for 80 and 100 °C by the Research Laboratory of the Sun Oil Company. The uncertainty in the values is estimated to be ±0.00008 for temperatures of 20, 25, and 30 °C and ±0.0002 for 80 and 100 °C. Values for other temperatures may be obtained by linear interpolation between the tabulated values.

Table 15.2.5.1. 1-Methylnaphthalene available from supplier (A)
Supplier (A), Sample 578–X, (99.97 ±0.03) mole per cent pure

Wavelength/nm	20 °C	25 °C	30 °C	80 °C	100 °C
667.81	1.60828	1.60592	1.60360	-	-
656.28	1.60940	1.60703	1.60471	1.5805	1.5710
589.26[a]	1.61755	1.61512	1.61278	1.5882	1.5786
546.07	1.62488	1.62240	1.62005	1.5952	1.5855
501.57	1.63513	1.63259	1.63022	-	-
486.13	1.63958	1.63701	1.63463	1.6092	1.5990
435.83	-	1.65627	1.65386	1.6274	1.6172

[a]Intensity-weighted mean of doublet, Sodium D_1, D_2

REFERENCE

1. Tilton, L. W. *J. Opt. Soc. Am.* **1942**, *32*, 371.

15.2.6. Toluene

Physical property: Optical Refraction (refractive index), n
Unit: Dimensionless
Recommended reference material: Toluene (C_6H_8)
Range of variables: 20 to 30 °C; wavelength range 435.83 to 667.81 nm
Physical state within the range: liquid
Contributors: A. Feldman, I. H. Malitson

Intended usage: Samples of toluene can be used to calibrate refractometers employed for measurements on solids and liquids. Details of methods and apparatus for use of this standard are given by Tilton (Ref. 1).

15. Optical Refraction

Sources of supply and/or methods of preparation: Samples of certified values of refractive index are available from suppliers (A) and (B).

Pertinent physicochemical data: Supplier (A) – The following values of refractive index, which apply to air-saturated material, were determined by the former Optical Instruments Section of the National Bureau of Standards, U.S.A. The uncertainty in these values is estimated to be ±0.00003.

Table 15.2.6.1. Supplier (A), Sample 211C-5S, (99.97 ±0.02) mole percent pure.

Wavelength/nm	20 °C	25 °C	30 °C
667.81	1.49180	1.48903	1.48619
656.28	1.49243	1.48966	1.48682
589.26[a]	1.49693	1.49413	1.49126
546.07	1.50086	1.49803	1.49514
501.57	1.50620	1.50334	1.50041
486.13	1.50847	1.50559	1.50265
435.83	1.51800	1.51506	1.51206

[a] Intensity-weighted mean of doublet, Sodium D_1, D_2

Supplier (B) – The following values of refractive index were determined at seven wavelengths and at temperatures near 20, 25, and 30 °C. Measurements were made on a precision spectrometer using the minimum deviation method. The toluene was contained in a water-jacketed prismatic cell mounted on the spectrometer table. A temperature controlled water bath maintained nearly constant temperature of the samples. Spectral lamps of mercury, cadmium, and helium were used for wavelength sources. The temperature coefficient was determined at each wavelength for all samples, and an average value of the temperature coefficient at each wavelength was used to determine the refractive index for that wavelength at 20, 25, and 30 °C.

The air temperature of the laboratory was maintained near 20 °C during all measurements. The certified values of refractive index, referenced to air at 20 °C for the three temperatures, are listed in Table 15.2.6.2. The refractive index values are considered to be accurate to $\pm 1 \times 10^{-4}$ at 20 °C, $\pm 2 \times 10^{-4}$ at 25 °C, and $\pm 4 \times 10^{-4}$ at 30°C. The refractive index value given for each wavelength at each temperature is the average of values determined for the four samples at that wavelength and temperature. These average values are considered to be representative of the refractive indices of this batch of toluene.

Table 15.2.6.2. Supplier (B), Sample SRM 211c

Wavelength/nm	20 °C	25 °C	30 °C
667.81	1.4918	1.4888	1.4858
643.83	1.4931	1.4901	1.4871
587.56	1.4970	1.4940	1.4910
546.07	1.5008	1.4978	1.4947
508.58	1.5052	1.5021	1.4991
479.99	1.5094	1.5063	1.5032
435.83	1.5180	1.5148	1.5116

REFERENCE

1. Tilton, L. W. *J. Opt. Soc. Am.* **1942**, *32*, 371.

15.2.7. Methylcyclohexane

Physical property: Optical Refraction (refractive index), n
Unit: Dimensionless
Recommended reference material: Methylcyclohexane (C_7H_{14})
Range of variables: 20 to 30 °C; wavelength range 435.83 to 667.81 nm
Physical state within the range: liquid
Contributors: A. Feldman, I. H. Malitson

Intended usage: Samples of methylcyclohexane can be used to calibrate refractometers employed for measurements on solids and liquids. Details of methods and apparatus for use of this standard are given by Tilton (Ref. 1).

Sources of supply and/or methods of preparation: Samples with certified values of refractive index are available from supplier (A).

Pertinent physicochemical data: The following values of refractive index, which apply to air-saturated material, were determined by the former Optical Instruments Section of the National Bureau of Standards, U.S.A. The uncertainty in these values is estimated to be ±0.00003.

REFERENCE

1. Tilton, L. W. *J. Opt. Soc. Am.* **1942**, *32*, 371.

Table 15.2.7.1. Supplier (A), Sample 218a-5S, (99.97 ±0.02) mole per cent pure.

Wavelength/nm	20 °C	25 °C	30 °C
667.81	1.42064	1.41812	1.41560
656.28	1.42094	1.41842	1.41591
589.26[a]	1.42312	1.42058	1.41806
546.07	1.42497	1.42243	1.41989
501.57	1.42744	1.42488	1.42233
486.13	1.42847	1.42590	1.42334
435.83	1.43269	1.43010	1.42752

[a]Intensity-weighted mean of doublet, Sodium D_1, D_2

15.2.8. Silicone liquids

Physical property: Optical Refraction (refractive index), n
Unit: Dimensionless
Recommended reference material: Two Silicone Liquids
Range of variables: 20 to 80 °C; wavelength range 435.83 to 667.81 nm
Physical state within the range: liquid
Contributors: A. Feldman, I. H. Malitson

Intended usage: Samples of silicone liquids can be used to calibrate refractometers employed for measurements on solids and liquids. Details of methods and apparatus for use of this standard are given by Tilton (Ref. 1).

Sources of supply and/or method of preparation: Samples with certified values of refractive index are available from supplier (B).

Pertinent physicochemical data: This standard consists of two silicone liquids that are chemically and thermally stable. These liquids are suitable for the calibration of refractometers. The homogeneity of each liquid was established by measuring the refractive indices of randomly selected samples. The refractive indices were measured by the classical minimum deviation method, using a water-jacketed hollow prism mounted on the table of a precision spectrometer. A thermistor sensor was immersed in the liquid during the measurements and temperature changes as small as 0.001 °C were monitored on a digital meter. Index determinations were made at or near 20, 40, 60 and 80 °C. Temperature coefficients, $\Delta n/\Delta T$, were computed and corrections were made to adjust to the exact temperatures that are listed in the tables. A statistical evaluation of the averaged index data for all samples yielded uncertainties in refractive index that are within $\pm 4 \times 10^{-5}$ refractive index units for SRM 1823 I, and within $\pm 5 \times 10^{-5}$ refractive index units for SRM 1823 II.

Table 15.2.8.1. Supplier (B), Sample SRM 1823 I

Wavelength/nm	20 °C	40 °C	60 °C	80 °C
667.81	1.51279	1.50465	1.49648	1.48811
656.28	1.51339	1.50524	1.49707	1.48871
643.85	1.51407	1.50593	1.49778	1.48939
589.26[a]	1.51767	1.50947	1.50119	1.49276
587.56	1.51780	1.50961	1.50139	1.49295
546.07	1.52140	1.51315	1.50485	1.49634
486.13	1.52859	1.52019	1.51176	1.50318
479.99	1.52949	1.52110	1.51273	1.50407
467.81	1.53145	1.52301	1.51454	1.50587
435.83	1.53751	1.52897	1.52043	1.51166

[a] Intensity-weighted mean of doublet, Sodium D_1, D_2

Table 15.2.8.2. Supplier (B), Sample SRM 1823 II.

Wavelength/nm	20 °C	40 °C	60 °C	80 °C
667.81	1.55330	1.54518	1.53693	1.52852
656.28	1.55403	1.54588	1.53757	1.52918
643.85	1.55487	1.54673	1.53842	1.53009
589.26[a]	1.55925	1.55106	1.54265	1.53418
587.56	1.55941	1.55119	1.54282	1.53444
546.07	1.56382	1.55555	1.54712	1.53867
486.13	1.57265	1.56425	1.55567	1.54702
479.99	1.57379	1.56536	1.55682	1.54814
467.81	1.57617	1.56779	1.55913	1.55042
435.83	1.58373	1.57519	1.56644	1.55763

[a] Intensity-weighted mean of doublet, Sodium D_1, D_2

REFERENCE

1. Tilton, L. W. *J. Opt. Soc. Am.* **1942**, *32*, 371.

15.2.9. Borosilicate glass

Physical property: Optical Refraction (refractive index), n
Unit: Dimensionless
Recommended reference material: Borosilicate glass
Range of variables: 20 °C; wavelength range 404.66 to 706.52 nm
Physical state within the range: solid
Contributors: A. Feldman, I. H. Malitson

Intended usage: Samples of borosilicate glass can be used to calibrate and check refractometers employed for measurements on solids and liquids. Details of methods and apparatus for use of this standard are given by Tilton (Ref. 1).

Sources of supply and/or methods of preparation: Samples with certified values of refractive index are available from supplier (B).

Pertinent physicochemical data: This standard was prepared from a selected portion of a commercial borosilicate glass that is homogeneous, stable, and has a high optical quality. It is designed for both the calibration of refractometers and the determination of refractive indexes of microscope immersion liquids. The standard consists of two rectangular slabs; a slab which is polished on two faces, intended for checking the performance of refractometers; and an unpolished slab, which can be broken into fragments, intended for microscopic determination of the refractive indexes of immersion liquids.

The homogeneity of the glass was established by measuring the refractive index of continguous rectangular slabs and prisms that were cut from bars of glass. The prisms were used to determine refractive index at 13 wavelengths. These indexes were measured by the classical minimum deviation method on a calibrated precision spectrometer. Statistical evaluation of averaged data for six prisms yielded uncertainties in refractive index that are within 5×10^{-6}. Each rectangular slab was measured at 2 wavelengths (589.26 nm and 643.85 nm) on a calibrated refractometer and agreed within 1×10^{-5} with the corresponding spectrometer values. Because of the established homogeneity of this glass, it is believed that the index variance at the other wavelengths is within the stated uncertainty. All refractive index measurements were carried out in a temperature controlled laboratory; the temperature varied not more than 0.3 °C. The values of refractive index are referred to 20 °C and a pressure of 101.325 kPa.

Information only: The temperature coefficient of refractive index, dn/dT, over a temperature range 20 to 80 °C $\approx +4.0 \times 10^{-6}$ K^{-1} at 589.26 nm. The density of the glass was (2.292 ± 0.001) g cm^{-3} at 24 °C.

Table 2.9.1. Supplier (B), Sample SRM 1820

Wavelength/nm	$n(20\ °C)$
706.52	1.48398
667.81	1.48499
656.28	1.48532
643.85	1.48569
589.26[a]	1.48755
587.56	1.48762
546.07	1.48939
508.58	1.49136
486.13	1.49275
479.99	1.49316
467.81	1.49404
435.83	1.49669
404.66	1.49994

[a] Intensity-weighted mean of doublet, Sodium D_1, D_2

REFERENCE

1. Tilton, L. W. *J. Opt. Soc. Am.* **1942**, *32*, 371.

15.2.10. Soda-lime glass

Physical property: Optical Refraction (refractive index), n
Unit: Dimensionless
Recommended reference material: Soda-lime glass
Range of variables: 20 °C; wavelength range 404.66 to 1082.97 nm
Physical state within the range: solid
Contributors: A. Feldman, I. H. Malitson

Intended usage: Samples of soda-lime glass can be used to calibrate and check refractometers employed for measurements on solids and liquids. Details of methods and apparatus for use of this standard are given by Tilton (Ref. 1).

Sources of supply and/or methods of preparation: Samples with certified values of refractive index are available from supplier (B).

Pertinent physicochemical data: This standard was prepared from a selected portion of a commercial soda-lime glass that is homogeneous, stable, and has a high optical quality. It is designed for both the calibration of refractometers and the determination of refractive indexes of microscope immersion liquids. The standard consists of two rectangular slabs; a slab which is polished on two faces, intended for checking the performance of refractometers; and an unpolished slab, which can be broken into fragments, intended for microscopic determination of the refractive indexes of immersion liquids.

The homogeneity of the glass was established by measuring the refractive index of each polished rectangular slab and five prisms cut from the glass. The prisms were used to determine refractive index at 15 wavelengths. These indexes were measured using the classical minimum deviation method on a calibrated precision spectrometer. Statistical evaluation of averaged data yielded uncertainties in refractive index that are within 9×10^{-6}. Each rectangular slab was measured at 2 wavelengths (589.26 nm and 643.85 nm) on a calibrated refractometer and agreed within 3×10^{-5} with the corresponding spectrometer values. Because of the homogeneity of this glass, it is believed that the index variance at the other wavelengths is within the stated uncertainty. All refractive index measurements were carried out in a temperature controlled laboratory; the temperature varied not more than 0.3 °C. The values of refractive index are referred to 20 °C and a pressure of 101.325 kPa.

Table 15.2.10.1. Supplier (B), Sample SRM 1822

Wavelength/nm	n(20 °C)
1082.97	1.507143
1013.98	1.508030
706.52	1.513723
667.81	1.514868
656.28	1.515244
643.85	1.515669
589.26[a]	1.517835
587.56	1.517914
546.07	1.520001
508.58	1.522337
486.13	1.524006
479.99	1.524503
467.81	1.525551
435.83	1.528761
404.66	1.532710

[a] Intensity-weighted mean of doublet, Sodium D_1, D_2

REFERENCE

1. Tilton, L. W. *J. Opt. Soc. Am.* **1942**, *32*, 371.

15.3. Contributors to the revised version

A. Feldman,
National Bureau of Standards,
United States Department of Commerce,
Gaithersburg, MD 20899
(U.S.A)

I. H. Malitson (retired),
National Bureau of Standards,
United States Department of Commerce,
Gaithersburg, MD 20899
(U.S.A)

15.4. List of Suppliers

A. API Standard Reference Materials,
Carnegie-Mellon University,
Schenley Park, Pittsburg,
Pennsylvania 15213
(U.S.A)

B. Office of Standard Reference Materials,
U.S. Department of Commerce,
National Bureau of Standards,
Gaithersburg, MD 20899
(U.S.A)

16

SECTION: REFLECTANCE

COLLATORS: H. FEUERBERG, D. GUNDLACH, H. TERSTIEGE

CONTENTS:

16.1. Introduction
    16.1.1. General considerations
    16.1.2. Commonly used methods
    16.1.3. Methods and possible difficulties

16.2. Reference materials for reflectance measurements
    16.2.1. Aluminium on glass
    16.2.2. First-surface gold mirror on a metallic substrate
    16.2.3. Polished glass with a specified refractive index
    16.2.4. Barium sulphate
    16.2.5. Pyroceram porcelain
    16.2.6. Opal glass
    16.2.7. Enameled iron discs
    16.2.8. Ceramic tiles
    16.2.9. Silicon carbide
    16.2.10. Flowers of sulphur
    16.2.11. Black porcelain enamel
    16.2.12. Second-surface aluminium mirror
    16.2.13. Polytetrafluoroethylene powder
    16.2.14. Ever-white

16.3. Contributors

16.4. List of Suppliers

16.1. Introduction

16.1.1. General considerations

If a material is exposed to optical radiation, some portion of the incident radiation is reflected, another portion absorbed, and often a third portion is transmitted (Ref. 1). These recommendations deal with the process of reflection and the characterization of the quantity called reflectance. Reflectance is the return of radiation by a specimen without a change in the wavelength. The reflected radiation can be regular, diffuse, or mixed.

Regular (specular) reflection is reflection without diffusion in accordance with the laws of optical reflection (as occurs with reflection by a mirror). Diffuse reflectance is reflection in which, on a macroscopic scale, there is no regular reflection. Reflectance is the ratio of the reflected flux ϕ_ρ to the incident flux ϕ:

$$\rho = \phi_\rho/\phi \qquad (16.1.1).$$

If there is a risk of ambiguity, then any term which involves a photometric characteristic should be preceded by the adjective 'luminous' and its symbol should have the subscript 'v' (*e.g.* luminous reflectance ρ_v) where as a radiant characteristic may be preceded by the adjective 'radiant' and its symbol given the subscript 'e' (*e.g.* radiant reflectance ρ_e). Thus, the spectral reflectance $\rho(\lambda)$ at wave length λ is defined by:

$$\rho(\lambda) = \phi_{e\lambda\rho}/\phi_{e\lambda} \qquad (16.1.2),$$

where $\phi_{e\lambda\rho}$ is the reflected spectral radiant flux and $\phi_{e\lambda}$ is the incident spectral radiant flux. The SI unit of the spectral radiant flux is W m^{-1}; the spectral reflectance is dimensionless because the reflectance is the ratio of two quantities of the same dimension.

When mixed reflection occurs, the (total) reflectance may be divided into two parts corresponding to the two modes of reflection: regular reflectance (ρ_r) and diffuse reflectance (ρ_d). The regular reflectance, ρ_r, is the ratio of the radiant flux (which after reflectance is proportional to the inverse square of the distance from the image of the source) to the incident flux. The diffuse reflectance, ρ_d, is the ratio of the radiant flux that has undergone diffuse reflectance to the incident flux. The radiance (luminance) factor, β, is the ratio of the radiance (luminance) of the sample to that of the perfect reflecting diffuser identically irradiated (illuminated). The conditions of irradiation and the view of the specimen must be specified in every instance; the CIE recommended geometries have been published (Ref. 2).

16. Reflectance

The basis of reflectance measurements and applications of reflectance are described in detail by Kortüm (Ref. 3). Measurements of the reflection properties of paper, textiles, and of building materials have been made for many years. The measurements have gained much importance during the last 20 years in the aerospace industry and more recently in energy conservation programs, especially those involving systems for the collection of radiation energy.

Reflectance reference materials are recommended by the CIE (Commission Internationale l´Eclairage) and by several Technical Committes of ISO (International Organization for Standardization), for example by ISO/TC 6 for brightness and reflectance measurements of paper, boards and pulps.

Reflectance is usually measured by comparing the reflecting properties of a sample with that of a reference material. Reference materials initially used for reflectance were freshly smoked magnesium or pressed magnesium oxide. However, this material was found to be inadequate because the reflectance values were found to vary with time due, in part, to the material aging. Magnesium oxide has therefore not been included in the following list of recommended materials. Barium sulphate has proved to have better reproducibility, its absolute reflectance varies only slightly from sample to sample, and it is most eminently suitable as a reference material for reflectance. For these reasons barium sulphate is recommended by the CIE Committee TC-2.3 (Ref. 1) and by ISO Committee TC/6 and by the German Standard Specification (Ref. 4). Barium sulphate is available with spectral reflectance (radiance) factors given for six wavelengths from 350 to 700 nm. Pressed polytetrafluoroethylene power is being increasingly used as an alternative to barium sulphate particularly in the UV region (Ref. 5).

Plates of Pyroceram glass ceramic, opalescent glass, enameled iron discs, and ceramic tiles with certified reflectances are convenient to use on a routine basis because they are easy to clean and are permanent in their optical behavior. These materials have to be calibrated using a barium sulphate sample. Regular reflectance values of suitable samples of black glass with a high refractive index can be calculated from the refractive index values. The reflectance of these materials should always be related to the perfect reflecting diffuser and corrections made for the deviations of the reflecting reference material from the behavior of a perfect reflecting diffuser. Fourteen reference materials are described in this section.

Absolute measurements of regular reflectance and of diffuse reflectance are frequently very difficult because the reflectance is not a specific property of a material since it depends on several parameters. In general, the reflectance depends on the following:

(i) The spectral composition of the radiation. Therefore, the spectral distribution of the radiation should be specified, for example, by naming the illuminant such as Standard Illuminant A or D65, or by its distribution temperature, or, for monochromatic radiation, by specifying the wavelength and bandwidth.

(ii) The state of polarization of the radiation. If the reflectance is not determined with unpolarized radiation, the state of polarization and the azimuth of the plane of polarization must be specified; the reflected radiation is usually partly polarized if the incident radiation is unpolarized.

(iii) The angle of incidence. If the reflectance is determined for any geometry other than normal by the use of a collimated beam, the geometrical distribution of the incident radiation must be specified.

(iv) The angular extent of the incident radiation. The value of the reflectance generally depends on the angular extent of the incident radiation. The reflectance approaches a constant value if the angular extent approached zero.

(v) The thickness of the sample. If necessary, the thickness of the sample must be given.

(vi) The temperature. The reflectance values given are assumed to be the values at 25 °C unless otherwise stated.

(vii) The state of the surface. The reflectance is normally given for a clean, dry sample, if not otherwise specified.

(viii) Other parameters. If other parameters affect the reflectance (for example illuminance or irradiance, UV-irradiation or IR-irradiation), they must be specified.

To determine the reflectance (the ratio of the reflected radiant flux to the incident flux) one has to measure both the reflected flux and the incident flux. To gain more precision in commercial reflectance measurements, the use of reference material is recommended. If the reflectance $\rho_S = \phi_{\rho S}/\phi_S$ of a sample has to be measured and the reflectance $\rho_{RM} = \phi_{\rho RM}/\phi_{RM}$ of the reference material is known and both are measured with the same instrument alternatively, then the incident flux on the sample and the reference material is the same so that

$$\phi_S = \phi_{RM} \qquad (16.1.3).$$

In this case, the reflectance ρ_s is:

$$\rho_S = \phi_{\rho S}/\phi_S = (\phi_{\rho S}/\phi_{\rho RM})\rho_{RM} \qquad (16.1.4).$$

The reflected fluxes of the sample and of the reference material are proportional to their readings on the instruments thus the reflectance of a sample is the ratio of the reading of the reflectance measurement of the sample to that of the reference material multiplied by the known reflectance of the reference material.

16.1.2. Commonly used methods

In the following tables, the most commonly used methods for measuring the reflectance and the appropriate reference materials are given. The survey should assist in finding the most appropriate method for a special purpose. More detailed descriptions of the methods are given later. Complex techniques required for special purposes which are used primarily in research work are not discussed.

16. Reflectance

Table 16.1.2.1. Spectral reflectance, properties, suitable instruments, and suitable reference materials

Property	Instrument	Recommendation	Use
$\rho(\lambda)$	Spectrophotometer with integrating sphere	4,5,6,10,13,14	Calibration of the photometric scale of spectral reflectance photometers
$\rho_d(\lambda)$	Spectrophotometer with integrating sphere fitted with gloss trap for regular reflectance	4,5,6,10,14	Calibration of the photometric scale of instruments for diffuse spectral reflectance
$\rho_r(\lambda)$	Spectrophotometer attached with device for for measuring regular reflectance	1,2,9	Calibration of the photometric scale of instruments for regular or total spectral reflectance

Table 16.1.2.2. Spectrally weighted reflectance, properties, instruments to be used, and suitable reference materials

Property	Instrument	Recommendation	Use
ρ	Photometer with integrating sphere	4,5,6,14	Calibration of the photometric scale of photometers
ρ_d	Photometer with integrating sphere fitted with gloss trap for regular reflectance	7,8	Calibration of color difference meters Calibration of tristimulus colorimeters
ρ_r	Photometer attached with device for measuring regular reflectance	1,2,3,9,11,12	Calibration of the photometric scale of instruments for measuring of regular reflectance (1,2) Calibration of Glossmeters (3)

Table 16.1.2.3. Suitable reference materials

No.	Material	Specified properties	Wavelength range of specified properties
1	Aluminium on glass	$\rho_r(\lambda)$	250 nm to 2500 nm
2	First-surface gold on a metallic substrate	$\rho_r(\lambda)$	600 nm to 2500 nm
3	Polished glass with specified refractive index	n, (ρ_r)	380 nm to 780 nm
4	Barium sulphate	$\beta(\lambda), \rho(\lambda)$	350 nm to 700 nm
5	Pyroceram	$\beta(\lambda), \beta$	380 nm to 780 nm
6	Opal glass	$\beta(\lambda), \beta, \rho(\lambda), \rho$	320 nm to 800 nm
7	Enameled iron discs	$\rho(\lambda), \rho$	380 nm to 760 nm
8	Ceramic tiles	$\beta(\lambda), \rho_d(\lambda), \rho(\lambda), \beta, \rho_d, \rho$	250 nm to 2500 nm
9	Silicon carbide	$\rho_r(\lambda)$	400 nm to 700 nm
10	Flowers of sulphur	$\beta(\lambda), \rho(\lambda)$	500 nm to > 2500 nm
11	Black porcelain enamel	$\rho_r(\lambda)$	250 nm to 2500 nm
12	Second surface Al mirror	$\rho_r(\lambda)$	250 nm to 2500 nm
13	Polytetrafluoroethylene powder	$\rho(\lambda)$	200 nm to 2500 nm
14	Ever-white	$\rho(\lambda)$	380 nm to 780 nm

Measurements relative to an absolute scale of reflectance made with any commercial instrument is only possible by the use of reference materials as the absolute values of the reflectance of the reference materials can only be measured with highly sophisticated instruments (Ref. 6).

16.1.3. Methods and possible difficulties

Different methods are recommended for the measurement of total reflectance, ρ, diffuse reflectance, ρ_d, and regular reflectance, ρ_r. For the measurement of the diffuse reflectance, integrating spheres with ports for excluding the regular part of the reflected light are used. The same spheres with closed ports can be used for measuring the total reflectance. The regular reflectance can be calculated from the values of ρ and ρ_r but are better measured in a separate arrangement.

When measuring reflectance with diffuse or direct illumination, different methods have to be used. Instruments mainly use direct or conical illumination for reflectance measurements. The main errors involved in reflectance measurements are:

(i) Errors in the photometric scale. Significant linearity errors may occur when only the zero and 100 per cent values of the scale are fixed. The 100 per cent value is usually calibrated with a reflectance standard, the zero value should be checked by mounting a light trap in place of the specimen. Reference materials with known reflectance can be used to check the linearity. For a more precise test, the light addition method should be used.

(ii) Straylight. Straylight appearing in addition to the measuring light is usually caused by imperfect or dirty optical surfaces.

(iii) Wavelength and Slitwidth. For measuring spectral reflectances, one has to take care that the wavelength scale is correct, that the influence of the slitwidths is taken into account, and that the dispersed light has no incorrect wavelengths superimposed.

REFERENCE

1. *Radiometric and Photometric Characteristics of Materials and their Measurement,* Publication CIE, No. 38 (TC-2-3), **1977**.
2. *Colorimetry,* Publication CIE, No. 15 (E-1.3.1), **1971**.
3. Kortüm, G. *Reflexionsspektroskopie,* Springer: Berlin, **1969**.
4. *Colour Measurement.* Reflectance Standards for Colorimetry and Photometry, DIN 5033, Part 9, **1970**.
5. Weidner, V. R.; Hsia, J. J. *Opt. Soc. Am.* **1981**, *71*, 856.
6. Erb, W. *Appl. Optics* **1975**, *14*, 493.

16.2. Reference materials for reflectance measurements

16.2.1. Aluminium on glass

Physical property: Regular spectral reflectance
Unit: Dimensionless
Recommended reference material: Aluminium on glass
Range of variables: 250 to 2500 nm, 25 °C
Physical state within the range: solid
Contributor to the first version: J. P. Cali
Contributors to the revised version: J. P. Cali, G. A. Uriano

Intended usage: This reference material is used for the calibration of measurement equipment and for the evaluation of the thermal radiation properties of materials.

Sources of supply and/or methods of preparation: Samples (Ref. 1) are available from supplier (E) as SRM Number 2003 A. The mirror is available as a 5.1 cm diameter disc.

Pertinent physicochemical data: The mirror is certified for near-normal (9°) regular reflectance at wavelengths ranging from 250 to 2500 nm. The uncertainty of the certified reflectance is given.

REFERENCE

1. National Bureau of Standards Special Publication 260-38, U.S. Department of Commerce: Washington, **1972**.

16.2.2. First-surface gold mirror on a metallic substrate

Physical property: Regular spectral reflectance
Unit: Dimensionless
Recommended reference material: First-surface gold mirror on a metallic substrate
Range of variables: 600 to 2500 nm, 25 °C
Physical state within the range: solid
Contributors: J. P. Cali, J. J. Hsia

Intended usage: This reference material is used for the calibration of measurement equipment and for the evaluation of the specular reflectance properties of materials.

Sources of supply and/or methods of preparation: Samples (Ref. 1) are available from supplier (E) as SRM Number 2011. The mirror is available as a 5.1 cm diameter disc.

Pertinent physicochemical data: The mirrors are 51 mm in diameter and the reflectance is calibrated at 50 nm intervals from 600 to 1000 nm, at 100 nm intervals from 1000 nm to 1300 mn, and at 250 mn intervals from 1500 mn to 2500 mn. The mirrors are also calibrated at the laser wavelengths 632.8 nm and 1060 nm. Certified values of reflectance are for 6° angle of incidence; but, in addition, non-certified values are also available for 30° and 45° angle of incidence.

16.2.3. Polished glass with a specified refractive index

Physical property: Regular reflectance
Unit: Dimensionless
Recommended reference material: Polished glass with a specified refractive index
Range of variables: 380 to 780 nm, 25 °C
Physical state within the range: solid

16. Reflectance

Contributors: H. Feuerberg, H. Terstiege

Intended usage: This reference material is used for the calibration of reflectometers employed for gloss assessment of plane surfaces (Refs. 1,2), and it can also be used with goniophotometers.

Sources of supply and/or methods of preparation: Samples are available from supplier (A).

Pertinent physicochemical data: Black glass with a theoretical refractive index of 1.567 gives a reflectometer value of 100 for 20°, 60° and 85° geometries. Black glasses with refractive indices lower than 1.567 which exhibit reflectometer values lower than 100 are suitable as reference materials (Ref. 3). The specular reflection for various angles of incident radiation are calculated from values of the refractive index by means of the Fresnel equation for unpolarized light.

REFERENCES

1. German Standard Specification DIN 67 530, **1972**.
2. ASM D523 - 67.
3. Czepluch, W. *Farbe und Lack* **1972**, *78*, 619.

16.2.4. Barium sulphate

Physical property: Radiance factor, reflectance
Unit: Dimensionless
Recommended reference material: Barium sulphate
Range of variables: 350 to 700 nm, 25 °C
Physical state within the range: solid
Contributors to the first version: H. Feuerberg, H. Terstiege
Contributors to the revised version: H. Feuerberg, H. Terstiege

Intended usage: This reference material is used for the calibration of spectral reflectance photometers.

Sources of supply and/or methods of preparation: Samples are available from suppliers (B), (C), (D), (M), (N), and (O).

Tablets of 25, 45, or 60 mm diameter are pressed from barium sulphate powder by means of a mechanical powder press. The tablets must be protected from dust and moisture and can be used for several weeks (Ref. 1). The powder can be pressed into tablets only once. Further details concerning the technique have been given (Ref. 2).

Pertinent physicochemical data: Each batch of this reference material issued by supplier (D) is certified (Ref. 3) by supplier (I). Six values of the spectral radiance factor, $\beta(\lambda)$, are given (Ref. 4) for d/0 and 45/0 geometries for the range 350 to 700 nm.

REFERENCES

1. Terstiege, H. *Lichttechnik* **1974**, *26*, 277.
2. *Radiometric and Photometric Characteristics of Materials and Their Measurements.* Publication CIE, No. 38 (TC-2-3), **1977**.
3. Korte, H.; Schmidt, M. *Lichttechnik* **1967**, *19*, 135.
4. German Standard Specification DIN 5033, Part 9, **1972**.

16.2.5. Pyroceram porcelain

Physical property: Reflectance
Unit: Dimensionless
Recommended reference material: Pyroceram glass ceramic
Range of variables: 380 to 780 nm, 25 °C
Physical state within the range: solid
Contributors to the first version: H. Feuerberg, H. Terstiege
Contributors to the revised version: H. Feuerberg, H. Terstiege

Intended usage: This reference material can be used for the calibration of spectral reflectance photometers.

Sources of supply and/or methods of preparation: Samples are available from supplier (K) with directions for use.

Pertinent physicochemical data: The spectral reflectance factors are measured with high accuracy by the use of barium sulphate as reference material.

16.2.6. Opal glass

Physical property: Reflectance
Unit: Dimensionless
Recommended reference material: Opal glass
Range of variables: 320 to 800 nm, 25 °C
Physical state within the range: solid
Contributors to the first version: F. J. J. Clarke, H. Feuerberg, H. Terstiege
Contributors to the revised version: H. Feuerberg, H. Terstiege

Intended usage: Opal glass is used for the calibration of spectral reflectance photometers, reflectometers or colorimeters.

Sources of supply and/or methods of preparation: Specimens are available from suppliers (E), (G), (H) and (K). Supplier (E) provides a reference material that is available in two sizes: SRM 2015 is 2.5 × 5.0 and SRM 2016 is 10 cm square. The 6° hemispherical reflectance is certified from 400 to 750 nm. These SRM's must be used with a black felt plate that is supplied with the unit. The white opal glass is very stable and can be easily cleaned.

Pertinent physicochemical data: Supplier (G) provides the 0/45 spectral radiance factor, the diffuse part of the 0/d spectral reflectance and the total 8/d spectral reflectance determined with respect to the perfect reflecting diffuser. For each of the three geometries the CIE colorimetric quantities (x, y, Y) are computed from the spectral values for CIE Standard Illuminants A, C and D65 (Refs. 1,2).

Supplier (H) supplies a rental service (period 2 or 3 weeks) of calibrated specimens which can be supplied with reflectance values for d/0 or 0/d geometries.

Supplier (K) provides a calibration certificate for spectral reflectance factors with each specimen but specimens have to be recalibrated after two or three months usage. Calibration can be done by the user with barium sulphate as a reference material or calibration can be done by supplier (A).

REFERENCES

1. Clarke, F. J. J.; Garforth, F. A.; Parry, D. National Physical Laboratory Report, MOM 13, National Physical Laboratory: Teddington, **1975**.
2. Budde, W. *J. Res. Nat. Bur. Std.* **1976**, *80A*, 585.

16.2.7. Enameled iron discs

Physical property: Reflectance
Unit: Dimensionless
Recommended reference material: Enameled iron discs
Range of variables: 380 to 760 nm, 25 °C
Physical state within the range: solid
Contributor: E. Juhász

Intended usage: These discs can be used for the calibration of spectral reflectance photometers.

Source of supply and/or methods of preparation: Supplier (F) provides a set of 16 discs: one white, one grey, four red, four yellow, three green and three blue all different.

Pertinent physicochemical data: The spectral reflectance factor has been determined with high accuracy relative to barium sulphate for the wavelength range 380 to 760 nm in 10 nm steps. Tristimulus values and chromaticity coordinates for the two illuminants A and C and geometries 0/d, d/0, 45/0 and 0/45 are calculated (Ref. 1).

REFERENCE

1. Dézsi, Gy.; Fillinger, L. *Mérésügyi Közlemények, Journal of the National Office of Measures, Hungary* **1968**, *9*, 34.

16.2.8. Ceramic tiles

Physical property: Radiance factor, diffuse reflectance, and total reflectance
Unit: Dimensionless
Recommended reference material: Ceramic tiles
Range of variables: 250 to 2500 nm, 25 °C
Physical state within the range: solid
Contributors to the first version: F. J. J. Clarke, G. A. Uriano
Contributors to the revised version: G. A. Uriano

Intended usage: These tiles can be used to calibrate reflection spectrophotometers, reflectometers (such as those used in the evaluation of solar energy materials), and colorimeters (Refs. 1 to 4).

Sources of supply and/or methods of preparation: White ceramic tiles are available from supplier (E) in size 5.1 × 5.1 cm or 3.8 × 7.6 cm under SRM numbers 2019 and 2020.

Tiles in sets of 12, measuring 100 mm × 100 mm, are available from supplier (G) and (S). In each set, three tiles are spectrally neutral greys for the investigation of errors of linearity or geometrical errors, and nine are spectrally selective colours for the investigation of waveband or wavelength calibration, scanning or recording mechanism errors or spectral response errors. Infrared measurements to 2 μm are also available.

Pertinent physicochemical data: The reflectance of each white tile from supplier (E) is measured for incidence at 6° from normal with a high precision reflectometer at 10 nm intervals from 250 to 2500 nm and certified. The 0/45 spectral radiance factor, the diffuse part of the 0/d spectral reflectance, and the total 8/d spectral reflectance of the tiles from supplier (G) have been determined with respect to the perfect reflecting diffuser. For each of three geometries, the CIE colorimetric quantities (x, y, Y) have been computed from the spectral values for CIE Standard Illuminants A, C and D65.

REFERENCES

1. Clarke, F. J. J.; Samways, P. R. National Physical Laboratory Report MC 2, National Physical Laboratory: Teddington, **1968**.
2. Clarke, F. J. J. *Printing Technol.* **1970**, *B*, 101.
3. Clarke, F. J. J. *Die Farbe* **1971**, *20*, 299.
4. Clarke, F. J. J.; Malken, F. J. *J. Soc. Dyers and Colorists* **1981**, *97*, 503.
5. Compton, J. A. *Color. Res. Appl.* **1984**, *9*, 15.

16.2.9. Silicon carbide

Physical property: Regular reflectance
Unit: Dimensionless
Recommended reference material: Polished slices of silicon carbide
Range of variables: 400 nm to 700 nm, 25 °C
Physical state within the range: solid
Contributor: D. Gundlach

Intended usage: This reference material is mainly used for the calibration of spectrometer in ore microscopy.

Sources of supply and/or methods of preparation: Samples are available from supplier (K) and (L). Polished slices of silicon carbide of about 5 mm diameter are fixed in resin and mounted in ring holders. The surface of the standard can be cleaned with tissue paper or cotton soaked with acetone. Residual moisture can be removed with pure alcohol.

Pertinent physicochemical data: The spectral regular reflectance is certified for near-normal incidence against air and immersion oil. Specification of the oil is according to German Standard DIN 58 884. In the wave length range 400 nm to 700 nm, two sets of 18 values for spectral regular reflectance against air and immersion oil are given. The relative standard deviation (referred to the mean reflectance) at a given wavelength is 1.5 per cent.

REFERENCES

1. *International Handbook of Coal Petrography, Chapter on Calibration of Reflection Measurement.* Centre National de la Recherche Scientifique: Paris, **1971**.
2. Piller, H. *Performance of Reflectance Standards,* Mineralogy and Materials New Bulletin for Quantitative Microscopic Methods **1972**, *1*, 4.
3. Piller, H. *Mineralogy and Materials New Bulletin for Quantitative Microscopic Methods* **1972**, *2*, 7.
4. Piller, H. *Microscope Photometry*, p. 125-128 Springer-Verlag: Berlin, **1977**.
5. Zeiss, C. Reflexions-Mikroskop-Photometrie, p. 12 Messungen an Kohlenanschliffen, **1980**.

16.2.10. Flowers of sulphur

Physical property: Radiance factor, reflectance
Unit: Dimensionless
Recommended reference material: Flowers of sulphur
Range of variables: 500 nm to 2500 nm, 25 °C
Physical state within the range: solid
Contributor: D. Gundlach

Intended usage: This reference material is used for the calibration of reflectance spectrometers in the near infrared region.

Sources of supply and/or methods of preparation: Tablets of 25, 45, or 60 mm diameter are pressed from flowers of sulphur by means of a mechanical powder press.

Pertinent physicochemical data: The certification of the radiance factor and the reflectance will be prepared.

REFERENCES

1. Tkachuk, R.; Kuzina, F. D. *J Appl. Optics* **1978**, *17*, 2817.
2. Erb, W.; Richter, W. *Optik.*, **1986**, *2*, 64.

16.2.11. Black porcelain enamel

Physical property: Reflectance
Unit: Dimensionless
Recommended reference materials: Black porcelain enamel
Range of variables: 250 to 2500 nm, 25 °C
Physical state within the range: solid
Contributor: G. A. Uriano

Intended usage: This reference material can be used in calibrating the reflectance scale of integrating sphere reflectometers such as those used in the evaluation of solar energy materials.

Sources of supply and/or methods of preparation: Reference materials are available from supplier (E) in sizes 5.1×5.1 cm or 2.5×2.5 cm as SRM numbers 2021 and 2022.

Pertinent physicochemical data: The certified values of the spectral reflectance for incidence at 6° from normal are measured at 10 nm intervals from 250 to 2500 nm with a high precision reflectometer and corrected by high accuracy measurements.

REFERENCE

1. Weidner, V. R.; Hsia, J. J. in the Certificate to the NBS Standard Reference Materials 2021 and 2022.

16.2.12. Second surface aluminum mirror

Physical property: Regular reflectance
Unit: Dimensionless
Recommended reference material: Aluminium mirror on quartz protected by second quartz plate
Range of variables: 250 to 2500 nm, 25 °C
Physical state within the range: solid
Contributor: G. A. Uriano

Intended usage: This reference material is used in calibrating the photometric scale of specular reflectometers.

Sources of supply and/or methods for preparation: The reference material is available from the supplier (E) as SRM numbers 2023, 2024, and 2025 in size 5.1 × 5.1 cm or 2.5 × 10.1 cm. The aluminium mirror is vacuum deposited on the surface of a 2 mm thick fused quartz plate and protected by a second fused-quartz plate attached to the first plate with epoxy cement.

Pertinent physicochemical data: The certified values of the spectral regular reflectance are given at 25 wavelengths as measured with a highly accurate specular reflectometer-spectrophotometer at angles of incidence of 6°. Information is also provided on the reflectance at 9 wavelengths for incidence at 6°, 30°, and 45° for vertically and horizontally polarized incident beams. Another reference material is available from supplier (E) as SRM 2025 having dimensions of 2.5 cm × 10.2 cm. The aluminium is deposited on the back surface of a 2 mm thick optical quality vitreous quartz plate with a wedge of 10 mrad (0.573°) between its long faces. The mirror is protected by a second similar plate cemented to the first plate so that the front and back surfaces are parallel.

REFERENCE

1. Weidner, V. R.; Hsia, J. J. in the Certificate to the NBS Standard Reference Materials 2023 and 2024.
2. Certificate for the NBS Standard Reference Material 2025.

16.2.13. Pressed polytetrafluorethylene (PTFE) powder

Physical property: Reflectance
Unit: Dimensionless
Recommended reference material: Pressed polytetrafluorethylene (PTFE) powder
Range of variables: 200 to 2500 nm, 25 °C
Physical state within the range: solid
Contributor: G. A. Uriano, J. J. Hsia

Intended usage: This material is used for the calibration of spectral reference photometers.

Sources of supply and/or methods for preparation: This material is available from supplier (J) and is marketed under the trade name of Halon.

Pertinent physicochemical data: Pressed polytetrafluoroethylene powder was first proposed as a reference white to the CIE by Grum and Saltzman and is being increasingly used as an alternative to barium sulphate. A full report of the properties has been given by Weidner and Hsia (Ref. 1) and by Weidner, Hsia, and Adams (Ref. 2).

REFERENCE

1. Weidner, V. R.; Hsia, J. J. *Opt. Soc. Am.* **1981**, *71*, 856.
2. Weidner, V. R.; Hsia, J. J.; Adams, B. *Appl. Opt.* **1985**, *24*, 2223.

16.2.14. Ever-white (opal glass)

Physical property: Reflectance
Unit: Dimensionless
Recommended reference material: Ever-white (opal glass)
Range of variables: 380 to 780 nm, 25 °C
Physical state within the range: solid
Contributor: G. A. Uriano, J. J. Hsia

Intended usage: Ever-white is used for the calibration of spectral photometers, reflectometers or colorimeters.

Sources of supply and/or methods for preparation: Specimens are available from supplier (P). Sizes of 4.0 cm square, 5.0 cm square, and 3.2 cm diameter are routinely available but sizes up to can be supplied to special order.

Pertinent physicochemical data: This is a recently developed opal, the neutrality and reflectance of this material being similar to Russian opal. The material is usually supplied with representative data for spectral diffuse and total reflectance over the wavelength range 380 nm to 780 mn, but samples can be individually calibrated on request in the 0°/45° geometry.

16.3. Contributors

J. P. Cali,
16405 Kipling Road,
Rochville, MD 20855 (U.S.A.).

F. J. J. Clarke,
National Physical Laboratory,
Teddington, Middlesex, TW11 OLW (UK).

H. Feuerberg,
Bundesanstalt fur Materialprüfung,
Unter den Eichen 87,
D-1000 Berlin 45,
(Federal Republic of Germany).

D. Gundlach,
Bundesanstalt fur Materialprüfung,
Unter den Eichen 87,
D-1000 Berlin 45,
(Federal Republic of Germany).

J. J. Hsia,
National Bureau of Standards,
U.S. Department of Commerce,
Gaithersburg, MD 20899 (U.S.A).

E. Juhász,
National Office of Measurements,
Némétölgyi út 37,
H-1124, Budapest XII (Hungary).

H. Terstiege,
Bundesanstalt fur Materialprüfung,
Unter den Eichen 87,
D-1000 Berlin 45,
(Federal Republic of Germany).

G. A. Uriano,
National Bureau of Standards,
U.S. Department of Commerce,
Gaithersburg, MD 20899 (U.S.A.).

16.4. List of Suppliers

(A) Bundesanstalt fur Materialprüfung,
Unter den Eichen 87,
D-1000 Berlin 45,
(Federal Republic of Germany).

(B) Eastman Kodak Company,
Rochester,
New York 14650 (U.S.A.).

(C) Hopkins and Williams Ltd.,
Freshwater Road,
Chadwell Heath, Essex (UK).

(D) E. Merck,
Postfach 4119,
D-6100 Darmstadt 2,
(Federal Republic of Germany).

(E) National Bureau of Standards,
U.S. Department of Commerce,
Office of Standard Reference Materials,
Room B 311 Chemistry
Gaithersburg, MD 20899 (U.S.A.).

(F) National Office of Measures,
Némétvölgyi út 37,
H-1124, Budapest XII (Hungary).

(G) National Physical Laboratory,
Division of Quantum Metrology,
Teddington, Middlesex TW11 OLW
(UK).

(H) National Research Council,
Ottawa, Ontario KIA OR6 (Canada).

(I) Physikalische-Technische Bundesanstalt,
Bundesallee 100,
Braunschweig D-3300,
(Federal Republic of Germany).

(J) Allied Chemical Co.,
Speciality Chemical Co.,
North Avenue East,
Elizabeth, NJ 07201 (U.S.A.).

(K) Carl Zeiss A.G.,
D-7082 Oberkochen/Wurtt,
(Federal Republic of Germany).

(L) Ernst Leitz Wetzlar GMBH,
D-6330 Wetzlar,
(Federal Republic of Germany).

(M) Union Chimique Belge RPL,
Grauwmeer,
B 3030 Leuven (Belgium).

(N) Wako Pure Chemical Industries, Ltd.,
3-10, Dohshuucho,
Higashi-Ku, Osaka 541 (Japan).

(O) VNIIM,
Moscow M-49,
Leninskii pr 9 (USSR).

(P) Yoneda Glass Bead Mfg. Co.,
969/2 Ohjicho,
Izumi, Osaka 594 (Japan).

(Q) National Institute of Metrology,
7, District 11. Heping Street,
Beijing (China).

(R) Laboratorie National d´Essais,
1, Rue Gaston Boissier,
75015 Paris (France).

(S) British Ceramic Research Association,
Queens Rd, Penkhull
Stoke-on-Trent ST4 7LQ (UK).

17

SECTION: WAVELENGTH AND TRANSMITTANCE

COLLATORS: R. W. BURKE, R. A. VELAPOLDI

CONTENTS:

- 17.1. Introduction
    - 17.1.1 General Considerations
    - 17.1.2 Definitions and Nomenclature
    - 17.1.3 Characteristics of Standards
- 17.2. Reference Materials for Wavelength
    - 17.2.1 Didymium Glass Filters
    - 17.2.2 Inorganic Solutions
    - 17.2.3 Discharge Lamps
    - 17.2.4 Lasers
    - 17.2.5 Polystyrene Film
    - 17.2.6 Organic Solutions
    - 17.2.7 Vapors
- 17.3. Reference Materials for Transmittance
    - 17.3.1 Neutral Density Glass Filters
    - 17.3.2 Semi-transparent Metal-Film Filters
    - 17.3.3 Inorganic Solutions/Calibrated Cuvette
    - 17.3.4 Rotating Sectors
- 17.4. Contributors
- 17.5. List of Suppliers

17.1. Introduction

17.1.1. General Considerations

The use of spectrometric methods for physicochemical and analytical studies has increased tremendously during the last several decades. With increasing frequency, many of these studies involve the measurement of the fundamental optical properties of materials. For such measurements to be meaningful, they must be accurate. A necessary operation in ascertaining their accuracy is the calibration of the accuracy of the instrumentation used. This calibration is best performed by the use of well-characterized and readily available reference materials.

In absorption spectrometry, or spectrophotometry, the transmittance of a sample constitutes an intrinsic property of that sample. Hence, the accurate measurement of this property is essential if transmittance is to be used to characterize meaningfully the material. The accuracy of the transmittance measurement can be assessed only after all the sources of systematic error are considered and their magnitudes determined. These errors can arise from limitations of the instrument, from limitations of the sample, and from instrument-sample interactions. Before the accuracy of a transmittance measurement can be rigorously established, the following parameters must be considered and independently evaluated: wavelength accuracy; accuracy of the transmittance scale; bandpass limitations; heterochromatic stray radiation; beam geometry and displacement errors; interreflections; polarization effects; optical pathlength; sample uniformity; and temperature dependence.

This proposal, an expansion of an earlier publication (Ref. 1), presents a summary of the reference materials that are recommended for checking two of the critical performance specifications of an absorption spectrometer - the accuracy of the wavelength scale and the accuracy of the transmittance scale. Also included is information on laboratories that will provide certified calibrations, on request, for customer supplied samples.

The calibration of the wavelength scale is discussed first, since accurate transmittance measurements at specified wavelengths can be obtained only after the *a priori* establishment of the wavelength accuracy of the instrument. Few Certified Reference Materials (CRM's) are available for wavelength calibration; thus both certified and recommended reference materials are listed in the wavelength section. On the other hand, CRM's for calibrating the transmittance scale of absorption spectrometers in the ultraviolet (UV), visible (VIS), and near infrared (IR) are available from sources in several countries; thus only transmittance standards that are CRM's are listed for use in the spectral range of 200 nm to 3.0 μm. Additionally, information is supplied for suggested reference materials (RM's) or calibration services for which CRM's are not available. Finally, only published references to calibration values of wavelength or transmittance are cited in this document.

17.1.2. Definitions and Nomenclature

Wavelength, λ, is defined as the distance in the direction of propagation of a periodic wave between two succesive points at which the phase is the same (at the same time). Its unit is the meter, m.

Names, symbols, and definitions used in this document follows closely that recommended by IUPAC Commission 1.5 (Ref. 2) and ISO (Ref. 3).

The majority of transmittance measurements in the UV, VIS, and near IR are made on materials whose spectra consist of absorption bands with half-height bandwidths (HHBW) of 20-50 nm. In these cases, a wavelength accuracy of 0.1 nm is sufficient to give a transmittance accuracy better than 0.01%, and is acceptable in most applications. Thus, various materials ranging from rare earth ions in glasses to line sources can be used in the above cases to calibrate the wavelength scale. Although the wavelengths of line sources are known to a higher degree of accuracy than necessary for the majority of applications, they are necessary for those cases requiring more stringent wavelength accuracy, *e.g.*, transmittance measurements on the sides of spectral bands or transmittance measurements or wavelength assignments of the maxima of sharp spectral features (Ref. 4).

Similarly, in the infrared many of the applications involve measurements of materials in condensed phases whose spectral bands tend to be broad (HHBW > 5 cm^{-1}). In these cases, assignment of peak wavenumbers to 0.2 to 0.5 cm^{-1} are generally sufficient and instruments with low to medium resolution (spectral slit widths of 10 to 1 cm^{-1}) can be used. Measurements of rotational fine structure and wavenumber assignment for the very sharp bands of small molecules in the vapour phase, however, require peak wavenumber assignments of 0.001 to 0.01 cm^{-1}. In these latter cases, instruments with medium to high resolution (spectral slit widths of 0.5 to 0.01 cm^{-1}) are required. Such instruments include optical and grating spectrometers, interferometers, and fourier transform infrared machines. Wavelengths in this document are expressed as follows: ultraviolet-visible (UV-VIS), nanometers (nm); infrared (IR), micrometers (μm) or wavenumbers (cm^{-1}), where wavenumber is the reciprical of wavelength.

The transmittance, τ, of a sample is defined as the ratio of the transmitted radiant or luminous flux to the incident flux under specified conditions of irradiation. Transmittance, as such, is a poorly defined property and cannot serve a useful purpose in spectrophotometry unless the constraints that have been delineated above are placed upon its measurement and use. Then, and only then, does the transmittance become a meaningful material property.

The spectral internal transmittance, $\tau_i(\lambda)$, of a sample is its transmittance exclusive of losses at boundary surfaces and interreflection effects between them. Its absorbance, A_i, is defined as the negative logarithm to the base ten of the internal transmittance, $A_i = -\log_{10} \tau_i(\lambda)$. Considerable confusion still exists in the definition of absorbance, particularly in the chemical literature, where it is frequently erroneously defined as $-\log_{10} \tau(\lambda)$, which was the same definition given to optical density. The use of the term optical density is now discouraged.

Mielenz has proposed an alternative term "transmittance density" for $-\log_{10} \tau(\lambda)$ as one of his recommendations for improving the consistency of spectrometry nomenclature used by optical physicists and chemists (Ref. 5). This new terminology has been incorporated in all of the certificates accompanying the transmittance standards provided by the National Bureau of Standards. In quantitative studies of radiation propagation through absorbing media it can be convenient to compute the adsorption index (k) which is the imaginary component of the complex refractive index and is related to A_i by $k = 2.30258 A_i \lambda / 4\pi l$.

In practice, the transmittance or transmittance density (optical density) is obtained when intensity measurements are made relative to air; the internal transmittance or absorbance is obtained, to a close approximation, when intensity measurements are made relative to a nonabsorbing solid or liquid reference sample of the same refractive index. The measurement is not exact, however, since the internal multiple reflection effect can be highly wavelength dependent because of anomalous dispersion changes in traversing a strong adsorption band (Ref. 7).

The basic requirement for the production of certified transmittance standards is the existence of instrumentation whose design permits the individual testing of the essential components and the testing of the complete instrument for all sources of systematic error. Activities of this type are primarily the responsibility of the National laboratories. Largely as a result of the work of Clarke (Ref. 8) at the National Physical Laboratory and of Mavrodineanu (Refs. 9, 10) and Mielenz (Ref. 11) at the National Bureau of Standards, instrumentation and procedures exist at these laboratories for measuring the transmittance of materials with a known accuracy. These two laboratories have used the capability to produce several kinds of Certified Reference Materials that are intended for verifying the accuracy of the transmittance scale of optical spectrophotometers, but not interferometers. In addition, the New Zealand Department of Scientific and Industrial Research has described the construction of a high-accuracy spectrophotometer that has been specially-built for the same purpose (Ref. 12). A reference spectrometer has also been described by the National Office of Measures (Hungary) (Ref. 13).

17.1.3. Characteristics of Standards

Reference materials for calibrating wavelength and transmittance scales may be solids, liquids, gases, or fabricated devices. Solid Reference Materials are frequently preferred because they are easy to handle and present few difficulties in being transported from one laboratory to another. Suitable liquid Reference Materials are often prepared from solid or liquid reagents. Gases and vapour Reference Materials are often difficult to transport and handle if they are not constituents of the atmosphere or are not contained in closed or sealed cells. Fabricated devices (*e.g.*, hollow cathode and low pressure discharge lamps, lasers, and rotating sectors) are often expensive and may require special equipment or modification of the instrument being calibrated.

For wavelength calibration, solid and liquid Certified Reference Materials (one of each) are available for use in the ultraviolet and visible regions of the spectrum. A broad range of recommended Reference Materials are available, however, to provide wavelength calibration in the UV, VIS, near IR, and IR. For those cases in which wavelength accuracies are not too stringent, (0.1 nm in the UV, VIS, and near IR and $2 - 10$ cm^{-1} in the IR) solutions, solids, low pressure and hollow cathode discharge lamps, lasers, and gases can be used. For those cases requiring more stringent wavelength accuracies, low pressure and hollow cathode discharge lamps, lasers, and small molecules in the vapour state must be used.

The IUPAC Commission on Molecular Structure and Spectroscopy, I.5., has made extensive recommendations (Refs. 14 to 18) for Reference Materials for wavelength measurements in the IR region. The culmination of these recommendations is an IUPAC sponsored book of wavenumbers for recommended Reference Materials for the calibration of IR spectrometers (Ref. 19). Commission 1.5 is in the process of preparing a new edition of this document. Tables of wavenumber values for a few materials from the compilation are listed in this document for low and medium resolution instruments. For additional Reference Materials for these instruments and especially for recommended reference materials for calibration of high resolution instruments, it is essential that this excellent book and the references therein be consulted. However, even these data do not satisfy the wavenumber accuracy requirements for a handful of applications. In these cases, an extremely accurate procedure based on measurements of absolute frequencies using heterodyne techniques has been developed. The extraordinarily accurate frequency values are related to the wavenumber by the speed of light, and wavelength accuracies of two to three orders of magnitude over what was previously obtainable are now possible in the IR (see for example reference 20 and those cited therein). Thus measurement procedures and Reference Materials are constantly evolving, driven by the demands of increased accuracy. Additionally, various laboratories (see Table 17.1.1.1) will provide wavelength calibration of submitted materials on request.

In the case of transmittance, both solid and liquid standards are available as Certified Reference Materials from a number of sources. Types of standards include the "optically neutral" absorbing glasses, thin metal-films on quartz, ampouled solutions, and a high purity, crystalline compound from which the user prepares appropriate solutions as needed.

The solid filters are more convenient to use and exhibit the more desirable spectral profile. In general, the smoother the spectral profile, the better suited the transmittance standard is for testing the photometric scale of a spectrophotometer because uncertainties associated with wavelength accuracy, bandwidth, and heterochromatic stray radiation are not generally significant. However, accuracy of the photometric scale determined by a CRM with a smooth spectral profile does not guarantee accurate transmittance values for samples with irregular spectral profiles because the effects of wavelength accuracy, bandwidth, and heterochromatic stray radiation may become significant.

Among the types of certified transmittance standards presently available, the absorbing glasses are the most satisfactory standards in general and have had the longest period of use. Their major limitation is that they do not transmit in the ultraviolet. The liquid standards that are now available were initially issued to fill the void that existed at the time for certified ultraviolet transmittance standards. In comparison to solid filters, their use requires more stringent restrictions on wavelength accuracy, spectral bandpass, and temperature control as well as increased analytical skills of the user. A calibrated cuvette (available as a Certified Reference Material) for which the internal pathlength and parallelism and flatness of the windows are known accurately, should also be used when measuring transmittances of highest accuracy.

Table 17.1.1.1. Sources of certified reference materials or certification services for calibration of wavelength (W) and transmittance (T) scales of spectrometers

Source	CRM's			Certification Services[a]
	Glasses	Metal Films	Liquids	
National Physical Laboratory (U.K.)	T	T	-	W,T
National Bureau of Standards (U.S.A.)	W,T	T	W,T	W,T
National Research Council (Canada)	-	-	-	W,T
New Zealand Department of Scientific and Industrial Res.	T	-	-	-
Physikalische-Technische Bundesanstalt (FRG)	-	-	-	W,T
Pye Unicam	T	-	-	-
Carl Zeiss	T	-	-	-
National Office of Measures Hungary	T	-	-	W,T

[a] Customer supplies sample

The thin-metal film filters are the most recent type of certified transmittance standard to be issued. They are prepared by vacuum-depositing thin films of nichrome, chromium, inconel, rhodium, etc. onto nonfluorescent, fused-quartz plates. Filters prepared in this manner exhibit excellent optical neutrality from 200 nm to 3 μm. However, a potential problem exists in the use of these metal-film filters in certain designs of conventional spectrophotometers because, in addition to absorption, they attenuate a significant proportion of the incident radiation by reflection. In some instruments, this reflected component will give problems because it can interact with either the irradiating or detecting optics to produce interreflection effects that are significantly larger than those that exist in the instrument during normal use. In such instances, the difference between the measured and certified transmittance cannot be used to correct the photometric scale. Clarke and co-workers (Ref. 21) have suggested that the high reflectance of the metal-film filters can be used to advantage to test for interreflective errors. To do so, however, requires the use of certified glass or liquid standards that have, at some wavelength, the same transmittance as the metal-film filter. Additional investigations and standards are needed to determine conditions under which the thin metal-film filters can be used to test all types of spectrophotometers and to provide directly corrections to the photometric scale.

A summary of the known sources for Certified Reference Materials for calibrating the wavelength and transmittance scales of absorption spectrometers is presented in Table 17.1.1.1. More detailed information on each of these standards is provided in the data sheets at the end of this section.

REFERENCES

1. Milazzo, G., (collator) Recommended Reference Materials for Realization of Physicochemical Properties, *Pure Appl. Chem.* **1977**, *49*, 661.
2. Names, Symbols, Definitions and Units of Quantities in Optical Spectroscopy. *Pure and Appl. Chem.* **1945**, *57*, 105.
3. Quantities and Units of Light and Related Electromagnetic Radiation, 150: Geneva, **1973**.
4. Mielenz, K. D. *Photoluminescence Spectrometry* in *Optical Radiation Measurements: Measurements of Photoluminescence*, Mielenz, K. D., editor, Vol. 3, Academic Press: New York, **1982**, pp. 1-88.
5. Mielenz, K. D., *Anal. Chem.* **1976**, *48*, 1093.
6. Crawford, B. L.; Goplen, T. G.; Swanson, D. *The Measurement of Optical Constants in the Infrared by Attenuated Total Reflection* in *Advances in Infrared and Raman Spectroscopy*, Clark, R. J. H.; Hester, R. E., editors, Vol. 4, Heyden: London, **1978**, pp. 47-83.
7. Young, R. P.; Jones, R. N. *Chemical Reviews* **1971**, *71*, 219.
8. Clarke, F. J. J., *Accuracy in Spectrophotometry and Luminescence Measurements*, National Bureau of Standards Special Publication 378, U.S. Department of Commerce:

Washington, **1973**, p. 1.
9. Mavrodineau, R., *Accuracy in Spectrophotometry and Luminescence Measurements*, National Bureau of Standards Special Publication 378, U.S. Department of Commerce: Washington, **1973**, p. 31.
10. Mavrodineanu, R., *Standardization in Spectrophotometry and Luminescence Measurements*, National Bureau of Standards Special Publication 466, U.S. Department of Commerce: Washington, **1977**, p. 127.
11. Mielenz, K. D., *App. Opt.* **1973**, *12*, 1630.
12. Bittar, A., Hamlin, J. D., *Physics and Engineering Laboratory Report No. 769*, Department of Scientific and Industrial Research, Private Bag, Lower Hutt, New Zealand, **1981**.
13. Fillinger, L.; Andor, Gy. *Mérés ès Automatika* **1983**, *31*, 369.
14. Tables of the calibration of grating spectrometers in the range 4300 − 600 cm^{-1}. *Pure Appl. Chem.* **1961**, *1*, 537.
15. Wavenumbers for the calibration of prism and small grating spectrometers (Resolution 0.5 − 10 cm^{-1}) in the range 3951 − 600 cm^{-1}. *Pure Appl. Chem.* **1961**, *1*, 601.
16. Tables for the calibration of moderately high resolution spectrometers in the range 600 − 1 cm^{-1}. *Pure Appl. Chem.* **1973**, *33*, 613.
17. Tables for the calibration of low resolution spectrometers in the range 600 − 15 cm^{-1}. *Pure Appl. Chem.* **1973**, *35*, 639.
18. Corrigendum, Appendix. Revised spectrum of indene-camphor-cyclohexanone for the calibration of low resolution spectrometers in the range 4000 − 600 cm^{-1}. *Pure Appl. Chem.* **1974**, *37*, 649.
19. Cole, A. R. H., IUPAC Commission on Molecular Structure and Spectroscopy, *Tables of Wavenumbers for the Calibration of Infrared Spectrometers*, 2nd Edition, Pergamon Press: Oxford, **1977**.
20. Pollack, C. R., Petersen, F. R., Jennings, D. A., Wells, J. S., Maki, A. G., *J. Mol. Spectroscopy* **1983**, *99*, 357.
21. Clarke, F. J. J., Downs, M. J., McGivern, W., *UV Group Bulletin* **1977**, *5*, 104.

17.2. Reference materials for wavelength

17.2.1. Didymium glass filters

Physical property: Wavelength in the visible region
Unit: Nanometers (nm)
Recommended reference materials: Didymium Glass Filters
Range of variables: Supplier A; Filter, 51 mm × 51 mm unmounted, or 10 *mm* × 31 *mm* mounted in cuvette-sized holders, batch or individually certified. Fifteen wavelengths certified at bandwidths between 1.5 and 10.5 nm.

Nominal wavelengths are 402, 431, 440, 445, 472, 478, 513, 529, 572, 585, 623, 629, 684, 739, and 748 nm.

Supplier B; Filter 12, 5 mm × 40 mm unmounted. Nominal wavelengths are 403, 431, 473, 513 and 684 nm.

Physical state within the range: Solid

Contributors: W. H. Venable, Jr., K. L. Eckerle, E. Juhász

Sources/Preparation: Supplier A; Office of Standard Reference Materials, National Bureau of Standards, Gaitherburg, MD 20899. Supplier B; National Office of Measures, H-1531 Budapest 126, Pf. 19, Hungary

Physicochemical data: See references 1 and 2.

REFERENCES

1. Venable, W. H., Jr., Eckerle, K. L., *Didymium Glass Filters for Calibrating the Wavelength Scale of Spectrophotometers*, NBS Special Publication 260-66, U.S. Department of Commerce: Washington, **1979**.
2. Fillinger, L.; Andor, Gy. *Mérés és Automatika* **1983**, *31*, 369.

17.2.2. Inorganic solutions

17.2.2.1. Holmium (III) solution as standard reference material

Physical property: Wavelength in the ultraviolet and visible regions
Unit: Nanometers (nm)
Recommended reference material: Holmium(III) ions in aqueous perchloric acid
Range of variables: Consists of 4% by mass solution of holmium(III) oxide dissolved in 100 g dm^{-3} perchloric acid solution placed in sealed quartz cuvetttes. Fourteen transmittance minima are certified over the nominal wavelength range 240 to 640 nm at (25 ± 5) °C and six spectral bandwidths.
Physical state within the range: Liquid
Contributors: V. R. Weidner, R. Mavrodineanu, K. D. Mielenz, R. A. Velapoldi, K. L. Eckerle, B. Adams

Sources/Preparation: Office of Standard Reference Materials, National Bureau of Standards, Gaithersburg, MD 20899

Physicochemical data: See reference 1 and table 17.2.2.1.1.

Table 17.2.2.1.1. Wavelength of transmittance minimum (nm) for holmium(III) solutions

Minimum No.	Wavelength nm
1	240.99[a]
2	249.83
3	278.15
4	287.01
5	333.47
6	345.55
7	361.36
8	385.45
9	416.07
10	b
11	467.82
12	485.28
13	536.54
14	640.51

[a] 25 °C, 0.1 nm spectral bandwidths, uncertainty in wavelength is ±0.1 nm at the 95% level.
[b] Splits into 2 minima for bandwidths less than 1 nm.

REFERENCE

1. Weidner, V. R.; Mavrodineanu, R.; Mielenz, K. D.; Velapoldi, R. A.; Eckerle, K. L.; Adams, B. *J. Res. Nat. Bur. Stds.* **1985**, *90*, 115.

17.2.2.2. Holmium (III) solution as laboratory intercomparison

Physical property: Wavelength in the ultraviolet and visible region
Unit: Nanometers (nm)
Recommended reference material: Holmium(III) ions in aqueous perchloric acid
Range of variables: Maxima occur in absorbance spectra over nominal wavelength range 240 to 640 nm.
Physical state within the range: Liquid
Contributors: R. A. Velapoldi, R. W. Burke.

Sources/Preparation: Suitable samples of holmium oxide (Ho_2O_3) are available from many suppliers. The following solutions have been prepared (1) 4 g Ho_2O_3 in 100 g of 1.4 mol dm^{-3}

perchloric acid solution and (2) 10 g of Ho_2O_3 in 100 cm^3 of 175 g dm^{-3} aqueous perchloric acid.

Physicochemical data: See references 1 to 4 and table 17.2.2.2.1.

Table 17.2.2.2.1

	Wavelength of maximum nm			
I[a]	II[b]	III[c]	IV[d]	V[e]
241.1 ± 0.05	241.15	241.1	241.0	241.1
249.7 ± 0.1	249.75	249.7	250.0	249.7
278.7 ± 0.1	278.2	278.7	277.8	278.2
287.1 ± 0.1	287.15	287.1	287.5	287.2
333.4 ± 0.1	333.5	333.4	333.3	333.3
345.5 ± 0.1	345.6	345.5	345.5	345.0
361.5 ± 0.1	361.5	361.5	361.0	361.2
385.4 ± 0.2	385.6	385.5	385.6	385.6
416.3 ± 0.2	416.2	416.3	416.0	416.6
450.8 ± 0.2	450.7	450.8	450.4	451.0
452.3 ± 0.2	452.0	-	-	-
467.6 ± 0.2	467.75	-	-	468.0
485.8 ± 0.2	485.25	485.8	485.2	485.2
536.4 ± 0.2	536.3	-	-	536.8
641.1 ± 0.2	640.5	-	-	-

[a] Solution (1) under "Preparation" above; spectrophotometer-Cary 14, (Ref. 1).
[b] Solution (1) under "Preparation" above; spectrophotometer-Cary 17, (Ref. 2).
[c] Solution (2) under "Preparation" above; spectrophotometer-Cary 118, (Ref. 3).
[d] Solution (2) under "Preparation" above; spectrophotometer-Beckman Acta CV, (Ref. 4).
[e] Solution (2) under "Preparation" above; spectrophotometer-Perkin-Elmer 200, (Ref. 4).

REFERENCES

1. McNeirney, J., Slavin, W., *Appl. Opt.* **1962**, *1*, 365.
2. Milazzo, G., *Pure and Appl. Chem.* **1977**, *49*, 661.
3. Burgess, C., U. V. Spectrometry Group Bulletin **1977**, *5*, 77.
4. Vinter, J. G., *Wavelength Calibration* in *Standardization in Absorption Spectrometry*, Vol. 1, Burgess, C.; Knowles, A., editors, Chapman and Hall: London, **1981**, pp. 111-120.

17.2.3. Discharge lamps

17.2.3.1. Hollow cathode lamps

Physical property: Wavelength in the ultraviolet, visible, and near infrared regions
Unit: Nanometers (nm)
Recommended reference material: Hollow cathode discharge lamps
Range of variables: Sharp maxima occur in emission spectra over nominal wavelength range 200 to 995 nm.
Physical state within the range: Manufactured device
Contributor: T. C. Rains

Sources/Preparation: Suitable devices available from many manufacturers

Physicochemical data: Wavelengths and lamps are only a few of those actually available; see table 17.2.3.1.1.

Table 17.2.3.1.1. Hollow cathode discharge lamps

Wavelength in air nm	Lamp	Wavelength in air nm	Lamp
202.582	Magnesium	520.844	Chromium
213.856	Zinc	550.649	Molybdenum
228.802	Cadmium	588.995	Sodium
253.652	Mercury	610.362	Lithium
285.213	Magnesium	640.847	Strontium
307.590	Zinc	670.784	Lithium
326.106	Cadmium	687.838	Strontium
343.489	Rhodium	705.994	Barium
400.875	Tungsten	766.490	Potassium
411.518	Vanadium	794.760	Rubidium
429.866	Barium	852.124	Cesium
451.331	Molybdenum	917.224	Cesium
479.992	Strontium	967.555	Titanium
499.107	Titanium	994.990	Rhenium

17.2.3.2. Low pressure lamps

Physical property: Wavelength in the ultraviolet, visible and near infrared regions
Unit: Nanometers (nm)
Recommended reference material: Low pressure discharge lamps
Range of Variables: Sharp maxima occur in emission spectra over nominal wavelength range 200 to 900 nm.
Physical state within the range: Manufactured device
Contributors: R. A. Velapoldi, R. W. Burke

Source: Suitable devices available from many manufacturers

Physicochemical data: See reference 1 and table 17.2.3.2.1.

Table 17.2.3.2.1.

Wavelength in air nm	Lamp	Wavelength in air nm	Lamp
186.96	Mercury	479.992	Cadmium
202.551	Zinc	491.604	Mercury
213.856	Zinc	501.568	Helium
228.802	Cadmium	508.582	Cadmium
253.652	Mercury	546.075	Mercury
275.278	Mercury	579.065	Mercury
296.728	Mercury	587.562	Helium
302.150	Mercury	643.847	Cadmium
334.148	Mercury	667.815	Helium
365.015	Mercury	692.947	Neon
388.865	Helium	724.517	Neon
404.656	Mercury	780.023	Rubidium
420.185	Rubidium	794.760	Rubidium
421.556	Rubidium	813.641	Neon
435.835	Mercury	852.110	Cesium
455.536	Cesium	894.350	Cesium
467.816	Cadmium		

REFERENCE

1. Harrison, G. R. *M. I. T. Wavelength Tables*, Vol. 2, M. I. T. Press: Cambridge, Mass. **1983**.

17.2.4. Lasers

17.2.4.1. Ultraviolet, visible, and near infrared

Physical property: Wavelength in the ultraviolet, visible, and near infrared regions
Unit: Nanometers (nm)
Recommended reference material: Lasers
Range of variables: Sharp maxima occur in emission spectra over nominal wavelength range 325 to 800 nm. The lasers can be gaseous-ion or solid state types.
Physical state within the range: Manufactured device
Contributors: R. A. Velpoldi, R. W. Burke

Sources/Preparation: Suitable devices available from many manufacturers

Physicochemical data:

Table 17.2.4.1.1. Lasers in ultraviolet, visible, and near infrared

Wavelength in air nm	Lamp	Wavelength in air nm	Lamp
325.030	Helium-Cadmium	528.692	Argon
351.115	Argon	578.213	Copper
363.789	Argon	627.818	Gold
441.570	Helium-Cadmium	632.817	Helium-Neon
454.508	Argon	647.089	Krypton
465.794	Argon	676.443	Krypton
476.489	Argon	694.3	Ruby
487.990	Argon	752.548	Krypton
496.512	Argon	799.298	Krypton
514.536	Argon		

17.2.4.2. Near infrared

Physical property: Wavelength in the near infrared
Unit: Micrometers (μm)
Recommended reference material: Laser
Range of variables: Sharp maxima occur in emission spectra over nominal wavelength range 1.0 to 3 μm.
  Physical state within the range: Manufactured device
Contributors: R. A. Velpoldi, R. W. Burke

Sources/Preparation: Suitable devices available from many manufacturers

Physicochemical data: See table 17.2.4.2.1

Table 17.2.4.2.1. Lasers in infrared region

Wavelength in air μ m	Lamp
1.0623	Neodynium-YAG
1.1523	Helium-Neon
3.3912	Helium-Neon

17.2.5. Polystyrene film

Physical property: Wavenumber (wavelength) in the infrared
Unit: Wavenumber (vacuum) cm^{-1}
Recommended reference material: Polystyrene film
Range of variables: Wavenumbers from 3000 to 700 cm^{-1}. Used for calibration of medium to low resolution spectrometers. In general, spectral slit widths in the range 2 to 10 cm^{-1} are required.
Physical state within the range: Solid
Contributor: G. Milazzo

Sources/Preparation: Suitable samples available from many manufacturers

Physicochemical data: See reference 1 and table 17.2.5.1.

Table 17.2.5.1. Polystyrene film

Wavenumber (vacuum) in cm^{-1}	Wavenumber (vacuum) in cm^{-1}
3027.1 ± 0.3	1583.1 ± 0.3
2924 ± 2.0	1154.3 ± 0.3
2850.7 ± 0.3	1069.1 ± 0.3
1944 ± 1.0	1028.0 ± 0.3
1871.0 ± 0.3	906.7 ± 0.3
1801.6 ± 0.3	698.9 ± 0.5
1601.4 ± 0.3	

REFERENCE

1. Cole, A. R. H., IUPAC Commission on Molecular Structure and Spectroscopy, *Tables of Wavenumbers for the Calibration of Infrared Spectrometers*, 2nd Edition, Pergamon Press: Oxford, **1977**.

17.2.6. Organic Solutions

Physical property: Wavenumber (wavelength) in the infrared
Unit: Wavenumber (vacuum) cm^{-1}
Recommended reference material: Indene (98.4%), camphor (0.8%), and cyclohexanone (0.8%), the percentages are by weight.
Ranges of variables: Wavenumbers from 4000 to 300 cm^{-1}. Used for calibration of medium to low resolution spectrometers. In general, spectral slit widths of 2 to 10 cm^{-1} are required.
Physical state within the range: Liquid
Contributor: G. Milazzo

Sources/Preparation: Samples are available from suppliers (A), (B), (C), (D), (E), and (F).

Physicochemical data: See references 1 and 2. Indene(98.4%), camphor(0.8%) and cyclohexanone (0.8%), the percentages are by weight. Values are given in table 17.2.6.1.

Table 17.2.6.1. Organic reference solution.

Band No.	Wavenumber[a] / cm^{-1}	Band No.	Wavenumber[a] / cm^{-1}	Band No.	Wavenumber[a] / cm^{-1}
1	3927.2 ± 1.0	36	2090.2	64	1122.4
2	3901.6	39	1943.1	66	1067.7 ± 1.0
3	3798.9	40	1915.3	67	1018.5
5	3660.6 ± 1.0	41	1885.1	69	947.2
8	3297.8 ± 1.0	42	1856.9	70	942.4
9	3139.5	44	1797.7 ± 1.0	71	914.7
10	3110.2	44α	1741.9	72	861.3
12	3025.4	44β	1713.4	73	830.5
15	2887.6	47	1661.8	74	765.3
17	2770.9	48	1609.8	76	718.1
19	2673.3	49	1587.5	77	692.6 ± 1.0
20	2622.3	51	1553.2	(1)[b]	592.1
21	2598.4 ± 1.0	53	1457.3 ± 1.0	(2)[b]	551.7
23	2525.5	54	1393.5	(3)[b]	521.4
28	2305.1	55	1361.1	(4)[b]	491.2 ± 1.0
29	2271.4	57	1312.4	(5)[b]	420.5
30	2258.7	58	1288.0	(6)[b]	393.1
33	2172.8	60	1226.2	(7)[b]	381.6
34	2135.8 ± 1.0	61	1205.1	(8)[b]	301.4
35	2113.2	62	1166.1		

[a] Values are accurate to ±0.5 cm^{-1} unless otherwise indicated.
[b] 1:1:1 indene: camphor: cyclohexanone mixture by weight; CsI cell thickness was 0.05 mm for these values.

Cell thickness: 0.2 mm for the regions 4000 to 3100, 2800 to 1500 cm^{-1}; 0.03 mm for the regions 3100 to 2800, 1650 to 1600, 1400 to 800 cm^{-1}; contact film for the region 800 to 690 cm^{-1}.

Band No. 44α is from camphor and Band No. 44β is from cyclohexanone. This mixture was also investigated by Lukasiewicz-Ziarkowska (Ref. 2) whose results differ from those in table 17.2.6.1 by 1.1 cm^{-1} at 4000 cm^{-1} and by 0.2 cm^{-1}; the differences vary linearly from 4000 to 600 cm^{-1}.

REFERENCES

1. Jones, R. N., Nadeau, A., *Can. J. Spect.* **1975**, *20*, 33. Data also compiled by Cole, A. R. H., IUPAC Commission on Molecular Structure and Spectroscopy, *Tables of Wavenumbers for the Calibration of Infrared Spectrometers*, 2nd Edition, Pergamon Press: Oxford, **1977**.
2. Lukasiewicz-Ziarkowska, Z., *Chem. Anal. (Warsaw)* **1969**, *14*, 1231; **1971**, *16*, 767.

17.2.7. Vapours

Physical property: Wavenumber (wavelength) in the infrared
Unit: Wavenumber (vacuum) cm^{-1}
Recommended reference material: Molecules in the vapour state
Range of variables: Wavenumbers from 4350 to 1 cm^{-1}. Used for calibration of high to medium resolution spectrometers. In general, spectral slit widths of 0.01 cm^{-1} or smaller are required
Physical state within the range: Gas
Contributors: R. A. Velpoldi, R. W. Burke

Physicochemical data: See reference 1 and table 17.2.7.1. Other values are given in reference 2.

Table 17.2.7.1. Reference materials in the vapour

Substance	Wavelength region covered cm^{-1}	Substance and band	Wavelength region covered cm^{-1}
CO	4350-4100	HCN	800-630
C_2H_2	4150-4015	CO_2	715-620
H_2O	4050-3480	N_2O	630-545
HCN	3410-3190	DCN	640-500
HCl	3130-2515	H_2O	625-310
HBr	2750-2265	H_2O	315-35
CO_2	2390-2290	H_2O	40-10
$^{13}CO_2$	2300-2225	C_2H_2	200-65
CO	2260-1990	HCl	325-20
DCl	2240-1890	HBr	260-16
DBr	1985-1630	HCN	120-3
HCN	1505-1330	CO	120-4
C_2H_2	1380-1285	N_2O	43-1
NH_3	1250-705		

REFERENCE

1. Cole, A. R. H., IUPAC Commission on Molecular Structure and Spectroscopy, *Tables of Wavenumbers for the Calibration of Infrared Spectrometers*, 2nd Edition, Pergamon Press: Oxford, **1977**.

2. Guelachvili, G.; Rao, K. N. *Handbook of Infrared Standards*, Academic Press: London, **1986**.

17.3. Reference materials for transmittance

17.3.1. Neutral density glass filters

17.3.1.1. Transmittance/optical density

Physical property: Transmittance/optical density
Unit: Dimensionless
Recommended reference material: Neutral density glass filters
Range of variables: Supplier A; Filters mounted in cuvette-style holders are available with nominal densities of 0.04, 0.15, 0.25, 0.5, 1.0, 1.5, 2.0, 2.5, 3.0; calibrations can be provided at any designated wavelength between about 380 nm and 2300 nm. From supplier B; Set of three unmounted filters in size of 12.5 mm ×40 mm, having nominal transmittances of 10, 20 and 30 percent; calibrations are available at any designated wavelength between 300 nm and 800 nm.
Physical state within the range: Solid
Contributors: J. F. Verrill, E. Juhász

Sources/Preparation: Supplier A; Division of Quantum Metrology, National Physical Laboratory, Teddington, Middlesex, United Kingdom TW11 OLW. Supplier B; National Office of Measures, H-1531 Budapest 126, Pf. 19, Hungary (see reference 1)

REFERENCE

1. Fillinger, L.; Andor, Gy. *Mérés és Automatika* **1983**, *31*, 369.

17.3.1.2. Transmittance/transmittance density

Physical property: Transmittance/transmittance density
Unit: Dimensionless

Recommended reference material: Neutral density glass filters
Range of variables: Sets consisting of three glass filters mounted in cuvette-type holders and having nominal transmittances of 10, 20 and 30 percent are stocked items. Calibration wavelengths are 440, 465, 546.1, 590 and 635 nm; other transmittances between 1 and 90 percent and for wavelengths up to 2500 nm are available on request.
Physical state within the range: Solid
Contributor: R. W. Burke

Sources/Preparation: Office of Standard Reference Materials, National Bureau of Standards, Gaithersburg, MD 20899

17.3.1.3. Absorbance

Physical property: Absorbance
Units: Dimensionless
Recommended reference material: Neutral density glass filters
Range of variables: Set of eight filters mounted in cuvette-type holders and having nominal absorbances of 0.1, 0.2, 0.5, 1.0, 1.5, 2.0, 2.5 and 3.0; calibrated relative to a clear glass plate at 546 nm.
Physical state within the range: Solid
Contributor: D. Irish

Sources/Preparation: Pye Unicam Ltd., Cambridge, England CB1 2PX

17.3.1.4. Transmittance/absorbance

Physical property: Transmittance/absorbance
Unit: Dimensionless
Recommended reference material: Neutral density glass filters
Range of variables: Set of six unmounted filters having nominal transmittances of 50, 25, 12, 3.5, 1.0 and 0.35 percent and calibrated against a clear plate at 550 nm.
Physical state within the range: Solid
Contributor: H. von Derschau

Sources/Preparation: Carl Zeiss, Oberkochen, West Germany

17.3.1.5. Transmittance

Physical property: Transmittance
Unit: Dimensionless
Recommended reference material: Neutral density glass filters
Range of variables: Filters mounted in cuvette-style holders are available having nominal transmittances of 10, 20, and 30 percent. Calibration wavelengths are 412, 463, 510, 600 and 710 nm.
Physical state within the range: Solid
Contributor: A. Bittar

Sources/Preparation: Physics and Engineering Laboratory, Department of Scientific and Industrial Research, Private Bag, Lower Hutt, New Zealand

17.3.2. Semi-transparent metal-film filters
17.3.2.1. Transmittance/optical density

Physical property: Transmittance/optical density
Unit: Dimensionless
Recommended reference material: Semi-transparent metallic films on fused quartz substrate
Range of variables: Nichrome film filters mounted in cuvette-style holders and having nominal densities of 0.25, 0.5, 1.0, 1.5 and 2.0 are stocked items; calibrations are available at any designated wavelengths between 200 nm and 2300 nm.
Physical state within the range: Solid
Contributor: J. F. Verrill

Sources/Preparation: Division of Quantum Metrology, National Physical Laboratory, Teddington, Middlesex, United Kingdom TW11 0LW

17.3.2.2. Transmittance/transmittance density

Physical property: Transmittance/transmittance density
Unit: Dimensionless
Recommended reference material: Semi-transparent metallic films on fused quartz substrates

Range of variables: One set consists of three filters mounted in cuvette-style holders. Two filters carry chromium films having nominal transmittances of 10 and 30 percent while the third filter is a clear quartz plate having 90 per cent transmittance. Calibrations are routinely provided at 250, 280, 340, 360, 400, 465, 500, 546.1, 590 and 635 nm; transmittances at other wavelengths between 200 nm and 2500 nm are available on request.

Physical state within the range: Solid
Contributor: R. W. Burke

Sources/Preparation: Office of Standard Reference Materials, National Bureau of Standards, Gaithersburg, MD 20899

17.3.3. Inorganic solutions

17.3.3.1. Ampouled solutions

Physical property: Absorbance
Unit: Dimensionless
Recommended reference material: Ampouled inorganic solutions
Range of variables: Three concentrations of an empirical inorganic mixture having nominal absorbances of 0.3, 0.6 and 0.9 are provided; calibration wavelengths are 302, 395, 512 and 678 nm; temperature coefficients and spectral bandpass dependence are specified.

Physical state within the range: Liquid
Contributor: R. W. Burke

Sources/Preparation: Office of Standard Reference Materials, National Bureau of Standards, Gaithersburg, MD 20899

17.3.3.2. Potassium dichromate

Physical property: Absorbance
Unit: Dimensionless
Recommended reference material: Crystalline potassium dichromate
Range of variables: Specific absorbances (absorptivities) are given at 235, 257 313, 345 and 350 nm for solutions prepared in 0.001 M perchloric acid; temperature coefficients and spectral bandpass requirements specified.

Physical state within the range: Liquid

Contributor: R. W. Burke

Sources/Preparation: Office of Standard Reference Materials, National Bureau of Standards, Gaithersburg, MD 20899

17.3.3.3. Quartz cuvette

Physical property: Pathlength
Unit: Millimeter
Recommended reference material: Calibrated quartz cuvette
Range of variables: Internal pathlength and parallelism of windows of 10 mm rectangular cuvette specified to ± 0.0005 mm.
Physical state within the range: Solid
Contributor: R. W. Seward

Sources/Preparation: Office of Standard Reference Materials, National Bureau of Standards, Gaithersburg, MD 20899

17.3.4. Rotating sectors

Physical property: Transmittance
Unit: Dimensionless
Recommended reference material: Rotating sectors
Range of variables: Transmittances from 0.04 to 0.96; for use in infrared where more conventional types of transfer standards cannot be used
Physical state within the range: Mechanical device
Contributor: G. Milazzo

Sources/Preparation: Widely available from suppliers of optical components

Physicochemical data: Such devices are suitable only for optical spectrometers and not interferometers. Also, they are not suitable for fourier transform infrared instruments. There can be problems with using sectors because of beating between the rotation frequency and the modulation frequency. The effect of stray light needs to be taken into account (Refs. 1, 2).

REFERENCE

1. Jones, R. N., *Pure and Appl. Chem.*, **1969**, *18*, 303.

2. Jones, R. N. Escolar, D.; Hawranek, J. P.; Neelakantan, P. *J. Mol. Spectroscopy* **1973**, *12*, 21.

17.4. Contributors

B. Adams,
Center for Radiation Research
National Bureau of Standards,
Gaithersburg, MD 20899 (U.S.A.)

A. Bitter,
Physics and Engineering Laboratory,
Department of Scientific and Industrial Research,
Private Bag, Lower Hutt (New Zealand)

R. W. Burke,
Center for Analytical Chemistry,
National Bureau of Standards,
Gaithersburg, MD 20899 (U.S.A.)

K. L. Eckerle,
Center for Radiation Research
National Bureau of Standards,
Gaithersburg, MD 20899 (U.S.A.)

D. Irish,
Pye Unicam Ltd.,
Cambridge, England CB1 2PX (UK)

E. Juhász,
National Office of Measures,
H-1531 Budapest 126, Pf. 19 (Hungary)

R. Mavrodineanu,
Center for Radiation Research
National Bureau of Standards,
Gaithersburg, MD 20899 (U.S.A.)

K. D. Mielenz,
Center for Radiation Research
National Bureau of Standards,
Gaithersburg, MD 20899 (U.S.A.)

G. Milazzo,
Instituto Superiore de Sanita,
Rome (Italy)

T. C. Rains,
Center for Analytical Chemistry,
National Bureau of Standards,
Gaithersburg, MD 20899 (U.S.A.)

A. R. Robertson,
National Research Council,
Division of Physics,
Ottawa, K1A OR6 (Canada)

R. W. Seward,
Office of Standard Reference Materials,
National Bureau of Standards,
Gaithersburg, MD 20899 (U.S.A.)

R. A. Velapoldi,
Center for Radiation Research
National Bureau of Standards,
Gaithersburg, MD 20899 (U.S.A.)

W. H. Venable, Jr.,
Center for Radiation Research,
National Bureau of Standards,
Gaithersburg, MD 20899 (U.S.A.)

J. F. Verrill,
Division of Quantum Metrology,
National Physical Laboratory,
Teddington, Middlesex, TW11 O1W (UK)

H. von Derschau,
Carl Zeiss,
Oberkochen (West Germany)

V. R. Weidner,
Center for Radiation Researach
National Bureau of Standards,
Gaithersburg, MD 20899 (U.S.A.)

17.5. List of Suppliers

A. Aldrich Chemical Co., Inc.,
   940 West Saint Paul Avenue,
   Milwaukee, WI 53233 (U.S.A.)

B. Aldrich Chemical Co., Ltd.,
   Old Brickyard, New Road,
   Gillingham, Dorset (UK)

C. Aldrich-Europe,
   c/o Janssen Pharmaceutica,
   B-2340 Beerse (Belgium)

D. Ega Chemie K. G,
   7924 Steinheim auf Albuch (FRG)

E. Perkin Elmer Corp,
   Norwalk, Connecticut (U.S.A.)

F. Polish Committee of Standardization
   and Measures,
   Division of Physicochemical Metrology,
   UL. Elecktoralna 2,
   PL 00-139, Warsaw (Poland)

G. National Physical Laboratory,
   Teddington, TW11 0LW (UK)

H. National Bureau of Standards,
   Gaithersburg, MD 20899 (U.S.A.)

I. National Research Council,
   Ottawa, K1A 0RG (Canada)

J. Pye Unicam Ltd.,
   Cambridge, England CB1 2PX (UK)

K. Carl Zeiss,
   Oberkochen (West Germany)

L. Department of Scientific and
   Industrial Research,
   Private Bag,
   Lower Hutt (New Zealand)

M. National Office of Measures,
   H-1531, Budapest 126, Pf. 19
   (Hungary)

18

SECTION: RELATIVE MOLECULAR MASS

COLLATOR: K. N. MARSH

CONTENTS:

18.1. Introduction

18.2. Reference materials for relative molecular mass measurements

   18.2.1. Reference materials for gel permeation chromatographic measurements

   18.2.2. Reference materials for light scattering photometric measurements

   18.2.3. Reference materials for colligative property measurements

   18.2.4. Reference materials for membrane osmometry and ultracentrifugation

18.3. Contributors

18.4. List of suppliers

18.1. Introduction

The relative molecular mass or molecular weight, M_r, of a substance is the ratio of the average mass per molecule of a specified isotopic composition of a substance to 1/12 of the mass of an atom of the nuclide ^{12}C. It is dimensionless. To avoid confusion with additional subscripts M_r will be replaced by M. Molecular weight may be measured by several methods, each with its characteristic range of application and need for calibration and test materials. Samples of synthetic polymers are heterogeneous with respect to molecular weight, and methods are available which yield the molecular weight distribution (MWD) or one of its averages.

Those averages can be defined in terms of n_i, the number of molecules of molecular weight M_i, or in terms of $f(M)$, the molecular weight distribution treated as a continuous function such that the mass fraction of molecular weight between M_1 and M_2 is given by

$$M = \int_{M_1}^{M_2} f(M)dM. \tag{18.1.1}$$

The number-average molecular weight is given by

$$M_n = \sum_i n_i M_i / \sum_i n_i$$
$$= 1 / \int [(1/M)f(M)dM]. \tag{18.1.2}$$

The weight average molecular weight is given by

$$M_w = \sum_i n_i M_i^2 / \sum_i n_i M_i$$
$$= \int M f(M)dM. \tag{18.1.3}$$

The Z-average molecular weight is given by

$$M_z = \sum_i n_i M_i^3 / \sum_i n_i M_i^2$$
$$= \int [M^2 f(M)dM]/M_w. \tag{18.1.4}$$

The summations and integrations are over all the material present.

The ranges of measurement given in table 18.1.1 are accessible with apparatus either available commercially or constructed simply. In favourable cases and with special equipment, the scope of the colligative property measurement methods can be extended considerably. Calibration materials are recommended for gel permeation chromatography, light-scattering photometry, and some colligative property measurement methods (vapour pressure osmometry, cryoscopy, and ebulliometry). Test materials may be useful in calibration and checking membrane osmometers and ultracentrifuges.

18. Relative Molecular Mass

Table 18.1.1. Methods of measurement, the quantities determined and their ranges and the upper temperature limit to the measurement

Method	Quantity measured	Molecular weight M_r	limit $t/°C$
Gel permeation chromatography	MWD	up to 10^7	140
Light-scattering photometry	M_w	10^4 to 10^7	160
Vapour pressure osmometry	M_n	up to 4×10^4	130
Cryoscopy	M_n	up to 5×10^4	165
Ebulliometry	M_n	up to 4×10^4	120
Membrane osmometry	M_n	2×10^4 to 10^6	130
Ultra-centrifugation	M_w, M_z	10^4 to 10^7	120

18.2. Reference materials for molecular weight measurements

18.2.1. Reference materials for gel permeation chromatographic measurements

In gel permeation, a dilute solution of the unknown is injected into a stream of solvent flowing continuously through columns containing microporous substrates; a detector, conventionally refractometric, provides a record of solute concentration against elution volume. Determination of molecular weight averages and distributions requires the relation of elution volume to molecular weight by means of reference materials. The calibration should extend over a wide range of molecular weight and can be established (Ref. 1) with reference materials of three kinds:

(a) A series of samples of narrow molecular weight distribution ($M_w/M_n < 1.1$) centered about values which span the range of interest and certified as to M_o, the molecular weight corresponding to the peak of the chromatogram; calibration consists in relating that value to the observed peak volume.

(b) A series of samples of certified number-average and/or weight-average molecular weight; calibration consists in finding the relation between molecular weight and elution volume which best generates those values from the observed chromatograms.

(c) A single sample of broad molecular weight distribution ($M_w/M_n > 3$) certified as to molecular weight distribution; calibration consists in matching an integrated chromatogram with the certified integral molecular weight distribution (Ref. 2).

Table 18.2.1.1. List of reference materials and suppliers

Material	Supplier	Calibration range in certified quantity	M_w/M_n	Certified Quantities
Polyethylene	C	$\sim 5 \times 10^3$ to 2×10^5	~ 3	M_n, M_w, M_z, MWD
	C	$\sim 10^4$	~ 1.2	M_n, M_w
	C	$\sim 3 \times 10^4$	~ 1.1	M_n, M_w
	C	$\sim 10^5$	~ 1.2	M_n, M_w
Poly(methyl methacrylate)	C	$\sim 10^5$	~ 1.1	M_n
Polystyrene	C	$\sim 2 \times 10^5$	~ 1.1	M_n, M_w
	C	$\sim 3 \times 10^5$	~ 2	M_w
	C	$\sim 4 \times 10^4$	~ 1.04	M_n, M_w
	C	$\sim 10^6$	~ 1.1	M_w

The table 18.2.1.1 lists reference materials available from national laboratories. Preferably, the reference material used should be a homologue of the sample to be analysed; but, in the absence of such a material, a universal calibration may be employed (Ref. 1).

The samples listed in table 18.2.1.2 are available without full details of their characterization.

Table 18.2.1.2. List of further reference materials

Material	Supplier	Calibration range of certified quantities	M_w/M_n	Certified Quantities
Polystyrene	A	5×10^2 to 2×10^6	1.06 to 1.15	M_n, M_w
	D	1×10^2 to 1.5×10^7	1.0 to 1.3	M_n, M_w, M_o
	F	8×10^2 to 1.8×10^6	1.05 to 1.13	M_n, M_w, M_o
	H	4×10^2 to 2×10^7	1.01 to 1.17	M_o
Polyethylene	D	5×10^2 to 1.2×10^5	1.1 to 6	M_n, M_w, M_o
	F	5×10^2 to 4×10^4	1.1 to 6	M_n, M_w, M_o
	G	1×10^4 to 5×10^5	~ 1.1	M_o
Polymethyl-	D	3×10^3 to 1.5×10^6	1.1 to 1.3	$M_n, M_w\, M_o$
methacrylate	F	7.7×10^3 to 8.4×10^5	1.07 to 1.17	M_n, M_w, M_o
Poly(α-methylstyrene)	D	1×10^3 to 1×10^6	1.1 to 1.3	$M_n, M_w\, M_o$
	F	5×10^3 to 3.5×10^5	1.1 to 1.2	M_n, M_w, M_o
Polyisoprene	D	1×10^3 to 3×10^6	1.04 to 1.15	$M_n, M_w\, M_o$
	F	1×10^3 to 3×10^6	1.04 to 1.15	M_n, M_w, M_o
	H	900		
Polytetrahydrofuran	D	1×10^3 to 5×10^5	1.1 to 1.15	M_n, M_w, M_o
	F	1×10^3 to 3×10^5	1.1 to 1.15	M_n, M_w, M_o
Polyvinylchloride	A	2.5×10^4 to 1.3×10^5	2.4 to 2.9	M_n, M_w
Polyethylene	D	106 to 2.2×10^4	1.0 to 1.06	M_n, M_w, M_o
glycol	F	106 to 1.9×10^4	1.0 to 1.34	M_n, M_w, M_o
Polypropyleneglycol	H	8×10^2 to 4×10^3		
Polysaccharide	D	6×10^3 to 8×10^5	1.06 to 1.14	M_n, M_w, M_o
Polyethylene	D	1.8×10^4 to 1×10^6	1.03 to 1.1	M_n, M_w, M_o
oxide	F	2×10^4 to 1×10^6	1.03 to 1.1	M_n, M_w, M_o
	H	2×10^4 to 1×10^6	1.02 to 1.14	M_w
Polystyrene sulphonate	D	1.8×10^3 to 1.2×10^6	1.1 to 1.25	M_n, M_w, M_o
Na salt	F	1.8×10^3 to 1.2×10^6	1.1 to 1.25	M_n, M_w, M_o
Poly(2-vinyl pyridine)	F	3×10^3 to 1.4×10^6		
Dextran	E	1×10^4 to 2×10^6		MWD

18.2.2. Reference materials for light-scattering photometric measurements

Methods of measurement

In light scattering photometry (Refs. 3, 4), a dilute solution of the unkown is illuminated monochromatically and the relative scattering irradiances are measured as a function of angle of scatter and concentration. The molecular weight of the solute (M_w) is related to the excess reduced irradiance $R(\theta, c)$ at a distance r from the center of the scattering volume $v(\theta)$ by the relation

$$R(\theta, c) = r^2 [E(\theta, c) - E(\theta, 0)] / E_o v(\theta) \qquad (18.2.2.1)$$

where $E(\theta, c)$ is the irradiance of light scattered through an angle θ by a solution of concentration c, and E_o is the incident irradiance.

Measurements are made by recording photomultiplier outputs proportional to $E(\theta, c)$ and $E(\theta, 0)$ at a range of values of θ and c, and calibration with reference materials is necessary to find $R(\theta, c)$; it is common practice to include r and v in the calibration constant. The scattering volume is a function of the refractive indices of the scattering medium and the thermostating liquid, if one is used, and a photometer should be calibrated under conditions as close to the experimental ones as possible.

Pure liquids are the preferred calibrants for work with a sensitive photometer near ambient temperatures; Rayleigh ratios $r^2 E(90°, 0)/E_o v(90°)$ are tabulated (Ref. 5) for a number of readily available liquids.

Solutions of 12-tungstosilicic acid in aqueous sodium chloride are convenient calibrants (Ref. 6) for work on aqueous media. For work on samples much above ambient temperatures, solutions of standard polymer samples are recommended. Only those for which the characterization is described fully should be used, and the published procedure should be adhered to closely.

Materials, data and suppliers

Rayleigh ratios for carbon disulphide, benzene, toluene, carbon tetrachloride, methyl ethyl ketone, and several aliphatic hydrocarbons are tabulated (Ref. 5). Data and procedures for the use of 12-tungstosilicic acid are available (Ref. 6). The above materials are available from laboratory suppliers. The following certified polymer samples are available from supplier (C).

Table 18.2.2.1. Reference materials available from the National Bureau of Standards

Material	Sample number	M_w, rsdm[a]
Polystyrene	NBS SRM 705	$179,300^b$, 0.4%
Polystyrene	NBS SRM 706	$247,800^b$, 0.4%
Linear Polyethylene (Ref. 7)	NBS SRM 1475	$52,000^b$, 4%
Polystyrene	NBS SRM 1478	$37,000^c$, 0.7%
Polystyrene	NBS SRM 1479	$1,050,000^b$, 4%
Linear Polyethylene (Ref. 8)	NBS SRM 1482	$13,600^b$, 1.0%
Linear Polyethylene (Ref. 8)	NBS SRM 1483	$32,000^b$, 2.3%
Linear Polyethylene (Ref. 8)	NBS SRM 1484	$119,600^b$, 1.8%
Poly(methyl methacrylate)	NBS SRM 1489	($M_n=115,000^d$, rsdm[a]=1.5%)

[a] relative standard deviation of the mean
[b] measured by light scattering
[c] measured by equilibrium ultracentrifugation
[d] measured by membrane osmometry

18.2.3. Reference materials for colligative property measurements

Reference materials are required to relate to molality the difference in freezing point between solvent and solution (in cryoscopy) (Refs. 9, 10), in boiling point between solvent and solution (in ebulliometry) (Refs. 11, 12), or in steady-state temperature between drops of solvent and solution in saturated solvent vapour (in vapour pressure osmometry) (Refs. 13, 14). For the first two methods, the calibration constant is calculable from the enthalpy of phase change of the solvent, but an empirical determination of the constant is preferred.

The reference material should be soluble, involatile, and neither associated nor dissociated under the experimental conditions. Evidence is accumulating (Refs. 15 to 17) that the calibration constants are dependent on molecular weight, and hence the reference material chosen should be close in molecular weight to the unknown. The following table lists reference materials with their molecular weights. All are available from laboratory suppliers; recommended samples of certified purity are available from the sources indicated.

Table 18.2.3.1. Reference materials for measurements using colligative properties

Material	Supplier	Molecular Weight
Resorcinol		110.1
Benzoic acid	B,C	122.1
Naphthalene		128.2
Camphor		152.1
Mannitol		182.1
Dimethylterephthalate		194.2
2,4-Dinitrochlorobenzene		202.6
Benzil		210.2
2,2-bis(ethylsulphonyl)propane		228.3
Dibenzyldisulphide		246.4
Hexachlorobenzene		284.8
Cholesterol	C	386.7
2,6,10,15,19,23-Hexamethyltetracosane		422.8
Hexatriacontane		507
Glyceryltristearate		891.5
Pentaerythrityl tetrastearate	F	1202

18.2.4. Reference materials for membrane osmometry and ultracentrifugation

Calibration is not required for these methods (Refs. 18, 19); however, the well-characterised polystyrene, NBS SRM 705, is recommended as a test material to check instrumental performance.

REFERENCES

1. Dawkins, J. V. *Brit. Polymer J.* **1972**, *4*, 87.
2. Swartz, T. D.; Bly, D. D.; Edwards, A. S. *J. Appl. Polymer Sci.* **1972**, *16*, 353.
3. Billmeyer, F.W. *Treatise on Analytical Chemistry.*, Vol. 5, Part 1, Kolthoff, I. M.; Elving, P. J., editors. Interscience: New York, **1964**.
4. Kratohvil, J. P. *Anal. Chem.* **1964**, *36*, 458R.
5. Coumou, D. J.; Mackor, E. L.; Hijmans, J. *Trans. Faraday Soc.* **1964**, *60*, 1539.

6. Krathovil, J. P.; Oppenheimer, L.E.; Kerker, M. *J. Phys. Chem.* **1966**, *70*, 2834.
7. Wagner, H. L.; Verdier, P. H., editors, *The Characterization of Linear Polyethylene SRM 1475*, National Bureau of Standards Special Publication 260-42, U.S. Department of Commerce: Washington, **1972**; *J. Res. Nat. Bur. Stand.* **1972**, *76A*, 137.
8. Verdier, P. H.; Wagner, H. L., ed. *The Characterization of Linear Polyethylene SRM'S 1482, 1483, and 1484*, National Bureau of Standards Special Publication 260-61, U.S. Department of Commerce: Washington, **1978**; *J. Res. Nat. Bur. Stand.* **1978**, *83*, 169.
9. Vofsi, D.; Katchalsky, A. *J. Polymer Sci.* **1957**, *26*, 127.
10. Newitt, E. J.; Kokle, V. *J. Polymer Sci.* **1966**, *4*, 705.
11. Dimbat, M.; Stross, F. H. *Anal. Chem.* **1958**, *29*, 219.
12. Lehle, R. S.; Majury, T. G. *J. Polymer Sci.* **1958**, *29*, 219.
13. Dohner, R. E.; Wachter A. H.; Simon, W. *Helv. Chim. Acta* **50**, 2193.
14. Adicoff, A.; Murback, W. *J. Anal. Chem.* **1967**, *39*, 302.
15. Bersted, B. H. *J. Appl. Polymer Sci.* **1973**, *17*, 1415.
16. Brzezinski, J.; Glowala, H.; Kornas-Calka, A. *Europ. Polymer J.* **1973**, *9*, 1251.
17. Glover, C. A. *American Chemical Society Advances in Chemistry Series*, Vol. 125, **1975**. p. 1.
18. Coll, H.; Stross, F. H. *Characterization of Macromolecular Structure*, McIntyre, D., editor. National Academy of Sciences: Washington, **1968**.
19. McCormick, H. W. *Polymer Fractionation*, Cantow, M. J. R., editor, Academic Press: New York, **1967**.

18.3. Contributors to the revised version

K. N. Marsh,
Thermodynamics Research Center,
Texas A&M University,
College Station, Texas 77843 (U.S.A.).

18.4. List of suppliers

A. Arro Laboratories Inc.,
   P.O. Box 686 Caton Farm Road,
   Joliet, IL 60434 (U.S.A.).

B. B.D.H. Chemicals Ltd.,
   Poole, Dorset, BH12 4NN (U.K.).

C. Office of Standard Reference Materials,
   U.S. Department of Commerce,
   National Bureau of Standards,
   Gaithersburg, MD 20899 (U.S.A.).

D. Polymer Laboratories, Ltd.
   Essex Road
   Church Stretton
   Shropshire SY6 6AX (U.K.)

E. Pharmacia Fine Chemicals AB,
   Box 175,
   S-75104 Uppsala (Sweden).

F. Pressure Chemical Co.,
   3419 Smallman Street,
   Pittsburgh, PA 15201 (U.S.A.).

G. Société Nationale des Pétroles d´Aquitaine,
   Tour Aquitaine,
   92080 Courbevoire (France).

H. Waters, Division of Millipore,
   34 Maple Street,
   Milford, MA 01757 (U.S.A.).

QD
455.2
R431
1987

RETURN CHEMISTRY LIB
TO → 100 Hildebrand
LOAN PERIOD

LIBRARY USE ONLY